Industrial Engineering: Management, Tools and Applications

Industrial Engineering: Management, Tools and Applications

Edited by **Michelle Vine**

Willford Press

New York

Published by Willford Press,
118-35 Queens Blvd., Suite 400,
Forest Hills, NY 11375, USA
www.willfordpress.com

Industrial Engineering: Management, Tools and Applications
Edited by Michelle Vine

International Standard Book Number: 978-1-68285-257-6 (Hardback)

Printed in China.

Contents

Preface

Industrial engineering incorporates concepts of various scientific disciplines under one umbrella such as safety engineering, management science, ergonomics, manufacturing engineering, etc. This book is a valuable compilation of topics such as decision making models, supply chain management, manufacturing optimization, quality engineering, operations research, etc. From theories to research to practical applications, case studies related to all topics of relevance to this field have been included in this book. It will prove to be an essential guide for all those who are engaged in the discipline of industrial engineering.

This book has been the outcome of endless efforts put in by authors and researchers on various issues and topics within the field. The book is a comprehensive collection of significant researches that are addressed in a variety of chapters. It will surely enhance the knowledge of the field among readers across the globe.

It gives us an immense pleasure to thank our researchers and authors for their efforts to submit their piece of writing before the deadlines. Finally in the end, I would like to thank my family and colleagues who have been a great source of inspiration and support.

Editor

1

A hybrid meta-heuristic algorithm for the vehicle routing problem with stochastic travel times considering the driver's satisfaction

Reza Tavakkoli-Moghaddam[1*], Mehdi Alinaghian[2], Alireza Salamat-Bakhsh[3] and Narges Norouzi[1]

Abstract

A vehicle routing problem is a significant problem that has attracted great attention from researchers in recent years. The main objectives of the vehicle routing problem are to minimize the traveled distance, total traveling time, number of vehicles and cost function of transportation. Reducing these variables leads to decreasing the total cost and increasing the driver's satisfaction level. On the other hand, this satisfaction, which will decrease by increasing the service time, is considered as an important logistic problem for a company. The stochastic time dominated by a probability variable leads to variation of the service time, while it is ignored in classical routing problems. This paper investigates the problem of the increasing service time by using the stochastic time for each tour such that the total traveling time of the vehicles is limited to a specific limit based on a defined probability. Since exact solutions of the vehicle routing problem that belong to the category of NP-hard problems are not practical in a large scale, a hybrid algorithm based on simulated annealing with genetic operators was proposed to obtain an efficient solution with reasonable computational cost and time. Finally, for some small cases, the related results of the proposed algorithm were compared with results obtained by the Lingo 8 software. The obtained results indicate the efficiency of the proposed hybrid simulated annealing algorithm.

Keywords: Vehicle routing problem, Stochastic travel times, Driver's satisfaction, Simulated annealing

Background

Vehicle routing problems (VRPs) have a significant role in logistics and distribution industries. VRPs include problems in which a fleet of vehicles presents service from one or several depots to different geographically located customers so that the total cost of transportation is minimized. It has been proven that vehicle routing problems are categorized as non-deterministic polynomial (NP)-hard problems (Lenstra and Rinnooy Kan 1981; Mingozzi et al. 1999) and that exact solution methods are not practical in a large scale due to high computational cost since most of the procedures have focused on using heuristics and meta-heuristics.

This problem was investigated mathematically by (Dantzig et al. 1994). The primitive model was developed by other researchers. Clarke and Wright (1994) solved VRPs by using a saving algorithm. Developing VRP models improves advanced solving methods based on mathematical programming, heuristics and meta-heuristics. Altinkemerk (1991) used a sweeping algorithm to develop a corresponding mathematical model. Laporte (1992) and Laporte et al. (1992) propagated a branch-and-bound algorithm. Tavakkoli-Moghaddam et al. (2005) proposed a multi-criteria vehicle routing problem considering soft time windows, and their problem is solved by simulated annealing. Also Tavakkoli-Moghaddam et al. (2006b) developed a mathematical model for VRP with backhauls by using memetic algorithm. Furthermore, Tavakkoli-Moghaddam et al. (2006b) presented a linear integer model of capacitated vehicle routing problems with the independent route length to minimize the heterogeneous fleet cost and maximize the capacity utilization, and their proposed model is solved by a hybrid simulated annealing (SA) based on the nearest neighborhood. Jozefowiez et al. (2009) developed a multi-objective vehicle routing problem by

* Correspondence: tavakoli@ut.ac.ir
[1]Department of Industrial Engineering, College of Engineering, University of Tehran, Tehran, Iran
Full list of author information is available at the end of the article

minimizing the route length and route balance by using a meta-heuristic method with traditional operators. A number of meta-heuristics, such as SA, tabu search and ant colony system, have been developed to solve more complicated problems containing more constraints (Fallahi et al. 2008; Bin et al. 2009). In general, the presented models for VRPs consider the cost function and propose the solution based on the cost minimization. However, there are other factors which have significant effect on the problem, such as drivers' satisfaction, which is investigated in this paper.

Since traveling time of the vehicle cannot be determined certainly due to various factors, such as traffic congestion, weather condition, vehicle failure and the like, this parameter is considered by a probability function. Obviously, increasing the total service time leads to decreasing the driver's satisfaction. In this paper, the total cost function of the vehicles is minimized, while the driver's satisfaction is increased by decreasing the probability of the service time. This problem is considered as a graph by n nodes, which represent customers, and connecting arcs show the route between customers. Each customer has a specific demand, and they should be served by only one vehicle. Due to its successful result in meta-heuristics, the model is solved by a hybrid simulated annealing (HSA) algorithm (Tavakkoli-Moghaddam et al. 2006a). The proposed algorithm shows much higher efficiency than the Lingo 8 software in large scale problems.

The rest of this paper is organized as follows. 'Problem definition' introduces the problem, and 'Model formulation' represents the model formulation. Problem-solving methodology is described in 'Hybrid simulation annealing', and computational results are discussed in 'Results and discussion'. Finally, a conclusion follows in the last section.

Methods

Problem definition

The traveling time for vehicles is a probability variable depending on the weather conditions, accident, failure of the vehicle and condition of the route. Before describing the problem, two definitions are presented based on the fuzzy theory (Zimmermann 1993).

Fuzzy number

Definition 2.1. A fuzzy set A in R (real line) is defined to be a set of ordered pairs $A = \{(x, \mu_A(x))|x\epsilon R\}$, where $\mu_A(x)$ is called the membership function for the fuzzy set.

Definition 2.2. A fuzzy set A is called normal if there is at least one point $x \in R$ with $\mu_A(x) = 1$.

Definition 2.3. A fuzzy set A on R is convex if, for any x,y R and any $\lambda \in [0,1]$, we have $\mu_A(\lambda x + (1-\lambda)y) \geq \min\{\mu_A(x), \mu_A(y)\}$.

Definition 2.4. A fuzzy number is a fuzzy set on the real line that satisfies the conditions of normality and convexity.

α-cut

One of the most important concepts of fuzzy sets is α-cut. Given fuzzy set A defined on X and any number $\alpha \in [0,1]$, the α-cut (i.e. ${}^{\alpha}A$) and the strong α-cut (i.e. ${}^{\alpha+}A$) are the crisp sets.

$$ {}^{\alpha}A = \{x|A(x) \geq \alpha\} \quad (1)$$

$$ {}^{\alpha+}A = \{x|A(x) > \alpha\} \quad (2)$$

Also, distributing companies are interested in the service time of each vehicle not exceeding from a specific time limit. Increasing of the service time leads to decreasing of the driver's working utility. It is supposed that the maximum time working utility of the drivers is achieved when service time is less than t_l in bound of $[0, t_u]$ and that increasing the service time affects linearly on the decreasing of the utility so that, for service time more than t_u, the utility level achieves its minimum. Also, a driver's utility function versus time is illustrated by a fuzzy number as shown in Figure 1. This function states that the driver's utility decreases linearly as traveling time increases.

Assumptions used in our presented model are as follows:

1. Travel time of each route is a normal variable with the specific mean and specific variance.
2. The driver's utility function depends on the travel time, and it decreases linearly with increasing the travel time.

In this way, a stochastic model is presented such that a vehicle travel time is restricted to time less than t_α with p percent probability in each level of satisfaction. t_α is the maximum length of tour in time for α level of satisfaction. To develop this model, the sum of the normal distribution function theorem is described in this section. Based on this theorem, if x_i is an independent variable of the normal distribution with mean μ_i's and variance σ_i's

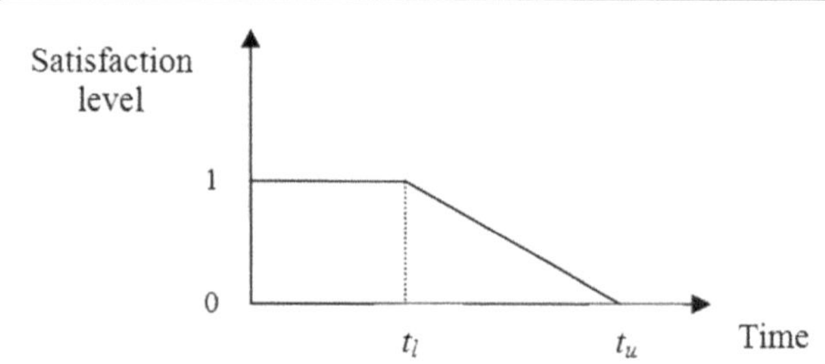

Figure 1 Driver's satisfaction versus working time.

by $x_i \sim N(\mu_i, \sigma_i^2)$, in which a_i is the constant number, y is a summation of x_i's by:

$$y = a_1x_1 + a_2x_2 + \ldots + a_nx_n$$

Then, y is also a normally distributed variable given by $y \sim N(\mu_y, \sigma_y^2)$, where $\mu_y = a_1\mu_1 + a_2\mu_2 + \ldots + a_n\mu_n$ and $\sigma_y^2 = a_1^2\sigma_1^2 + a_2^2\sigma_2^2 + \ldots + a_n^2\sigma_n^2$.

It is now supposed that t_{ij}^v represents the travel time of vehicle v from node i to node j, in which t_{ij}^v is a sequence of the normal independent variables and follows the normal distribution of $N(\mu_{ij},\sigma_{ij}^2)$. t_i^v is the required time for servicing customer (node) i by vehicle v and has normal distribution with μ_i and σ_i^2 as its parameters. Also, decision variable x_{ij} is defined such that:

$$\begin{cases} x_{ij}^v = 1 \\ x_{ij}^v = 0 \end{cases}$$

If vehicle v travels through route (i, j).

Otherwise,t_α is defined as the maximum time in each α-level of the satisfaction function. For all of the drivers, t_α is supposed to be the same. It is ideal that the total service time for each vehicle to be less than t_α.

$$\sum_{i=1}^{n} t_i^v \sum_{j=1}^{n} x_{ij}^v + \sum_{i=1}^{n}\sum_{j=1}^{n} t_{ij}^v x_{ij}^v \le t_\alpha; \quad v = 1, 2, \ldots, NV, \; i \ne j \tag{3}$$

where NV is the total number of vehicles.

Expanding Equation 3 for each vehicle (v) leads to the following equation.

$$\begin{aligned} &t_1\left(x_{12}^v + x_{13}^v + \ldots + x_{1N}^v\right) \\ &+ \ldots + t_N\left(x_{N1}^v + x_{N3}^v + \ldots + x_{NN-1}^v\right) \\ &+ \left(t_{12}^v x_{12}^v + t_{13}^v x_{13}^v + \ldots + t_{1N}^v x_{1N}^v\right) + \ldots \\ &+ \left(t_{N2}^v x_{N2}^v + t_{N3}^v x_{N3}^v + \ldots + t_{NN-1}^v x_{NN-1}^v\right) \le t_\alpha. \end{aligned} \tag{4}$$

Thus, we consider p percent probability by:

$$P\left(\sum_{i=1}^{n} t_i^v \sum_{j=1}^{n} x_{ij}^v + \sum_{i=1}^{n}\sum_{j=1}^{n} t_{ij}^v x_{ij}^v \le t_\alpha\right) \ge \alpha; \quad v = 1, 2, \ldots, NV \tag{5}$$

Now, by defining the normal distribution of the travel time as parameter A, we have:

$$\begin{aligned} A = \sum_{i=1}^{n}\sum_{j=1}^{n} t_{ij}^v x_{ij}^v \sim N[&\sum_{i=1}^{n}\mu(t_i^v)\sum_{j=1}^{n} x_{ij}^v + \sum_{i=1}^{n}\sum_{j=1}^{n}\mu\left(t_{ij}^v\right)x_{ij}^v, \\ &\sum_{i=1}^{n}\sigma^2(t_i^v)\sum_{j=1}^{n}\left(x_{ij}^v\right)^2 + \sum_{i=1}^{n}\sum_{j=1}^{n}\sigma^2\left(t_{ij}^v\right)x_{ij}^v] \; v = 1, 2, \ldots, NV. \end{aligned} \tag{6}$$

It can be concluded that

$$P(A \le t_\alpha) \ge \alpha \Rightarrow P\left(\frac{A - \mu_A}{\sqrt{\sigma_A^2}} \le \frac{t_\alpha - \mu_A}{\sqrt{\sigma_A^2}}\right) \ge \alpha. \tag{7}$$

This equation is expanded in the presented model as follows:

$$\frac{t_\alpha^v - \left(\sum_{i=1}^{n}\mu(t_i^v)\sum_{j=1}^{n} x_{ij}^v + \sum_{i=1}^{n}\sum_{j=1}^{n}\left(t_{ij}^v\right)x_{ij}^v\right)}{\sqrt{\sum_{i=1}^{n}\sigma^2(t_i^v)\sum_{j=1}^{n}\left(x_{ij}^v\right)^2 + \sum_{i=1}^{n}\sum_{j=1}^{n}\sigma^2\left(t_{ij}^v\right)^2 x_{ij}^v}} \ge Z_\alpha; \quad v = 1, 2, \ldots, NV. \tag{8}$$

By this approach, the VRP can be extended in order to consider the limitation of the specific travel time and the distribution time of vehicles.

Model formulation

In our presented nonlinear-integer model, customers are introduced by nodes connected to each other by arcs. This graph contains n nodes in which each node has its demand d_i (depot is located in node 1). The total number of vehicles is shown by NV, and each vehicle has the capacity of K_v. t_i^v is the required time for servicing the customer (i.e. node) i by vehicle v, and t_{ij}^v is the required time for traveling between customers i and j by vehicle v. Also, C_{ij} is the travel cost (i.e. length of the arc) between customers i and j. S is the collection of nodes defined as $S = \{i | i = 1, \ldots, n\}$.

Now, the problem can be mathematically formulated as follows.

$$z = \min \sum_{i=1}^{n}\sum_{j=1}^{n}\sum_{v=1}^{NV} c_{ij} x_{ij}^v \tag{9}$$

subject to:

$$\sum_{i=1}^{n}\sum_{v=1}^{nv} x_{ij}{}^v = 1; \; j = 1, 2, \ldots, n \tag{10}$$

$$\sum_{j=1}^{n}\sum_{v=1}^{nv} x_{ij}{}^v = 1; \; i = 1, 2, \ldots, n \tag{11}$$

$$\sum_{i=1}^{n} x_{ip}^v - \sum_{v=1}^{nv} x_{pi}^v = 0; \; p = 1, 2, \ldots, n; \quad v = 1, 2, \ldots, NV \tag{12}$$

$$\frac{t_{ij}^{v} - \left(\sum_{i=1}^{n} E\left(t_{i}^{v}\right) \sum_{j=1}^{n} x_{ij}^{v} + \sum_{i=1}^{n}\sum_{j=1}^{n} \left(t_{ij}^{v}\right) x_{ij}^{v}\right)}{\sqrt{\sum_{i=1}^{n} \sigma^{2}\left(t_{i}^{v}\right) \sum_{j=1}^{n} \left(x_{ij}^{v}\right)^{2} + \sum_{i=1}^{n}\sum_{j=1}^{n} \sigma^{2}\left(t_{ij}^{v}\right)^{2} x_{ij}^{v}}} \geq Z_{\alpha};$$

$$v = 1, 2, \ldots, N \tag{13}$$

$$\sum_{i=2}^{n} d_i \left(\sum_{j=1}^{n} x_{ij}^{v}\right) < k_v; \ v = 1, 2, \ldots, NV \tag{14}$$

$$\sum_{j=1}^{N} x_{1j}^{k} = 1 \ \ v \in NV \tag{15}$$

$$\sum_{k=1}^{K} \sum_{j \in S} \sum_{j \notin S} x_{ij}^{k} \leq |S| - r(S) \quad \forall S \subseteq A - \{1\} \quad S \neq \phi \tag{16}$$

$$x \in S \tag{17}$$

In the above model, Equation 9 minimizes the total travel cost and total travel time. Constraints (10) and (11) ensure that each customer location is serviced from only one vehicle. Constraint (12) states that, if a vehicle arrives at a node (i.e. customer location), it should leave it, and by this way, the route continuity is ensured. Constraint (13) states that standard normally distributed function of the driver's transit time should be larger than z_{α}. Constraint (14) imposes that the vehicle capacity does not exceed from its limit. Constraint (15) states that the depot is the first location of each vehicle. Finally, constraint (16) eliminates the sub-tours.

Hybrid simulated annealing

Simulated annealing is the process of physical annealing with solids in which a crystalline solid is heated and then allowed to cool very slowly until it achieves its most regular possible crystal lattice configuration (i.e., its minimum lattice energy state) and, thus, is free of crystal defects. If the cooling schedule is sufficiently slow, the final configuration results in a solid with such superior structural integrity. Simulated annealing establishes the connection between this type of thermo-dynamic behavior and the search for global minima for a discrete optimization problem. Furthermore, it provides an algorithmic means for exploiting such a connection (Koskosidis et al. 1992; Kirkpatrick et al. 1983; Hajek 1985; Gidas 1985; Glover and Kochenberger 2003).The SA parameters are as follows:

Epoch length (EL): the number of accepted solutions in each temperature for achieving to equilibrium.
MTT: maximum number of consecutive temperature trails.
T_0: initial temperature.
α: rate of the current temperature decrease (cooling schedule).
X: a feasible solution.
$C(X)$: objective function value of solution X.
L: counter for the number of accepted solutions in each temperature;
K: counter for the number of consecutive temperature trails, where T_K is equal to the temperature in iteration K.

Initial solution generation

To generate initial solutions, which is the first step in any meta-heuristic approach, a creative approach is applied as follows.

1) Create a new route: Select randomly a node (i.e. r) which has not been served. Allocate a new vehicle (i.e. v) with maximum capacity to that from the depot.
2) Determine neighborhood nodes: Find the nearest unserved node neighborhoods (i.e. node k) such that, if vehicle v services k after r, the stochastic service time constraint and capacity of the vehicle are not violated for reaching node k. If such node is found, then vehicle v services k after r ($x_{vrk} = 1$).
3) Complete the route: Repeat step 2 until vehicle v returns to the depot.
4) Complete the service: Repeat above steps until all of the nodes are served.
5) Determine the optimal kind of vehicles: After running the algorithm, the optimal kind of vehicles can be determined by assuming that all of the vehicles are available with their maximum capacity.

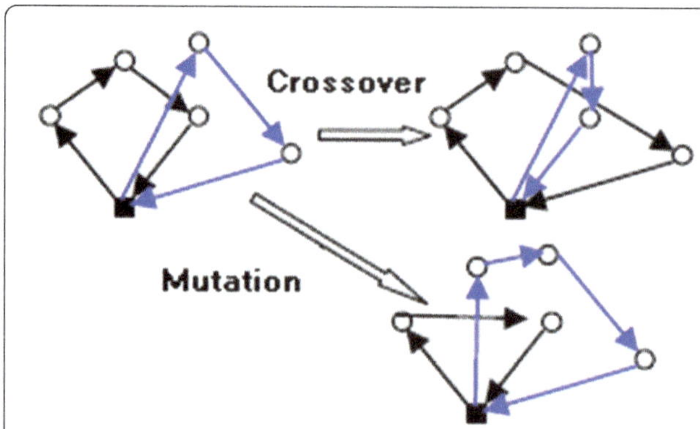

Figure 2 Crossover and mutation operators.

Table 1 Comparison of results obtained by the HSA for ten small-scale cases

Numbers of Problems	Numbers of Customes	Numbers of Vehicles	Objective function value											
			HSA						Lingo 8					
			α=0	α=0.25	α=0.5	α=0.75	α=1	CPU.time	α=0	α=0.25	α=0.5	α=0.75	α=1	CPU.time
VRP1	5	2	4250	4257	4269	4269	4363	00	4250	4257	4269	4269	4363	03
VRP2	7	2	5316	5374	5473	5501	5536	12	5316	5374	5473	5501	5536	03.3
VRP3	7	3	6369	6376	6383	6496	7012	47	6369	6376	6383	6496	7012	05.67
VRP4	8	3	5476	5483	5499	5512	5528	54	5476	5483	5499	5512	5528	05.82
VRP5	9	2	6873	6879	6891	6905	6925	76	6873	6879	6891	6905	6925	06.54
VRP6	9	3	8023	8056	8168	8174	8192	143	8023	8056	8168	8174	8192	06.98
VRP7	11	2	10367	10381	10393	10659	10896	237	10367	10381	10393	10659	10881	06.31
VRP8	11	3	14368	14382	14697	14697	14987	362	14368	14382	14697	14697	14987	07.27
VRP9	12	2	15698	15717	15764	15764	15925	476	15680	15711	15742	15759	15890	08.11
VRP10	12	3	17698	17698	17705	17705	17798	681	17698	17698	17700	17697	17780	08.67

α, α-cut; CPU, central processing unit; HSA, hybrid simulated annealing; VRP, vehicle routing problem.

Feasible solution space search

Two efficient genetic operators were designed for searching in a feasible solution space and obtaining neighborhood solution, as follows:

1) Crossover operator: This operator randomly selects two different routes of two vehicles of the feasible solution. Then two nodes of these routes are replaced with each other while considering vehicle capacity and stochastic transit times limit.
2) Mutation operator: This operator randomly selects two routes of two vehicles of the feasible solution. Then, one node is deleted from one route, and it is added to another route while considering vehicle capacity and stochastic transit times limit (Tavakkoli-Moghaddam et al. 2005, 2006b)

Also, a schematic diagram of the two genetic operators used in the proposed algorithm is represented in Figure 2.

The main steps of the proposed HSA embedded with genetic operators are as follows.

Step 1: Generate an initial solution according to the initial solution generation algorithm.
Step 2: Generate a neighborhood solution for X_1 (i.e. X_2) and estimate its corresponding objective function value, E_2.
Step 3: If $E_1 \geq E_2$, then X_2 is a new solution of the given problem; otherwise, regenerate a random number, Rnd (0, 1).
Step 4: If Rnd (0, 1) < exp $\{-(E_2\text{-}E_1)/\text{KT}\}$, then accept X_2; otherwise, reject X_2, and the new solution is X_1. Also, k is a constant, and it considers 1 in problems.
Step 5: Repeat the algorithm several times for T temperature.
Step 6: Decrease T by a relation such as $(n + 1) = \eta \times T(n)$, $0 < \eta < 1$, in which η is the cooling schedule.

Table 2 Computational results for large-scale cases

Problems	Customers	Vehicles	Objective function value					CPU time (s)
			α = 0	α = 0.25	α = 0.5	α = 0.75	α = 1.0	
VRP1	20	4	18968	18976	18990	18995	19014	4
VRP2	25	4	19354	19369	19387	19394	19406	4
VRP3	30	5	20365	20391	20409	20431	20434	7
VRP5	35	5	22157	22161	22169	22177	22190	7
VRP6	40	6	24698	24698	24710	24726	24759	7
VRP7	45	6	26874	26894	26902	28187	29036	8
VRP8	50	7	31265	31286	31297	31297	31306	11
VRP9	55	7	32169	32212	32236	32263	32298	12
VRP10	60	8	33698	33705	33742	33775	33796	13

α, α-cut; CPU, central processing unit; VRP, vehicle routing problem.

Step 7: Terminate the algorithm when the convergence criterion is satisfied.

Results and discussion

The computational results are obtained from the proposed solution approach in this section. All solution procedures are coded in the visual basic, and all test problems are run using the Intel Dual Core 1.66-GHz compiler and 1 GB of RAM. In the proposed HSA, the travel route and the travel cost are estimated for each service. Hence a small-sized problem is generated by using the uniform distribution, and it is solved with the Lingo 8 software. To solve this problem by the proposed HSA, the related parameters are set as EL = 100, MTT = 100, $T_0 = 100$ and $\eta = 0.99$.

In addition, the stochastic travel time of a fleet for the different route is estimated with the confident coefficient of 95% and in several satisfaction levels. The results of ten cases in small sizes are shown in Table 1. The comparison of the Lingo software with the proposed algorithm shows that this algorithm has good capability to solve the proposed problem in less computational time than Lingo. Furthermore, increasing the size of the problem increases the solution time of Lingo exponentially, while it does not affect the solution time of the proposed HSA. The efficiency of this algorithm is shown for ten large-scale cases in Table 2.

As illustrated in these tables, the results show that increasing the driver's satisfaction grows the routing cost, resulting in the increase of the total cost of a distributing company. For this reason, increasing the driver's satisfaction and decreasing the costs are considered as objective functions which are in conflict with each other. Improving each objective leads to the deterioration of the other objective. Thus, definition of an appropriate satisfaction level can be considered as a decision management. Paying attention to Pareto set and following its data with regard to management utility, appropriate solutions can be obtained.

Conclusions

Vehicle routing problems are categorized as NP-hard problems in which exact solution methods, such as solving with Lingo, are not practical in large scale due to high computational cost. For this kind of problems, using meta-heuristic algorithms are efficient. In this paper, a hybrid simulated annealing algorithm was proposed to solve vehicle routing problems. The total travel time was limited to a definite probability percent, and also, other constraints, such as capacity and time distribution restrictions, are considered while the total cost of the transportation was minimized. Based on the obtained results, the proposed HSA can be used for obtaining high-quality solutions with a reasonably computational cost.

Competing interests

The author(s) declare that they have no competing interests.

Author details

[1]Department of Industrial Engineering, College of Engineering, University of Tehran, Tehran, Iran. [2]Department of Industrial and Systems Engineering, Isfahan University of Technology, Isfahan, Iran. [3]Department of Industrial Engineering, Tehran South Branch, Islamic Azad University, Tehran Iran.

Authors' contributions

RTM managed the study and was responsible for integrating and revising the manuscript. MA participated in the design of the study, contributed to the development of goals and managed the data. AS analyzed the data and prepared the figures. NN conducted the statistical analysis, participated in the problem solving approach and drafted the manuscript. All authors read and approved the final manuscript.

References

Altinkemerk GB (1991) Parallel savings based heuristic for the delivery problem. Operations Research 39:456–469
Bin Y, Zhong-Zhen Y, Baozhen Y (2009) An improved ant colony optimization for vehicle routing problem. European Journal of operational Research 196:171–176
Clarke C, Wright JQ (1994) Scheduling of vehicle from a central depot to a number of delivery points. Operations Research 12:568–581
Dantzig GB, Fulderson R, Johnson SM (1994) Solution of a large scale traveling salesman problem. Management Science 6:80–91
Fallahi A, Pprins C, Calvo R (2008) A memetic algorithm and tabu search for the multi-comparment vehicle routing problem. Computer & operations Research 35:1725–1741
Gidas B (1985) Non-stationary markov chains and convergence the annealing algorithm. Journal of Statistical Physics 39:73–131
Glover F, Kochenberger GA (2003) Handbook of meta-heuristics. Kluwer Academic Publishers, London
Hajek B (1985) Cooling schedules for optimal annealing. University of Illions Champaign, Urbana, Department of Electrical Engineering
Jozefowiez N, Semet F, Talbi E-G (2009) An evolutionary algorithm for the vehicle routing problem with route balancing. European Journal of Operational Research 195(3):761–769
Kirkpatrick F, Gelatt C, Vecci M (1983) Optimization by simulated annealing. Science 2:117–133
Koskosidis Y, Powell W, Solomon M (1992) An optimization-based heuristic for vehicle routing and scheduling with soft time window constraints. Transportation Science 26:69–85
Laporte G (1992) The vehicle routing problem: an over view of exact and approximate algorithm. European Journal of Operational Research 59:345–358
Laporte G, Mercure H, Nobert Y (1992) A branch and bound algorithm for a class of asymmetrical vehicle routing problems. Journal of Operational Research Society 43:469–481
Lenstra JK, Rinnooy Kan AHG (1981) Complexity of vehicle and scheduling problem. Networks 11:221–227
Mingozzi A, Giorgi S, Baldacci R (1999) Solution of a large scale traveling salesman problem. Management Science 6:80–91
Tavakkoli-Moghaddam R, Safaei N, Gholipour Y (2006a) A hybrid simulated annealing for capacitated vehicle routing problems with the independent route length. Applied Mathematics and Computation 176:445–454
Tavakkoli-Moghaddam R, Safaei N, Shariat MA (2005) A multi-criteria vehicle routing problem with soft time windows by simulated annealing. Journal of Industrial Engineering - Int 1(1):28–36
Tavakkoli-Moghaddam R, Saremi AR, Ziaee MS (2006b) A memetic algorithm for a vehicle routing problem with backhauls. Applied Mathematics and Computation 181:1049–1060
Zimmermann HJ (1993) Fuzzy set theory and its applications, 2nd edn. Kluwer Academic Publishers, London

2

A population-based algorithm for the railroad blocking problem

Masoud Yaghini*, Masoud Seyedabadi and Mohamad M Khoshraftar

Abstract

Railroad blocking problem (RBP) is one of the problems that need an important decision in freight railroads. The objective of solving this problem is to minimize the costs of delivering all commodities by deciding which inter-terminal blocks to build and by specifying the assignment of commodities to these blocks, while observing limits on the number and cumulative volume of the blocks assembled at each terminal. RBP is an NP-hard combinatorial optimization problem with billions of decision variables. To solve the real-life RBP, developing a metaheuristic algorithm is necessary. In this paper, for the first time, a new genetic algorithm-based solution method, which is a population-based algorithm, is proposed to solve the RBP. To evaluate the efficiency and the quality of solutions of the proposed algorithm, several simulated test problems are used. The quality and computational time of the generated solutions for the test problems with the proposed genetic algorithm are compared with the solutions of the CPLEX software. The results show high efficiency and effectiveness of the proposed algorithm.

Keywords: Railroad blocking problem, Genetic algorithm, Budget design problem

Background

Railroad blocking problem (RBP) is an important subject for freight railroad companies because its solution can reduce some extra costs. To reach a destination, a load on its route may traverse on many classification yards from its origin to its destination. In these places, the arriving load, which is usually composed of different individual shipments, is regrouped to be carried by departing trains. However, because the reclassification process is difficult and capital is intensive, this classification always produces some delay in the movement of these loads. Therefore, to avoid loads from being regrouped at every classification yard they pass through, it is better to group various shipments together to form a block. A block may contain individual shipments whose origin or destination may or may not be the origin or destination of the whole block. These shipments are regrouped only after reaching the final destination of the block.

The RBP is a network budget design problem (BDP). In the BDP, the flow cost of a set of commodities through a network is minimized while observing budget constraints on the fixed costs of the links used. BDP is related to several other problems, such as the fixed-charge network design problem and the multicommodity flow problem. The fixed-charge network design problem is different because the fixed costs for the arcs appear in the objective function. This problem is known as nondeterministic polynomial time-hard (NP-hard) combinatorial optimization problem (Magnanti and Wong 1984; Minoux 1989). No polynomial-time optimization techniques are known to solve the NP-hard problem. The multicommodity flow problem is a BDP with fixed binary arc selection variables. The interested reader of the network design problem should consult the surveys by Magnanti and Wong (1984), Minoux (1989), and the annotated bibliography of Balakrishnan et al. (1997), Crainic (2000, 2006, 2009).

We can define the RBP as follows. To reduce the intermediate handlings of shipments as they travel over the railroad network, a set of shipments is classified (or grouped together) at a railroad yard to create a group of cars called block. The aim of the RBP is to choose which blocks to put together at each yard and to determine the sequences of blocks that deliver each shipment to minimize the total miles that the shipments travel and

* Correspondence: yaghini@iust.ac.ir
School of Railway Engineering, Iran University of Science and Technology, Tehran 16846-13114, Iran

their intermediate handlings (or reclassifications). In fact, the RBP aims to identify the classification plan for all shipments at all yards in the network to minimize the total shipment cost, i.e., to create a blocking plan which dictates which blocks should be built at each individual yard and which traffic should be assigned to each block as they travel from origins to destinations.

The RBP in real-life is a very large-scale problem with billions of decision variables. To solve it, we need to design the underlying blocking network, which is a virtual network that is overlaid on the physical network, and to route different commodities, where each set of railcars with the same origin-destination pair of nodes defines a separate commodity on the blocking network to minimize the system-wide transportation costs. Multicommodity flow and network-design problems are among the most difficult combinatorial problems in research operations. When they comprise a few hundred design variables, we try solving them optimally. However, the real-life RBP often contain over a million design variables and hundreds of billions of flow variables. Modeling and solving them, using commercial software, is nearly impossible.

A pioneer effort to place railroad modeling within a hierarchical structure is provided in the work of Assad (1980). The author offers a three-tiered hierarchy of models, including strategic, tactical, and operational levels. Differences in these levels are about planning horizon, required investment, and modeling methods (Assad 1980). RBP is one of the main problems that need a decision in rail transportation planning, which is made in the tactical level of railroad operation.

Central to railroad operations is the operating plan that dictates the movement of shipments (railcar loads), crews, and locomotives over the railroad's network. Each railroad company has a service design department that is responsible for creating operating plans to enable efficient movement of shipments (Figure 1; Ahuja et al. 2007).

The blocking plan determines how to aggregate a large number of shipments into blocks of shipments as they travel from origins to destinations. Train scheduling consists of designing train routes, days of operation, timings, and routing of blocks on trains to minimize the total system wide cost; this cost includes car hire, crew, and locomotive costs. An operating plan dictates the flow of three important railroad assets: crews, locomotives, and railcars. A well-designed operating plan can reduce the costs of railroads significantly (Ahuja et al. 2007).

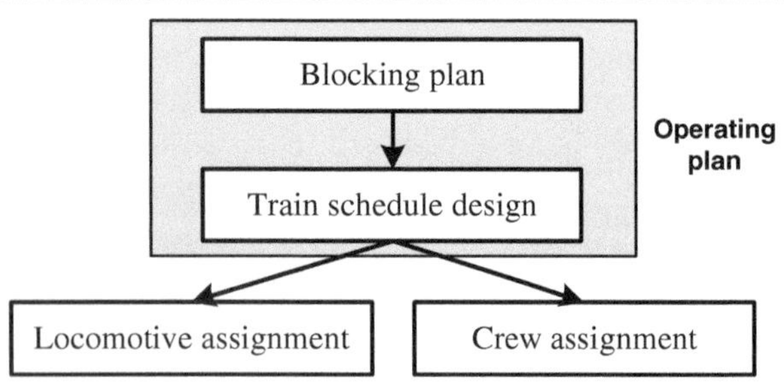

Figure 1 An operating plan consists of a blocking plan and a train schedule. Modified from Ahuja et al. (2007).

Greedy heuristic methods produce solutions that are far from optimal and/or cannot handle real-life complexities of the blocking problem. As a result, no railroad company uses any optimization-based approach to determine blocking plans, but they rely on manual decision-making processes to make and change these plans. This manual process is very time intensive and does not save substantially on transportation costs. Consequently, we believe that there are major opportunities for cost saving using an optimization-based approach that can determine near-optimal blocking plans.

In this research, we used a population-based metaheuristic (p-metaheuristics) for solving RBP. The p-metaheuristics share many common concepts. They could be viewed as an iterative improvement in a population of solutions. First, the population is initialized; a new population of solutions is then generated. Finally, this new population is integrated into the current one, using some selection procedures. The search process is stopped when a given condition is satisfied (stopping criterion) (Talbi 2009).

We use a mathematical model for RBP wherein its decision variables identify a blocking scheme and a commodity assignment to this block. For the first time, we proposed a new genetic algorithm-based solution method for the model, and several genetic search operators are used to find optimal solutions. To implement the proposed algorithm, we utilized the Java programming language (Oracle Corporation, Redwood City, CA, USA), and to validate the proposed algorithm, we made several blocking problems randomly and compared the proposed algorithm results with the results of the CPLEX software (IBM Corporation, Armonk, NY, USA) for these problems. The evaluating of results revealed the effectiveness and efficiency of the proposed solution method for RBP.

The outline of this paper is as follows. In the 'Literature review' subsection, related work is reviewed. 'The railroad blocking problem' subsection introduces the railroad blocking problem. Two mathematical models that include nod-arc and path-based formulations are presented in 'The Mathematical formulation' subsection of the 'Methods' section. In 'The proposed algorithm' subsection, the proposed algorithm is described. A small example is solved with the proposed algorithm in the 'An example' subsection. The next two sections presented are 'Results and discussion' and 'Conclusions', respectively.

Literature review

Previous surveys by Cordeau et al. (1998) and Crainic (2000, 2002) reviewed optimization models for rail transportation.

One of the first models for blocking belongs to Bodin et al. (1980) who suggested a nonlinear mixed integer programming formulation of the problem. The model is a multicommodity flow problem with additional side constraints that simultaneously determine the optimal blocking strategies for all the classification yards in a railroad system.

Assad (1983) proposed a solution approach for a problem defined on a line network. The cars are received at the first yard in arbitrary order and must be separated as they proceed along the line to allow each successive yard to extract the traffic destined for it.

Van Dyke (1986, 1988) described an interactive, heuristic approach that attempts to improve an existing blocking plan by solving a series of shortest-path problems on a network whose arcs represent available blocks.

Keaton (1989) proposed a nonlinear mixed integer program (MIP) model and a heuristic method based on Lagrangian relaxation for the combined problem of blocking, train routing, and makeup. In a subsequent paper by Keaton (1992), pure strategy constraints for blocking and maximum transit times for each origin-destination pair are also considered.

Huntley et al. (1995) developed a simulated annealing approach, and Gorman (1998) proposed a hybrid tabu search and genetic algorithm for constructing operating plans including blocking, train connections, and block-to-train assignments.

One of the significant researches on RBP has been done by Newton (1996) and Newton et al. (1998). They modeled the RBP as a network design model and formulated it as an MIP. They developed column generation and branch-and-price algorithms to solve it. Barnhart et al. (2000) used the same formulation as Newton et al. (1998) and proposed the Lagrangian relaxation technique to decompose the problem into two subproblems. Their approaches focused on determining a near-optimal solution. However, the running times are not scalable as the problem increases in size; this prohibits their use in practice.

Ahuja et al. (2007) developed an algorithm using a technique known as very large-scale neighborhood (VLSN) search that is able to solve the problem to near optimality using 1 to 2 h of computer time on a standard workstation computer. Jha et al. (2008) presented two formulations for the block-to-train assignment problem: an arc-based formulation and a path-based formulation. The latter is generally smaller than the former, and it can better handle practical constraints. They also proposed exact and heuristic algorithms based on the path-based formulation.

Zhu et al. (2009) attempted to combine blocking, makeup, and empty car distribution in an integrated dynamic model. To do this, they built a two-layer network, one for block flows and another for car flows. The objective function includes two fixed costs on open service and block, and a variable cost on all blocks. The model has three types of capacity constraints on maximum cars in each train, maximum car on each block, and maximum block being built in each yard at each time. They used a tabu search metaheuristic to solve the proposed model. Yang et al. (2010) proposed a freight transportation model with mixed uncertainty of randomness and fuzziness. Table 1 represents the well-known research in this field.

The railroad blocking problem

In this section, we describe the RBP and rail operations in general. In the RBP, the physical rail network (the railroad terminals and tracks) is already defined. The blocking network to be constructed is a virtual network that is overlaid on the physical network. The blocks are virtual express arcs which a commodity may take to have uninterrupted service between two terminals that are not necessarily connected by a physical link (Newton 1996).

On railroads, a typical general merchandise shipment may pass through many terminals on its route from origin to destination (OD). At these terminals, the incoming traffic is reclassified (sorted and grouped together) to be placed on outgoing trains. On average, each reclassification results in a 1-day delay for the shipment. In addition, the classification process is labor and capital intensive since many workers and large quantities of equipment are needed to sort the traffic, and construction and maintenance of large terminals are necessary to handle the sorting task. To prevent shipments from being reclassified at every terminal they pass through, several shipments may be grouped together to form a block. A block has associated with it an OD pair, which may or may not be the OD pair of any of the individual cars contained in the block. Once a shipment is placed in a block, it is not reclassified until it reaches the destination of that block (Newton 1996).The railroads develop the blocking plans that dictate which blocks should be built at each individual terminal and which traffic should be assigned to each block.

The objective of the RBP is to develop a feasible blocking plan which, in conjunction with other operating policies, minimizes the total cost of delivering the commodities. These costs are usually broken down into car-handling costs associated with handling (or blocking) a car and car-mile costs associated with the movement of a car. To solve the RBP, we must decide which blocks to include in the blocking plan and which blocks to use to

Table 1 The prominent railroad car blocking research

Authors	Problem type	Planning horizon	Objective function	Model structure	Solution approach
Bodin et al. (1980)	Blocking	Tactical	Minimum operating and delay costs	Nonlinear MIP	Heuristic
Assad (1983)	Blocking	Operational	Minimum total classification	Shortest path	Dynamic programming
Van Dyke (1986)	Blocking	Tactical	Minimum operating costs	Shortest path	Heuristic
Newton (1996)	Blocking	Tactical	Minimum operating costs	NDP with node budget	Dantzig-Wolfe decomposition
Barnhart et al. (2000)	Blocking	Tactical	Minimum operating costs	NDP with node budget	Lagrangian relaxation
Ahuja et al. (2007)	Blocking	Tactical	Minimum operating costs	Linear MIP	VLSN

deliver each commodity (Newton 1996). An example of a rail system with four terminals is shown in Figure 2.

We define commodities as origin-destination pairs of terminals. We assume that the only commodities are A $\rightarrow$ B with 100 cars, A $\rightarrow$ C with 80 cars, and A $\rightarrow$ D with 90 cars. Terminal A has a volume capacity of 270 cars, which is its minimum since 270 cars originate there. Terminals B and C can each block 90 cars. Terminal A can build two blocks, whereas each of the other terminals can only build one block.

At each terminal, limited classification resources restrict the number of cars (or car volume), which can be classified, and the number of blocks. In terms of our graph, the car volume constraints would be the upper bounds on the flow through each node, and the number of block constraints would be limits on the out-degree of each node. One performance metric of a blocking plan is the total number of classifications necessary to deliver all commodities. We consider five different blocking plans for this example.

With blocking plan 1, at terminal A, all three commodities are sorted to move to terminal B. The resulting string (block) of cars may have the commodities intermingled. At terminal B, commodity A $\rightarrow$ B has reached its destination and leaves the system. Commodities A $\rightarrow$ C and A $\rightarrow$ D are blocked to be move to terminal C. At terminal C, commodity A $\rightarrow$ C has reached its destination and leaves the system, and commodity A $\rightarrow$ D is blocked to move to terminal D. Using this blocking plan, the 100 cars for commodity A $\rightarrow$ B use one block, (A, B), the 80 cars for commodity A $\rightarrow$ C use two blocks, (A, B) and (B, C), and the 90 cars for commodity A $\rightarrow$ D use three blocks, (A, B), (B, C), (C, D). Therefore, for blocking plan 1, there are 530 classifications; however,

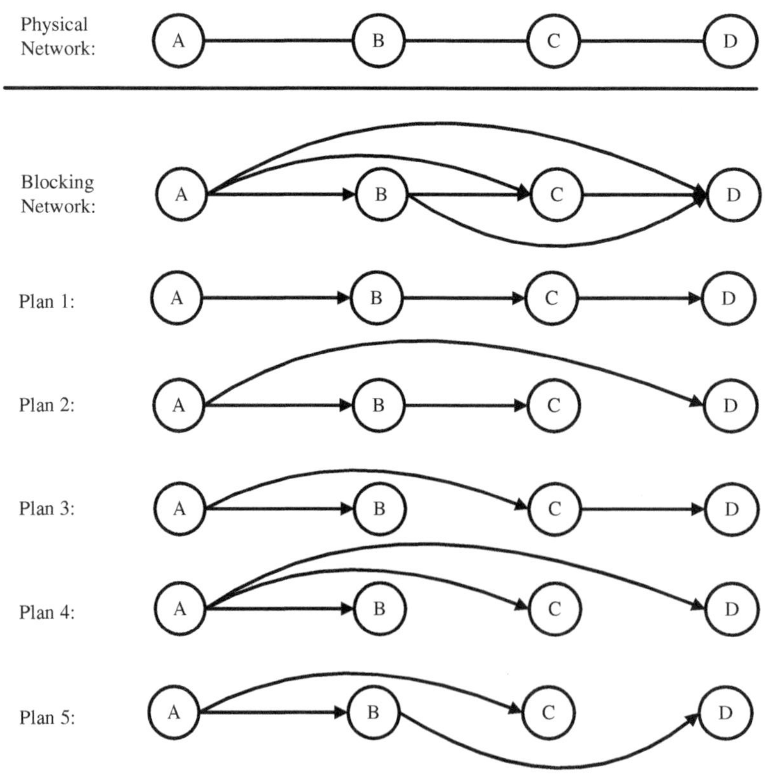

Figure 2 Physical rail system and five blocking plans.

this plan is not feasible because it requires blocking 170 cars at terminal B that has a maximum volume of 90 cars.

Blocking plan 2 provides a direct block from terminal A to terminal D. Cars which travel in this block will still move along the physical track through terminals B and C, but since they are presorted for terminal D, they do not require classification resources at B or C. To deliver the three commodities using this blocking plan requires 350 classifications, since the 90 cars of commodity A → D now require only one block, (A, D). This blocking plan is feasible since it observes the limits on number and total volume of the blocks built at each terminal.

Similar calculations show that blocking plans 3, 4, and 5 require 360, 270, and 360 classifications, respectively. However, blocking plan 4 is infeasible since it requires three blocks for terminal A, which has a maximum capacity of two blocks. Commodities are always carrying directly to their destination, as in blocking plan 4.

Therefore, blocking plan 2 with 350 handlings is optimal with respect to the total number of classifications required. However, if the number of cars which could be blocked at B is reduced to less than 80 (or equivalently, the volume of commodity A → C were more than 80), then blocking plan 2 would be infeasible and blocking Plan 3 with 360 handlings would be optimal.

Methods

The mathematical formulation

The RBP can be modeled as a network design problem where the nodes represent the railroad terminals, and the arcs represent potential blocks (Newton 1996). In this section, we present a node-arc MIP for the RBP. The parameters and variables used in this model are as follows:

- Parameters
  - $G = (N, A)$:the graph with terminal set N and potential blocks set A.
  - K:the set of all commodities k designated by an origin–destination pair of nodes.
  - v^k:the volume of commodity k.
  - orig(k):the origin terminal for commodity k.
  - dest(k):the destination terminal for commodity k.
  - orig(a):the origin of potential block a.
  - dest(a):the destination of potential block a.
  - u_a:the capacity of potential block a.
  - c_a:the per unit cost of flow on arc a.
  - $B(i)$:the number of blocks which may originated at terminal i.
  - $V(i)$:the volume which may be classified at terminal i.
- Variables
  - $x_a^k = 1$ if the commodity k is flowing on block a; 0 otherwise.
  - $y_a^k : y_a = 1$ if block a is included in the blocking network; 0 otherwise.

- Node-arc formulation

$$\min \sum_{k \in K} \sum_{a \in A} c_a v^k x_a^k \tag{1}$$

st.

$$\sum_{\substack{a \in A \\ \text{orig}(a) = i}} x_a^k - \sum_{\substack{a \in A \\ \text{dest}(a) = i}} x_a^k = \begin{cases} 1 & \text{orig}(k) = i \\ -1 & \text{dest}(k) = i \forall i \in N, k \in K \\ 0 & \text{otherwise} \end{cases} \tag{2}$$

$$\sum_{k \in K} v^k x_a^k \le u_a y_a \forall a \in A \tag{3}$$

$$\sum_{\substack{a \in A \\ \text{orig}(a) = i}} y_a \le B(i) \forall i \in N \tag{4}$$

$$\sum_{k \in K} \sum_{\substack{a \in A \\ \text{orig}(a) = i}} v^k x_a^k \le V(i) \forall i \in N \tag{5}$$

$$y_a \in \{0, 1\} \forall a \in A \tag{6}$$

$$x_a^k \in \{0, 1\} \forall a \in A, k \in K \tag{7}$$

The objective is to minimize the sum of the costs of delivering each commodity using the blocking network formed by blocks for which $y_a = 1$. For each terminal, constraints (2) are balance equations for the flow of each commodity. For each potential block, constraints (3) prevent flow on blocks, which are not built and enforce the upper bound u_a on flow for blocks, which are built. The constraints (4) enforce the terminal limit $B(i)$ for the sum of the blocks which leave the terminal. The constraints (5) model the volume of cars, which may be classified at each terminal.

We also present the path-based formulation for the RBP. The parameters and variables used in this model are as follows:

- Parameters$Q(k)$:
  - the set of legal paths for commodity k.
  - orig(k):the origin node for commodity k.
  - dest(k):the destination node for commodity k.
  - PC_q^k :the path cost for flowing one unit of commodity k on path q.

δ_a^q :the binary design parameter; $\delta_a^q = 1$ if qth path contain arc (block) ath 0 otherwise.

- Variablesy_a:
  the binary design variable for including arc (block) a:$y_a = 1$ if arc a is selected; 0 otherwise.
  f_q^k :the binary flow variable; f_q^k : = 1 if commodity k flows on path q 0 otherwise.

The path-based mathematical model for RBP is explained as flowing equations:

- Path based formulation

$$\min \sum_{k\in Q} \sum_{q\in Q(k)} PC_q^k v^k f_q^k \tag{8}$$

st.

$$\sum_{k\in Q} \sum_{q\in Q(k)} f_q^k v^k \delta_a^q - u_a y_a \leq 0 \forall a \in A \tag{9}$$

$$\sum_{q\in Q(k)} f_q^k = 1 \forall a \in A \tag{10}$$

$$\sum_{\substack{a \in A \\ orig(a) = i}} y_a \leq B(i) \forall a \in A \tag{11}$$

$$\sum_{k\in K} \sum_{q\in Q(k)} \sum_{\substack{a \in A \\ orig(a) = i}} v^k f_q^k \delta_a^q \leq V(i) \forall i \in N \tag{12}$$

$$f_q^k \in \{0,1\} \forall q \in Q(k), k \in K \tag{13}$$

$$y_a \in \{0,1\} \forall a \in A \tag{14}$$

Equation 8 is the objective function of this model; it minimizes total costs of delivering each commodity on paths because classification must be done on all commodities at its origin and destination. In this model, classification cost is not considered at that place for each commodity. Constraints (9) ensure that flow is only allowed on arcs (blocks) included in the network. Constraints (10) ensure that all of each commodity is delivered. Constraints (11) limit the number of blocks, which may be built at each node, which is modeled by the node-budget constraints. Constraints (12) limit the volume of cars, which may be classified at each terminal.

The proposed algorithm

This section introduces the proposed p-metaheuristic algorithm for solving the RBP. The proposed solution method is based on genetic algorithms (GAs). GAs are one of the most famous p-metaheuristic algorithms which have been developed by J. Holland in the 1970s (University of Michigan, USA) to understand the adaptive processes of natural systems; Then, they have been applied to optimization and machine learning in the 1980s. A GA usually applies a crossover operator to two solutions that plays a major role, plus a mutation operator that randomly modifies the individual contents to promote diversity. GA uses a probabilistic selection that is originally the proportional selection. The replacement (survivor selection) is generational, i.e., the parents are replaced systematically by the offsprings. The crossover operator is based on the n-point or uniform crossover while the mutation is bit flipping. A fixed probability p_m (p_c) is applied to the mutation (crossover) operator (Talbi 2009). GAs have several advantages. Instead of single solution optimization, GAs use a population of solution simultaneously, so the probability of getting stuck in local minima is prevented; in comparison with other p-metaheuristic methods, GAs have high-quality solutions and good convergence speed.

General structure of the proposed algorithm

The proposed algorithm has several stages. At the first stage, solutions (feasible and infeasible) are encoded into the chromosome. The idea of this stage is simulation and good representation of all possible solutions for the next stage. Encoding solutions into chromosomes is important to other stages. In the proposed algorithm, to find solutions in a reasonable time, each possible solution is encoded into strings of integer numbers (chromosome). To construct the initial population, each possible solution is created randomly, but at least one feasible solution is added to the initial population. There are two approaches to find a feasible initial solution. If the car blocking problem is implemented for a railway which has a specified blocking plan for each O-D pairs, this blocking plan could be used as initial feasible solution. However, if a car blocking problem is implemented for a railway without a specified blocking plan for each O-D pairs, it needs a heuristic algorithm to initialize feasible solutions. To present a feasible solution, one must consider two problem constraints (i.e., number of block and car volume). Figure 3 represents the process of this heuristic.

To do a comparison among chromosomes, we use a fitness function. A fitness value is assigned to each chromosome based on this function. This stage of the proposed algorithm for solving the problem is extremely important. Because it establishes a measure to indicate superiority or inferiority of each possible solution, the measure could attribute a real number to each possible solution; this number represents its degree of dominance, e.g., suppose f is a fitness function, and X_1 and X_2 are two chromosomes of a population and $f(X_1) = 3$ and

Step 1. **O-D Pair Selection:** If there are unselected O-D pairs, choose an unselected O-D pairs randomly, otherwise go to *Step 7*;
Step 2. **Path Selection:** Select one legal path of the O-D pairs, randomly;
Step 3. **Assignment:** Among all possible commodities, assign those which could be delivered through this selected path;
Step 4. **Update:** Update number of block and car volume for the selected O-D pairs and its intermediate station;
Step 5. **Feasibility:** If this path is a feasible path then go to *Step 1*, if not, go to *Step 2* and select another legal path. Otherwise if all path are infeasible go to *Step 6*;
Step 6. **Cancellation:** Unselect all O-D pair which this origin or destination is a part of it and go to *Step 1*;
Step 7. **End:** the end of method.

Figure 3 The heuristic steps for finding a feasible solution.

$f(X_2) = 5$, then X_2 is better than X_1, or it has a greater fitness value.

After evaluating the fitness value of each chromosome of the initial population, the proposed GA randomly chooses two chromosomes according to the roulette wheel method; the chromosomes with greater fitness value are selected as the parents. In constructing a new generation of chromosomes, crossover and mutation operators are utilized. The crossover operator first selects two parents by roulette wheel selection method and chooses a place on each parent as crossover point randomly. To make the first (second) child, the algorithm swaps the content of the left (right) side of each parent with another one. The mutation operator selects a chromosome from the population randomly and then selects one point on the chromosome randomly, and the content of its gene is changed according to the mutation probability (p_m); this process is continued for all selected chromosome gene. After producing a new generation of chromosomes, the next generation is identified. In the stage of producing the next generation, the proposed algorithm keeps k as the best feasible solution of the current generation to include in the next generation. If m is the number of initial population, in order to create the next generation, the m and k chromosomes of new generation are initially selected randomly, and then, k is selected as the best feasible solution of the current population. Their union makes the next generation. The algorithm checks two termination conditions simultaneously: (1) if the iteration number does exceed N and (2) if after L iterations of the algorithm, the best solution so far does not improve. Figure 4 is a pseudocode for the proposed algorithm.

Algorithm: the GA algorithm for solving RBP
Begin
  Read data;
  Create paths for all commodities;
  Create initial population;
  Identify *K* best feasible solution;
  Repeat
    Do parent selection;
    Do recombination;
    Do mutation;
    Replace current population with new population;
    Evaluate fitness of chromosomes;
    Find feasible solutions;
    Update K best feasible solutions based on new population;
    Replace *K* best feasible solutions in the new population;
  Until termination condition is met;
End

Figure 4 The pseudocode for the proposed algorithm.

Figure 5 Physical rail network with four stations.

Parameter tuning

The metaheuristics are sensitive to the value of their parameters. The parameters are the configurable components of a metaheuristic algorithm. Parameter tuning may allow a larger flexibility and robustness to the algorithm but requires a careful initialization. Those parameters may have a great influence on the efficiency and effectiveness of the search. It is not obvious to define *a priori* which parameter setting should be used. The optimal values for the parameters depend mainly on the problem, and even the instance to deal with, and on the search time that the user wants to spend in solving the

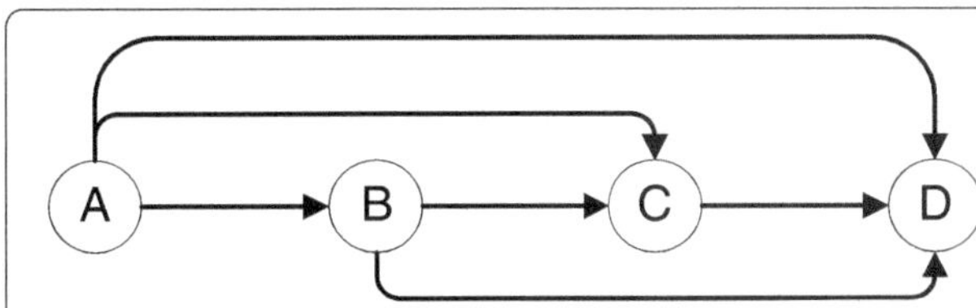

Figure 6 The blocking network.

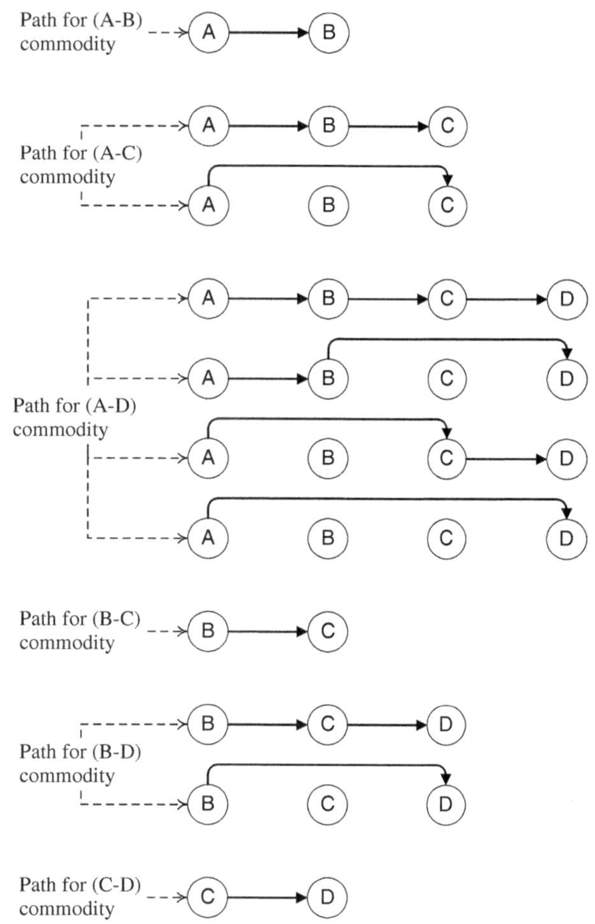

Figure 7 Legal path for each demand.

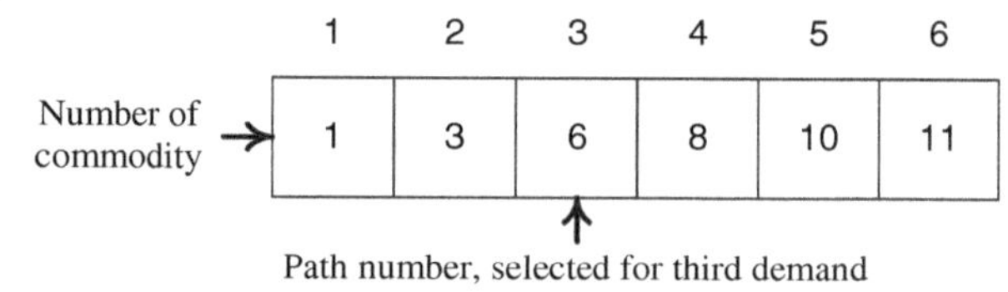

Figure 8 A schematic of the proposed chromosomes.

problem. A universally optimal parameter value set for a given metaheuristic does not exist.

One main step of this research is fine parameter tuning of algorithm. In the proposed algorithm, there are three main parameters which include the number of individuals of a population, crossover, and mutation rate. We use a trial-and-error procedure to obtain locally optimal values for the parameters. To tune parameters, we generated three instances with different sizes. For tuning, only one parameter changed at that time, while all other parameters didn't change. Proper values of the parameters were then determined through running the algorithm 10 times over different values of the parameters and through calculating the average of the goal function for these 10 runs. The criteria for modifying parameters are the quality of solutions and time spent to find them. The final value for the parameter is as follows: the population size is $6 \times$ the number of commodity, cross over rate is 0.7, and mutation rate is (1/the number of commodity).

An example

In the proposed method, each gene represents one commodity, and the value of this gene is the legal

Table 2 Comparison results of the proposed algorithm with CPLEX

Problem			CPLEX		GA algorithm		
Name	Number of terminals	Number of commodities	Objective value	CPU time (s)	Objective value	CPU time (ms)	Percentage error
SP1	5	3	5640	1000	5640	15	0%
SP2	5	5	11400	1000	11400	16	0%
SP3	5	10	20880	2000	20880	46	0%
SP4	10	5	11100	3000	11100	93	0%
SP5	10	10	23940	3000	23940	140	0%
SP6	10	31	60720	5000	60720	951	0%
SP7	20	105	8586	14000	8656	1,717	1%
SP8	20	195	13676	15000	13898	4,041	1.6%
SP9	30	133	10298	18000	10672	4,150	3.6%
SP10	30	292	21604	23000	22770	16,099	4.9%
SP11	50	220	15922	38000	16702	15,803	4.8%
SP12	50	521	39700	110000	39804	81,242	0.3%

blocking path number (on the blocking network) which this commodity can select. To determine this legal path for each commodity, a comprehensive preprocess is needed. In the preprocess procedure, the legal paths for each commodity, cost of each path, number and type of block on each path are determined.

Figure 5 represents a physical rail network with four stations and six demands (A → B, A → C, A → D, B → C, B → D, and C → D). Demand A → B is named (1),. . ., and demand C → D is named (6), respectively.

On the first step, the blocking network is made with respect to the physical rail network and demand. Figure 6 represents the blocking network for the presented physical rail network. Each arc on this network represents a legal block from each station.

The next step is to compose legal paths of each demand as shown in Figure 7, and then delivering cost of cars on a legal path, which contains the shipment cost of cars on the physical path and reclassification cost at the intermediate terminal (expect origins and destinations), is computed.

According to this legal path, the chromosomes of the proposed algorithm are created. Figure 8 represents a schematic of these chromosomes that we used to solve the problem.

To do a comparison among chromosomes, a fitness function is used as Equation 15, which X_i is a chromosome of the population, K is set of commodity, $c_k^{x_i}$ is the cost of selected path at kth gene of X_i chromosome, and v^k is the volume of kth commodity:

$$f(X_i) = \sum_{k \in K} v^k c_k^{X_i} \tag{15}$$

Note that the proposed algorithm has a possibility to produce infeasible solutions (chromosomes). Hence, after producing an infeasible chromosome, a penalty term is considered; this term is added to the fitness value of infeasible solutions as an extra cost.

To create a new generation, we select some chromosomes as parents to produce children. This selection must give greater chances not only to chromosomes with lower cost, but also to chromosomes with greater cost because they may have some genes with high quality. To do this, we implemented the roulette wheel method. Suppose a circular plane that is able to circle around its central point. On this plane, consider some sections; the number of these section is equal to the number of the population, and the area of this section has a direct relationship with the selection probability of an individual chromosome. Beside this plane is a fixed point; the plane spins around its central point to stop. A section that is in front of this fixed point is selected

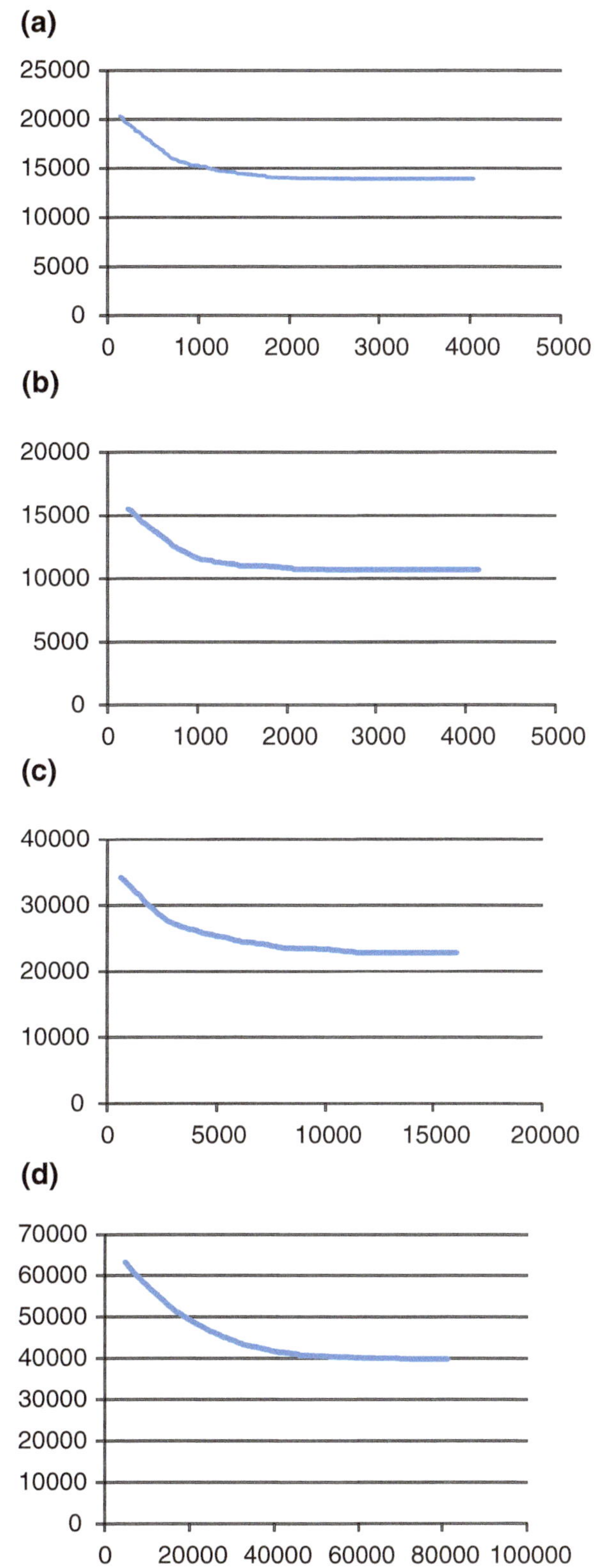

Figure 9 Time-cost graphs for (a) SP8, (b) SP9,(c) SP10, and (d) SP12.

when the plane stops spinning. To simulate the roulette wheel action, according to Equation 16, each chromosome of the current generation initially gets a number between 0 to 1. X_i is a chromosome; m is the number of population, and sumfitness $= \sum_{m}^{j=1} f(x_i)$.

$$(1-(f(X_i)/\text{sumfitness})/(m-1)). \quad (16)$$

A random number between 0 to 1 named r is selected, and n is selected as Equation 17.

$$\sum_{n-1}^{i=1}(1-(f(X_i)/\text{sumfitness})/(m-1)) \le r \le \sum_{m}^{i=1}(1-(f(X_i)/\text{sumfitness})/(m-1)). \quad (17)$$

After the parent selection, we use a one point crossover to produce a new generation. After selecting two parents from $[1, L-1]$ interval, this operator creates a random number (with uniform distribution) named k; L is the length of the chromosome. The gene at $k+1$, $k+2,\ldots, L-1$, and L from the first parent is replaced by the corresponding gene at the second parent to create two new strings. After the crossover action, the mutation operator changes some genes of the chromosome. The mutation operator on chromosome X_i first creates a random number between 0 and 1; if this number is less than the mutation probability, then the algorithm changes the content of this gene. To change the value of gene kth of the chromosome, from interval $[1, |Q(k)|]$, an integer number is initially created randomly ($|Q(k)|$ is the number of legal path for commodity k). The algorithm then swaps the generated number with the previous one at gene kth of the chromosome.

Results and discussion

The proposed algorithm is implemented with the use of the Java programming language, and a personal computer with Intel® Pentium® processor (2M cache, 2 GHz; Santa Clara, CA, USA), 32-bit Windows 7 Ultimate operating system (Microsoft Corporation, Redmond, WA, USA), and 2-GB installed memory (RAM) is used to achieve all of the results.

Capacity to classification operations in each terminal, commodity volume, commodities' origin and destination, and physical location of each terminal were all under consideration. Through the process, the main issue was to create a problem similar to real-world conditions as possible. The generated problems were solved using the proposed algorithm, which were implemented with Java programming language and CPLEX 11.1 software, separately. The CPLEX software is designed for large and complex problems. It is capable to solve linear programming, quadratic programming, and integer programming problems. Each one of the simulated problems is solved using the proposed algorithm 10 times. Table 2 lists the average of the objective values and the average of CPU time spent for finding the solutions by the proposed algorithm.

The last column shows the error percentages. As illustrated in Table 2, the proposed algorithm found optimum solutions for problems SP1 to SP12 in a short time. The algorithm has superior performance than the CPLEX software with respect to time spent to generate the solutions. The error percentage for the first six of these problems is 0%, and for the six remaining is approximately near 0%.

Figure 9a,b,c,d represents convergence behavior of the proposed algorithm for some test problems to find optimal solution during the successive iteration. In this figure, (1) time was measured in one thousandth second (millisecond), and (2) the solid line represents the GA solutions.

The proposed algorithm checks two termination conditions simultaneously. In the first termination condition, if after 50 iterations there is not any improvement in the best-so-far solution, the algorithm stops. In the second termination condition, if the number of iterationsexceeds5,000 iterations, the algorithm stops.

Conclusions

This paper, for the first time, introduced an approach based on genetic algorithm metaheuristic to solve the railroad blocking problem in a short reasonable time and high accuracy. The seven simulated problems were solved using the CPLEX software and the proposed algorithm, the solution accuracy, and the CPU time spent to find the solutions used to make comparison. The results show high efficiency and effectiveness of the proposed algorithms. It is possible to make significant savings in time and costs of operations through applying solutions produced by the algorithm. Finding feasible solutions for the algorithm is a challenging task with respect to great solution space. Future research on this paper could have two directions; one direction is to develop a heuristic algorithm to find initial feasible solution to speed up convergence, and the other direction is to use a hybrid genetic algorithm and ant colony optimization to enable the algorithm for solving more large-scale problems and decrease solving time of algorithm.

Competing interests

The authors declare that they have no competing interests.

Authors' contributions

MY has proposed the genetic algorithm for the railroad blocking problem. MS has worked on the modeling and the literature review; he also performed numerical experiments. MMK has generated the test problems and implemented the proposed genetic algorithm by Java programming language. All authors read and approved the final manuscript.

References

Ahuja RK, Jha KC, Liu J (2007) Solving real-life railroad blocking problems. Interfaces 37(5):404–419

Assad AA (1980) Modeling of rail networks: toward a routing/makeup mode. Transp Res 14B:101–114

Assad AA (1983) Analysis of rail classification policies. Infor 21:293–314

Balakrishnan A, Magnanti TL, Mirchandani P (1997) Network design. In: Maffioli F, Martello S (eds) Dell 'Amico M. Annotated Bibliographies in Combinatorial Optimization. John Wiley, New York, pp 311–334

Barnhart C, Jin H, Vance PH (2000) Railroad blocking: a network design application. Oper Res 48:603–614

Bodin LD, Golden BL, Schuster AD, Romig W (1980) A model for the blocking of trains. Transp Res 14B:115–120

Cordeau JF, Toth P, Vigo D (1998) A survey of optimization models for train routing and scheduling. Transp Sci 32:380–404

Crainic TG (2000) Service network design in freight transportation. Eur J Oper Res 122(2):272–288

Crainic TG (2002) A survey of optimization models for long-haul freight transportation. In: Hall RW (ed) Handbook of Transportation Science, 2nd edn. Kluwer, Boston

Crainic TG (2006) Long-haul freight transportation. International Series in Operations Research & Management Science 56:451–516

Crainic TG (2009) Service design models for rail intermodel transportation. Lecture Notes in Economics and Mathematical Systems 619(1):53–67

Gorman MF (1998) An application of genetic and tabu searches to the freight railroad operating plan problem. Ann Oper Res 78:51–69

Huntley CL, Brown DE, Sappington DE, Markowics BP (1995) Freight routing and scheduling at CSX transportation. Interfaces 25(3):58–71

Jha KC, Ahuja RK, Guvenc S (2008) New approaches for solving the block-to-train assignment problem. Networks 51(1):48–62

Keaton MH (1989) Designing optimal railroad operating plans: lagrangian relaxation and heuristic approaches. Transp Res 23B:415–431

Keaton MH (1992) Designing railroad operating plans: a dual adjustment method for implementing lagrangian relaxation. Transp Sci 26:263–279

Magnanti TL, Wong RT (1984) Network design and transportation planning: models and algorithms. Transp Sci 18:1–55

Minoux M (1989) Network synthesis and optimum network design problems: models, solution methods and applications. Networks 19:313–360

Newton HN (1996) Network design under budget constraints with application to the railroad blocking problem. Auburn University, USA, Ph.D. dissertation

Newton HN, Barnhart C, Vance PH (1998) Constructing railroad blocking plans to minimize handling costs. Transp Sci 32:330–345

Talbi EG (2009) Metaheuristics: from Design to Implementation. Wiley, New Jersey

Van Dyke CD (1986) The automated blocking model: a practical approach to freight railroad blocking plan development. Transportation Research Forum 27:116–121

Van Dyke CD (1988) Dynamic management of railroad blocking plans. Transportation Research Forum 29:149–152

Yang L, Gao Z, Li K (2010) Railway freight transportation planning with mixed uncertainty of randomness and fuzziness. Appl Soft Comput 11(1):778–792

Zhu E, Crainic TG, Gendreau M (2009) Integrated Service Network Design for Freight Transportation. CIRRELT: Interuniversity Research Centre on Enterprise Networks. Logistics and Transportation, Montreal, Canada, Working Paper

3

Locomotive assignment problem with train precedence using genetic algorithm

Siamak Noori* and Seyed Farid Ghannadpour

Abstract

This paper aims to study the locomotive assignment problem which is very important for railway companies, in view of high cost of operating locomotives. This problem is to determine the minimum cost assignment of homogeneous locomotives located in some central depots to a set of pre-scheduled trains in order to provide sufficient power to pull the trains from their origins to their destinations. These trains have different degrees of priority for servicing, and the high class of trains should be serviced earlier than others. This problem is modeled using vehicle routing and scheduling problem where trains representing the customers are supposed to be serviced in pre-specified hard/soft fuzzy time windows.

A two-phase approach is used which, in the first phase, the multi-depot locomotive assignment is converted to a set of single depot problems, and after that, each single depot problem is solved heuristically by a hybrid genetic algorithm. In the genetic algorithm, various heuristics and efficient operators are used in the evolutionary search. The suggested algorithm is applied to solve the medium sized numerical example to check capabilities of the model and algorithm. Moreover, some of the results are compared with those solutions produced by branch-and-bound technique to determine validity and quality of the model. Results show that suggested approach is rather effective in respect of quality and time.

Keywords: Locomotive assignment problem, Vehicle routing and scheduling, Fuzzy time windows, Genetic algorithm

Background

The rail transportation industry has many problems that can be modeled by mathematical programming and solved using soft computing techniques. But the research in railroad scheduling has experienced a slow growth (Ahuja et al. 2005). A growing interest for using optimization techniques in railroad problems has appeared in the operation research literature (see, e.g., Brannlund et al. (1998) and Cordeau et al. (1998)). One of the most important problems in rail transportation industry in view of operating costs is locomotive assignment or locomotive routing and scheduling problem. Because the considerable cost usually is paid by rail companies for operating the locomotives according to the properly assignment plan that has a direct impact on operating cost, punctuality and performance, which in turn affect customers' satisfaction, finding a way to design a proper locomotive assignment and scheduling policy can be an important decision in railroad companies. The planning stage in railroad scheduling is divided to strategic, tactical and operational planning. These stages are in accordance with the length of the respective planning horizon and the temporal impact and relevance of the decision. According to these levels of planning, locomotive scheduling is the final stage of the railroad scheduling and depends on incoming requests, but with this assumption, engines and crews are rostered and appropriate cars are available. It must be mentioned that this paper is a complementary version of our recent researches (Ghoseiri and Ghannadpour 2009a; Ghoseiri and Ghannadpour 2010) about the locomotive assignment problem. Ghoseiri and Ghannadpour (2009a) suggested the solving of the general locomotive planning (or scheduling) that is one of the most attractive topics in operation research by vehicle routing problem. Ghoseiri and Ghannadpour (2010) extended the previous research and tried to study the multi-depot

* Correspondence: noori@iust.ac.ir
Department of Industrial Engineering, Iran University of Science and Technology, Narmak, Tehran 16846-13114, Iran

homogenous locomotive assignment with more real-life assumptions. This model was formulated by vehicle routing problem with time windows (VRPTW) and solved heuristically by an efficient hybrid genetic algorithm. This paper, in continuation of previous researches, tries to consider the different degrees of priority of trains for servicing using the concept of fuzzy time windows.

As mentioned earlier, few references can be found in the literature regarding operation research for assignment of locomotives to trains. A version of the problem, where a single locomotive must be assigned to each train, no deadhead is allowed, and no maintenance requirements are taken into account was addressed by Forbes et al. (1991). In an earlier paper, Florian et al. (1976) introduced an integer programming model based on a multi-commodity network for the case where several locomotives can be assigned to each train. Ziarati et al. (1997) extended Florian et al. (1976) formulation to include almost all the operational constraints encountered at CN North America. Ahuja et al. (2005) presented the real-life locomotive scheduling faced by CSX transportation (Jacksonville, Florida), a major US railroad company. They considered the planning version of the locomotive scheduling model, where there were multiple types of locomotives and needed to decide the set of locomotives to be assigned to each train. A Benders decomposition approach to the locomotive assignment problem can be found in Cordeau et al. (2000). In a subsequent paper, Cordeau et al. (2001a) extended the previous work with various real-life constraints, such as maintenance. Rouillon et al. (2006) presented an efficient backtracking mechanism that can be added to this heuristic branch-and-price approach. Moreover, Vaidyanathan et al. (2008b) developed new formulations for the locomotive planning problem (LPP). In this paper, new constraints were added to the planning problem desired by locomotive directors, and additional formulations necessary to transition solutions of models to practice were developed. Also, two formulations namely *consist formulation* and *hybrid formulation* were suggested for this generalized LPP.

In this area, Vaidyanathan et al. (2008a) developed robust optimization methods to solve the LPP. In this paper, two major sets of constraints were considered and should have satisfied by each locomotive route: (1) *locomotive fueling constraints*, which requires that every unit visit of a fueling station be conducted at least once for every F miles of travel, and (2) *locomotive servicing constraints*, which require that every unit visit of a service station be conducted at least once for every S miles of travel. This problem was formulated as an integer programming problem on a suitably constructed space-time network, and it was shown that this problem is NP-complete. Other important locomotive assignment papers can be found in Cordeau et al. (2001b), Fioole et al. (2006), Lingaya et al. (2002), etc.

As mentioned earlier, the model that is considered in this paper is presented using the vehicle routing and scheduling problem in which the trains are supposed to be serviced in pre-specified hard/soft fuzzy time windows. This problem is an important variant of vehicle routing problem (VRP) with adding time window constraints to the model in which a set of vehicles with limited capacity is to be routed from a central depot to a set of geographically dispersed customers with known demands and predefined time windows in order that fleet size of vehicles and total traveling distance are minimized, and capacity and time windows constraints are not violated. Usually, in real world VRPs, many side constraints appear. Because of many applications of different kinds of VRP problems, many researchers have focused to develop solution approaches for these problems. We can find useful techniques for the general VRP in Bräysy and Gendreau (2001), Laporte and Semet (1999), and Pisinger and Ropke (2007). In this area, Czech and Czarnas (2002) solved VRPTW with simulated annealing, and Gambardella et al. (1999) applied multiple ant colony system for VRPTW. Alvarenga et al. (2007), Berger and Barkaoui (2003), Ghoseiri and Ghannadpour (2009b), and Tan et al. (2006) used genetic algorithm for VRPTW. Other very good techniques and applications of VRPTW can be found in Cerda and Dondo (2007), Crevier et al. (2007), Irnich et al. (2006), Kim et al. (2006), Li et al. (2005), and Tan et al. (2007). As mentioned earlier, this paper uses the concept of fuzzy approach to consider the different degrees of priority of trains for servicing. This concept is known as fuzzy time windows in the literature of VRP models in which the customers hop to be served at desired time if possible. The most important studies in this area can be found in Sheng et al. (2006) and Cheng et al. (1996). According to literature, the suggested model based on VRPTW with fuzzy concept is NP-hard, and due to the intrinsic difficulty of the problem, heuristic methods are most promising for solving it. In this paper, among the number of solving methods which were explored, genetic algorithm is examined in greater depth and combined with other heuristics to solve the suggested model.

The remaining parts of paper are organized as follows: 'Locomotive assignment with train precedence' section defines the locomotive assignment problem with fuzzy time windows. 'Solution procedure' section introduces the hybrid genetic search algorithm to solve the problem. 'Numerical example and results analysis' section discusses the model validation and computational complexity of the proposed method, and 'Conclusion' section provides the concluding remarks.

Results and discussion

This section describes computational experiments carried out to investigate the performance of the proposed GA. The algorithm was coded in MATLAB 7 and run on a PC with 1.6-GHz CPU and 512-MB memory. In this section, a complete randomly generated medium size problem is considered as a numerical example. Before solving and analyzing this problem, the validity and quality of suggested method should be checked. So, a few small- and medium-sized problems are created and solved by branch-and-bound (B&B) technique. The solutions yielded by the exact optimization technique are compared with those of the hybrid algorithm in general approach. Table 1 summarizes the results.

The times that are reported in column CPU time (Loco-GA) are the times to find the best solutions of each test problem during the 600 generations. In comparison with the exact solution of branch-and-bound technique, the hybrid GA algorithm has a zero percentage of error and drastically reduces the CPU time for the generated test problems. Also, in order to determine computational complexity of the algorithm, 15 instance problems are solved ranging from 10 to 100 trains which the produced results are illustrated in Figure 1. With reference to results, the computational complexity of the algorithm falls between n^2 and $nlog\ (n)$, i.e., the hybrid genetic algorithm solve a NP-hard problem in a polynomial time computation that is an advantage for the algorithm.

Now, a randomly generated medium size problem should be considered. It includes 80 nodes and 40 trains per day in a weekly planning horizon. Ten trains are considered high class, and the fuzzy time windows with different widths are considered for them. These trains have to be serviced without any delay, and their satisfaction for receiving services should be maximized. A maximum permissible operating time of 18 h is defined for all the trains running in the planning horizon, and maintenance time for each locomotive is assumed to be 6 h. The locomotive speed is variable at different points of route in accordance with a normal probability distribution from 45 to 65 km/h. At first, this problem is solved in which all trains are considered normal, and they have classical time windows. Then, 10 trains are selected randomly, and they are supposed to be high class with different fuzzy time windows. So, at first stage, nine locomotives are needed to service all trains with classical time windows. The total traveling time of this solution is 64.93 h; total traveling distance is 3,386.4 km; and the total waiting time is 8.4 h. Also, the order of servicing is shown in Figure 2.

With reference to Figure 2, each locomotive according to its constructed route services to the trains en route. For instance, the sixth locomotive starts its journey at depot and hauls trains 21, 39, 28 and 5 to their destinations and then returns to the central depot for its routine daily maintenance. In fact, the sixth locomotive goes to the origin node of the 21st train zone and hauls the train to its destination, and then goes to the origin node of the 39th train zone and keeps on going, i.e., the sixth route is as follows:

$$\{O_{21}, D_{21}, O_{39}, D_{39}, O_{28}, D_{28}, O_5, D_5\}.$$

At the second stage, ten trains are selected randomly, and they were considered as high priority trains. These trains are highlighted in Figure 2. According to these changes, the locomotive assignment model was solved again, and the produced result is shown in Figure 3.

According to this figure, nine locomotives are needed for serving all trains. The total traveling time of this solution is 65.93 h, total traveling distance is 3586.7 km and the total waiting time is 6.15 h. Also, the order of servicing is changed to maximize the satisfaction rate of

Table 1 Comparing the results with exact solutions

Problem number	Input size (nodes)	CPU time		Fitness value		Percentage of error
		Loco-GA	B&B	Loco-GA	B&B	
1	4	0.10 s	2 s	51.25	51.25	0%
2	8	0.14 s	21 s	67.5	67.5	0%
3	8	0.23 s	56 s	93.80	93.80	0%
4	12	0.29 s	9 min	59.45	59.45	0%
5	12	0.25 s	8.5 min	126.90	126.90	0%
6	16	2 s	———[a]	65.7	65.7	0%
7	16	6.6 s	———[a]	154.34	———	———
8	16	5 s	———[a]	130.55	———	———
9	20	3.1 s	———[a]	70.35	———	———
10	20	3.3 s	———[a]	202.85	———	———

[a]The optimizer is unable to solve the problem within 24 h of running.

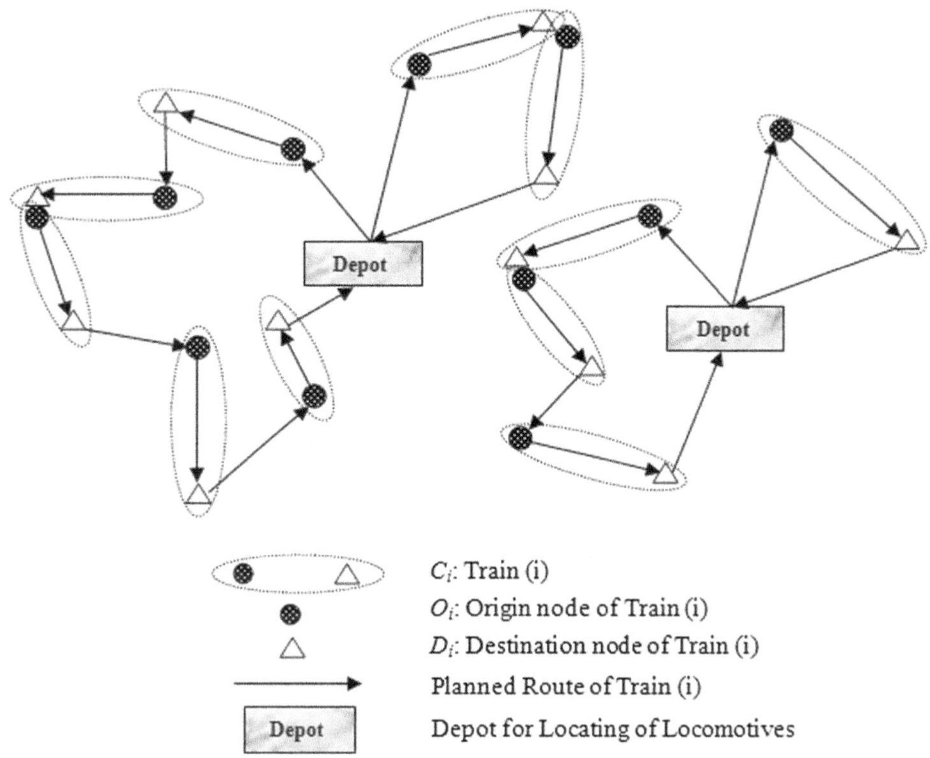

Figure 1 Typical routes with and without waiting time.

trains with high degree of priority. Considering the ten high class trains, the summation of satisfaction rate for this solution is 30.7, while this rate for the solution of Figure 2 is 27.1. So, the algorithm tries to change the servicing arrangement of trains to earn more value of satisfaction. For example, for train 7 which is considered as high priority train, the satisfaction rate was 0.27 in previous solution and increased to 0.61 in the new solution by changing to arrival time of locomotive and servicing arrangement. Hence, the suggested algorithm for serving the trains tries to maximize the satisfaction rate of high priority trains considering the other objectives, namely, traveling time, distance and waiting time.

Moreover, an operator deletion-retrieval strategy is executed to probe the efficiency of the inner working of the suggested method. According to this strategy, genetic operators are eliminated one at a time and each time. Algorithm is put into run, and convergence behavior is studied and compared with the operator retrieved. Figure 4 summarizes the analysis of the operator's effect on convergence behavior of the hybrid genetic algorithm for a test problem.

According to Figure 4, the hill climbing operator has the main role to improve the chromosomes obtained through crossover and mutation, and it works highly efficient. Figure 4 shows that all the inner components of the hybrid genetic algorithm work properly and indicate good behavior of convergence towards the best solutions.

As mentioned earlier, this paper used an adaptive mutation probability scheme, which changed the mutation probability as the standard deviation of the population

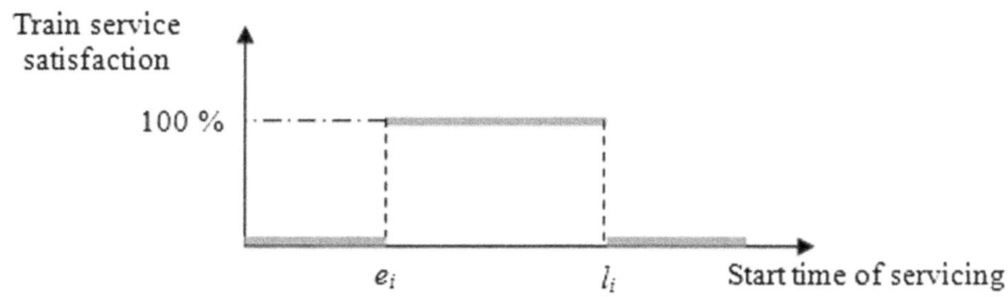

Figure 2 Best solution of problem in 600 generations (classical time windows).

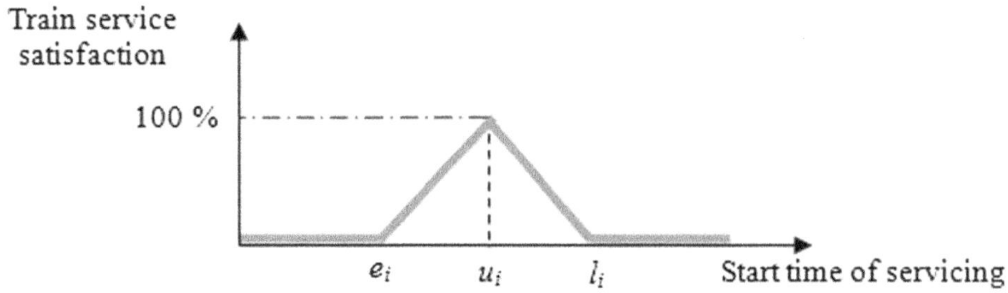

Figure 3 Best solution of problem in 600 generations (ten high class trains).

fitness changes. This scheme was proposed to maintain and control the diversification of population in each generation. Figure 5 illustrates the behavior of this scheme in the first 200 generations of a test problem. According to this figure, the mutation probability (upside schema) should be increased when the diversification of population is decreased. Also, this probability should be decreased to maintain the population diversification when the standard deviation of the population fitness changes is increased.

As mentioned earlier, this paper tried to check the effectiveness of other crossover operators, namely, route exchange crossover (REX) versus the heuristic and merge crossover (HMX). To check the efficiency of these operators, a comparison is undertaken between these operators and partially mapped crossover (PMX) which is the common operator for sequential problem like TSP and VRP. Figure 6 illustrates this comparison.

Unlike PMX and HMX, which were used in our recent research (Ghoseiri and Ghannadpour 2010), the REX that is suggested in this paper produces better solution and works properly than them. Eventually, produced results show that suggested approach is quiet effective in respect of quality and time.

Conclusion

This paper presented the locomotive assignment problem which is very important for railway companies, in view of high cost of operating locomotives. This problem was to determine the minimum cost assignment of homogeneous locomotives located in some central depots to a set of pre-scheduled trains in order to provide sufficient power to pull the trains from their origins to their destinations. These trains had different degrees of priority for servicing, and the high class of trains should have serviced earlier than others. This problem was modeled using vehicle routing and scheduling problem where trains performed as customers, and they should have serviced in pre-specified hard/soft fuzzy time windows.

A two-phase approach was used in which, in the first phase, the multi-depot locomotive assignment was converted to a set of single depot problems, and after that, each single depot problem was solved heuristically by a hybrid genetic algorithm. In the genetic algorithm, various heuristics and efficient operators were used in the evolutionary search.

In suggested algorithm, the push-forward insertion heuristic (PFIH) was used to determine the initial solution, and λ-interchange mechanism was used for neighborhood search and improving method. Moreover, new crossover operator was suggested and it was shown that it worked more properly than the old operators.

The suggested algorithm was applied to solve the medium-sized numerical example to check capabilities of the model and algorithm. Moreover, some of the results were compared with those solutions produced by branch-and-bound technique to determine validity and

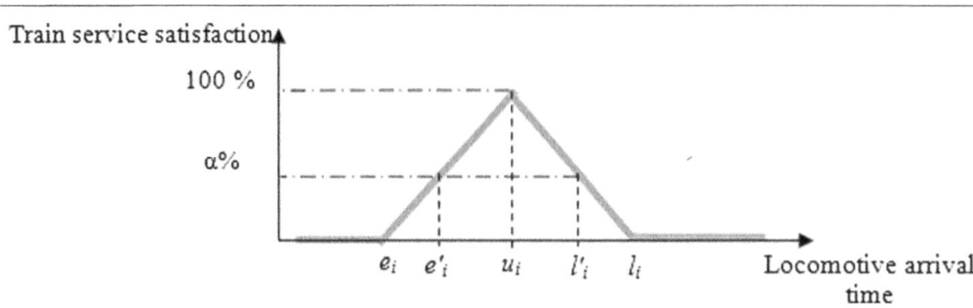

Figure 4 Inner working of method under the operation of different operators.

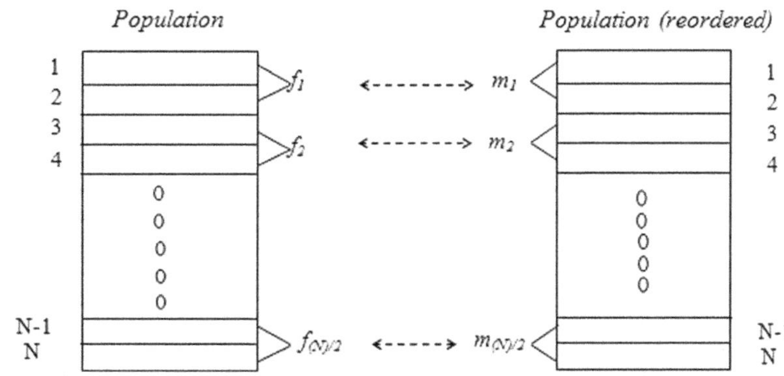

Figure 5 Inner working of method under the operation of different operators.

quality of the model. Results indicated that this algorithm was efficient and solved the problem in a polynomial time.

Methods

Locomotive assignment with train precedence

The problem includes a set of homogeneous locomotives, a set of depots where locomotives are initially located, and a set of pre-scheduled trains with different degrees of priority (e.g., normal passenger trains, high speed trains, freight trains, etc.). As mentioned earlier, this problem is modeled by VRPTW. In this model, the trains act as customers of a VRPTW that should be serviced in their time windows. It is assumed that, according to a pre-planned schedule, origin and destination nodes for each of the trains are known during the T-days planning horizon. It is worth noting that trains (customers) in this network have two coordinates of origin and destination. For each train (C_i), there exists an origin-destination pair node (O_i, D_i). Distance between the destination node of train i and the origin node of train j is considered to be the distance between C_i and C_j. Therefore, the network is asymmetrical which means that the distance between train i and train j is not equal to the distance between train j and train i. Figure 7 is a typical output of this problem.

Two parameters are assigned to each arc: $D_{D(i)O(j)}$ as the distance between train i and j, and $t_{D(i)O(j)}$ as the travel time between train i and j. In this network, the set of depots (e.g., P depots) are considered as central zones that provide the neighboring zones (current customers or trains) with locomotives.

Each locomotive k starts its journey from a depot and reaches to the origin of train i and hauls the train to its destination. Afterward, it is decided for locomotive k whether it should return to its home depot or be dispatched to the origin node of another train. The factor which forces the locomotive to return to its home depot is the maximum allowable operating time. It is assumed that total operating time for each locomotive is less than or equal to the pre-determined maximum operating time. The maximum allowable daily operating time for locomotive k is calculated as follows:

$$r_k = 24 - z_k, \tag{1}$$

where, z_k is the daily routine maintenance and service time on locomotive k. Therefore, as long as the

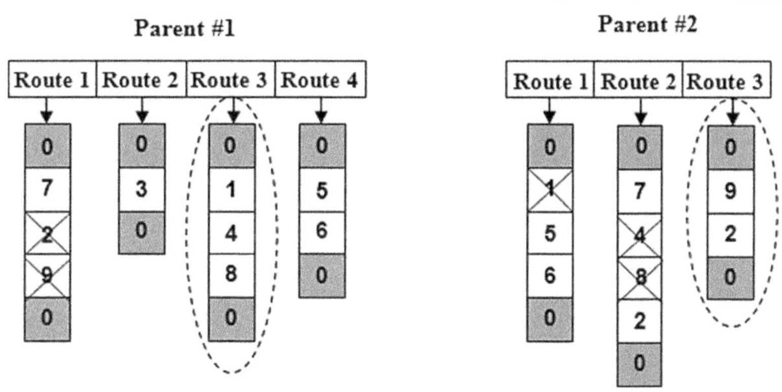

Figure 6 Comparison between REX, HMX, and PMX.

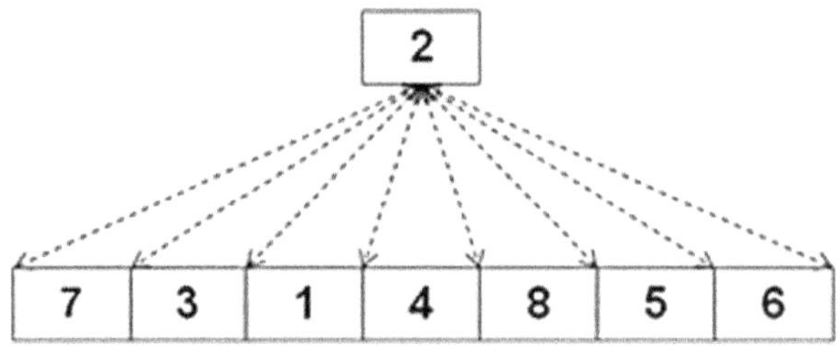

Figure 7 Typical output for the locomotive routing and scheduling problem.

locomotive is allowed to operate, it will proceed to its journey in the network to service the customers, and after that, it will return to its home depot. As mentioned earlier, not exceeding the maximum travel time is considered as a condition of feasibility of the solution. Another basic and important constraint in this problem is the time window assigned to each train, so that the train must be serviced in this time window; based on the punctuality and operating class of railroads, time windows width can be tightened or loosened for the trains. It is assumed that, if locomotive k arrives at train i before the earliest time of its service initiation, e_i, it must wait until e_i which represents idleness and undesirable state for locomotive k, and it will appear as a penalty term in the cost function. But if locomotive k arrives at train i after the latest time of its service initiation, l_i, due to the delay in servicing, it will increase the cost function as another penalty term. In this paper, it is assumed that delay in service is not allowed to the trains. In other words, assigning locomotives without delay in service is considered as another condition of feasibility of the solution. So, the starting time of each train is calculated as follows:

$$t_i = \begin{cases} e_i & \text{if} \quad t_{i-1}+f_{i-1}+r_{i-1,i} \leq e_i \\ t_{i-1}+f_{i-1}+r_{i-1,i} & \text{Otherwise} \end{cases}, \tag{2}$$

where f_i is the service time for train i that is equal to the travel time of train i between its origin and destination nodes. Also $r_{i-1,i}$ is the travel time between the train $i-1$ and i.

As mentioned earlier, this paper uses the fuzzy time windows for considering the different degrees of train priority for servicing. In classical time window, each train has the same satisfaction during the time window for servicing. This time window is shown in Figure 8.

This classical time window cannot meet the need of the real life, and it does not reflect the train's preference for receiving the services, as each train might prefer to be serviced at a certain time within the time window. Figure 9 shows the typical fuzzy time window that can reflect the train's preferences for servicing.

According to this figure, the classical time window is changed to the triple time window $[e_i, u_i, l_i]$. For example, if a train is served at its desired time, the grade of its satisfaction is 1; otherwise, the grade of satisfaction gradually decreases along with the increase of difference between the arrival time of locomotive and desired time. The grade of satisfaction will be 0 (no satisfaction) if the arrival time falls outside the time interval. The membership function of train i or $\mu_i(t_i)$ represents the grade of satisfaction when the start of service time, t_i, is defined as follows:

$$\mu_i(t_i) = \begin{cases} 0 & t_i < e_i \\ \dfrac{t_i - e_i}{u_i - e_i} & e_i \leq t_i \leq u_i \\ \dfrac{l_i - t_i}{l_i - u_i} & u_i \leq t_i \leq l_i \\ 0 & t_i > l_i \end{cases} \tag{3}$$

This paper uses the fuzzy time windows to represent the different degrees of trains' priority, e.g., between freight and passenger trains or different classes of passenger trains. In this case, the low class of trains is considered to be having the classical time windows. It means that the satisfaction of these trains is full during the time window. So, the trains with high priority have the narrow fuzzy time windows, and the desired time (u_i) is nearest to the earliest arrival time of train i (e_i) in which this fuzzy time window is tolerable for lateness but less tolerable for earliness. These two attributes (narrow window and less tolerable for earliness) represent that the high priority trains want to be received as locomotive as soon as possible to start the travel. If it is desirable for the decision maker, the servicing of some trains with at least satisfaction rate α%, the α-cut approach can be used for changing the fuzzy time windows to classical one with acceptable satisfaction rate. Figure 10 shows this point. According to this figure for these trains, the time window $[e_i, u_i, l_i]$ is changed to $[e'_i, l'_i]$ based on Equation 4:

$$\begin{aligned} e'_i &= \alpha(u_i - e_i) + e_i \\ l'_i &= -\alpha(l_i - u_i) + l_i \end{aligned} \tag{4}$$

Moreover, in this problem, it is assumed that K_0 locomotives are initially located at each depot, and then the

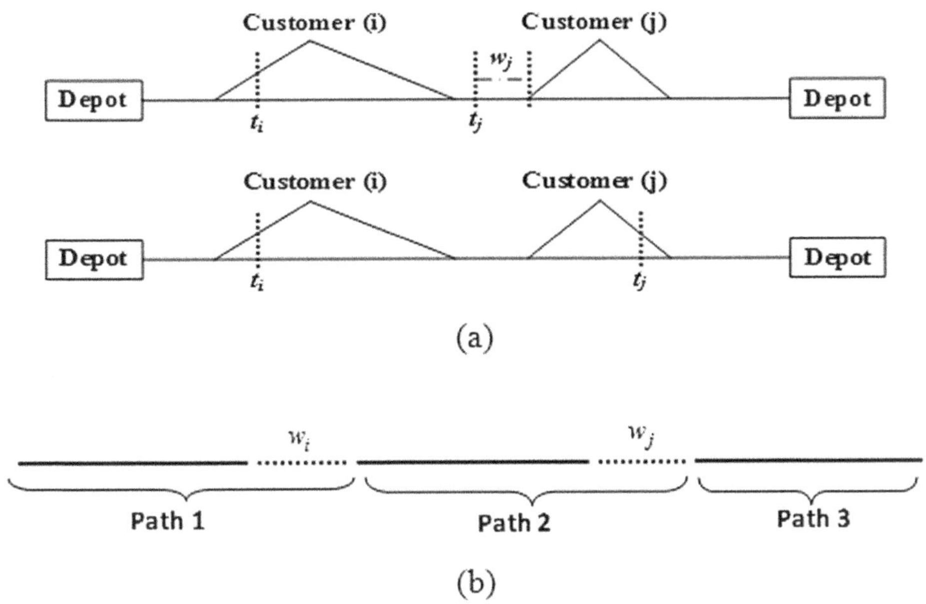

Figure 8 The classical time window for each train.

optimum numbers of locomotives are determined for each depot by solving the model. It is worth noting that the number of set routes in network is equal to the number of locomotives for each period of time.

Finally, this paper tries to design the proper assignment policy that the total locomotive assignment cost is minimized by making a tradeoff between travel time, travel distance, and waiting time, and considering the different degrees of priority (by maximizing the satisfactions of trains in service). The solution procedure is described in the following section.

Solution procedure

In this paper, the cluster-first, route-second approach is used to convert the multi-depot locomotive assignment to P single depot problems (for more details, see Ghoseiri and Ghannadpour (2010)). By this approach, at first stage, trains are assigned to each depot to obtain P clusters, and after that each cluster is solved by the suggested hybrid genetic algorithm. Genetic algorithm is a class of adaptive heuristics based on the drawing concept of evaluation – 'survival of the fitness,' and it has been developed by Holland (1975) at the University of Michigan

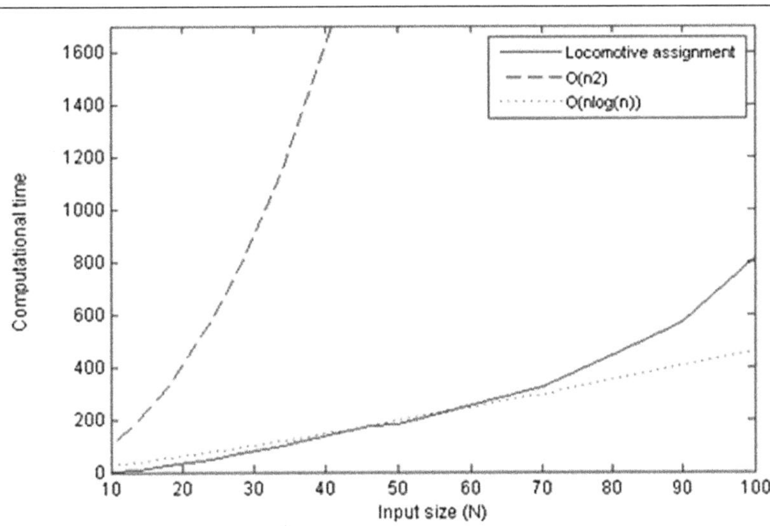

Figure 9 The fuzzy time window for trains.

Locomotive 1	[	0	10	12	18	32	37	0	-	]
Locomotive 2	[	0	6	16	19	22	26	7	0	]
Locomotive 3	[	0	27	30	34	17	13	0	-	]
Locomotive 4	[	0	2	31	23	25	36	0	-	]
Locomotive 5	[	0	24	9	14	1	11	0	-	]
Locomotive 6	[	0	21	39	28	5	0	-	-	]
Locomotive 7	[	0	8	4	35	15	33	0	-	]
Locomotive 8	[	0	29	3	38	0	-	-	-	]
Locomotive 9	[	0	40	20	0	-	-	-	-	]

Figure 10 **Fuzzy time window with at least satisfaction rate (α%).**

in 1975. A GA starts with a set of chromosomes referred to as initial population. Each chromosome represents a solution to the problem, and the initial population is either randomly generated or generated by heuristics. A selection mechanism will then be used to select the prospective parents based on their fitness computed by evaluation function. The selected parent chromosomes will then be recombined via the crossover operator to create potential new population. The next step will be to mutate a small number of newly obtained chromosomes in order to introduce a level of randomness that will preserve the GA from converging to a local optimum. The GA will then reiterate through this process until a pre-defined number of generations have been produced, or until there was no improvement in the population, which would mean that the GA has found a near optimal solution, or until a pre-defined level of fitness has been reached. The use of the hybrid GA is designed as follows.

Chromosome representation

A solution to the problem is represented by an integer string of length N, where N is the number of customers which needed to be served. All routes are encoded together, with no special route termination characters in between; chromosomes are decoded back into routes based on the feasibility conditions, namely, maximum allowable operating time and servicing without delay time.

Initial population

Part of the population is initialized using modified PFIH method and λ-interchange mechanism, and part is initialized randomly. The PFIH method was first introduced by Solomon (1987) to create an initial route configuration for each single depot problem in a time unit of the planning horizon (for more details, see Solomon (1987)). This paper uses the modified PFIH method according to defined problem that cost function for inserting a customer into a new route is as follows:

$$Cost(C_i) = -\alpha\, t_{(0)O(i)} + \beta\, l_i + \gamma\Big(\frac{|\theta_{O(i)} - \theta_{D(j)}|}{360} \times t_{(0)O(i)}\Big) \tag{5}$$

where $\theta_{o(i)}$ and $\theta_{D(j)}$ are the polar angle of the train in question and the last visited train in the last formed route, respectively. $t_{(0)O(i)}$ is the travel time between the home depot and train i. l_i is the latest arrival time at node i. Therefore, the unrouted train with the lowest cost is selected as the first train to be visited. Once the first train is selected for the current route, the heuristic selects from the set of unrouted trains the train j^* which minimizes the total insertion cost between every edge $\{k, l\}$ in the current route without violating the time and maximum route time constraints. The insertion cost function is according to relation (6):

$$Cost(C_i) = \delta\, D_K + \varphi\, W_K + \eta\, O_K + \kappa\, T_K + \kappa'\, T'_K - \theta\, S_K \tag{6}$$

where D_k is total distance traveled by locomotive k; W_k is the total travel time consumed by locomotive k; O_k is the over timing of locomotive k; T_k, T'_k and S_k are respectively the summation of the tardiness time, waiting time, and satisfaction rate for all the trains visited by locomotive k. The constant coefficients of δ, ϕ, η, κ, κ', η and θ are the weight factors. Exceeded amount of maximum operating time for each locomotive is according to relation (7):

$$O_K = \max\{0, W_K - r_K\} \tag{7}$$

Waiting and delay times imposed on each locomotive k is equal to sum of the waiting and delay times according to Equations 8 and 9 for trains on route. In this relation, a_{ik} is the arrival time of locomotive k at origin node of train i:

$$T'_K = \sum_i \max\{0, e_i - a_{ik})\tag{8}$$

$$T_K = \sum_i \max\{0, a_{ik} - l_i\}\tag{9}$$

Moreover, the total satisfaction rates for trains which are on locomotive route k (S_k) can be calculated by relation (3).

This paper uses a *λ-interchange mechanism* to move trains between routes to generate neighborhood solution for the problem (for more details see Ghoseiri and Ghannadpour (2010)). In one version of the algorithm called *global best*, the whole neighborhood is explored, and the best move is selected. In the other version, *first best*, the first admissible improving move is selected if one exists; otherwise, the best admissible move is implemented.

Selection

Candidates for mating are selected using the tournament selection. In tournament selection, two identical (through differently ordered) copies of the population are kept. In every generation, we compare adjacent chromosomes in one copy of the population pair by pair and select the chromosome with best fitness value, and then we proceed with the second copy of the population to select the other half of the selected population. So, the candidates for mating are selected by this schema, and they recombined by crossover. This procedure is illustrated in Figure 11.

As shown in Figure 11, two identical copies of the population with size N are maintained at every generation and ranked arbitrary. For each population, adjacent chromosomes are compared, and the solutions with better chromosomes qualify to be potential parent. After comparing all pairs in two populations, $N/2$ 'fathers', namely, $f_1, \ldots f_{(N)/2}$ and $N/2$ 'mothers', namely, $m_1, \ldots m_{(N)/2}$ are created, and each f_i and m_i are mated subsequently. In this procedure, the superior chromosomes are given priority in mating, but average entities have chance of being selected too.

Crossover and mutation

The classical crossover (one-point crossover, n-point crossover…) is not appropriate for this sequencing problem. Their use may cause the offspring not to have a valid sequence due to duplication and omission of vertices. So, this paper uses the heuristic and merge crossovers for recombination phase. The heuristic crossover deals with distances between nodes, for example, a random cut was made on two chromosomes, and the genes will be compared immediately after the cut. The first gene is chosen randomly, and it is considered for two chromosomes. The following genes will be the one which is geographically closer to previous gene. The merge crossover operates on the basis of time precedence, defined by the time windows corresponding to each node. Similarly, the first gene is chosen randomly, and the following genes will be the one which time window comes earlier (for more details see Ghoseiri and Ghannadpour (2010)).

Moreover, this paper tries to check the effectiveness of other crossover operators namely REX versus the previous operator. REX selects a random route from each parent. Next, for a given parent, the trains in the chosen route from the opposite parent are removed. Since each

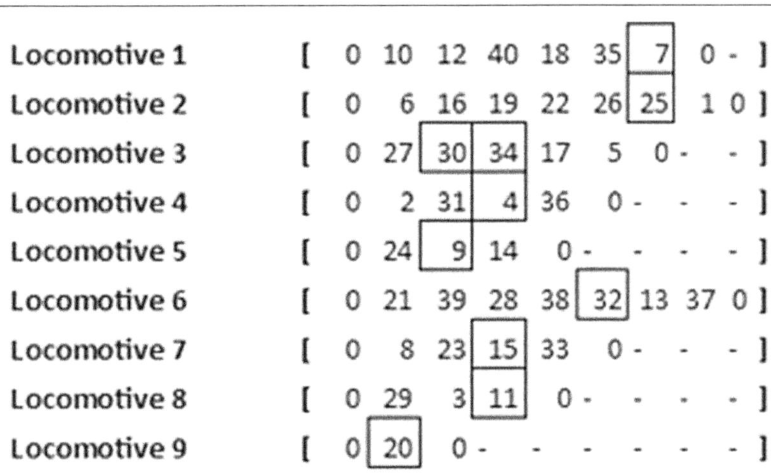

Figure 11 Selection procedure.

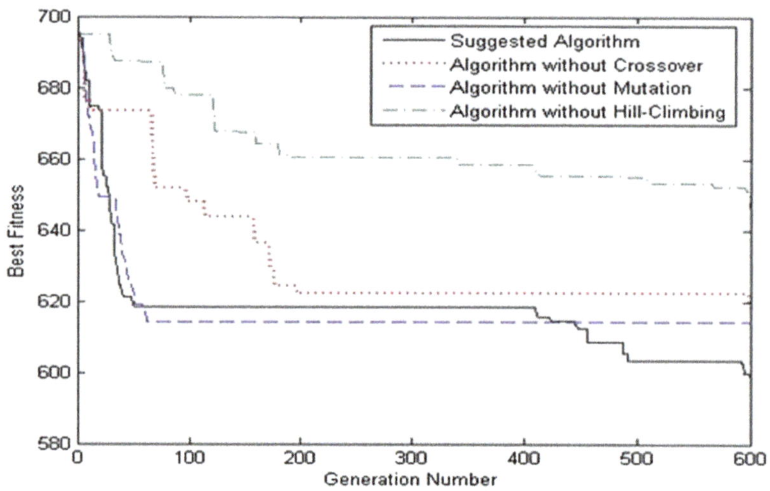

Figure 12 Route exchange crossover.

chromosome should contain the entire train numbers, the next step is to locate the best possible locations for the removed trains in the corresponding children. This procedure is illustrated in Figure 12.

According to this figure, route (3) from parent #1 is selected randomly, and the trains on this route are removed from the routes of parent #2. This process is done similarly for another parent. Hereinafter, for each parent, the best location of removed trains is determined by the insertion procedure one at a time. This procedure for finding the best location of train (2) on the parent #1 is illustrated in Figure 13.

The mutation schemes used are swap node and swap sequence. This paper uses an adaptive mutation probability scheme, which changes the mutation probability as the standard deviation of the population fitness changes, described as:

$$
\begin{aligned}
P_{mutation} &= 0.5 \times \left(1 - \frac{S}{S_{\max}}\right) + 0.06 \\
S &= \sqrt{\frac{\sum_{i=0}^{N-1} (x_i - \bar{x})^2}{N-1}}
\end{aligned}, \quad (10)
$$

where x_i is the fitness of a chromosome; $\bar{x}$ is the average fitness; N is the population size; S is the standard deviation of the population fitness; and $S_{\max}$ is the maximum standard deviation among all the generated populations.

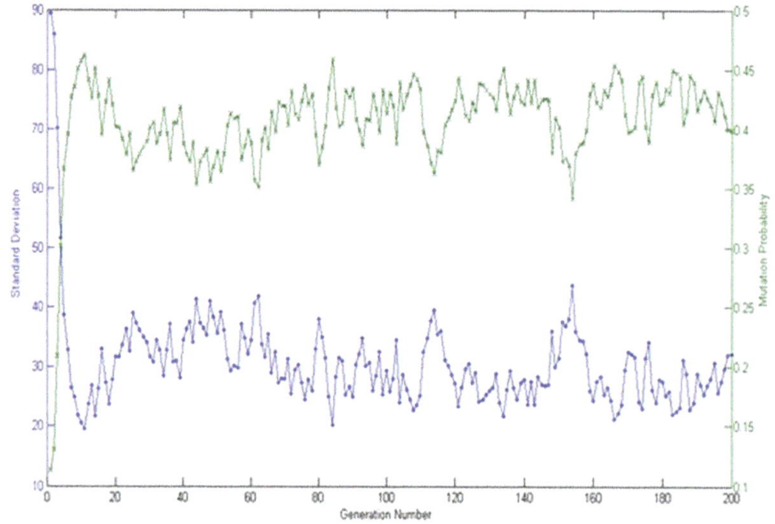

Figure 13 Finding the best location of train (2).

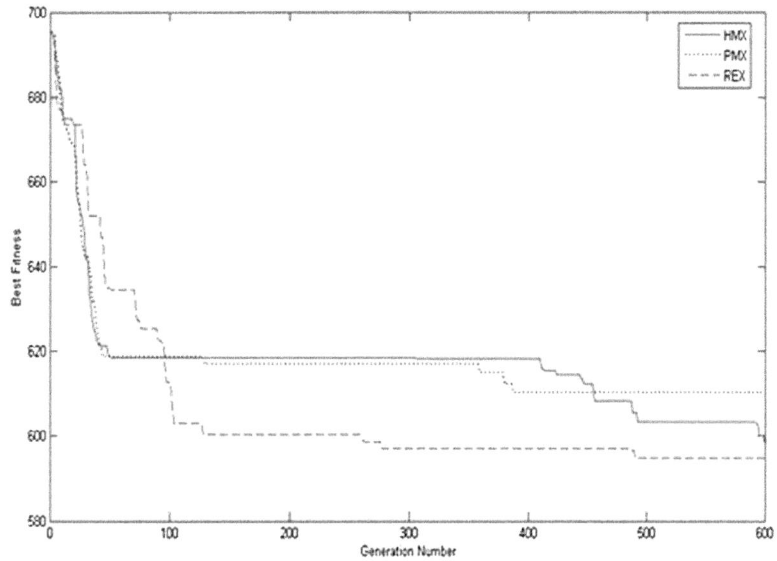

Figure 14 Typical routes with and without waiting time.

Fuzzy satisfaction operator

As mentioned earlier, the different degrees of priority are considered by fuzzy time windows. So, except to previous operators that focused on travel time and travel distance, we need special operator to improve the satisfaction rate of each high priority trains. This operator is used to determine the best arrival time for each train to maximize the total grade of satisfaction without increasing the waiting time but replacing the waiting time of each train between others. So, this operator scans the feasible schedule from left to right and tries to find out a possible forward push within the route. This push will increase the total degree of satisfaction along the route without violating the feasibility conditions. In general, this operator is applied on the chromosomes which have the following characteristics:

1. This operator is applied on the route which has at least one train with non-zero waiting time.
   Figure 14a shows the typical routes with and without waiting time.
   It must be mentioned that each route is divided to some sections (each section is named *path*) according to the number of trains with waiting time, and this operator is applied on each path that has one train with waiting time. This procedure is shown in Figure 14b, and according to it, the fuzzy operator is applied on paths 1 and 2.
2. The sum of the slope of satisfaction function at locomotive arrival time for each train on the path should be larger than zero. In this case, a possible forward push will cause the increase of total grade of satisfaction on the path.

So, this operator is applied on each path with above characteristics. In each step, the feasible forward push should be calculated, and the arrival time of each train should be pushed according to it. The feasible forward push in each step is shown below:

$$Push = \min(\Delta_i, w), \tag{11}$$

where

$$\Delta_{(i)} = \begin{cases} u_i - t_i \mathrm{ife_i} < t_i < u_i \\ l_i - t_i \mathrm{ifu_i} < t_i < l_i \end{cases}. \tag{12}$$

After this push, the section of path from the train* to end of the path is considered, and the above characteristics are checked again. So, if the characteristics are okay, the new feasible forward push will be calculated, and the operator will apply on this section again. This procedure is repeated until the new feasible forward push could not be found. It must be mentioned that, the train* is the train that the previous minimum push is found on it.

Hill climbing and recovery

Also the hill climbing is used in order to improve the chromosomes obtained through crossover and mutation. Hill climbing is a scheme for randomly selecting a portion of the population and improving them by a few iterations of removal and reinsertion. At the end, to additionally improve the quality of the population, the worst portion of the population will be replaced with the best of the parent population.

Competing interests

Both authors declare that they have no competing interests.

Authors' contributions

SN conceived of the study and participated in its design and coordination. SFG participated in the design of the study and performed the statistical analysis. All authors read and approved the final manuscript.

Authors' information

SN is the associate professor of Industrial Engineering Department in Iran University of Science and Technology (IUST) since October 1990. The teaching experiences are mainly in project management, human resource management, organizational behaviour and HRM, and engineering economic and preventive maintenance. SFG is a PhD student of Industrial Engineering Department in Iran University of Science and Technology (IUST). He specializes research in the field of transportation, railway engineering, vehicle routing problem, and engineering economic and preventive maintenance.

Acknowledgments

The authors would like to acknowledge and thank the editors and reviewers of this manuscript that their valuable input and correspondences helped this manuscript to improve.

References

Ahuja RK, Liu J, Orlin JB, Sharma D, Shughart LA (2005) Solving real-life locomotive scheduling problems. Transp Sci 39(4):503–517

Alvarenga GB, Mateus GR, de Tomi G (2007) A genetic and set partitioning two-phase approach for the vehicle routing problem with time windows. Computers & Operation Research 34:1561–1584

Berger J, Barkaoui M (2003) A hybrid genetic algorithm for the capacitated vehicle routing problem. In: Cantú-Paz E (ed) GECCO'03 proceedings of the 2003 international conference on genetic and evolutionary computation: Part I, Chicago, IL, USA, July 12-16, 2003. Springer, Heidelberg, pp 646–656

Brannlund V, Lindberg PO, Nou A, Nilsson JE (1998) Railway timetabling using Lagrangian relaxation. Transp Sci 32(4):358–369

Bräysy O, Gendreau M (2001) Metaheuristics for the vehicle routing problem with time windows. Internal Report STF42 A01025, SINTEF Applied Mathematics. Department of Optimization, Norway

Cerda J, Dondo R (2007) A cluster-based optimization approach for the multi-depot heterogeneous fleet vehicle routing problem with time windows. Eur J Oper Res 176:1478–1507

Cheng R, Gen M, Tozawa T (1996) Vehicle routing problem with fuzzy due-time using genetic algorithm. Japan Society for Fuzzy Theory and System 7 (5):1050–1061

Cordeau JF, Toth P, Vigo D (1998) A survey of optimization models for train and scheduling. Transp Sci 32(4):988–1005

Cordeau JF, Soumins F, Desrosiers J (2000) A Bender decomposition approach for the locomotive and car assignment problem. Transp Sci 34(4):133–149

Cordeau JF, Desauliniers G, Lingaya N, Soumis F, Desrosiers J (2001a) Simultaneous locomotive and car assignment at VIA Rail Canada. Transportation Research Part B 35:767–787

Cordeau JF, Soumins F, Desrosiers J (2001b) Simultaneous assignment of locomotives and cars to passenger trains. Operation Research 49:531–548

Crevier B, Cordeau JF, Laporte G (2007) The multi-depot vehicle routing problem with inter-depot routes. Eur J Oper Res 176:756–773

Czech ZJ, Czarnas P (2002) Parallel simulated annealing for the vehicle routing problem with time windows. In: Proceeding of the 10th Euromicro workshop on parallel, distributed and network-based processing, Canary Islands, Spain, January 9, 2002. IEEE, NY, USA, pp 376–383

Fioole PJ, Kroon L, Maroti G, Schrijver A (2006) A rolling stock circulation model for combining and splitting of passenger trains. European Journal of Operation Research 174:1281–1297

Florian M, Bushell G, Ferland J, Guerin G, Nastansky L (1976) The engine scheduling problem in a railway network. Infor 14:121–138

Forbes MA, Holt JN, Walts AM (1991) Exact solution of locomotive scheduling problem. Journal of the Operation Research Society 42:825–831

Gambardella LM, Taillard E, Agazzi G (1999) MACS-VRPTW: A multiple ant colony system for vehicle routing problems with time windows. In: Corne D, Dorigo N, Glover F (eds) New Ideas in Optimization. McGraw-Hill, Columbus, pp 63–76

Ghoseiri K, Ghannadpour SF (2009a) Locomotive routing problem using a hybrid genetic algorithm. Journal of Transportation Research 5(3):259–273

Ghoseiri K, Ghannadpour SF (2009b) Hybrid genetic algorihtm for vehicle routing and scheduling problem. Journal of Applied Science 9(1):79–87

Ghoseiri K, Ghannadpour SF (2010) A hybrid genetic algorithm for multi-depot homogenous locomotive assignment with time windows. Appl Soft Comput 10:53–65

Holland JH (1975) Adaptation in natural and artificial system. The University of Michigan Press, Ann Arbor, MI

Irnich S, Funke B, Grunert T (2006) Sequential search and its application to vehicle routing problems. Computer & Operations Research 33:2405–2429

Kim BI, Kim S, Sahoo S (2006) Waste collection vehicle routing problem with time windows. Comput Oper Res 33:3624–3642

Laporte G, Semet F (1999) Classical Heuristics for the Vehicle Routing Problem. Les Cahiers du GERAD, G-98-54. Group for Research in Decision Analysis, Montreal, Canada, http://www.gerad.ca/en/publications/cahiers_rech.php?r_numero=&r_titre=Classical+Heuristics+for+the+Vehicle+Routing+Problem&r_titre_ancien=&r_auteurs=&r_journal_paru=&r_paru_mois=&r_paru_annee=&r_rev_mois=&r_rev_annee=&r_resume=&tri=num&rech=Search Accessed 06 December 2010

Li F, Golden B, Wasil E (2005) Very large scale vehicle routing: new test problems algorithms and results. Comput Oper Res 32:1165–1179

Lingaya N, Cordeau JF, Desaulniers G, Desrosiers J, Soumis F (2002) Operational car assignment at VIA Rail Canada. Transportation Research Part B 36:755–778

Pisinger D, Ropke S (2007) A general heuristic for vehicle routing problem. Computer & Operations Researches 34:2403–2435

Rouillon S, Desaulniers G, Soumis F (2006) An extended branch-and-bound method for locomotive assignment. Transportation Research Part B 40:404–423

Sheng HM, Wang JC, Huang HH, Yen DC (2006) Fuzzy measure on vehicle routing problem of hospital materials. Expert Systems with Application 30:367–377

Solomon MM (1987) Algorithms for vehicle routing problem and scheduling problems with time window constraints. Operation research **35**:254–365

Tan KC, Chew YH, Lee LH (2006) A hybrid multiobjective evolutionary algorithm for solving vehicle routing problem with time windows. Comput Optim Appl 34:115–151

Tan KC, Cheong CY, Goh CK (2007) Solving multiobjective vehicle routing problem with stochastic demand via evolutionary computation. Eur J Oper Res 177:139–813

Vaidyanathan B, Ahuja RK, Orlin JB (2008a) The locomotive routing problem. Transp Sci 42:492–507

Vaidyanathan B, Ahuja RK, Liu J, Shughart LA (2008b) Real-life locomotive planning: New formulations and computational results. Transportation Research Part B: Methodological 42:147–168

Ziarati K, Soumis F, Desrosiers J, Gelinas S, Saintonge A (1997) Locomotive assignment with heterogeneous consists at CN North America. European Journal of Operation Research 97:281–292

4

A goal geometric programming problem (G^2P^2) with logarithmic deviational variables and its applications on two industrial problems

Payel Ghosh[1*] and Tapan Kumar Roy[2]

Abstract

A very useful multi-objective technique is goal programming. There are many methodologies of goal programming such as weighted goal programming, min-max goal programming, and lexicographic goal programming. In this paper, weighted goal programming is reformulated as goal programming with logarithmic deviation variables. Here, a comparison of the proposed method and goal programming with weighted sum method is presented. A numerical example and applications on two industrial problems have also enriched this paper.

Keywords: Goal programming, Geometric programming, Pareto optimality, Nonlinear programming

Background

The earliest goal programming formulation was introduced by Charnes et al. (1955). Later, Charnes and Cooper (1977), Ijiri (1965), Lee (1972), and Ignizio (1976) are the contributors of goal programming for which goal programming became a useful tool in multi-criteria decision-making (MCDM) problem. The updated presentations of goal programming have been discussed by Tamiz et al. (1998), Lee and Olson (2000), Jones and Tamiz (2002), and Ignizio and Romero (2003). Methodologies of goal programming such as weighted goal programming, min-max goal programming, lexicographic goal programming have been discussed in the study of Romero (2004). Except for these three methods, another method, the logarithmic goal programming, is introduced (Wang et al. 2005).

Our proposed method is goal geometric programming with logarithmic deviational variables. In goal programming formulation with logarithmic deviational variables, we use geometric programming for solving because there are lots of real-life situations and many engineering applications where equations may be nonlinear. For a special type of nonlinear programming problem, geometric programming is a very useful tool. Since we use geometric programming method to solve a nonlinear goal programming problem, therefore, the degree of difficulty has a great role in this context. The degree of difficulty of the proposed method is lesser than that of other methods such as goal geometric programming using weighted sum method.

The concept of taking multiplicative deviational variables as an objective function is not new. Previously, Verma (1990) and a paper entitled 'Goal geometric programming problem (G^2P^2) with product method' by Ghosh and Roy (2012) used this concept. In this paper, we have started with additive deviational variables as the objective function which were then converted into multiplicative deviational variables as objective function using the logarithmic concept. The method of conversion is given in the form of 'Result 1'.

The arrangement of the paper is as follows: the background of the study followed by the goal programming model are presented. A result is presented together with its proof (Result 1), and the model of weighted goal programming with logarithmic deviational variables is then presented. The sections for goal geometric programming model with logarithmic deviational variables and its solution procedure are followed by a theorem on the model of weighted goal programming with logarithmic deviational variables and its proof (Result 2). Next, a numerical

*Correspondence: ghoshpayel86@yahoo.com
[1] Department of Mathematics, Adamas Institute of Technology, Barasat, North 24 Parganas 700126, West Bengal, India
Full list of author information is available at the end of the article

example and applications on lightly loaded bearing problem, optimal production, and marketing planning are presented. Finally, the conclusions of the study is presented.

Goal programming

A multi-objective programming can be written as follows:

$$\text{Find } X = (x_1, x_2, \ldots, x_n)^T \tag{1}$$

$$\text{so as to minimize} f_{10}(X) = \sum_{i=1}^{P_{10}} C_{10i} \prod_{k=1}^{n} x_k^{a_{koi}} \text{ with target } C_{10},$$

$$\text{minimize} f_{20}(X) = \sum_{i=1}^{P_{20}} C_{20i} \prod_{k=1}^{n} x_k^{a_{koi}} \text{ with target } C_{20},$$

$$\text{minimize} f_{m0}(X) = \sum_{i=1}^{P_{m0}} C_{m0i} \prod_{k=1}^{n} x_k^{a_{koi}} \text{ with target } C_{m0},$$

$$\text{subject to} f_r(X) = \sum_{i=1}^{P_r} C_{ri} \prod_{k=1}^{n} x_k^{a_{ki}} \leq C_r;\ r = 1, 2, \ldots, q$$

$$x_k > 0;\ k = 1, 2, \ldots, n.$$

C_{j0i} and C_{ri} are positive real numbers $\forall\, j, r, i$, and a_{k0i}, a_{ki} are real numbers $\forall\, k, i$.

P_{j0} = Number of terms present in j0th objective function,

P_r = Number of terms present in rth constraint,

C_r = Boundary value of rth constraint,

The multi-objective programming model contains m, the number of minimizing objective functions; q, the number of inequality type constraints; and n, the number of strictly positive decision variables.

Result 1. As mentioned, the goal programming model may be reduced to the following form:

$$\text{Minimize} \prod_{j=1}^{m} u_{j0}^{+} \prod_{r=1}^{q} v_r^{+}$$

subject to

$$f_{j0}(X)/u_{j0}^{+} \leq C_{j0}, j = 1, 2, \ldots, m,$$

$$f_r(X)/v_r^{+} \leq C_r, r = 1, 2, \ldots, q,$$

$$x_k > 0, k = 1, 2, \ldots, n; u_{j0}^{+}, v_r^{+} > 1,$$

with the conditions

$$f_{j0}(X) > 0,\ C_{j0} > 0,\ C_r > 0.$$

Proof. In the multi-objective programming model (1), objective functions are minimized and have target values, e.g., minimize $f_{j0}(X)$ with target value C_{j0}, i.e., minimize $\log(f_{j0}(X))$ with target value $\log(C_{j0})$. According to the method of goal formulation, positive deviation should be minimized.

Similarly, in model (1), constraints are of $\leq$ type. Thus, positive deviations should also be minimized. Therefore, when

$$f_r(X) \leq C_r,$$

then

$$\log(f_r(X)) \leq \log(C_r).$$

The goal formulation is as follows:

$$\text{Minimize} \sum_{j=1}^{m} d_{j0}^{+} + \sum_{r=1}^{q} d_r^{+} \tag{2}$$

subject to

$$\log(f_{j0}(X)) + d_{j0}^{+} - d_{j0}^{-} = \log(C_{j0}); j = 1, 2, \ldots, m,$$

$$\log(f_r(X)) + d_r^{+} - d_r^{-} = \log(C_r); r = 1, 2, \ldots, q,$$

$$x_k > 0, k = 1, 2, \ldots, n\ d_{j0}^{+}, d_{j0}^{-}, d_r^{+}, d_r^{-} > 0,$$

$$d_{j0}^{+} \times d_{j0}^{-} = 0; d_r^{+} \times d_r^{-} = 0.$$

d_{j0}^{+} = Positive deviation of objective function,

d_{j0}^{-} = Negative deviation of objective function,

d_r^{+} = Positive deviation of constraint,

d_r^{-} = Negative deviation of constraint.

However, with a logarithmic change of deviational variables $d_{j0}^{+} = \log(u_{j0}^{+})$, $d_{j0}^{-} = \log(u_{j0}^{-})$, $d_r^{+} = \log(v_r^{+})$, $d_r^{-} = \log(v_r^{-})$, we can turn model (2) into the following problem:

$$\text{Minimize} \left(\log \prod_{j=1}^{m} u_{j0}^{+} \prod_{r=1}^{q} v_r^{+} \right) \tag{3}$$

subject to

$$\log(f_{j0}(X) \cdot u_{j0}^{-}/u_{j0}^{+}) = \log(C_{j0}), j = 1, 2, \ldots, m,$$

$$\log(f_r(X) \cdot v_r^{-}/v_r^{+}) = \log(C_r), r = 1, 2, \ldots, q,$$

$$x_k > 0, k = 1, 2, \ldots, n; u_{j0}^{+}, u_{j0}^{-}, v_r^{+}, v_r^{-} > 1,$$

which is obviously equivalent to the following goal programming form with logarithmic deviational variables:

$$\text{Minimize} \prod_{j=1}^{m} u_{j0}^{+} \prod_{r=1}^{q} v_r^{+} \tag{4}$$

subject to

$$f_{j0}(X) \cdot u_{j0}^{-}/u_{j0}^{+} = C_{j0}, j = 1, 2, \ldots, m,$$

$$f_r(X) \cdot v_r^{-}/v_r^{+} = C_r, r = 1, 2, \ldots, q,$$

$$x_k > 0, k = 1, 2, \ldots, n; u_{j0}^{+}, u_{j0}^{-}, v_r^{+}, v_r^{-} > 1.$$

The goal programming formulation where the constraints are in inequality form the following:

$$\text{Minimize} \prod_{j=1}^{m} u_{j0}^{+} \prod_{r=1}^{q} v_r^{+} \tag{5}$$

subject to

$$f_{j0}(X)/u_{j0}^{+} \leq C_{j0}, j = 1, 2, \ldots, m,$$

$$f_r(X)/v_r^{+} \leq C_r, r = 1, 2, \ldots, q,$$

$$x_k > 0, k = 1, 2, \ldots, n; u_{j0}^{+}, v_r^{+} > 1,$$

hence the result. □

Results and discussion

Weighted goal programming with logarithmic deviational variables

According to model (1), all of the objective functions are minimized. If the decision maker wants to get a much more minimized value for any particular objective function or wants to satisfy strictly the constraints, then weight factors (priorities) are introduced. In goal programming formulation with logarithmic deviational variables, weights (priorities) are given with the deviational variable. Hence, the weighted goal programming formulation becomes the following:

$$\text{Minimize } \prod_{j=1}^{m}(u_{j0}^{+})^{W_{j0}}\prod_{r=1}^{q}(v_r^{+})^{W_r} \tag{6}$$

subject to

$$f_{j0}(X)/u_{j0}^{+} \le C_{j0}, j = 1, 2, \ldots, m,$$

$$f_r(X)/v_r^{+} \le C_r, r = 1, 2, \ldots, q,$$

$$x_k > 0, k = 1, 2, \ldots, n; u_{j0}^{+}, v_r^{+} > 1.$$

Here, W_{j0} values are the weights for objective functions and W_r values are the weights for the constraints.

Solutions of goal programming (Romero 1991), even those of weighted goal programming and lexicographic goal programming (Miettinen 1999), are pareto optimal. Here, we prove a result which also shows that goal programming with logarithmic deviation gives pareto optimal solutions.

Result 2. The following is the solution of weighted goal programming with logarithmic deviation:

$$\text{Minimize } \prod_{i=1}^{k}(u_i^{+})^{w_i}$$

subject to

$$\left(\sum_{r=1}^{p} C_{mr}\prod_{l=1}^{n} x_l^{a_{lr}}\right)_i (u_i^{+})^{-1} \le \overline{C_i}, i = 1, 2, \ldots, k,$$

$$X \in S, u_i^{+} > 1, \ i = 1, 2, \ldots, k,$$

which comes from the following goal programming model:

$$\text{Minimize} f_i(X) = \left(\sum_{r=1}^{p} C_{mr}\prod_{l=1}^{n} x_l^{a_{lr}}\right)_i \text{with target } \overline{C_i},$$

$$i = 1, 2, \ldots, k; \ X \in S$$

is pareto optimal if u_i^{+} for each function $f_i(X)$ to be minimized has a value greater than 1 at the optimum.

Proof. If $x^* \in S$ with a positive deviation vector, then let $(u_i^{+})^* \ (> 1)$ be the solution of the following weighted goal programming problem:

$$\text{Minimize } \prod_{i=1}^{k}(u_i^{+})^{w_i} \tag{7}$$

subject to

$$\left(\sum_{r=1}^{p} C_{mr}\prod_{l=1}^{n} x_l^{a_{lr}}\right)_i (u_i^{+})^{-1} \le \overline{C_i}, i = 1, 2, \ldots, k,$$

$$X \in S, u_i^{+} > 1, \ i = 1, 2, \ldots, k.$$

If possible, let x^* be not pareto optimal, then there exists a vector x^0 with a positive deviational variable $(u_i^{+})^0 (> 1)$ such that

$$\left(\sum_{r=1}^{p} C_{mr}\prod_{l=1}^{n}(x_l^0)^{a_{lr}}\right)_i \le \left(\sum_{r=1}^{p} C_{mr}\prod_{l=1}^{n}(x_l^*)^{a_{lr}}\right)_i ; \tag{7.1}$$

$$\forall i = 1, 2, \ldots, k$$

and

$$\left(\sum_{r=1}^{p} C_{mr}\prod_{l=1}^{n}(x_l^0)^{a_{lr}}\right)_j < \left(\sum_{r=1}^{p} C_{mr}\prod_{l=1}^{n}(x_l^*)^{a_{lr}}\right)_j ; \tag{7.2}$$

$$\text{for at least one } j.$$

or

$$\frac{\left(\sum_{r=1}^{p} C_{mr}\prod_{l=1}^{n}(x_l^*)^{a_{lr}}\right)_j}{\left(\sum_{r=1}^{p} C_{mr}\prod_{l=1}^{n}(x_l^0)^{a_{lr}}\right)_j} > 1.$$

Let

$$\frac{\left(\sum_{r=1}^{p} C_{mr}\prod_{l=1}^{n}(x_l^*)^{a_{lr}}\right)_j}{\left(\sum_{r=1}^{p} C_{mr}\prod_{l=1}^{n}(x_l^0)^{a_{lr}}\right)_j} = \beta > 1. \tag{7.3}$$

We set

$$(u_i^{+})^0 = (u_i^{+})^* \ (> 1) \ for \ i = 1, 2, \ldots, k \tag{7.4}$$

and

$$(u_j^{+})^0 = \max(1, \ (u_j^{+})^*/\beta) \ge 1 \text{ and } i \ne j. \tag{7.5}$$

Here, $(u_i^{+})^0$ is the positive deviational variable corresponding to x^0, $i = 1, 2, \ldots, k$. From (7.1),

$$\left(\sum_{r=1}^{p} C_{mr}\prod_{l=1}^{n}(x_l^0)^{a_{lr}}\right)_i ((u_i^{+})^0)^{-1} \le \left(\sum_{r=1}^{p} C_{mr}\prod_{l=1}^{n}(x_l^*)^{a_{lr}}\right)_i \times ((u_i^{+})^0)^{-1}$$

$$\le \left(\sum_{r=1}^{p} C_{mr}\prod_{l=1}^{n}(x_l^*)^{a_{lr}}\right)_i ((u_i^{+})^*)^{-1} \text{ using (7.4)}$$

$$\le \overline{C_i} \text{ as } x^* \text{ be the solution of (7), i.e.,}$$

$$\left(\sum_{r=1}^{p} C_{mr}\prod_{l=1}^{n}(x_l^0)^{a_{lr}}\right)_i ((u_i^{+})^0)^{-1} \le \overline{C_i} \text{ for}$$

$$i = 1, 2, \ldots, k, \text{ but } i \ne j. \tag{7.6}$$

From 7.5,

$$(u_j^+)^0 = \max\left(1, \frac{(u_j^+)^*}{\beta}\right),$$

Thus,

$$(u_j^+)^0 = \frac{(u_j^+)^*}{\beta}, \text{ if } \frac{(u_j^+)^*}{\beta} > 1 \tag{7.7}$$

$$= 1 \text{ if } \frac{(u_j^+)^*}{\beta} \leq 1 \tag{7.8}$$

□

Case 1

$\frac{(u_j^+)^*}{\beta} > 1$, then using (7.7),

$$\left(\sum_{r=1}^{p} C_{mr} \prod_{l=1}^{n} (x_l^0)^{a_{lr}}\right)_j ((u_j^+)^0)^{-1} = \left(\sum_{r=1}^{p} C_{mr} \prod_{l=1}^{n} (x_l^0)^{a_{lr}}\right)_j \times ((u_j^+)^*)^{-1}\beta$$

$$= \left(\sum_{r=1}^{p} C_{mr} \prod_{l=1}^{n} (x_l^*)^{a_{lr}}\right)_j ((u_j^+)^*)^{-1} \leq \overline{C_j}$$

Therefore,

$$\left(\sum_{r=1}^{p} C_{mr} \prod_{l=1}^{n} (x_l^0)^{a_{lr}}\right)_j ((u_j^+)^0)^{-1} \leq \overline{C_j} \tag{7.9}$$

Thus, x_0 satisfies the constraints of (7). From (7.7), $(u_j^+)^0 = \frac{(u_j^+)^*}{\beta} < (u_j^+)^*$.

Since $\beta > 1$ and $(u_j^+)^* > 1$,

using (7.4),

$(u_i^+)^0 \leq (u_i^+)^*, \; \forall\, i = 1, 2, \ldots, k.$

Case 2

$\frac{(u_j^+)^*}{\beta} \leq 1$, then using (7.8),

$$\left(\sum_{r=1}^{p} C_{mr} \prod_{l=1}^{n} (x_l^0)^{a_{lr}}\right)_j ((u_j^+)^0)^{-1} = \left(\sum_{r=1}^{p} C_{mr} \prod_{l=1}^{n} (x_l^0)^{a_{lr}}\right)_j = \frac{(\sum_{r=1}^{p} C_{mr} \prod_{l=1}^{n} (x_l^*)^{a_{lr}})_j}{\beta}.$$

$$\text{Using (7.3), } \leq \frac{(\sum_{r=1}^{p} C_{mr} \prod_{l=1}^{n} (x_l^*)^{a_{lr}})_j}{(u_j^+)^*}; \tag{7.10}$$

$$\text{from (7.8), } \leq \overline{C_j}$$

Thus, x^0 satisfies the constraints of (7), and from (7.8),

$$(u_j^+)^0 = 1 < (u_j^+)^*. \tag{7.11}$$

Therefore, from (7.4) and (7.11), $(u_i^+)^0 \leq (u_i^+)^* \; \forall\, i = 1, 2, \ldots, k$. Thus, $\forall$ positive weights W_i, $(i = 1, 2, \ldots, k)$

$$((u_i^+)^0)^{W_i} \leq ((u_i^+)^*)^{W_i} \text{ or } \prod_{i=1}^{k} ((u_i^+)^0)^{W_i} \leq \prod_{i=1}^{k} ((u_i^+)^*)^{W_i}. \tag{7.12}$$

Thus, from (7.6), (7.9), (7.10), and (7.12), we have seen that x^0 is a solution of (7), which contradicts the fact that x^* is a solution of (7). Hence, x^* is pareto optimal.

Goal geometric programming model with logarithmic deviational variables and its solution procedure

Linear goal programming is a very commonly used tool of the MCDM problem. However, nonlinear goal programming is very rare in this context. In many engineering problems, as well as problems of science, there are nonlinear equations to optimize. To solve that type of nonlinear goal programming problem, the geometric programming method can be used. Hence, we can turn model (6) into a goal geometric programming form as in the following:

$$\text{Minimize} \prod_{j=1}^{m} (u_{j0}^+)^{W_{j0}} \prod_{r=1}^{q} (v_r^+)^{W_r} \tag{8}$$

subject to

$$f_{j0}(X)(u_{j0}^+)^{-1}/C_{j0} \leq 1, \; j = 1, 2, \ldots, m,$$

$$f_r(X)(v_r^+)^{-1}/C_r \leq 1, \; r = 1, 2, \ldots, q,$$

$$x_k > 0, \; k = 1, 2, \ldots, n; \; u_{j0}^+, \; v_r^+ > 1.$$

The corresponding dual geometric programming of model (8) can be written as follows:

$$\text{Maximize } d(\delta) = \left[\left(\frac{1}{\delta_{10}}\right)^{\delta_{10}} \prod_{j=1}^{m} \prod_{i=1}^{P_{j0}} \left(\frac{C_{j0i}}{C_{j0}\delta_{ji}}\right)^{\delta_{ji}} \prod_{r=1}^{q} \prod_{i=1}^{P_r} \times \left(\frac{C_{ri}}{C_r\delta_{ri}}\right)^{\delta_{ri}} \prod_{j=1}^{m} \lambda_j(\delta)^{\lambda_j(\delta)} \prod_{r=1}^{q} \lambda_r(\delta)^{\lambda_r(\delta)}\right]$$

such that

$$\delta_{10} = 1, \; W_{j0}\delta_{10} - \sum_{i=1}^{P_{j0}} \delta_{ji} = 0, \; j = 1, 2, \ldots, m;$$

$$W_r\delta_{10} - \sum_{i=1}^{P_r} \delta_{ri} = 0, \; r = 1, 2, \ldots, q;$$

$$\sum_{j=1}^{m}\sum_{i=1}^{P_{j0}} a_{k0i}\delta_{ji} - \sum_{r=1}^{q}\sum_{i=1}^{P_r} a_{ki}\delta_{ri} = 0;\ k = 1, 2, \ldots, n;$$

$$\lambda_j(\delta) = \sum_{i=1}^{P_{j0}} \delta_{ji},\ j = 1, 2, \ldots, m;$$

$$\lambda_r(\delta) = \sum_{i=1}^{P_r} \delta_{ri},\ r = 1, 2, \ldots, q.$$

Numerical example

A multi-objective goal programming problem is:

$$\text{Minimize} f_1(x_1, x_2) = x_1^{-1}x_2^{-2} \text{ with target value } 4, \quad (9)$$

$$\text{Minimize} f_2(x_1, x_2) = 2x_1^{-2}x_2^{-3} \text{ with target value } 50,$$

subject to $x_1 + x_2 \leq 1,\ x_1, x_2 > 0.$

In goal geometric programming model with logarithmic deviational variables, model (8) can be written as follows:

$$\text{Minimize } u^{W_1}v^{W_2} \quad (10)$$

subject to

$$x_1^{-1}x_2^{-2}u^{-1} \leq 4,$$

$$2x_1^{-2}x_2^{-3}v^{-1} \leq 50,$$

$$x_1 + x_2 \leq 1,\ x_1, x_2 > 0,\ u, v > 1.$$

Illustration

Degree of difficulty = 5 − (4 + 1) = 0, dual of (10) is given by the following:

$$\text{Maximize } d(\delta) = \left[\left(\frac{1}{\delta_{10}}\right)^{\delta_{10}}\left(\frac{1}{4\delta_{11}}\right)^{\delta_{11}}\left(\frac{2}{50\delta_{21}}\right)^{\delta_{21}}\left(\frac{1}{\delta_{31}}\right)^{\delta_{31}} \times \left(\frac{1}{\delta_{32}}\right)^{\delta_{32}}\lambda_1(\delta)^{\lambda_1(\delta)}\lambda_2(\delta)^{\lambda_2(\delta)}\lambda_3(\delta)^{\lambda_3(\delta)}\right]$$

such that

$$\delta_{10} = 1, \quad (10.1)$$

$$W_1\delta_{10} - \delta 11 = 0, \quad (10.2)$$

$$W_2\delta_{10} - \delta 21 = 0, \quad (10.3)$$

$$-\delta_{11} - 2\delta_{21} + \delta_{31} = 0, \quad (10.4)$$

$$-2\delta_{11} - 3\delta_{21} + \delta_{32} = 0, \quad (10.5)$$

$$\lambda_1(\delta) = \delta_{11},\ \lambda_2(\delta) = \delta_{21},\ \lambda_3(\delta) = \delta_{31} + \delta_{32}.$$

Solving (10.1) to (10.5), we get the following:

$$\delta_{10} = 1,\ \delta_{11} = W_1,\ \delta_{21} = W_2,\ \delta_{31} = W_1 + 2W_2,$$

$$\delta_{32} = 2W_1 + 3W_2,\ \lambda_1(\delta) = W_1,\ \lambda_2(\delta) = W_2,$$

$$\lambda_3(\delta) = 3W_1 + 5W_2.$$

From primal dual relation

$$\frac{x_1^{-1}x_2^{-2}u^{-1}}{4} = \frac{\delta_{11}}{\lambda_1(\delta)} = 1 \text{ or } u = \frac{1}{4x_1x_2^2},$$

$$\frac{2x_1^{-2}x_2^{-3}v^{-1}}{50}\ \frac{\delta_{21}}{\lambda_2(\delta)} = 1 \text{ or } v = \frac{2}{50x_1^2x_2^3},$$

$$x_1 = \frac{W_1 + 2W_2}{3W_1 + 5W_2},\ x_2 = \frac{2W_1 + 3W_2}{3W_1 + 5W_2}.$$

Solving from primal dual relation for different values of weights, we get the optimal values of the decision variables which are given in Table 1.

From the table, we see that each deviation (u_i, v_i) has values greater than 1 when minimized. Thus, according to our theorem, the solutions are pareto optimal.

Again, we have solved the mentioned example in goal geometric programming with weighted sum method. Here, we have compared the results of the mentioned example in equal weights solved in two different methods: goal geometric programming with weighted sum method and goal geometric programming with logarithmic deviational variables which are given in Table 2.

From the comparison, it is clear that in both methods, the optimum values of the first and second objectives are almost the same. We have solved the same example in both processes where we have used geometric programming to solve a nonlinear goal programming problem. The advantage of the proposed method lies in the method of solution, i.e., geometric programming where degree of difficulty is less than the degree of difficulty of the previous process (goal geometric programming with weighted sum method). For this reason, the solution procedure of this process becomes easier than that of the previous.

Application on lightly loaded bearing problem

A lightly loaded bearing is to be designed to minimize the linear combination of frictional moment and angle of twist of the shaft and the temperature rise of the oil while carrying a load of 1,000 lb, and the angular velocity of the shaft is to be greater than 100 rad s^{-1}. Assume that 1 in-lb of frictional moment in the bearing is equal to 0.0025 rad of the angle of twist. The following are the goals:

> Priority 1: Linear combination of frictional moment, angel of twist of the shaft, and temperature rise of the oil should be minimized and near 10.
> Priority 2: Angular velocity of the shaft per 100 rad s^{-1} should be minimized and near 0.2.

In formulating the mentioned goal programming problem and finding the dimension of the bearing that is to be built for this purpose, it should be done in such a way that it can carry the maximum load.

Table 1 Optimal values of decision variables using G^2P^2 with logarithmic deviational variables

W_1	W_2	Primal variables	Dual variables	First objective (f_1^*)	Second objective (f_2^*)
0.9	0.1	$x_1^* = 0.34375$,	$\Delta_{01} = 1, \Delta_{11} = 0.9$	6.754896	59.88756
		$x_2^* = 0.65625$,	$\Delta_{21} = 0.1, \Delta_{31} = 1.1$		
		$u^* = 1.688724$,	$\Delta_{32} = 2.1$		
		$v^* = 1.197751$			
0.8	0.2	$x_1^* = 0.3529412$	$\Delta_{01} = 1, \Delta_{11} = 0.8$,	6.767218	59.26442
		$x_2^* = 0.6470588$,	$\Delta_{21} = 0.2, \Delta_{31} = 1.2$,		
		$u^* = 1.691804$	$\Delta_{32} = 2.2$		
		$v^* = 1.185288$			
0.7	0.3	$x_1^* = 0.36111$	$\Delta_{01} = 1, \Delta_{11} = 0.7$	6.784354	58.81286
		$x_2^* = 0.638889$,	$\Delta_{21} = 0.3, \Delta_{31} = 1.3$		
		$u^* = 1.696088$,	$\Delta_{32} = 2.3$		
		$v^* = 1.176257$			
0.6	0.4	$x_1^* = 0.3684211$	$\Delta_{01} = 1, \Delta_{11} = 0.6$	6.804563	58.48684
		$x_2^* = 0.6315789$,	$\Delta_{21} = 0.4, \Delta_{31} = 1.4$		
		$u^* = 1.701141$	$\Delta_{32} = 2.4$		
		$v^* = 1.169737$			
0.5	0.5	$x_1^* = 0.375$	$\Delta_{01} = 1, \Delta_{11} = 0.5$	6.826667	58.25422
		$x_2^* = 0.625$,	$\Delta_{21} = 0.5, \Delta_{31} = 1.5$		
		$u^* = 1.70667$,	$\Delta_{32} = 2.5$		
		$v^* = 1.165084$			
0.4	0.6	$x_1^* = 0.3809524$	$\Delta_{01} = 1, \Delta_{11} = 0.4$	6.849852	58.09201
		$x_2^* = 0.6190476$,	$\Delta_{21} = 0.6, \Delta_{31} = 1.6$		
		$u^* = 1.712463$,	$\Delta_{32} = 2.6$		
		$v^* = 1.16184$			
0.3	0.7	$x_1^* = 0.3863636$	$\Delta_{01} = 1, \Delta_{11} = 0.3$	6.873558	57.98348
		$x_2^* = 0.6136364$,	$\Delta_{21} = 0.7, \Delta_{31} = 1.7$		
		$u^* = 1.718389$,	$\Delta_{32} = 2.7$		
		$v^* = 1.159670$			
0.2	0.8	$x_1^* = 0.3913043$	$\Delta_{01} = 1, \Delta_{11} = 0.2$	6.897392	57.91620
		$x_2^* = 0.6086957$,	$\Delta_{21} = 0.8, \Delta_{31} = 1.8$		
		$u^* = 1.724348$	$\Delta_{32} = 2.8$		
		$v^* = 1.158324$			
0.1	0.9	$x_1^* = 0.3958333$	$\Delta_{01} = 1, \Delta_{11} = 0.1$	6.921084	57.88086
		$x_2^* = 0.6041667$,	$\Delta_{21} = 0.9, \Delta_{31} = 1.9$		
		$u^* = 1.730271$	$\Delta_{32} = 2.9$		
		$v^* = 1.157617$			

Solution Let R (in.) be the radius of the journal and L (in.) be the half length of the bearing, T be the temperature rise of the oil, and frictional moment of the bearing $(M) = \frac{8\pi\mu\omega R^2 L}{\sqrt{1-n^2}\, c}$ where ω is the angular velocity of the shaft, μ is the viscosity of the oil (lubricant), n is the eccentricity ratio, and c is the radial clearance.

The angle of twist of the shaft $(\phi) = \frac{S_e l}{GR}$, where S_e is the shear stress, l is the length between the driving point and rotating mass, and G is the shear modulus. The temperature rise of the oil in the bearing is given by $T = \frac{0.045\mu\omega R^2}{c^2 n\sqrt{1-n^2}}$. For the given data, $\frac{c}{R} = 0.0015$, $n = 0.9$, $\mu = 10^{-6}$ lb s in.$^{-2}$, $l = 10$ in., $S_e = 30,000$ psi, and $G = 12 \times 10^6$ psi.

Table 2 Comparison of optimal solutions in two different methods

Method	W_1	W_2	First objective (f_1^*)	Second objective (f_2^*)
Goal geometric programming with weighted sum	0.5	0.5	6.919487	57.88240
Goal geometric programming with logarithmic deviational variables	0.5	0.5	6.826667	58.25422

Hence, linear combination of frictional moment, angle of twist of the shaft, and temperature rise of the oil equals

$$0.038\omega R^2 L + 0.025R^{-1} + 0.592RL^{-3} \text{ with target value } 10 \quad (11.1)$$

and angular velocity

$$\omega \geq 100 \text{ rad s}^{-1}. \quad (11.2)$$

From the given data in the chart of 'Dimensionless performance parameters for full journal bearing' $\omega R^{-1}L^3 = 11.6$, i.e., $\omega = 11.6R/L^3$.

As per the assumption that 1 in. lb of frictional moment in bearing is equal to 0.0025 rad angle of twist, Equation 11.1 becomes $Z_1 = 0.44R^3L^{-2} + 10R^{-1} + 0.592RL^{-3}$ with the target value of 10.

Equation 11.2 becomes $Z_2 = 8.62R^{-1}L^3$ with the target value of 0.2. Hence, the model of lightly loaded bearing problem in G^2P^2 with logarithmic deviational variable is as follows:

$$\text{Minimize } u^{W_1}v^{W_2} \quad (11.3)$$

subject to

$$0.44R^3L^{-2}u^{-1} + 10R^{-1}u^{-1} + 0.592RL^{-3}u^{-1} \leq 10,$$

$$8.62R^{-1}L^3v^{-1} \leq 0.2,$$

$$u, v > 1, R, L > 0.$$

In solving with the use of geometric programming method where the degree of difficulty is $5-(4+1) = 0$, we get the optimal values of the radius of the journal (R) and half length of the bearing (L) which are given in Table 3. From the table, we have seen that each deviation (u, v) has values greater than 1. Thus, the solution is pareto optimal.

Application on optimal production and marketing planning

Consider a manufacturer who produces a single product where the demand is affected by the selling price. Let P be the selling price per unit, α be the price elasticity to the demand, M be the marketing expenditure per unit, and γ be the marketing expenditure elasticity to the demand (Sadjadi et al. 2005). Assume that demand $D = KP^{-\alpha}M^{\gamma}$, where K is the predetermined constant and production cost C, which is inversely related to production lot size (units) Q, i.e., $C = rQ^{-\beta}$, where r is the predefined constant for unit production cost and β is the lot size elasticity of production unit cost. Again, let μ and a be the production rate and the setup cost of production, respectively. We assume the production rate μ to vary with the demand D proportionally. Hence, $\mu = uD$ where $u > 1$. There are some restrictions on variables such as α, γ, and β. The equation $\alpha > 1$ indicates that D increases at a diminishing rate as P decreases. The equation $0 < \beta < 1$ is almost the same as α and $0 < \gamma < 1$.

We want to minimize the equation (Marketing cost + Production cost + Setup cost + Holding cost), which is subject to some constraint that total revenue should be bigger. These are the following goals:

Priority 1: Total revenue should be greater than 0.1386×10^5,
Priority 2: (Marketing cost + Production cost + Setup cost + Holding cost) should be minimized and near 0.692791.

Thus, the model is as follows:

$$\text{Minimize } MD + CD + \frac{aD}{Q} + iC\left(1 - \frac{D}{\mu}\right)\frac{Q}{2} \text{ with target value } 0.692791$$

subject to

$$PD \geq 0.1386 \times 10^5$$

$$P, M, Q > 0.$$

Let $\widehat{u} = 1 - \frac{1}{u}$, and from assumptions and consideration, the above model becomes the following:

$$\text{Minimize } KP^{-\alpha}M^{\gamma+1} + rKP^{-\alpha}M^{\gamma}Q^{-\beta} + aKP^{-\alpha}M^{\gamma}Q^{-1} \quad (12.1)$$

Table 3 Optimal values of radius of the journal (R) and half length of the bearing (L)

W_1	W_2	Dual variables	Primal variables	First objective (Z_1)	Second objective (Z_2)
0.6	0.4	$\Delta_{01} = 1$,	$R^* = 0.97395$,	35.2027	0.2485057
		$\Delta_{11} = 0.075$,	$L^* = 0.3039405$,		
		$\Delta_{12} = 0.175, \Delta_{13} = 0.35$,	$u^* = 3.520266$,		
		$\Delta_{21} = 0.4$.	$v^* = 1.24253$.		

Table 4 Optimal values of price per unit (P), production lot size (Q), marketing expenditure per unit (M)

W_1	W_2	Dual variables	Primal variables	Objective function (Z) (US$)	Total revenue (US$)
0.4	0.6	$\Delta_{01} = 1$,	$P^* = 72.38344$,	129.7589	9098297
		$\Delta_{11} = 0.0072$,	$M^* = 0.1189591$,		
		$\Delta_{12} = 0.3163636$,	$Q^* = 83.07024$,		
		$\Delta_{13} = 0.03643$,	$u^* = 200.7293$,		
		$\Delta_{14} = 0.04$,	$v^* = 9.098298$.		
		$\Delta_{21} = 0.6$.			

$$+ ir\widehat{u}\frac{Q^{-\beta+1}}{2} \quad \text{with target value } 0.692791$$

subject to

$$KP^{-\alpha}M^{\gamma} \geq 0.1386 \times 10^5$$

$$P,\ M,\ Q > 0.$$

Consider the following data: $\alpha = 2.5$, $\beta = 0.01$, $\gamma = 0.03$, $r = 5$, $K = 10^6$, $a = 50$, $i = 0.1$, $\widehat{u} = 0.7$, and converting the model (12.1) according to the goal geometric programming model, we have the following:

$$\text{Minimize (Z) } 10^6P^{-2.5}M^{1.03} + 5 \times 10^6P^{-2.5}M^{0.03}Q^{-0.01} + 50 \times 10^6P^{-2.5}M^{0.03}Q^{-1} + \frac{0.1 \times 0.7 \times 5}{2} \times Q^{0.99} \quad \text{with target value } 0.692791 \tag{12.2}$$

subject to

$$0.1386 \times 10^5 \times 10^{-6}P^{1.5}M^{-0.03} \leq 1$$

$$P,\ M,\ Q > 0.$$

Transforming the model (12.2) into G^2P^2 with logarithmic deviation variables, we get the following:

$$\text{Minimize } u^{W_1}v^{W_2} \tag{12.3}$$

subject to

$$\frac{10^6P^{-2.5}M^{1.03}u^{-1}}{0.692791} + \frac{5 \times 10^6P^{-2.5}M^{0.03}Q^{-0.01}u^{-1}}{0.692791} + \frac{50 \times 10^6P^{-2.5}M^{0.03}Q^{-1}u^{-1}}{0.692791} + \frac{\frac{0.1\times0.7\times5}{2}Q^{0.99}u^{-1}}{0.692791} \leq 1$$

$$0.1386 \times 10^5 \times 10^{-6}P^{1.5}M^{-0.03}v^{-1} \leq 1$$

$$P,\ M,\ Q > 0, u, v > 1.$$

Solving with the use of geometric programming method where the degree of difficulty is $6 - (5 + 1) = 0$, we get the optimal values of decision variables, e.g., price per unit (P), production lot size (Q), and marketing expenditure per unit (M), which are given in Table 4.

Here, we have also observed from the table that each deviation (u, v) has values greater than 1. Thus, the solution is pareto optimal.

Conclusions

The aim of this paper was to introduce a new approach to solve a nonlinear goal programming problem. The geometric programming approach is the best tool to solve nonlinear programming problems as compared with the other approach (Khun-Tucker conditions) that is already discussed in this paper. We have used logarithmic deviational variables in the goal programming model instead of the commonly used addition of deviational variables. Also, the applications on lightly loaded bearing problem and optimal production and marketing planning shows the efficiency of this method. Two applications have two different aims. In the first application, the decision maker has given more priority to the first objective function, whereas in the second application, priority is given to the second objective. Further, this method could be more applicable in imprecise environment rather than in precise environment.

Competing interests

The authors declare that they have no competing interests.

Authors' contributions

PG and TKR have discussed a new approach to solve a nonlinear goal programming problem. The model formulation of goal programming using logarithmic deviational variables is rare. Instead of using Kuhn-Tucker conditions to solve nonlinear programming, they have used geometric programming. They have also shown that geometric programming gives a better result than Kuhn-Tucker conditions. Both authors read and approved the final manuscript.

Authors' information

PG is an assistant professor at the Department of Mathematics in Adamas Institute of Technology. She received her master's of science degree from Bengal Engineering and Science University in 2009. Her research interests are optimization and fuzzy mathematics. TKR is a professor at the Department of Mathematics in Bengal Engineering and Science University. His research interests are in the areas of fuzzy mathematics, fuzzy optimization, and inventory control. He received his bachelor's degree in mathematics from Burdwan University in 1977. He completed his master's degree in mathematics from Jadavpur University in 1986 and his PhD from Vidyasagar University in 1999.

Acknowledgements
The authors are thankful to the reviewers for their suggestions on improving the quality of this paper as well as to the language editor whose suggestions helped improve the presentation of the paper. The authors are grateful to Mr. Saroj Kumar Rana, assistant professor at the Department of Mechanical Engineering, Adamas Institute of Technology, for the help in the application part of this paper.

Author details
[1]Department of Mathematics, Adamas Institute of Technology, Barasat, North 24 Parganas 700126, West Bengal, India. [2]Department of Mathematics, Bengal Engineering and Science University, Shibpur, Howrah 711103, West Bengal, India.

References

Charnes, A, Cooper WW: **Goal programming and multiple objective optimization.** *European J Oper Res* 1977, **1:**39–54.
Charnes, A, Cooper W, Ferguson R: **Optimal estimation of executive compensation by linear programming.** *Manag Sci* 1955, **1:**138–151.
Ghosh, P, Roy TK: **Goal geometric programming problem (G^2P^2) with product method.** In *International Conference of IMBIC on Mathematical Sciences for Advancement of Science and Technology*. Kolkata, 2012. 21–23 December 2012.
Ignizio, JP: *Goal programming and extensions*. Lexington: Lexington Series, DC Heath and Company; 1976.
Ignizio, JP, Romero C: **Goal programming.** In *In: Bidgoli H (ed) Encyclopedia of information systems, vol 2.* San Diego, Academic Press; 2003. pp 489–500.
Ijiri, Y: *Management goals and accounting for control.* Amsterdam: North Holland; 1965.
Jones, DF, Tamiz M: **Goal programming in the period 1990–2000.** In *In: Ehrgott M , Gandibleux X (eds) Multi-criteria optimization: state of the art annotated bibliographic surveys.* Dordrecht, Kluwer; 2002. pp 129–170.
Lee, SM: *Goal programming for decision analysis*. Philadelphia: Auerbach Publishers; 1972.
Lee, SM, Olson D: **Goal programming.** In *In: Gal T, Stewart TJ, Hanne, T (eds) Multicriteria decision making: advances in MCDM models, algorithms, theory, and applications*. Boston: Kluwer Academic Publishers; 2000.
Miettinen, K: *Nonlinear multi-objective optimization*. Boston: Kluwer Academic Publishers; 1999.
Romero, C: *Handbook of critical issues in goal programming*. Oxford: Pergamon Press; 1991.
Romero C: **Continuous optimization : a general structure of achievement function for a goal programming model.** *Eur J Oper Res* 2004, **153:**675–686.
Sadjadi, SJ, Oroujee M, Aryanezhad MB: **Optimal production and marketing planning.** *Comput Optimization Appl* 2005, **30:**195–203.
Tamiz, M, Jones DF, Romero C: **Goal programming for decision making: an overview of the current state-of-the-art.** *Eur J Oper Res* 1998, **111:**569–581.
Verma, RK: **Fuzzy geometric programming with several objective functions.** *Fuzzy Sets Syst* 1990, **35**(1):115–120.
Wang, YM, Yang JB, Xu DL: **A two-stage logarithmic goal programming method for generating weights from interval comparison matrices.** *Fuzzy Sets Syst* 2005, **152:**475–498.

5

Forecasting time and place of earthquakes using a Semi-Markov model (with case study in Tehran province)

Ramin Sadeghian

Abstract

The paper examines the application of semi-Markov models to the phenomenon of earthquakes in Tehran province. Generally, earthquakes are not independent of each other, and time and place of earthquakes are related to previous earthquakes; moreover, the time between earthquakes affects the pattern of their occurrence; thus, this occurrence can be likened to semi-Markov models. In our work, we divided the province of Tehran into six regions and grouped the earthquakes regarding their magnitude into three classes. Using a semi-Markov model, it proceeds to predict the likelihood of the time and place of occurrence of earthquakes in the province.

Keywords: Semi-Markov model, Earthquake occurrence, Forecasting, Transition matrix

Background

Forecasting the time and place of earthquakes is considered to be very important in science (Sadeghian 2007). Many researchers in the fields of mathematics, physics, and geology have tried to forecast the time and place of earthquake occurrence. Most of these are one-dimensional studies, looking at either the time or place of occurrence, and some of them have studied the prediction of occurrence as two-dimensional (Papazachos 1992; Rikitake 1976; Kelleher et al. 1973; Nanjo et al. 2006; Qin et al. 1999). While forecasting time or place alone cannot help prevent devastation caused by earthquakes, it is helpful if both dimensions are looked at together; it is useful to consider the dimension of magnitude as the third dimension along them. Some researchers have explained that the occurrence of earthquakes can be considered as memoryless models such as Markov models (Altinok and Kolcak 1999; Patwardhan et al. 1980; Nava et al. 2005). Some researchers have studied on earthquake prediction using a Markov chain model (such as Di Luccio et al. 1997; Console et al. 2002; Console 2001), and few of them have considered the model as semi-Markov (Altinok and Kolcak 1999; Patwardhan et al. 1980). These papers using a semi-Markov model don't have either much analysis or any evaluation on obtained results. Altinok and Kolcak (1999) and Patwardhan et al. (1980) didn't use scientific techniques for selecting used regions and magnitudes, while we apply a clustering method for regions and magnitudes. Altinok and Kolcak (1999) didn't discuss about selecting time unit. Nava et al. (2005) used a Markov model for the validation of earthquake occurrence, while the proposed model is a semi-Markov model. Patwardhan et al. (1980) and Altinok and Kolcak (1999) didn't forecast time, place, and magnitude of earthquake occurrence simultaneously, while we obtain joint functions for forecasting all dimensions simultaneously. The paper here tries to improve weaknesses of the papers and apply different methods for determining regions, magnitudes, and time unit. Using a semi-Markov model to forecast time, place, and magnitude of occurrence of earthquakes in the province, it is possible to improve the precautionary measures. Some advantages of semi-Markov models in comparison to other models were explained by Patwardhan et al. (1980). Because energies saved under faults are freed when an earthquake occurs, the time and magnitude of the earthquake within a specified region depend on the previous earthquake in the region. Consequently, earthquake occurrence can be explained by Markov models. Note that future periods can be forecasted by previous periods in Markov models. Nava et al. (2005) used a Markov model for the validation of earthquake,

Correspondence: ramin_sadeghian@yahoo.com
Islamic Azad University, South Tehran Branch, Industrial Engineering Faculty, Tehran, Iran

while the proposed model is a semi-Markov model. Also, Sadeghian and Jalali-Naini (2008a,b) applied a semi-Markov model and probability distributions on the field of earthquakes. In the next sections, we describe definitions of a semi-Markov model and its necessary parameters such as holding time, core matrix, interval transition probability matrix, and their analyses. A semi-Markov model is used for Tehran province. Next, the data corresponding to the province and its magnitudes during 1966 to 2004 are clustered, and then, both region-to-region and magnitude-to-magnitude transition matrices are obtained. Forecasting is performed by matrices obtained during next 3 years.

Case description

Semi-Markov model

Consider a continuous time chain $\{X(t)\}_{t\in[0,\infty)}$, taking value in a discrete state space S:

1. $T_n(n\geq 1)$ is the time of the nth transition after $t = 0$. (Note that the chain does not have to change states at each transition, as at time T_4 in Figure 1).
2. The duration between two consecutive transitional epochs T_{n-1}, T_n is defined as the chain's nth holding time.
3. If the chain's sample path is right continuous (i.e., the time is considerable as piecemeal), then we write $X_n \equiv X(T_n)$.

The chain of $\{X(t)\}_{t\in[0,\infty)}$ is said to be semi-Markov if its development after each transition is independent of its behavior before that time. This means that the distribution of its holding time $T_n - T_{n-1}$ is independent of its behavior before T_{n-1} but may be a function of $X_{n-1} = X(T_{n-1}), X_n = X(T_n)$. If $X_{n-1} = i, X_n = j$, we denote the holding time $T_n - T_{n-1}$ by t_{ij}.

If all holding times in a semi-Markov chain are equal to a constant value, the chain can be studied as a discrete-time Markov chain. To describe it completely, we need only to have all transition probabilities.

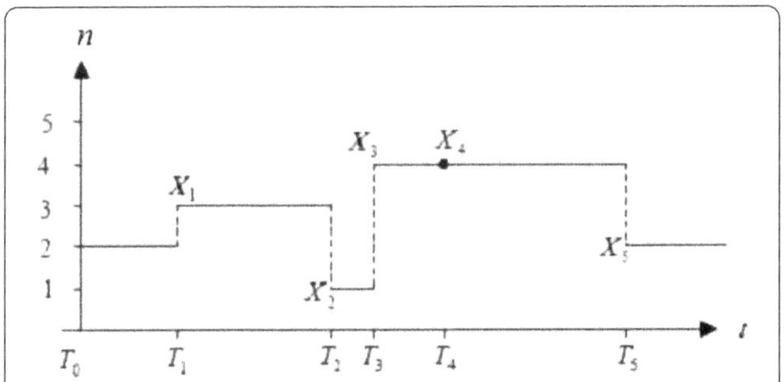

Figure 1 A continuous time chain.

If a semi-Markov chain has only one state, all its holding times can only be a function of this one state; hence, they are independent and identically distributed. The chain therefore can be studied as a renewal stream, with a renewal at each transition; we only need its holding time distribution as the life distribution of the stream (Jalali-Naini 1997; Minh 2001). See previous studies of Kulkarni (1995), Cotea and Stein (2006), Bradley et al. (2006), Bradley and Wilson (2005), and Jenamani et al. (2003) for some general references about semi-Markov models.

Transition matrix

The transition probability G_{ij} is the probability that a semi-Markov process which has entered state i on its last transition will enter state j on its next transition. These probabilities must satisfy the conditions:

$$G_{ij}\geq 0\ ;\quad i = 1,\ldots,N,\ j = 1,\ldots,N$$
$$\sum_{j=1}^{N} G_{ij} = 1\ ;\quad i = 1,\ldots,N \tag{1}$$

where N *is* the total number of states in the system (Minh 2001; Altinok and Kolcak. 1999; Kulkarni 1995; Jenamani et al. 2003).

Holding time

Before the transition from state i to state j, the process remains in state i for a time t_{ij}. The holding times t_{ij} are positive, integer-valued, random variables. All holding times are finite, and each is at least equal to one time unit. The probability mass function T_{ij} in t_{ij} is called the holding time mass function for a transition from state i to j:

$$\Pr\{t_{ij} = m\} = T_{ij}(m);\quad m = 1,\ldots,n \tag{2}$$

Here, n is the number of time intervals. We must specify both the holding time mass functions and the transition probabilities to describe a discrete-time semi-Markov process completely (see Minh 2001; Altinok and Kolcak 1999; Kulkarni 1995; Jenamani et al. 2003).

Core matrix

The ijth element of the core matrix $C(m)$ is the probability of the joint event in which a system that entered state i at time 0 makes its next transition to state j and makes that transition after a holding time m (Altinok and Kolcak 1999).

$$\begin{aligned} C_{ij}(m) &= \Pr((X_n = j|X_{n-1} = i), (t_{ij} = m))\\ &= \Pr(X_n = j|X_{n-1} = i).\Pr(t_{ij} = m)\\ &= G_{ij}.T_{ij}(m)\Rightarrow C_{ij}(m) = G_{ij}T_{ij}(m);\quad i,j\\ &= 1,\ldots,N, m = 1,\ldots,n \end{aligned} \tag{3}$$

We denote the above equation in congruent matrix multiplication form by,

$$C(m) = G \otimes T(m), \quad (4)$$

where the operator $\otimes$ denotes multiplication of corresponding elements. If we sum the elements of $C(m)$ across the i throw, we obtain the waiting time mass function $w_i(m)$ for the ith state:

$$\sum_{j=1}^{N} C_{ij}(m) = \sum_{j=1}^{N} G_{ij} T_{ij}(m) = w_i(m), \quad (5)$$

Indeed, $w_i(m)$ is the probability that waiting time for the ith state is equal to m.

The cumulative probability distribution of the waiting time is obtained from

$$LEw_i(n) = \sum_{m=1}^{n} w_i(m), \quad (6)$$

$LEw_i(m)$ is the probability that the waiting time for the ith state is less than or equal to n. The complementary of LEw_i is then given by

$$Gw_i(n) = \sum_{m=n+1}^{\infty} w_i(m), \quad (7)$$

$Gw_i(m)$ is the probability that waiting time for the ith state is greater than n.

Interval transition probability matrix

The most important statistics of the semi-Markov process are the interval transition probabilities. The probability of a transition from state i to state j in the interval $(0, n)$ requires that the process makes at least one transition during that interval. The process could have made its first transition from state i to some state at time m, where $(0 \le m \le n)$, and then by the sum of a succession of transitions, it could have made its way to state j at time n.

$$F(n) = GW(n) + \sum_{m=0}^{n} G \otimes T(m) F(n-m) = GW(n) + \sum_{m=0}^{n} C(m) F(n-m); \quad n = 0, 1, 2, \ldots \quad (8)$$

Here, $GW(n)$ is a diagonal matrix with its ith element equal to $Gw_i(n)$. The interval transition probability $F(n)$ is obtained by a recursive procedure. Because $T(0)$ is equal to 0, $F(n)$ is obtained for the interval $1 \le m \le n$. In the case $n = 0$, $F(n)$ is equal to the Kronecker Delta (Equation 9) or identity matrix.

$$F_{ij}(0) = \begin{cases} 1 & i = j \\ 0 & i \neq j \end{cases} \quad (9)$$

In the earthquake phenomenon, these probabilities can be used for studying earthquake hazards (Altinok and Kolcak 1999; Patwardhan et al. 1980) and evaluating seismic hazard (Nava et al. 2005).

Discussion and evaluation

Application of the model

Tehran province is selected as an area of investigation. This area is bounded by longitudes 50.42°E, 53.16°E and latitudes 35.23°N, 36.33°N (Figure 2). Earthquake data are obtained from a catalog in the USGS website (USGS 2011).

After filtering and removing unsuitable data, the data wherein 44 earthquakes occurred from 1966 to 2004 are used. Note that if there are sufficient data such as a large number, accuracy, and long time interval, we expect that the accuracy of forecasting is increased and that the proposed model is validated mathematically.

We consider two transitions: from region-to-region and magnitude-to-magnitude.

Region-to-region transition

In the first stage, regions are chosen as states. We divided Tehran province into six regions $R_1, R_2, R_3, R_4, R_5, R_6$ (Figure 2). The transition probability matrix of region-to-region transitions was determined in Equation 10. Selecting 1 year for time unit, the largest time interval for the region-to-region transitions is found to be 5 years; therefore, holding time mass functions $T_R(m)$ are obtained by taking into account the time lapse between successive earthquakes (Equation 11). Also, interval transition probabilities $F_R(n)$ are calculated using Equation 8. They are shown in Equations 11 and 12.

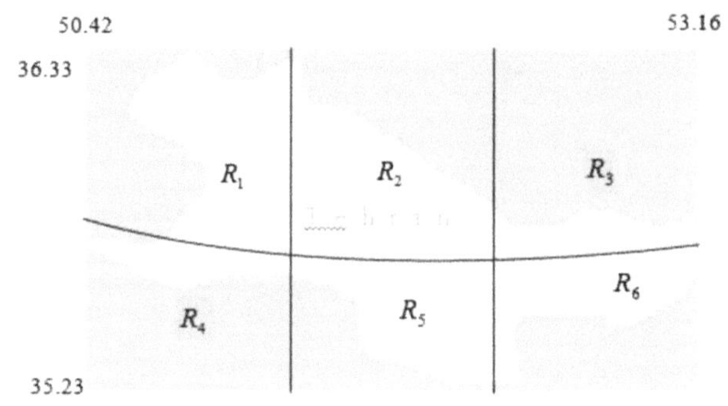

Figure 2 Map of Tehran province.

$$G_R = \begin{matrix} R_1 \\ R_2 \\ R_3 \\ R_4 \\ R_5 \\ R_6 \end{matrix} \begin{pmatrix} 0 & 0 & .67 & 0 & .33 & 0 \\ .125 & .375 & .375 & .125 & 0 & 0 \\ 0 & .17 & .33 & 0 & .08 & .42 \\ 0 & 1 & 0 & 0 & 0 & 0 \\ .25 & .25 & 0 & 0 & 0 & .5 \\ 0 & .06 & .22 & 0 & .11 & .61 \end{pmatrix} \tag{10}$$

$$T_R(1) = \begin{pmatrix} 0 & 0 & .5 & 0 & 1 & 0 \\ 1 & 1 & 1 & 1 & 0 & 0 \\ 0 & 1 & 1 & 0 & 0 & .8 \\ 0 & 1 & 0 & 0 & 0 & 0 \\ 0 & 1 & 0 & 0 & 0 & .5 \\ 0 & 1 & .5 & 0 & .5 & .91 \end{pmatrix}$$

$$T_R(2) = \begin{pmatrix} 0 & 0 & 0 & 0 & 0 & 0 \\ 0 & 0 & 0 & 0 & 0 & 0 \\ 0 & 0 & 0 & 0 & 0 & .2 \\ 0 & 0 & 0 & 0 & 0 & 0 \\ 0 & 0 & 0 & 0 & 0 & .5 \\ 0 & 0 & .25 & 0 & .5 & .09 \end{pmatrix}$$

$$T_R(3) = \begin{pmatrix} 0 & 0 & .5 & 0 & 0 & 0 \\ 0 & 0 & 0 & 0 & 0 & 0 \\ 0 & 0 & 0 & 0 & 0 & 0 \\ 0 & 0 & 0 & 0 & 0 & 0 \\ 1 & 0 & 0 & 0 & 0 & 0 \\ 0 & 0 & .25 & 0 & 0 & 0 \end{pmatrix} \tag{11}$$

$$T_R(4) = \begin{pmatrix} 0 & 0 & 0 & 0 & 0 & 0 \\ 0 & 0 & 0 & 0 & 0 & 0 \\ 0 & 0 & 0 & 0 & 0 & 0 \\ 0 & 0 & 0 & 0 & 0 & 0 \\ 0 & 0 & 0 & 0 & 0 & 0 \\ 0 & 0 & 0 & 0 & 0 & 0 \end{pmatrix}$$

$$T_R(5) = \begin{pmatrix} 0 & 0 & 0 & 0 & 0 & 0 \\ 0 & 0 & 0 & 0 & 0 & 0 \\ 0 & 0 & 0 & 0 & 1 & 0 \\ 0 & 0 & 0 & 0 & 0 & 0 \\ 0 & 0 & 0 & 0 & 0 & 0 \\ 0 & 0 & 0 & 0 & 0 & 0 \end{pmatrix}$$

$$F_R(1) = \begin{pmatrix} .335 & 0 & .335 & 0 & .33 & 0 \\ .125 & .375 & .375 & .125 & 0 & 0 \\ 0 & .17 & .494 & 0 & 0 & .336 \\ 0 & 1 & 0 & 0 & 0 & 0 \\ 0 & .25 & 0 & 0 & .5 & .25 \\ 0 & .06 & .11 & 0 & .055 & .775 \end{pmatrix}$$

$$F_R(2) = \begin{pmatrix} .335 & .14 & .165 & 0 & .165 & .195 \\ .089 & .329 & .368 & .047 & .041 & .126 \\ .021 & .14 & .345 & .021 & .018 & .455 \\ .125 & .375 & .375 & .125 & 0 & 0 \\ .031 & .109 & .121 & .031 & .264 & .444 \\ .008 & .088 & .193 & .008 & .113 & .59 \end{pmatrix}$$

$$F_R(3) = \begin{pmatrix} .017 & .083 & .49 & .017 & .093 & .3 \\ .1 & .24 & .334 & .041 & .043 & .242 \\ .025 & .137 & .33 & .018 & .056 & .434 \\ .089 & .329 & .368 & .047 & .041 & .126 \\ .274 & .119 & .168 & .014 & .052 & .373 \\ .014 & .117 & .262 & .011 & .112 & .484 \end{pmatrix}$$

$$F_R(4) = \begin{pmatrix} .1 & .142 & .331 & .01 & .036 & .381 \\ .06 & .193 & .356 & .03 & .054 & .307 \\ .03 & .133 & .35 & .017 & .073 & .397 \\ .1 & .24 & .334 & .041 & .043 & .242 \\ .114 & .11 & .281 & .015 & .15 & .33 \\ .035 & .129 & .274 & .015 & .096 & .451 \end{pmatrix}$$

$$F_R(5) = \begin{pmatrix} .055 & .128 & .325 & .018 & .08 & .394 \\ .058 & .17 & .348 & .025 & .054 & .342 \\ .033 & .13 & .29 & .017 & .155 & .375 \\ .06 & .193 & .356 & .03 & .054 & .307 \\ .111 & .144 & .264 & .014 & .107 & .36 \\ .051 & .132 & .288 & .016 & .086 & .427 \end{pmatrix} \tag{12}$$

Magnitude-to-magnitude transitions

In the second stage, magnitudes are chosen as states. Similar to the previous stage, we apply the AGNES method (Ghazanfari and Rezaei 2006) for the clustering of magnitudes, and three clusters on magnitudes as states are established in the whole regions as follows:

$$\begin{aligned} &M1 : \text{MB} < 4.5 \\ &M2 : 4.5 \leq \text{MB} \leq 5 \\ &M3 : 5 < \text{MB} \end{aligned} \tag{13}$$

where MB is a unit of the magnitude of earthquakes; it is called as body wave magnitude.

The transition matrix of magnitude-to-magnitude transitions is given in Equation 17. The holding time mass functions $T_M(m)$and interval transition probabilities

$F_M(n)$ are calculated using Equation 2 through Equation 8. These are shown in Equations 15 and 16.

$$G_M = \begin{matrix} M_1 \\ M_2 \\ M_3 \end{matrix} \begin{pmatrix} .5 & .3 & .2 \\ .45 & .5 & .05 \\ .167 & .5 & .333 \end{pmatrix} \tag{14}$$

$$T_M(1) = \begin{pmatrix} 1 & 1 & .75 \\ .67 & .7 & 0 \\ 0 & 1 & 1 \end{pmatrix} \quad T_M(2) = \begin{pmatrix} 0 & 0 & .25 \\ .11 & .2 & 0 \\ 1 & 0 & 0 \end{pmatrix}$$

$$T_M(3) = \begin{pmatrix} 0 & 0 & 0 \\ .22 & 0 & 1 \\ 0 & 0 & 0 \end{pmatrix} \quad T_M(4) = \begin{pmatrix} 0 & 0 & 0 \\ 0 & 0 & 0 \\ 0 & 0 & 0 \end{pmatrix}$$

$$T_M(5) = \begin{pmatrix} 0 & 0 & 0 \\ 0 & .1 & 0 \\ 0 & 0 & 0 \end{pmatrix} \tag{15}$$

$$F_M(1) = \begin{pmatrix} .55 & .3 & .15 \\ .302 & .698 & 0 \\ 0 & .5 & .5 \end{pmatrix} \quad F_M(2) = \begin{pmatrix} .365 & .435 & .2 \\ .322 & .633 & .045 \\ .318 & .515 & .167 \end{pmatrix}$$

$$F_M(3) = \begin{pmatrix} .326 & .51 & .164 \\ .38 & .486 & .134 \\ .359 & .538 & .103 \end{pmatrix} \quad F_M(4) = \begin{pmatrix} .347 & .507 & .146 \\ .336 & .514 & .15 \\ .37 & .495 & .135 \end{pmatrix}$$

$$F_M(5) = \begin{pmatrix} .347 & .51 & .143 \\ .328 & .526 & .146 \\ .346 & .507 & .147 \end{pmatrix} \tag{16}$$

Forecasting

In this section, time and magnitude of earthquake occurrence are forecasted for each region in the next 3 years. Note that some forecasting methods based on probability approaches don't have sufficient accuracy, and others are more accurate. Matrices P_1 through P_6 show region-to-region transition probabilities by time of occurrence without considering the dimension of magnitude. In other words, the sum of each column is equal to 1 (see Tables 1, 2, 3, 4, 5, and 6). For example, the *ij*th element of matrix P_1 shows the probability of earthquake occurrence in region R_i during year *j* if the last earthquake has occurred in region R_1. Similarly, matrices P_7 through P_9 show magnitude-to-magnitude transition probabilities by time of occurrence without considering the spatial dimension (see Tables 7, 8, 9). In these matrices, the sum of each column is also equal to 1.

Here, a semi-Markov model for considering the time of earthquake occurrence is applied. This method requires time, place, and magnitude of previous

Table 1 Matrix *P1*

		1	2	3	4	5
P1 =	*R1* to *R1*	0.335	0.335	0.017	0.1	0.055
	R1 to *R2*	0	0.14	0.083	0.142	0.128
	R1 to *R3*	0.335	0.165	0.49	0.331	0.325
	R1 to *R4*	0	0	0.017	0.01	0.018
	R1 to *R5*	0.33	0.165	0.093	0.036	0.08
	R1 to *R6*	0	0.195	0.3	0.381	0.394

Table 2 Matrix *P2*

		1	2	3	4	5
P2 =	*R2* to *R1*	0.125	0.089	0.1	0.06	0.058
	R2 to *R2*	0.0375	0.329	0.24	0.193	0.17
	R2 to *R3*	0.375	0.368	0.334	0.356	0.348
	R2 to *R4*	0.125	0.047	0.041	0.03	0.025
	R2 to *R5*	0	0.041	0.043	0.054	0.054
	R2 to *R6*	0	0.126	0.242	0.307	0.342

Table 3 Matrix *P3*

		1	2	3	4	5
P3 =	*R3* to *R1*	0	0.021	0.025	0.03	0.033
	R3 to *R2*	0.17	0.14	0.137	0.133	0.13
	R3 to *R3*	0.494	0.345	0.33	0.35	0.29
	R3 to *R4*	0	0.021	0.018	0.017	0.017
	R3 to *R5*	0	0.018	0.056	0.073	0.155
	R3 to *R6*	0.336	0.455	0.434	0.397	0.375

Table 4 Matrix *P4*

		1	2	3	4	5
P4 =	*R4* to *R1*	0	0.125	0.089	0.1	0.06
	R4 to *R2*	1	0.375	0.329	0.24	0.193
	R4 to *R3*	0	0.375	0.368	0.334	0.356
	R4 to *R4*	0	0.125	0.047	0.041	0.03
	R4 to *R5*	0	0	0.041	0.043	0.054
	R4 to *R6*	0	0	0.126	0.242	0.307

Table 5 Matrix *P5*

		1	2	3	4	5
P5 =	*R5* to *R1*	0	0.031	0.274	0.114	0.111
	R5 to *R2*	0.25	0.109	0.119	0.11	0.144
	R5 to *R3*	0	0.121	0.168	0.281	0.264
	R5 to *R4*	0	0.031	0.014	0.015	0.014
	R5 to *R5*	0.5	0.264	0.052	0.15	0.107
	R5 to *R6*	0.25	0.444	0.373	0.33	0.36

Table 6 Matrix *P6*

		1	2	3	4	5
P6 =	*R6* to *R1*	0	0.008	0.014	0.035	0.051
	R6 to *R2*	0.06	0.088	0.117	0.129	0.132
	R6 to *R3*	0.11	0.193	0.262	0.274	0.288
	R6 to *R4*	0	0.008	0.011	0.015	0.016
	R6 to *R5*	0.055	0.113	0.112	0.096	0.086
	R6 to *R6*	0.775	0.59	0.484	0.451	0.427

Table 7 Matrix *P7*

		1	2	3	4	5
P7 =	*M1* to *M1*	0.55	0.365	0.326	0.347	0.347
	M1 to *M2*	0.3	0.435	0.51	0.507	0.51
	M1 to *M3*	0.15	0.2	0.164	0.146	0.143

Table 8 Matrix *P8*

		1	2	3	4	5
P8 =	*M2* to *M1*	0.302	0.322	0.38	0.336	0.328
	M2 to *M2*	0.698	0.633	0.486	0.514	0.526
	M2 to *M3*	0	0.045	0.134	0.15	0.146

Table 9 Matrix *P9*

		1	2	3	4	5
P9 =	*M3* to *M1*	0	0.318	0.359	0.37	0.346
	M3 to *M2*	0.5	0.515	0.538	0.495	0.507
	M3 to *M3*	0.5	0.167	0.103	0.135	0.147

earthquakes. If the previous earthquake is within region r_0 and its magnitude is m_0, then the probability of occurrence of the next earthquake within region r_1 and magnitude m_1 after d time periods is a conditional probability defined as follows:

$$P(R = r_1, M = m_1 | r_0, m_0, d) = P(r_1, m_1 | r_0, m_0, d)$$

$$\underset{\text{magnitude-to-magnitude transition probabilities}}{\overset{\text{region-to-region transition probabilities are independent from}}{=\!=\!=\!=\!=\!=\!=\!=\!=\!=}}$$

$$P(r_1 | r_0, d).P(m_1 | m_0, d) = F_{R_{r_0\,r_1}}(d).F_{M_{m_0\,m_1}}(d)$$

Using the above equations and the list of data, the last earthquake in Tehran province occurred at 28 May 2004 within region R_1 with a magnitude of 6.3 (MB). Consider the current time as November 2006; in this way, the proposed model forecasts that the most probable occurrence of an earthquake within Tehran province within region R_3 (northeast of Tehran province) with a magnitude of M_2 ($4.5 \leq MB \leq 5$) from May 2006 to May 2007 is 0.2636 (26.36%).

Conclusion

This article attempts to forecast the likelihood of all earthquakes within Tehran province in the next years by a semi-Markov model. The proposed model forecasts that the most probable occurrence of an earthquake within Tehran province within region R_3 (northeast of Tehran province) with a magnitude of M_2 ($4.5 \leq \text{MB} \leq 5$) from May 2006 to May 2007 is 0.2636 (26.36%). According to our forecasting, an earthquake occurred in Roudehen (northeast of Tehran province) with a magnitude of 4.2 MB on 20 December 2006. This can be considered as the evaluation of the proposed model.

The average magnitude of earthquake occurrence in Tehran province is 3.1 MB yearly. The most number of earthquake occurrences have occurred in intervals ($4 < \text{MB} \leq 4.5$). It includes 32% all of the events. Here, in this paper, the spatial, temporal, and magnitude dimensions are considered, while in past researches, these three dimensions were considered less.

As a conclusion, the amount of forecast errors shows that MAPE (Haj Shirmohammadi 2003) in this model is equal to 4.45%, while in other models, it is more than 10% (Papazachos 1992; Karakaisis 1994).

The obtained results during the next 50 days show that in other models, 42% of earthquakes have been correctly forecasted. The places of 16% of earthquake occurrences have been correctly forecasted, but not their magnitudes. Sixteen percent of the forecasted earthquakes occurred in places adjacent to the forecasted ones, but their magnitudes were true, and 26% of earthquakes could not have been forecasted at all.

In this model, 56% of earthquakes have been correctly forecasted in all the three dimensions. The places of 19% of earthquake occurrences have been correctly forecasted, but not their magnitudes. Nineteen percent of the forecasted earthquakes occurred in places adjacent to the forecasted ones, but their magnitudes were true, and 6% of earthquakes could not have been forecasted at all.

The mentioned statistics above show that this model in comparison to others can be applied in a more accurate forecasting.

References

Altinok Y, Kolcak D (1999) An application of the semi-Markov model for earthquake occurrences in North Anatolia, Turkey. Journal of Balkan Geophysical Society 2:90–99

Bradley JT, Wilson HJ (2005) Iterative convergence of passage-time densities in semi-Markov performance models. Performance Evaluation 60:237–254
Bradley JT, Dingle NJ, Harrison PG, Knottenbelt WJ (2006) Distributed computation of transient state distributions and passage time quantiles in large semi-Markov models. Future Generation Computer Systems 22:828–837
Console R (2001) Testing earthquake forecast hypotheses. Tectonophysics 338:261–268
Console R, Pantosti D, D'Addezio G (2002) Probabilistic approach to earthquake prediction. Ann Geophys 45(6):723–731
Cotea MJ, Stein WE (2006) A stochastic model for a visit to the doctor's office. Mathematical and Computer Modeling 45:309–323
Di Luccio F, Console R, Imoto M, Murru M (1997) Analysis of short time-space range seismicity pattern in Italy. Ann Geophys 40:4
Ghazanfari M, Rezaei M (2006) Introduction on fuzzy theory and its applications., IUST Press
Haj Shirmohammadi A (2003) Principle of production and inventory planning and control. Arkan Press, Isfahan
Jalali-Naini SG (1997) Stochastic processes and its applications in biology, agriculture, management and industries. Statistics and Information Office Press, Tehran
Jenamani M, Mohapatra PKJ, Ghose S (2003) A stochastic model of e-customer behavior. Electronic Commerce Research and Applications 2:81–94
Karakaisis GF (1994) Long-term earthquake prediction in Iran based on the time- and magnitude-predictable model. Physics of the Earth and Planetary Interiors 83:129–145
Kelleher J, Sykes L, Oliver J (1973) Possible criteria for predicting earthquake locations and their application to major plate boundaries of the Pacific and the Caribbean. Journal of Geophysical Research 78(14):2547–2585
Kulkarni VG (1995) Modeling and analysis of stochastic systems. Chapman & Hall, London
Minh DL (2001) Applied probability models. Duxbury, Thomson Learning Press, California, pp 225–246
Nanjo KZ, Holliday JR, Chen C, Rundle JB, Turcotte DL (2006) Application of a modified pattern informatics method to forecasting the locations of future large earthquakes in the central Japan. Tectonophysics 424:351–366
Nava FA, Herrera C, Frez J, Glowacka E (2005) Seismic hazard evaluation using Markov chains: application to the Japan area. Pure and Applied Geophysics 162:1347–1366
Papazachos BC (1992) A time- and magnitude- predictable model for generation of shallow earthquakes in the Aegean area. Pure and Applied Geophysics 138(2):287–308
Patwardhan AS, Kulkarni RB, Tocher D (1980) A semi-Markov model for characterizing recurrence of great earthquakes. Bulletin of the Seismological Society of America 70(1):323–347
Qin C, Papadimitriou EE, Papazachos BC, Karakaisis GF (1999) On the validity of the regional time and magnitude predictable model in China. Ann Geophys 42(5)
Rikitake T (1976) Recurrence of great earthquakes at subduction zones. Tectonophysics 35:335–362
Sadeghian R (2007) What ever you don't know about earthquakes. Shooka Press, Tehran
Sadeghian R, Jalali-Naini GR (2008a) A new probability density function in earthquake occurrences. Journal of Industrial Engineering International 4(6):1–6
Sadeghian R, Jalali-Naini GR (2008b) Applying semi-Markov models for forecasting the triple dimensions of next earthquake occurrences. Dynamics of Continuous, Discrete and Impulsive Systems, Series B. Applications & Algorithms 15(3b):395–416
USGS (2011) Earthquake hazards program http://neic.usgs.gov/neis/epic/epic.html

6

A multi-period distribution network design model under demand uncertainty

Babak H Tabrizi* and Jafar Razmi

Abstract

Supply chain management is taken into account as an inseparable component in satisfying customers' requirements. This paper deals with the distribution network design (DND) problem which is a critical issue in achieving supply chain accomplishments. A capable DND can guarantee the success of the entire network performance. However, there are many factors that can cause fluctuations in input data determining market treatment, with respect to short-term planning, on the one hand. On the other hand, network performance may be threatened by the changes that take place within practicing periods, with respect to long-term planning. Thus, in order to bring both kinds of changes under control, we considered a new multi-period, multi-commodity, multi-source DND problem in circumstances where the network encounters uncertain demands. The fuzzy logic is applied here as an efficient tool for controlling the potential customers' demand risk. The defuzzifying framework leads the practitioners and decision-makers to interact with the solution procedure continuously. The fuzzy model is then validated by a sensitivity analysis test, and a typical problem is solved in order to illustrate the implementation steps. Finally, the formulation is tested by some different-sized problems to show its total performance.

Keywords: Supply chain, Distribution network design, Uncertain demand, Fuzzy logic

Introduction

Moving towards competitive markets compels the manufacturers to increase their qualifications so that they can best fulfill the customers' needs. The supply chain management (SCM) concept arose as an effective managerial tool to enhance the customers' satisfaction within the last 20 years in particular (Azaron et al. 2008; Jayaraman and Ross 2003). Generally, a supply chain (SC) can be defined as a set of plants, distribution centers, and customers in which the products are completed once they are transmitted downstream the chain and delivered to the customers (Altiparmak et al. 2006; Blackhurst et al. 2004).

Distribution network design (DND) is one of the primitive principles in establishing a successful SC. It is associated with the category of production-distribution and facility location-allocation problems (Altiparmak et al. 2009). The researchers investigate the distribution networks under two distinctive categories including strategic and tactical levels. The first level concerns with whatever affects the entire network configuration (e.g., the numbers, capacities, and locations), and the second level deals with whatever affects the aggregate quantities (e.g., material handling, processing, and distribution) (Santoso et al. 2005). Consequently, considering the two planning levels simultaneously is very fruitful in enhancing the network performance.

A DND problem can also incorporate five aspects according to the necessity (please refer to Tang (2006) for more description):

- Network configuration (NC)
- Product assignment (PA)
- Customer assignment (CA)
- Production planning (PP)
- Transportation planning (TP)

Regarding the above, the proposed model covers the first four aspects at the same time.

Some other issues are expected in the model structure likewise (e.g., the number of consideration periods, sources, echelons, and commodities). For example, Amiri (2006) suggested a multi-capacity distribution

* Correspondence: babaktabrizi@ut.ac.ir
Department of Industrial Engineering, Faculty of Engineering, University of Tehran, P.O. Box: 11155-4563, Tehran, Iran

network in deterministic environments so that both strategic and tactical levels were included. Gupta and Maranas (2003) concentrated on the tactical level in a multi-commodity, multi-echelon, multi-period problem. The authors considered demand as the uncertain parameter. Selim and Ozkarahan (2006) developed a fuzzy multi-echelon, multi-commodity, multi-capacity problem by considering both levels. They introduced the demand as the uncertain part likewise. Liu et al. (2006) introduced a multi-commodity DND problem and dealt with the potential uncertainties by the combination of gray and fuzzy factors. Azaron et al. (2008) proposed a multi-commodity, multi-source model and solved it by the goal attainment technique. However, they extended the uncertainty and vagueness to all three demand-, supply-, and process-side risks. Gumus et al. (2009) applied a neuro-fuzzy technique for demand uncertainty in a multi-echelon condition. Xu et al. (2009) developed a multi-echelon, multi-source problem where the demand, supply, and process follow a fuzzy nature. They applied a heuristic solution approach called spanning tree technique that originated from genetic algorithm. Peidro et al. (2010) proposed a multi-echelon, multi-period, multi-commodity, multi-source mathematical modeling under fuzzy programming so that all three mentioned sides were uncertain. Compared with the mentioned researches, our developed model lies in a three-echelon, multi-source, multi-period, multi-commodity, single-capacity category in which the two levels are planned. The environment under consideration faces a demand-side risk which is tackled with by the fuzzy mathematical modeling. A brief overview on DND literature is depicted in Table 1.

The rest of the paper is organized as follows: model formulation is presented in the next section. Solution methodology is described in the third section, which also demonstrates how the fuzzy steps can be implemented. The fourth section pertains to the computational study, and finally, the conclusions and future research interests are discussed in the 'Conclusions' section.

Model formulation

In this section, the mixed-integer mathematical model is defined and the fuzzy approach is described. Before we construct the mathematical formulation, the assumptions, indices, parameters, and decision variables are introduced as follows:

1. Assumptions
   - Demand parameter follows a fuzzy nature.
   - A given plant or distributor can be open or close in each horizon.
   - The number of practicing facilities cannot exceed the predefined upper bound.
   - The processing capacity of each plant and distributor is known with respect to the commodities.
   - The facility capacity is fixed during the total horizon periods.
   - Each tier can supply its needs from more than one member of the upper tier.
2. Indices
   - C: set of commodities ($c \in C$)
   - P: set of plants ($p \in P$)
   - J: set of distributors ($j \in J$)
   - I: set of customers ($i \in I$)
   - H: set of planning horizons ($h \in H$)
3. Parameters
   - FC_{ph}: annual fixed cost of plant p in period h
   - FC'_{jh}: annual fixed cost of distributor j in period h
   - LC_{pchj}: logistics cost for supplying commodity c in period h from plant p to distributor j
   - LC'_{jchi}: logistics cost for supplying commodity c in period h from distributor j to customer i
   - d_{ich}: demand of customer i for commodity c in period h
   - α_{pc}: capacity of plant p for commodity c
   - α'_{jc}: capacity of distributor j for commodity c
   - P_{Upper}: the upper bound for the number of open plants
   - J_{Upper}: the upper bound for the number of open distributors
4. Decision variables
   - x_{pchj}: the proportion of commodity c in period h transported from plant p to distributor j
   - y_{jchi}: the proportion of commodity c in period h transported from distributor j to customer i
   - Z_{ph}: the binary variable that takes 1 if plant p is open in period h and 0, otherwise
   - Z'_{jh}: the binary variable that takes 1 if distributor j is open in period h and 0, otherwise
   - LQ_{chj}: the left quantity of commodity c in period h for distributor j

Now, the model can be developed by Equations 1 to 8, as follows:

$$\text{Min} = \begin{array}{l} \sum_{p=1}^{P}\sum_{c=1}^{C}\sum_{h=1}^{H}\sum_{j=1}^{J} LC_{pchj}x_{pchj}\alpha'_{jc} + \\ \sum_{j=1}^{J}\sum_{c=1}^{C}\sum_{h=1}^{H}\sum_{i=1}^{I} LC'_{jchi}y_{jchi}d_{ich} + \\ \sum_{p=1}^{P}\sum_{h=1}^{H} FC_{ph}Z_{ph} + \sum_{j=1}^{J}\sum_{h=1}^{H} FC'_{jh}Z'_{jh} \end{array} \quad (1)$$

Table 1 An overview on DND literature

Articles	Aspects					Features					Nature	Uncertain parameter(s)	Solution approach
	NC	PA	CA	PP	TP	Product	Stage	Period	Source	Capacity			
Jayaraman and Ross (2003)	***	***	***			Multiple	Three	Single	Single	Single	Deterministic	-	Simulated annealing
Santoso et al. (2005)	***	***	***	***		Multiple	Two	Single	Multiple	Single	Stochastic	Demand, supply, and process	Sample average approximation scheme and accelerated Benders' decomposition
Altiparmak et al. (2006)	***		***	***		Single	Multiple	Single	Single	Single	Deterministic	-	Genetic algorithm
Selim and Ozkarahan (2006)	***	***		***		Multiple	Three	Single	Single	Multiple	Fuzzy	Demand	Fuzzy multi-objective programming
Goh et al. (2007)	***		***			Single	Two	Single	Multiple	Single	Stochastic	Demand and supply	Heuristic algorithm
Chen et al. (2007)		***	***	***		Multiple	Three	Multiple	Multiple	Single	Fuzzy	Demand	Fuzzy programming
You and Grossmann (2008)	***	***		***		Multiple	Multiple	Multiple	Single	Single	Probabilistic	Demand	ε-Constraint
Cakir (2009)	***	***		***	***	Multiple	Two	Single	Multiple	Single	Deterministic	-	Benders' decomposition
Altiparmak et al. (2009)	***	***	***	***		Multiple	Multiple	Single	Single	Single	Deterministic	-	Genetic algorithm
Georgiadis et al. (2011)	***	***	***	***		Multiple	Three	Single	Multiple	Single	Stochastic	Demand	Branch-and-bound
Cintron et al. (2010)	***		***			Single	Three	Single	Single	Multiple	Stochastic	Demand	Goal programming
Hajiaghaei-Keshteli (2011)	***		***			Single	Two	Single	Multiple	Single	Deterministic	-	Genetic algorithm and artificial immune algorithm
Cardona-Valdes et al. (2011)	***		***	***	***	Single	Three	Single	Multiple	Single	Stochastic	Demand	L-shaped algorithm
Rezapour and Farahani (2010)	***		***			Single	Three	Single	Multiple	Multiple	Deterministic	-	A modified projection method
Park et al. (2010)	***		***			Single	Three	Single	Single	Single	Deterministic	-	A two-phase heuristic algorithm
Our proposed model	***	***	***	***		Multiple	Three	Multiple	Multiple	Single	Fuzzy	Demand, supply, and process	Fuzzy programming

S.t.:

$$\sum_{j-1}^{J} y_{jchi} \le 1 \quad ;\forall i, c, h \tag{2}$$

$$\sum_{j-1}^{J} x_{pchj} a'_{jc} \le Z_{ph} a_{pc} \quad ;\forall p, h, c \tag{3}$$

$$\mathrm{LQ}_{chj} = \sum_{p=1}^{P} x_{pchj} a'_{jc} - \sum_{i=1}^{I} y_{jchi} d_{ich} + \mathrm{LQ}_{c(h-1)j} \quad ;\forall c, h, j \tag{4}$$

$$\sum_{i=1}^{I} y_{jchi} d_{ich} \le \mathrm{LQ}_{chj} \le Z'_{jh} a'_{jc} \quad ;\forall j, c, h \tag{5}$$

$$\sum_{p=1}^{P} Z_{ph} = P_{\text{Upper}} \quad ;\forall h \tag{6}$$

$$\sum_{j=1}^{J} Z'_{jh} = J_{\text{Upper}} \quad ;\forall h \tag{7}$$

$$Z_{ph}, Z'_{jh} \in \{0.1\}, \quad 0 \le x_{pchj}, y_{jchi} \le 1. \tag{8}$$

Considering the presented formulation, Equation 1 is the objective function (OF) which consisted of logistic and annual fixed costs. In other words, the first two parts deal with the total sum of logistic costs (i.e., variable costs including processing, transportation, etc.). It is calculated for both the practicing plants and distributors separately. This type of cost is obtained from the commodity volume multiplied by the associated capacity. For example, the logistic cost of plants is determined by the corresponding unit logistic cost, to supply a given commodity from a plant to a distributor in a specific period, multiplied by the satisfied quantities. The values of responded needs are obtained from the multiplication of the given capacity by the satisfied proportion. The same calculations can be presented for distributors likewise. Furthermore, the fixed cost is realized for each of the plants and distributors with respect to the extant cost once the facility is practicing in the corresponding period. Equation 2 shows the fraction of satisfied demands for each of the commodities in each planning horizon with respect to the existing customers. However, a potential customer can acquire its demanded commodities from a set of distributors instead of focusing on a single one. Equation 3 guarantees that the transportation quantity from plants to distributors cannot surpass the given plant capacity, if it is open in the mentioned period. Equation 4 denotes that the left quantity for each commodity and distributor in each period can be calculated by the sum of transported quantities to customers subtracted from the sum of transported quantities to a distributor added to the left quantity in the previous period. Equation 5 is the same as Equation 3 but concerns the distributors and affirms that the output of each distribution center does not exceed its extant capacity, restricted by the remaining commodities. Moreover, the volume of the left commodities cannot exceed the distributor capacity. Equations 6 and 7 state that the number of open plants and distribution centers in each of the planning horizons should not exceed their upper bounds (i.e., the maximum allowed numbers), respectively. Furthermore, Equation 8 deals with the nature of existing variables.

The schematic performance of our multi-commodity, multi-period, multi-source network is depicted in Figure 1 for a given planning horizon.

Solution methodology

Since we presumed that the developed distribution network practices in a vague environment, demand uncertainty can be represented by fuzzy numbers efficiently. In order to solve the model, we applied the fuzzy approach

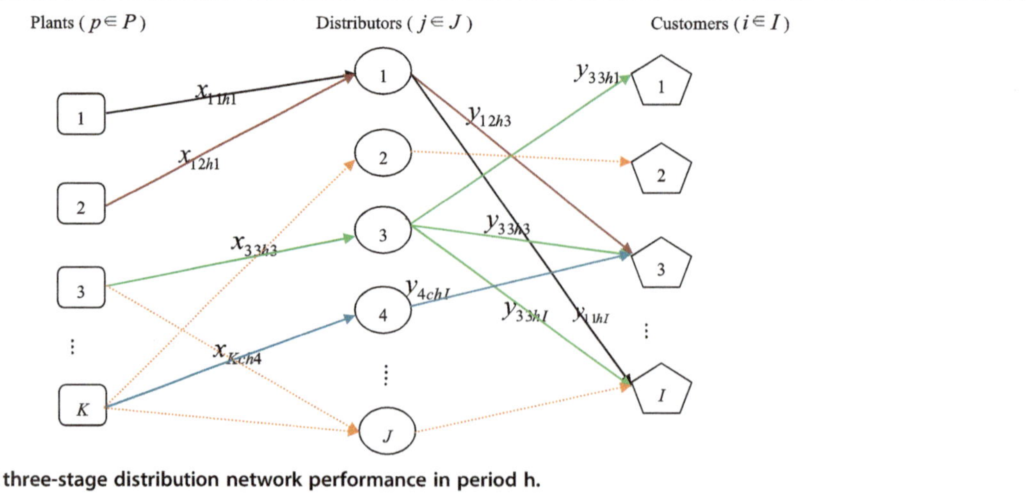

Figure 1 The proposed three-stage distribution network performance in period h.

suggested by Jimenez et al. (2007) which consists of a three-phase interaction algorithm.

Converting the fuzzy model into its crisp equivalent

In this section, we show the main framework of the implementation method of fuzzy approach proposed by Jimenez et al. (2007). Regarding the risk threatening estimated demands, all OF coefficients, technological multipliers, and right-hand side coefficients may take a vague quantity. For instance, consider the following problem in which $\tilde{C}' = (\tilde{c}'_1, \tilde{c}'_2, \ldots \tilde{c}'_n)$, $\tilde{A}'' = \left[\tilde{a}''_{ij}\right]_{m \times n}$, and $\tilde{B}' = (\tilde{b}'_1, \tilde{b}', \ldots \tilde{b}'_m)^t$ are the fuzzy parameters representing the OF coefficients, technological multiplies, and right-hand side coefficients, respectively. Therefore, the fuzzy mathematical model can be represented by Equation 9:

$$\begin{aligned} &\text{Min} = \tilde{C}'^t x \\ &\text{S.t.} : \tilde{A}'' x \geq \tilde{B}'_i \qquad ; i = 1, 2, \ldots, m \\ &x \in R^n, \geq 0. \end{aligned} \tag{9}$$

The uncertainty in the nature of fuzzy problems makes the decision-makers (DMs) find a solution so that both feasibility and optimality conditions can be satisfied efficiently. Consequently, there has been a great effort on determining different methods which could fulfill the two above characteristics (e.g., refer to Rommelfanger and Slowinski (1998)). Some papers have also discussed the ranking methods besides their justifications. We applied the method introduced by Jimenez (1996) to our problem.

It is assumed that the applied fuzzy parameters follow a trapezoidal nature in order to provide a broader range of potential values. Figure 2 shows the trapezoidal fuzzy number n, in which the α-cut equals the feasibility degree for a specific decision.

Considering the above definitions, the corresponding parts of the OF and constraints (i.e., the parts that include

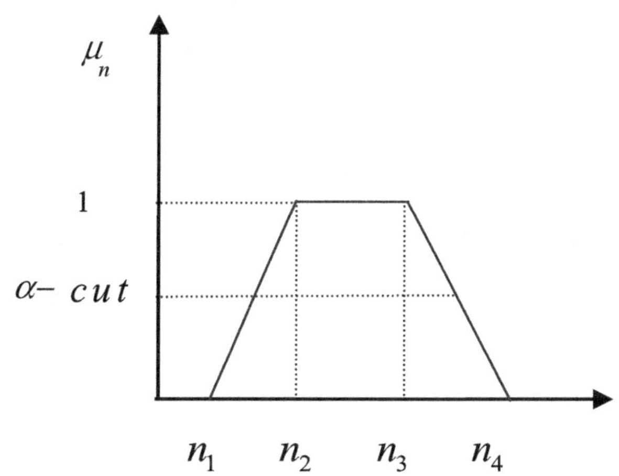

Figure 2 The membership function of trapezoidal fuzzy number *n*.

demand parameter) can be transformed to their relevant crisp values due to Equations 10 to 12, respectively:

$$\text{MinEI}(\tilde{C}')x = \left[\frac{1}{2}(c'_1 + c'_2) + \frac{1}{2}(c'_3 + c'_4)\right]x \tag{10}$$

$$\left[(1-\alpha)\frac{a''_1 + a''_2}{2} + \alpha\frac{a''_3 + a''_4}{2}\right]x \leq \alpha\frac{b'_1 + b'_2}{2} + (1-\alpha)\frac{b'_3 + b'_4}{2}. \tag{11}$$

$$\left[\left(1-\frac{\alpha}{2}\right)\frac{a''_1 + a''_2}{2} + \frac{\alpha}{2}\frac{a''_3 + a''_4}{2}\right]x \leq \frac{\alpha}{2}\frac{b'_1 + b'_2}{2} + \left(1-\frac{\alpha}{2}\right)\frac{b'_3 + b'_4}{2} \tag{12}$$

$$\left[\left(1-\frac{\alpha}{2}\right)\frac{a''_3 + a''_4}{2} + \frac{\alpha}{2}\frac{a''_1 + a''_2}{2}\right]x \geq \frac{\alpha}{2}\frac{b'_3 + b'_4}{2} + \left(1-\frac{\alpha}{2}\right)\frac{b'_1 + b'_2}{2}$$

It should be noted that EI stands for the expected interval for the corresponding fuzzy number and can be computed by finding the half point of the given fuzzy number. We can therefore calculate the expected value of a triangular or any other shaped fuzzy numbers by considering the mentioned concept likewise. Equation 11 can be applied for less than or equal type constraints, and Equation 12 can be applied for the equality type constraints likewise. However, the crisp form of the proposed model can be rewritten as Equations 13 to 16 based on Equations 2, 3, 6, 7, and 8:

$$\begin{aligned} \text{Min} = {} & \sum_{p=1}^{P}\sum_{c=1}^{C}\sum_{h=1}^{H}\sum_{j=1}^{J} \text{LC}_{pchj} x_{pchj} a'_{jc} + \\ & \sum_{j=1}^{J}\sum_{c=1}^{C}\sum_{h=1}^{H}\sum_{i=1}^{I} \text{LC}'_{jchi} y_{jchi} \left(\frac{d^1_{ich} + d^2_{ich} + d^3_{ich} + d^4_{ich}}{4}\right) \\ & + \sum_{p=1}^{P}\sum_{h=1}^{H} \text{FC}_{ph} Z_{ph} + \sum_{j=1}^{J}\sum_{h=1}^{H} \text{FC}'_{jh} Z'_{jh} \end{aligned} \tag{13}$$

S.t.:

$$\begin{aligned} \text{LQ}_{chj} \leq & \sum_{p=1}^{P} x_{pchj} a'_{jc} - \sum_{i=1}^{I} y_{jchi} \left[\frac{\alpha}{2}\left(\frac{d^1_{ich} + d^2_{ich}}{2}\right) + \left(1-\frac{\alpha}{2}\right)\left(\frac{d^3_{ich} + d^4_{ich}}{2}\right)\right] \\ & + \text{LQ}_{c(h-1)j}; \forall c, h, i \end{aligned} \tag{14}$$

$$\begin{aligned} \text{LQ}_{chj} \geq & \sum_{p=1}^{P} x_{pchj} a'_{jc} - \sum_{i=1}^{I} y_{jchi} \left[\frac{\alpha}{2}\left(\frac{d^3_{ich} + d^4_{ich}}{2}\right) + \left(1-\frac{\alpha}{2}\right)\left(\frac{d^1_{ich} + d^2_{ich}}{2}\right)\right] \\ & + \text{LQ}_{c(h-1)j}; \forall c, h, i \end{aligned} \tag{15}$$

$$\sum_{i=1}^{I} y_{jchi} \left[(1-\alpha)\left(\frac{d^1_{ich} + d^2_{ich}}{2}\right) + \alpha\left(\frac{d^3_{ich} + d^4_{ich}}{2}\right)\right] \leq \text{LQ}_{chj} \quad ; \forall j, c, h \tag{16}$$

Interaction resolution method

In this section, we explain the three levels of the resolution method in which the DM is involved interactively.

Firstly, the linear fuzzy model transformed into its crisp mode should be solved regarding each different value of α-cuts independently. In order to make the comparison more clearly, the solutions are compared against three quantitative measures as follows:

- Fixed cost: it can be calculated by the sum of annual fixed costs of practicing facilities in planning horizons.
- Logistics cost: it can be calculated by the sum of cost of supplying commodities from plants to customers in the given periods.
- Average customers' satisfaction: it can be determined by Equation 17 so that the average satisfied demands are determined as follows:

$$\sum_{h=1}^{H} \frac{\left(\sum_{j=1}^{J} y_{jchi} d_{ich} / d_{ich}\right)}{H} \quad ;\forall c, i. \tag{17}$$

Secondly, a decision vector should be obtained so that the DM is satisfied with respect to two conflicting factors (i.e., *feasibility* degree and *global satisfaction*). In this phase, the DM is asked to provide an aspiration level z added to the relevant tolerance threshold t for the values obtained by the three measures introduced above. The DM's satisfaction level can be assessed by means of a fuzzy set $\tilde{z}$ whose membership function is depicted by Equation 16 or 17 for a 'less is better' or 'much is better' case, respectively. In other words, the first two measures exemplify the 'less is better' state and the last measure typifies the 'much is better' one. However, λ is projected in a descending and ascending manner for Equations 18 and 19, respectively:

$$\mu_{\tilde{z}}(\mathrm{OF}) = \begin{array}{ll} 1, \text{if} & \mathrm{OF} \leq z \\ \lambda \in \{0,1\} \text{if} & z \leq \mathrm{OF} \leq z+t \\ 0, \text{if} & \mathrm{OF} \geq z+t \end{array} \tag{18}$$

$$\mu_{\tilde{z}}(\mathrm{OF}) = \begin{array}{ll} 1, \text{if} & \mathrm{OF} \geq z \\ \lambda \in \{0,1\} \text{if} & z-t \leq \mathrm{OF} \leq z \\ 0, \text{if} & \mathrm{OF} \leq z-t\,. \end{array} \tag{19}$$

Since we have defined three measures $\lambda_k (k = 1, 2, 3)$ for our DND problem, we are required to specify the associated weights ω_k, with regard to the DM's opinion by one of the multi-criteria decision-making tools (e.g., AHP, ANP, or TOPSIS). At the end of the second level, the global satisfaction degree Ω can be determined by Equation 20:

$$\Omega = \lambda_1 \omega_1 + \lambda_2 \omega_2 + \lambda_3 \omega_3. \tag{20}$$

Finally, we should reach a favorite balanced solution considering the feasibility degree and the global satisfaction degree at the third level. This is carried out by using Equation 21 that consists of two fuzzy sets whose membership functions represent the magnitude the DM accepts for each feasibility degree ρ_α and satisfaction degree ρ_Ω importance:

$$P = \theta \rho_\alpha + (1-\theta) \rho_\Omega \qquad \text{if } \rho_\alpha, \rho_\Omega \neq 0 \text{ and } 0, \text{otherwise}. \tag{21}$$

Note that P is the joint acceptation index obtained by the linear combination of the two mentioned degrees and $\theta \in [0,1]$ points to the relative importance the DM assigns to feasibility in comparison with global satisfaction.

Computational study

In this section, the developed fuzzy model is tested by the sensitivity analysis, and then a numerical example is solved to show its applicability and efficiency in real-world problems. Finally, the total performance of the model is investigated by different-sized problems. The calculations are run by Lingo 8 solver on a PC with characteristics of Intel®, Pentium® 4, central processing unit (CPU) 3.20 GHz, and 2,046 MB of RAM.

Sensitivity analysis

Applying the sensitivity analysis helps us understand how accurate the problem performance is. This is demonstrated by changes in model outputs according to manipulations in parameters. In other words, it is assumed that the input data may vary independently. Hence, we have to investigate the potential changes in values of model variables. To do so, a typical example whose data have been created randomly is considered here. The results are depicted in Table 2.

Implementation of interactive phases

This section is presented in order to show how the model can be implemented with respect to the three interactive solution steps. To do so, the DM is required to specify the aspiration level of each three introduced measures $\mu\tilde{z}(\mathrm{OF})$ as follows. It should be noted that the measures' quantity is calculated according to the lower and upper bound results, obtained from different cuts:

- Fixed cost: z = US$17,650,000 and $z + t$ = US $18,800,000.

Table 2 The results of fuzzy sensitivity analysis

Variation source	Fixed costs (US$)	Logistic costs (US$)	OF (US$)	Average customers' satisfaction (%)
Fixed costs	18,095,016	947,500	19,042,516	89.2
Logistic costs	17,650,000	765,014	18,415,014	83.1
Demand	17,923,000	1,100,540	19,023,540	85.7
Capacities	17,025,000	945,000	17,970,000	86.2

Table 3 The comparison results between feasibility sets and deterministic state

α-cut	OF (US$)	CPU time (s)	Iteration	Fixed cost (US$)	Logistic cost (US$)	Average customers' satisfaction (%)
0	18,595,000	14	52,741	17,650,000	945,000	78.6
0.1	18,602,400	13	47,856	17,665,500	952,400	81.4
0.2	18,784,700	15	58,996	17,823,000	961,700	86.2
0.3	18,937,300	14	44,258	17,965,000	972,300	85.7
0.4	19,204,020	21	68,439	18,220,000	984,020	88.9
0.5	19,558,500	26	89,477	18,575,000	983,500	94.5
0.6	19,510,750	27	71,644	18,522,000	988,750	92.8
0.7	19,810,600	34	83,355	18,800,000	1,010,600	92.1
0.8	19,844,330	31	67,195	18,800,000	1,044,330	87.7
0.9	19,844,050	50	85,008	18,800,000	1,044,050	85.5
1	19,860,010	63	107,414	18,800,000	1,053,070	77.3
Deterministic	19,789,840	19	104,220	18,800,000	1,060,010	81.2

- Logistic cost: z = US$945,000 and $z + t$ = US $1,060,010.
- Average customers' satisfaction: $z - t$ = 77.3% and z = 94.5%.

The global satisfaction degree can be determined by weighting the measures which is done here by considering equal importance (i.e., $\omega_1 = \omega_2 = \omega_3$).

Afterwards, it is time to specify the joint acceptance indicator, shown by Equation 19. The results of our typical example are demonstrated in Tables 3 and 4, respectively, in which the former shows the output for each feasibility cut and the latter shows the final quantity regarding the feasibility and global satisfaction simultaneously. Note that θ is assumed to be 0.5 to give equal weights to both feasibility and global conditions.

Referring to Table 3, it can be observed that the fuzzy formulation led to lower costs for all cuts compared with the deterministic mode. However, cuts greater than 0.6 resulted in an equal value for the annual fixed cost. In order to compare the results with respect to the elapsed solution time (i.e., the CPU time), it is apparent that the deterministic mode needs a rather small time although the lower cuts do the same. However, it indicates an approximately ascending order, so the last cuts required much more time that the first ones. On the other hand, the average customers' satisfaction measure showed the third lowest value for the deterministic mode. However, the measure increases as the cut magnitude goes up until it reaches the maximum value and then starts to decrease.

Table 4 The best cut with respect to the DM's opinion

α	ρ_α	Ω	ρ_Ω	P
0	0	0.02	0	0
0.1	0.05	0.09	0.1	0.075
0.2	0.15	0.26	0.33	0.24
0.3	0.25	0.32	0.4	0.325
0.4	0.4	0.47	0.55	0.475
0.5	0.55	0.71	0.78	0.765
0.6	0.7	0.65	0.7	0.7
0.7	0.85	0.81	0.93	0.89
0.8	0.9	0.82	0.95	0.925
0.9	0.95	0.77	0.86	0.905
1	1	0.66	0.72	0.86

Considering Table 4, it is obvious that α = 0.8 yielded the most acceptable result for the final solution. The DM's opinions can be seen for each of the different feasibility and satisfaction degrees likewise.

Total performance of the model

In order to understand the total performance of the mathematical formulation, some different-sized problems are solved and compared with respect to the elapsed solution time. The results are shown in Table 5.

Table 5 The results when considering different-sized problems

Problem size $p \times c \times j \times i \times h$	CPU time (s)	OF (US$)	Average customers' satisfaction (%)	Variables	Constraints
2 × 2 × 4 × 6 × 2	68	16,855,025	81	418	420
3 × 2 × 5 × 8 × 2	217	17,120,250	75	622	754
3 × 2 × 5 × 9 × 3	873	23,864,700	86	1,216	1,436
3 × 3 × 6 × 9 × 3	1,307	25,004,610	79	1,484	1,668
4 × 3 × 6 × 10 × 2	4,876	93,324,080	80	1,560	1,610
5 × 3 × 10 × 12 × 3	9,654	26,080,950	76	3,044	3,152
5 × 4 × 10 × 15 × 4	14,662	35,775,940	78	5,696	5,842

It is clear that because of intrinsic intricacy, it takes a long time for the model to respond as the dimension enlarges. These problems belong to the NP-hard category, and the solution time increases exponentially as the problem size rises, in other words. For example, in Table 5, there is a noticeable difference in CPU time while even one of the problem elements changes to a higher value. This can be easily observed as the required solution time gets multi-fold once the considered problems become larger. For instance, the second problem is larger than the first one just as one unit for the number of plants and two units for the number of distribution centers, but its elapsed time is 3.19 times as much as that of the first one. Like the previous comparison, the third problem needed a more remarkable time than the second one while the number of considered customers and planning horizons has been added by one unit. That same analysis can be carried out for the rest of the problems as well. Consequently, it is suggested that intelligent computation approaches be applied for large-sized cases in particular. This makes the mathematical model be applied efficiently for any real market conditions.

Conclusions

The noticeable increase in competition amongst the market holders makes the SCM a necessary tool in fulfilling the customers' needs. DND is of great importance as it establishes the first steps in handling the network. Consequently, a multi-period, multi-commodity, multi-source DND problem was considered in this paper. We extended our formulation to an uncertain environment, due to receiving demands, by fuzzy mathematical programming. The fuzzy model consisted of a three-interaction resolution method, so it included the DMs throughout the solution procedure. Afterwards, it was validated by a sensitivity analysis, and then a numerical example was solved in order to give a picture of how the method steps can be implemented. The consideration of different-sized problems was also carried out and indicated to the intrinsic complexity.

The presented model focused on increasing the network design efficiency so that it can help more to the drawn decisions. Considering that demand uncertainty causes the disruptions, it threatens the network (e.g., the Bullwhip effect) to decrease to a great extent as the practitioners are well equipped by demand fluctuations *a priori*. In fact, this issue helps in the necessity of the least changes in executive plans.

For future studies, it is suggested that the presented model be solved by more effective tolls like meta-heuristic or exact methods and even compared with each other. The model can also be enriched by added concepts like the inventory control management. Furthermore, the problem can be considered by other backgrounds like economic viewpoints at the same time.

Competing interests

The authors declare that they have no competing interests.

Authors' contributions

BHT carried out the problem formulation and its solution, and JR participated in its design and coordination. Both authors read and approved the final manuscript.

Acknowledgments

We would like to express our appreciation to the University of Tehran for the financial support for this study (grant number 8108023/1/18). The authors are also grateful to all persons who helped them in developing this paper.

References

Altiparmak F, Gen M, Lin L, Karaoglan I (2009) A steady-state genetic algorithm for multi-product supply chain network design. Comput Ind Eng 56:521–537

Altiparmak F, Gen M, Lin L, Paksoy T (2006) A genetic algorithm approach for multi-objective optimization of supply chain networks. Comput Ind Eng 51:196–215

Amiri A (2006) Designing a distribution network in a supply chain system: formulation and efficient solution procedure. Eur J Oper Res 171:567–576

Azaron A, Brown KN, Tarim SA, Modarres M (2008) A multi-objective stochastic programming approach for supply chain design considering risk. Int J Prod Econ 116:129–138

Blackhurst J, Wu T, O'Grady P (2004) Network-based approach to modelling uncertainty in a supply chain. Int J Prod Res 42(8):1639–1658

Cakir O (2009) Benders decomposition applied to multi-commodity, multi-mode distribution planning. Expert Syst Appl 36(4):8212–8217

Cardona-Valdes Y, Alvarez A, Ozdemir D (2011) A bi-objective supply chain design problem with uncertainty. Transport Res C-Emer 19(5):821–832

Chen CL, Yuan TW, Lee WC (2007) Multi-criteria fuzzy optimization for locating warehouses and distribution centers in a supply chain network. J Chin Inst Chem Eng 38:393–407

Cintron A, Ravindran AR, Ventura JA (2010) Multi-criteria mathematical model for designing the distribution network of a consumer goods company. Comput Ind Eng 58(4):584–593

Georgiadis MC, Tsiakis P, Longinidis P, Sofioglou MK (2011) Optimal design of supply chain networks under uncertain transient demand variations. Omega 39(3):254–272

Goh M, Lim JYS, Meng F (2007) A stochastic model for risk management in global supply chain networks. Eur J Oper Res 182(1):164–173

Gumus AT, Guneri AF, Keles S (2009) Supply chain network design using an integrated neuro-fuzzy and MILP approach: a comparative design study. Expert Syst Appl 36(10):12570–12577

Gupta A, Maranas CD (2003) Managing demand uncertainty in supply chain planning. Comput Chem Eng 27(8–9):1219–1227

Hajiaghaei-Keshteli M (2011) The allocation of customers to potential distribution centers in supply chain networks: GA and AIA approaches. Appl Soft Comput 11(2):2069–2078

Jayaraman V, Ross A (2003) A simulated annealing methodology to distribution network design and management. Eur J Oper Res 144:629–645

Jimenez M, Arenas M, Bilbao A, Guez MV (2007) Linear programming with fuzzy parameters: an interactive method resolution. Eur J Oper Res 177:1599–1609

Jimenez M (1996) Ranking fuzzy numbers through the comparison of its expected intervals. Int J Fuzziness Knowledge-Based Syst 4(4):379–388

Liu D, Huang D, Chen Y (2006) Distribution inventory cost optimization under grey and fuzzy uncertainty. Wuhan University J Nat Sci 11(5):1238–1242

Park S, Lee T-E, Sung CS (2010) A three-level supply chain network design model with risk-pooling and lead times. Transport Res E 46(5):563–581

Peidro D, Mula J, Jimenez M, Botella MDM (2010) A fuzzy linear programming based approach for tactical supply chain planning in an uncertainty environment. Eur J Oper Res 205(1):65–80

Rezapour S, Farahani RZ (2010) Strategic design of competing centralized supply chain networks for markets with deterministic demands. Adv Eng Softw 41 (5):810–822

Rommelfanger H, Slowinski R (1998) Fuzzy linear programming with single or multiple objective functions. In: Slowinski R (ed) Fuzzy sets in decision analysis, operations research and statistics. Kluwer, Norwell, pp 179–207

Santoso T, Ahmed S, Goetschalckx M, Shapiro A (2005) A stochastic programming approach for supply chain network design under uncertainty. Eur J Oper Res 167(1):96–115

Selim H, Ozkarahan I (2006) Application of fuzzy multi-objective programming approach to supply chain distribution network design problem. In: Gelbukh A, Reyes-Garcia CA (eds) MICAI 2006: advances in artificial intelligence. 5th Mexican international conference on artificial intelligence, Apizaco, Mexico, November 13–17, 2006 proceedings. Lecture notes in artificial intelligence (Lecture notes in computer science). Springer, Berlin, pp 415–425

Tang CS (2006) Perspectives in supply chain risk management. Int J Prod Econ 103(2):451–488

Xu J, He Y, Gen M (2009) A class of random fuzzy programming and its application to supply chain design. Comput Ind Eng 56:937–950

You F, Grossman IE (2008) Design of responsive supply chains under demand uncertainty. Comput Che Eng 32:3090–3111

7

Identifying the change time of multivariate binomial processes for step changes and drifts

Seyed Taghi Akhavan Niaki* and Majid Khedmati

Abstract

In this paper, a new control chart to monitor multi-binomial processes is first proposed based on a transformation method. Then, the maximum likelihood estimators of change points designed for both step changes and linear-trend disturbances are derived. At the end, the performances of the proposed change-point estimators are evaluated and are compared using some Monte Carlo simulation experiments, considering that the real change type presented in a process are of either a step change or a linear-trend disturbance. According to the results obtained, the change-point estimator designed for step changes outperforms the change-point estimator designed for linear-trend disturbances, when the real change type is a step change. In contrast, the change-point estimator designed for linear-trend disturbances outperforms the change-point estimator designed for step changes, when the real change type is a linear-trend disturbance.

Keywords: Multi-binomial processes, Maximum likelihood estimator, Multi-attribute processes, Step change, Linear-trend disturbance, Root/power transformation

Background

Enabling process engineers and practitioners to monitor the state of processes by distinguishing between common and special causes of variation, control charts have been one of the most widely used statistical process control tools in industries. When a control chart generates a signal due to an out-of-control state of a process, process engineers initiate a search to identify the root causes of process variation. However, the signaling time is different from the first time that a change manifests itself into the process (change point), and in most cases there exists a considerable time lag between them. Therefore, knowing the exact time of the change would simplify the search for root cause identification and corrective action implementation in order to improve the quality of processes.

A binomial distribution with parameters n and p is used to model the number of non-conforming items in a sequence of n trials for an attribute quality characteristic, in which p denotes the process fraction non-conforming. In the case of multi-attribute processes, the multivariate binomial distribution with parameters n and p_k ($k = 1, 2,\ldots, q$) can be used to model the number of non-conforming items in a sequence of n trials for each attribute quality characteristic, in which p_k denotes the process fraction non-conforming of the kth quality characteristic. The problem of estimating the time of a change in a univariate binomial distribution for different change types of step change, linear-trend disturbance (drift), multiple-step changes, monotonic change, and other change types defined as a combination of the above has been addressed in the literature. In a step-change disturbance, the parameter of the process shifts to an out-of-control level and remains in that level until corrective measures are taken. Although a step change can occur because of tool breakage, the parameter of a process may gradually drift. In other words, linear-trend disturbances can occur as a result of tool wear or equipment depreciation. Moreover, a process parameter may change due to multiple-step changes. A change in several parameters at different times or a change in one parameter at different times can be the results of this type of change. In monotonic changes, however, the only assumption is that the change type can be described as belonging to a family of monotonic change types.

* Correspondence: Niaki@sharif.edu
Department of Industrial Engineering, Sharif University of Technology, P.O. Box 11155-9414, Azaadi Ave., Tehran 1458889694, Iran

Samuel and Pignatiello (2001) derived the maximum likelihood estimator (MLE) of a step-change point for binomial processes. Evaluating the estimator for different shift sizes, they concluded that it provides accurate results. In the case of a step change, Perry and Pignatiello (2005) compared the MLE of the process fraction non-conforming with the built-in change-point estimators of the binomial CUSUM and EWMA control charts described by Page (1954) and Nishina (1992), respectively. They showed that the MLE provides better results than the other two estimators do. Perry et al. (2007) derived the MLE of a monotonic change point of the process fraction non-conforming and compared it with the MLE of a step-change point derived by Samuel and Pignatiello (2001). They concluded that using their proposed estimator is better when the type of change is only known to be monotonic. Zandi et al. (2011) proposed an MLE to estimate the time of a linear-trend change in the process fraction non-conforming and compared it to the MLEs derived for step and monotonic change disturbances proposed by Samuel and Pignatiello (2001) and Perry et al. (2007), respectively.

To the best of the authors' knowledge, there are only two research works available in the literature to estimate the change point of multi-attribute processes. In the first work, Niaki and Khedmati (2012a) proposed an MLE of a step-change point for multivariate Poisson processes and evaluated its performances using some numerical experiments. They showed that their estimator leads into good results for all process dimensions. In the second work, Niaki and Khedmati (2012b) derived an MLE of a linear-trend change in multivariate Poisson processes and compared its performances with those of an MLE of a step change for both step-change and linear-trend disturbances. However, there is no research work concerning the change-point estimation of multivariate binomial processes, whereas in many situations, monitoring processes containing several quality characteristics of attribute type, each following a binomial distribution, is a need. Therefore, in this paper, we first propose a new multi-attribute control chart to monitor multi-attribute processes, and then, we derive the MLEs of both a step-change and a linear-trend disturbance in the process fraction non-conforming. Following a signal from the proposed multi-attribute control chart, the MLEs are applied to estimate the step-change and linear-trend disturbance points.

The rest of this paper is organized as follows: In the next section, a new multi-attribute control chart is proposed based on a transformation method to monitor multi-attribute processes. In the 'Multivariate binomial process modeling and MLE derivation' section, the model of the process behavior is described and the MLEs of a step-change and a linear-trend disturbance are derived. The performances of the two change-point estimators are evaluated and are compared in the 'Results and discussion' section, in the presence of both step-change and linear-trend disturbances. Finally, the paper is concluded in the 'Conclusions' section.

The proposed multi-attribute control chart

As a generalization of Shewhart-type control charts, most of multivariate control charts have been constructed based upon the multivariate normal distribution. Despite the existence of some skewness in multi-attribute distributions, the normal distribution is symmetric with zero skewness. As a result, using normal-based control charts with zero skewness for multi-attribute processes causes some problems. The main problem is the skewness of these processes in which inattention leads to poor results in distinguishing the in- and out-of-control conditions. To overcome this problem, the root and the power transformation methods proposed by Niaki and Abbasi (2007) and Abbasi et al. (2011), respectively, are employed to almost eliminate the skewness. The following subsection is devoted to describe the method.

The root/power transformation methods

To eliminate the skewness involved in multi-attribute processes, Niaki and Abbasi (2007) first proposed the root transformation method, in which the skewness of the transformed attributes becomes almost zero. In this transformation, using the bisection method, a power within (0,1) is found for each attribute in the vector $\mathbf{X} = [X_1, X_2, \ldots, X_q]'$ to transform it into a new attribute vector $\mathbf{Y}{=}\mathbf{X}^{\mathbf{r}} = \left[X_1^{r_1}, X_2^{r_2}, \ldots, X_q^{r_q}\right]'$ such that the transformed attribute vector contains attributes of almost zero skewness. However, this transformation method is only applicable for right-skewed distributions. Hence, in another work, Abbasi et al. (2011) extended the root transformation method and proposed a power transformation technique for left-skewed distributions as well. In the power transformation method, one first subtracts the observations from their minimum value and then finds a power within (0,1) for each attribute in the vector $[\mathbf{X} - \mathrm{Min}(\mathbf{X})] = [(X_1 - \mathrm{Min}(X_1)), (X_2 - \mathrm{Min}(X_2)), \ldots, (X_q - \mathrm{Min}(X_q))]'$ such that the transformed attribute vector $\mathbf{Y} = [\mathbf{X} - \mathrm{Min}(\mathbf{X})]^{1/r} = \left[(X_1 - \mathrm{Min}(X_1))^{1/r_1}, \ldots, (X_q - \mathrm{Min}(X_q))^{1/r_q}\right]'$ contains attributes of almost zero skewness.

Applying the above transformation methods, an attribute vector with zero skewness is obtained. Consequently, one may assume that the transformed vector approximately follows a multivariate normal distribution, and hence, the use of normal-based control charts is recommended. Moreover, to make sure the transformation works well, the Jarque and Bera (1987) test statistic given in Equation 1 is used to test the normality of the transformed attributes:

$$\mathrm{JB} = M\left(b_1^2/6 + b_2^2/24\right), \tag{1}$$

where M is the number of observations, b_1 is the sample skewness, and b_2 is the sample kurtosis. This statistic follows an asymptotic chi-square distribution with 2 degrees of freedom and is used to test the null hypothesis that the transformed attributes are from normal distributions.

Applying this transformation method, attributes with almost zero skewness that approximately follow multivariate normal distribution are obtained. Therefore, the well-known Hotelling T^2 control chart can be used to monitor the multi-attribute processes.

The multi-attribute T^2 control chart based on transformed data

The Hotelling T^2control chart was first introduced by Hotelling (1947) to come up with the correlation between the quality characteristics in monitoring multivariate processes. This control chart has been designed based on multivariate normal distribution. We use this control chart to monitor transformed attribute vector with approximately multivariate normal distribution.

Consider the observation vector $\mathbf{X_{ij}} = [X_{ij1}, X_{ij2}, \ldots, X_{ijq}]'$ coming from a multivariate binomial distribution with in-control parameter vector $[\mathbf{N}', \mathbf{P_0}'] = [(n_1, n_2, \ldots, n_q)', (p_1, p_2, \ldots, p_q)']$ and in-control correlation matrix $\mathbf{\Sigma_0}$, in which X_{ijk} represents the jth observation in the ith subgroup of the kth quality characteristic. We apply the root/power transformation method in order to transform the vector $\mathbf{X_{ij}}$ to $\mathbf{Y_{ij}} = [Y_{ij1}, Y_{ij2}, \ldots, Y_{ijq}]'$ that approximately follows a multivariate normal distribution with the mean vector $\boldsymbol{\mu}_\mathrm{Y}$ and the covariance matrix $\mathbf{Cov}_\mathrm{Y}$.

For the transformed vector, the sample mean vector for each subgroup of size n is obtained using

$$\overline{\mathrm{Y}_i} = \frac{1}{n}\sum_{j=1}^{n} Y_{ij} \ ; \ i = 1, 2, \ldots, m. \tag{2}$$

Then, for each subgroup the T^2 statistic is calculated by

$$T_i^2 = n\left(\overline{\mathbf{Y_i}} - \overline{\overline{\mathbf{Y}}}\right)' \hat{\mathbf{Cov}}_\mathrm{Y}^{-1}\left(\overline{\mathbf{Y_i}} - \overline{\overline{\mathbf{Y}}}\right), \tag{3}$$

in which $\overline{\overline{\mathrm{Y}}} = \left[\overline{\overline{Y}}_1, \overline{\overline{Y}}_2, \ldots, \overline{\overline{Y}}_q\right]' = \frac{1}{mn}\left[\sum_{i=1}^{m}\sum_{j=1}^{n} Y_{ij1}, \sum_{i=1}^{m}\sum_{j=1}^{n} Y_{ij2}, \ldots, \sum_{i=1}^{m}\sum_{j=1}^{n} Y_{ijq}\right]'$ is the estimated in-control mean vector of the process.

The statistic in Equation 3 is plotted on a T^2 control chart with an upper control limit (UCL) of

$$\mathrm{UCL} = \frac{q(m-1)(n-1)}{mn - m - q + 1} F_{\alpha:q,mn-m-q+1}, \tag{4}$$

where for large number of subgroups may be estimated by $\chi^2_{\alpha;q}$ (see Lowry and Montgomery (1995) for more details). An out-of-control signal is detected when the T^2 statistic exceeds UCL.

Multivariate binomial process modeling and MLE derivation

In monitoring processes using the T^2 control chart, as long as the plotted points fall below UCL, the process is assumed in-control. However, when one or more points exceed UCL, the chart signals a change in the parameters of the process and the process is assumed out-of-control. The most important problem is that the chart signals after a considerable amount of time after the change point. Therefore, knowing when the change had actually occurred would substantially improve the diagnostic procedure.

The MLEs of the change points, designed for step changes and linear-trend disturbances are derived in the next two subsections.

The proposed change-point estimator for a step-change disturbance

In this section, we assume that the real change type presented in a multivariate binomial process is a step change and derive the MLE of a change point in the process fraction non-conforming. We also assume that the process is initially in-control with known parameter vector $[\mathbf{N}', \mathbf{P_0}']$. Following an unknown point in time τ_{SC} (the step-change point), the process moves to an out-of-control state with parameter vector $[\mathbf{N}', \mathbf{P_1}']$, where $\mathbf{P_1}$ is given in Equation 5 and $\boldsymbol{\delta}$ is the vector of unknown magnitude of the change. An element $\delta > 1$ in $\boldsymbol{\delta}$ denotes an increase in the corresponding process fraction non-conforming, while $\delta < 1$ denotes an improvement:

$$\mathbf{P}_1 = \boldsymbol{\delta}\mathbf{P}_0. \tag{5}$$

The control chart genuinely (not false alarm) signals the above change with a delay at time S, when the T_S^2 statistic exceeds the UCL of the T^2 control chart. Accordingly, the subgroup averages $\overline{Y_1}, \overline{Y_2}, \ldots, \overline{Y_{\tau_{SC}}}$ come from the in-control process, while the subgroup averages $\overline{Y_{\tau_{SC}+1}}, \overline{Y_{\tau_{SC}+2}}, \ldots, \overline{Y_S}$ come from the out-of-control process. As a result, the MLE of the proposed change-point estimator, denoted by $\hat{\tau}_{SC}$, is obtained as

$$\hat{\tau}_{SC} = arg \underset{0 \le t < S}{\mathrm{Min}} \left\{ \begin{array}{l} \frac{t}{2}\mathrm{Ln}\left|\mathbf{Cov}_{0,Y}\right| + \frac{n}{2}\sum_{i=1}^{t}\left(\overline{\mathbf{Y}}_i - \boldsymbol{\mu}_{0,Y}\right)' \mathbf{Cov}_{0,Y}^{-1} \\ \times\left(\overline{\mathbf{Y}}_i - \boldsymbol{\mu}_{0,Y}\right) + \frac{S-t}{2}\mathrm{Ln}\left|\hat{\mathbf{Cov}}_{1,Y}\right| \\ + \frac{n}{2}\sum_{i=t+1}^{S}\left(\overline{\mathbf{Y}}_i - \hat{\boldsymbol{\mu}}_{1,Y}\right)' \hat{\mathbf{Cov}}_{1,Y}^{-1}\left(\overline{\mathbf{Y}}_i - \hat{\boldsymbol{\mu}}_{1,Y}\right) \end{array} \right\}, \tag{6}$$

where $\hat{\boldsymbol{\mu}}_{1,Y} = \sum_{i=\tau_{SC}+1}^{S} \overline{\mathbf{Y}}_i / S - \tau_{SC}$ and $\hat{\mathbf{Cov}}_{1,Y} = \frac{n}{S-\tau_{SC}} \sum_{i=\tau_{SC}+1}^{S} \left(\overline{\mathbf{Y}}_i - \hat{\boldsymbol{\mu}}_{1,Y}\right)' \left(\overline{\mathbf{Y}}_i - \hat{\boldsymbol{\mu}}_{1,Y}\right)$ (see Appendix 1 for details). The MLE of the change point $\hat{\tau}_{SC}$ is the value for τ_{sc} that minimizes Equation 6.

The proposed change-point estimator for a linear-trend disturbance

Assuming again that the process is initially in-control with known parameter vector $[\mathbf{N}', \mathbf{P}_0']$, in a linear-trend disturbance when the T_S^2 statistic exceeds the UCL of the T^2 control chart, it is assumed that a change has occurred at an unknown point in time τ_{LT} (the change point under linear trend). As a result the parameter vector of the out-of-control process becomes $[\mathbf{N}', \mathbf{P}_1']$, where $\mathbf{P}_1$ at different points in time takes a value based on Equation 7 and vector $\boldsymbol{\beta}$ contains the magnitude (or slope) of the linear-trend disturbance:

$$\mathbf{P}_{1i} = \mathbf{P}_0 + \boldsymbol{\beta}(i - \tau_{LT}) \; ; \quad i = \tau_{LT} + 1, \tau_{LT} + 2, \ldots . \tag{7}$$

Based on the above descriptions, the subgroup averages $\overline{Y_1}, \overline{Y_2}, \ldots, \overline{Y_{\tau_{LT}}}$ come from the in-control process, while the subgroup averages $\overline{Y_{\tau_{LT}+1}}, \overline{Y_{\tau_{LT}+2}}, \ldots, \overline{Y_S}$ come from an out-of-control process. As a result, the MLE of the change point, $\hat{\tau}_{LT}$, is the value of τ_{LT} that minimizes Equation 8 and is obtained as

$$\hat{\tau}_{LT} = arg \underset{0 \le t \le S}{\mathrm{Min}} \left\{ \begin{array}{l} \frac{t}{2}\mathrm{Ln}\left|\mathbf{Cov}_{0,Y}\right| + \frac{n}{2}\sum_{i=1}^{t}\left(\overline{\mathbf{Y}}_i - \boldsymbol{\mu}_{0,Y}\right)' \mathbf{Cov}_{0,Y}^{-1} \\ \times\left(\overline{\mathbf{Y}}_i - \boldsymbol{\mu}_{0,Y}\right) + \frac{S-t}{2}\mathrm{Ln}\left|\hat{\mathbf{Cov}}_{1,Y}\right| \\ + \frac{n}{2}\sum_{i=t+1}^{S}\left(\overline{\mathbf{Y}}_i - \boldsymbol{\mu}_{0,Y} - \hat{\boldsymbol{\beta}}(i-t)\right)' \hat{\mathbf{Cov}}_{1,Y}^{-1} \\ \times\left(\overline{\mathbf{Y}}_i - \boldsymbol{\mu}_{0,Y} - \hat{\boldsymbol{\beta}}(i-t)\right) \end{array} \right\}, \tag{8}$$

where

$$\hat{\boldsymbol{\beta}} = \sum_{i=\tau_{LT}+1}^{S} (i - \tau_{LT})\left(\overline{\mathbf{Y}}_i - \boldsymbol{\mu}_{0,Y}\right) \Big/ \sum_{i=\tau_{LT}+1}^{S} (i - \tau_{LT})^2$$

and

$$\hat{\mathbf{Cov}}_{1,Y} = \frac{n}{S - \tau_{LT}} \sum_{i=\tau_{LT}+1}^{S} \left(\overline{\mathbf{Y}}_i - \boldsymbol{\mu}_{0,Y} - \hat{\boldsymbol{\beta}}(i - \tau_{LT})\right)' \times \left(\overline{\mathbf{Y}}_i - \boldsymbol{\mu}_{0,Y} - \hat{\boldsymbol{\beta}}(i - \tau_{LT})\right)$$

(see Appendix 2 for details).

In the next section, the performances of the proposed change-point estimators are evaluated and are compared using some simulation experiments. The MATLAB 7 software (Math Works, Inc., Natick, MA, USA) is utilized to perform all the programming works of this research.

Performance evaluation and comparison

Monte Carlo simulations are used to evaluate and to compare the performances of the proposed change-point estimators designed for step changes and linear-trend disturbances, when the actual changes are of step changes and linear-trend disturbances in the process fraction nonconforming of a multivariate binomial process.

Change-point estimators applied to step-change disturbances

The change point is simulated to occur at $\tau = 100$. Thus, the first 100 subgroups are randomly generated from an in-control multivariate binomial distribution with parameter vector $[\mathbf{N}', \mathbf{P}_0']$, using the normal to anything (NORTA) method (Cario and Nelson 1997; Niaki and Abbasi 2008). The in-control subgroup averages exceeding the UCL of the proposed control chart are false alarms. Consequently, these subgroups are discarded and are replaced with other in-control subgroups. This is replicated until all of the in-control subgroups are below UCL. Following subgroup 100, the parameter vector is changed based on a step-change disturbance given in Equation 5 where the subgroups are randomly generated from an out-of-control process with parameter vector $[\mathbf{N}', \mathbf{P}_1']$. The subgroup averages are plotted on the proposed control chart until a signal is generated, and at this time, the accuracy and the precision measures of the change-point estimators are computed.

Simulation experiment 1

Consider a multivariate binomial process with two correlated attributes, an in-control parameter vector $[\mathbf{N}', \mathbf{P}_0'] = [(20, 30)', (0.2, 0.15)']$ and a correlation between the two attributes of 0.25. For subgroups $i = 1, 2, \ldots, 100$, independent observations are first generated from the in-control process using the NORTA method. After subgroup 100, observations are generated from an out-of-control process with parameter vector $[\mathbf{N}', \mathbf{P}_1']$. Obtaining a signal from the control chart, the change points are estimated using Equations 6 and 8 for 1,000 simulation runs. Then, the averages, the standard deviations, and the mean square errors

(MSEs) of the estimates in addition to the expected time when the control chart signals a change ($E(T)$) are calculated and are reported in Table 1. Table 1 also contains the precision performances of the estimated change points, in which $\hat{P}(|\hat{\tau}-\tau|=0)$, $\hat{P}(|\hat{\tau}-\tau|\leq 1)$, $\hat{P}(|\hat{\tau}-\tau|\leq 2)$, $\hat{P}(|\hat{\tau}-\tau|\leq 3)$, and $\hat{P}(|\hat{\tau}-\tau|\leq 4)$ are denoted by ***P0, P1, P2, P3,*** and ***P4***, respectively. The estimated probabilities in parentheses correspond to $\hat{\tau}_{SC}$.

Based on the results in Table 1, the change-point estimator designed for step changes outperforms the one designed for linear-trend disturbances in terms of both the accuracy and the precision measures, when the true change type is a step-change disturbance. In other words, for almost all of the shift magnitudes, $\hat{\tau}_{SC}$ is considerably closer to the actual change point than $\hat{\tau}_{LT}$ is. Furthermore, MSE($\hat{\tau}_{SC}$) is smaller than MSE($\hat{\tau}_{LT}$) for all of the shift values.

Simulation experiment 2

Consider a multivariate binomial process with five correlated attributes. The sample size and the process fraction non-conforming of the process are $[\mathbf{N}', \mathbf{P}_0'] = [(25, 25, 25, 25, 25)', (0.4, 0.27, 0.15, 0.35, 0.21)']$ with an in-control correlation matrix of

$$\Sigma_0 = \begin{pmatrix} 1.00 & 0.15 & 0.21 & 0.12 & 0.35 \\ 0.15 & 1.00 & 0.41 & 0.18 & 0.24 \\ 0.21 & 0.41 & 1.00 & 0.26 & 0.14 \\ 0.12 & 0.18 & 0.26 & 1.00 & 0.32 \\ 0.35 & 0.24 & 0.14 & 0.32 & 1.00 \end{pmatrix}.$$

Using the NORTA method, the first 100 subgroups are randomly generated from the in-control process. Following subgroup 101, observations are generated from an out-of-control process based on the step-change disturbance given in Equation 5. The subgroup averages are then plotted on the proposed control chart until a signal is generated. Then, the averages, the standard deviations, the mean square errors, and the precision performances of the change-point estimates for 1,000 simulation runs are reported in Tables 2 and 3 for $\delta < 1$ and $\delta > 1$, respectively. Similar to Table 1, the precision performances of $\hat{\tau}_{SC}$ in Tables 2 and 3 are shown in parentheses.

Again, the results in Tables 2 and 3 show that $\hat{\tau}_{SC}$ outperforms $\hat{\tau}_{LT}$ for almost all shift magnitudes, since $\hat{\tau}_{SC}$ is closer to the true change point than $\hat{\tau}_{LT}$ is. Moreover, MSE($\hat{\tau}_{SC}$) is smaller than MSE($\hat{\tau}_{LT}$), and $\hat{\tau}_{SC}$ is more precise than $\hat{\tau}_{LT}$.

Based on the results obtained from the above two simulation experiments, one can conclude that when the real change type is a step-change disturbance, the change-point estimator designed for step changes outperforms the change-point estimator designed for linear-trend disturbances for almost all of the shift magnitudes and process dimensions.

Change-point estimators applied to linear-trend disturbances

Assuming a linear-trend disturbance, in this section, the performances of the two proposed change-point estimators designed for step-change and linear-trend disturbances are

Table 1 Estimated change points and precision performances of two different MLEs in simulation experiment 1

δ	$E(T)$	$E(\hat{\tau}_{LT})$	$E(\hat{\tau}_{SC})$	MSE($\hat{\tau}_{LT}$)	MSE($\hat{\tau}_{SC}$)	*P0*	*P1*	*P2*	*P3*	*P4*
(0.7,0)	115.14	93.21 (10.61)	99.45 (4.87)	158.6	24.0	0.100 (0.410)	0.254 (0.675)	0.357 (0.785)	0.430 (0.852)	0.501 (0.897)
(0.8,0)	143.58	82.60 (24.54)	100.68 (6.84)	904.5	47.3	0.039 (0.246)	0.101 (0.455)	0.157 (0.589)	0.198 (0.668)	0.236 (0.726)
(0.9,0)	215.07	75.55 (50.83)	107.96 (26.27)	3178.6	752.8	0.010 (0.069)	0.037 (0.187)	0.068 (0.271)	0.089 (0.324)	0.103 (0.376)
(1.1,0)	203.86	83.38 (56.50)	110.94 (29.96)	3465.4	1016.4	0.019 (0.086)	0.053 (0.181)	0.076 (0.263)	0.097 (0.337)	0.123 (0.389)
(1.2,0)	139.63	84.40 (23.72)	100.57 (7.69)	805.4	59.4	0.044 (0.212)	0.117 (0.430)	0.176 (0.568)	0.219 (0.642)	0.255 (0.713)
(1.5,0)	103.74	97.41 (5.99)	98.04 (7.00)	42.6	52.9	0.175 (0.374)	0.502 (0.679)	0.695 (0.823)	0.790 (0.874)	0.846 (0.909)
(0,0.7)	112.84	94.76 (8.55)	99.53 (3.90)	100.6	15.4	0.094 (0.394)	0.264 (0.659)	0.407 (0.806)	0.496 (0.877)	0.559 (0.910)
(0,0.8)	141.67	83.63 (23.49)	100.32 (6.92)	819.1	47.9	0.043 (0.256)	0.116 (0.473)	0.167 (0.609)	0.207 (0.695)	0.254 (0.755)
(0,0.9)	208.11	76.95 (53.36)	106.72 (26.66)	3375.5	755.4	0.014 (0.073)	0.046 (0.213)	0.073 (0.289)	0.098 (0.346)	0.124 (0.407)
(0,1.1)	197.60	81.32 (49.78)	108.53 (25.28)	2824.9	711.1	0.027 (0.085)	0.059 (0.214)	0.085 (0.289)	0.106 (0.342)	0.134 (0.400)
(0,1.2)	134.80	86.73 (23.09)	100.92 (6.79)	709.0	46.9	0.037 (0.227)	0.123 (0.439)	0.188 (0.574)	0.239 (0.656)	0.298 (0.717)
(0,1.5)	103.44	98.02 (4.13)	98.44 (5.43)	20.9	32.0	0.181 (0.369)	0.514 (0.675)	0.755 (0.840)	0.834 (0.894)	0.883 (0.933)
(0.7,0.7)	107.88	96.39 (6.16)	99.08 (4.45)	50.9	20.7	0.133 (0.445)	0.373 (0.711)	0.523 (0.840)	0.621 (0.900)	0.704 (0.931)
(0.8,0.8)	126.99	88.45 (17.01)	99.75 (5.62)	422.5	31.7	0.055 (0.340)	0.152 (0.580)	0.232 (0.719)	0.295 (0.789)	0.351 (0.839)
(0.9,0.9)	199.75	71.63 (43.21)	103.98 (17.51)	2670.2	322.3	0.016 (0.113)	0.048 (0.249)	0.078 (0.336)	0.103 (0.390)	0.122 (0.451)
(1.1,1.1)	176.76	78.02 (41.05)	103.80 (18.74)	2166.4	365.3	0.022 (0.117)	0.060 (0.251)	0.088 (0.348)	0.116 (0.427)	0.140 (0.476)
(1.2,1.2)	122.09	90.83 (15.92)	100.04 (7.64)	337.2	58.3	0.077 (0.254)	0.187 (0.510)	0.264 (0.637)	0.328 (0.721)	0.374 (0.783)
(1.5,1.5)	102.10	97.60 (5.33)	97.26 (7.72)	34.1	67.1	0.158 (0.276)	0.550 (0.596)	0.798 (0.796)	0.855 (0.851)	0.895 (0.884)

Experiment has a step-change disturbance; standard deviations are shown in parentheses.

Table 2 Estimated change points and precision performances of two different MLEs in simulation experiment 2

δ	$E(T)$	$E(\hat{\tau}_{LT})$	$E(\hat{\tau}_{SC})$	MSE$(\hat{\tau}_{LT})$	MSE$(\hat{\tau}_{SC})$	P0	P1	P2	P3	P4
(0.7,0,0,0,0)	103.22	97.18 (3.43)	97.39 (3.69)	19.8	20.4	0.081 (0.150)	0.238 (0.306)	0.434 (0.511)	0.639 (0.668)	0.792 (0.828)
(0.9,0,0,0,0)	161.33	87.38 (38.82)	106.32 (18.37)	1665.3	377.1	0.013 (0.163)	0.059 (0.331)	0.099 (0.444)	0.148 (0.525)	0.180 (0.587)
(0,0.7,0,0,0)	107.52	97.42 (5.76)	99.13 (3.54)	39.9	13.2	0.081 (0.262)	0.240 (0.471)	0.423 (0.639)	0.584 (0.753)	0.698 (0.874)
(0,0.9,0,0,0)	189.00	93.83 (56.78)	112.29 (28.54)	3259.3	965.5	0.020 (0.125)	0.050 (0.230)	0.072 (0.300)	0.101 (0.355)	0.123 (0.414)
(0,0,0.7,0,0)	121.58	93.90 (15.98)	101.50 (6.61)	292.5	45.9	0.059 (0.278)	0.149 (0.466)	0.245 (0.601)	0.340 (0.705)	0.408 (0.784)
(0,0,0.9,0,0)	225.12	106.74 (72.81)	126.97 (47.98)	5342.1	3028.0	0.008 (0.052)	0.036 (0.135)	0.062 (0.203)	0.079 (0.261)	0.099 (0.308)
(0,0,0,0.7,0)	104.69	97.29 (4.03)	98.26 (3.33)	23.7	14.1	0.099 (0.220)	0.274 (0.410)	0.450 (0.590)	0.640 (0.748)	0.784 (0.868)
(0,0,0,0.9,0)	172.99	88.14 (43.42)	110.03 (22.37)	2023.8	600.8	0.017 (0.136)	0.053 (0.258)	0.086 (0.362)	0.114 (0.430)	0.141 (0.485)
(0,0,0,0,0.7)	112.98	95.69 (9.32)	100.31 (4.57)	105.4	21.0	0.085 (0.311)	0.229 (0.530)	0.355 (0.664)	0.456 (0.768)	0.570 (0.844)
(0,0,0,0,0.9)	206.54	93.68 (58.44)	118.83 (42.39)	3451.2	2150.0	0.014 (0.100)	0.028 (0.200)	0.055 (0.279)	0.093 (0.340)	0.113 (0.392)
(0.7,0.7,0.7, 0.7,0.7)	101.65	96.05 (4.32)	96.32 (3.98)	34.3	29.4	0.026 (0.037)	0.110 (0.131)	0.262 (0.310)	0.487 (0.548)	0.726 (0.752)
(0.9,0.9,0.9, 0.9,0.9)	144.72	88.20 (27.67)	104.20 (12.74)	904.5	179.9	0.031 (0.232)	0.092 (0.430)	0.155 (0.528)	0.211 (0.608)	0.249 (0.669)

Experiment has a step-change disturbance when $\delta < 1$; standard deviations are shown in parentheses.

evaluated and are compared in multivariate binomial processes with two and five correlated quality characteristics.

Considering the change point at $\tau = 100$, the first 100 subgroups are randomly generated from an in-control multivariate binomial distribution with parameter vector $[\mathbf{N}', \mathbf{P}_0']$, using the NORTA method. Following subgroup 100, the parameter vector is changed according to a linear-trend disturbance given in Equation 7, and as a result, subgroups are randomly generated from an out-of-control process with parameter vector $[\mathbf{N}', \mathbf{P}_1']$. The subgroup averages are plotted on the proposed control chart until a signal is generated. At this time, the accuracy and the precision measures of the change-point estimators are computed.

Simulation experiment 3

Consider the process in simulation experiment 1. Using the NORTA method, the first 100 subgroup averages are generated from the in-control process. After subgroup 100, observations are generated from an out-of-control process. Following a signal by the proposed control chart, the change points are estimated using Equations 6 and 8 for 1,000 simulation runs. Then, the averages, the standard deviations, the mean square errors, and the precision performances of the change-point estimates in addition to the expected time when the control chart signals a change are summarized in Table 4.

The results in Table 4 show that when the real change type is a linear-trend disturbance, the proposed MLE for linear-trend disturbances outperforms the one designed for step-change disturbances in terms of both the accuracy and the precision for almost all values of $\boldsymbol{\beta}$. That is, $\hat{\tau}_{LT}$ is closer to the true change point than $\hat{\tau}_{SC}$ is.

Simulation experiment 4

Considering the process described in simulation experiment 2, 100 subgroups are first randomly generated from the in-control process. Then, the parameter vector is changed based on a linear-trend disturbance given in Equation 7. Consequently, observations are generated from an out-of-control process. Following a signal from the control chart, the averages, the standard deviations, the mean square errors, and the precision performances of the change-point estimates in 1,000 simulation runs are reported in Table 5.

Based on the results in Table 5, in the presence of a linear-trend disturbance, $\hat{\tau}_{LT}$ outperforms $\hat{\tau}_{SC}$ in terms of both the accuracy and the precision measures. In other words, in addition to the better precision performances of $\hat{\tau}_{LT}$, the estimated change points by $\hat{\tau}_{LT}$ are closer to the real change point than $\hat{\tau}_{SC}$ for almost all values of $\boldsymbol{\beta}$.

Table 3 Estimated change points and precision performances of two different MLEs in simulation experiment 2

δ	$E(T)$	$E(\hat{\tau}_{LT})$	$E(\hat{\tau}_{SC})$	$MSE(\hat{\tau}_{LT})$	$MSE(\hat{\tau}_{SC})$	*P0*	*P1*	*P2*	*P3*	*P4*
(1.1,0,0,0,0)	161.49	88.15 (38.26)	108.03 (21.11)	1602.9	509.5	0.028 (0.160)	0.067 (0.304)	0.111 (0.412)	0.142 (0.468)	0.176 (0.532)
(1.2,0,0,0,0)	112.17	97.08 (9.28)	100.55 (4.99)	94.7	25.2	0.073 (0.248)	0.219 (0.475)	0.369 (0.615)	0.487 (0.738)	0.588 (0.821)
(1.5,0,0,0,0)	101.13	95.80 (3.09)	96.85 (2.99)	27.1	26.1	0.006 (0.008)	0.034 (0.044)	0.193 (0.203)	0.424 (0.438)	0.685 (0.699)
(0,1.1,0,0,0)	188.57	91.61 (53.06)	115.46 (34.11)	2883.7	1401.3	0.012 (0.099)	0.051 (0.221)	0.082 (0.302)	0.103 (0.374)	0.124 (0.428)
(0,1.2,0,0,0)	126.57	93.47 (19.43)	102.87 (10.71)	419.8	122.8	0.041 (0.195)	0.129 (0.393)	0.195 (0.522)	0.275 (0.631)	0.347 (0.716)
(0,1.5,0,0,0)	101.87	96.38 (3.57)	96.62 (2.51)	25.9	17.7	0.034 (0.061)	0.154 (0.172)	0.329 (0.326)	0.550 (0.563)	0.759 (0.767)
(0,0,1.1,0,0)	213.58	112.39 (72.25)	129.51 (50.21)	5369.7	3389.9	0.010 (0.049)	0.031 (0.119)	0.056 (0.166)	0.082 (0.225)	0.098 (0.269)
(0,0,1.2,0,0)	150.77	92.23 (36.35)	107.34 (17.01)	1380.8	342.7	0.027 (0.142)	0.083 (0.293)	0.134 (0.393)	0.179 (0.480)	0.218 (0.559)
(0,0,1.5,0,0)	105.68	97.68 (4.42)	98.73 (3.55)	25.1	14.2	0.119 (0.202)	0.289 (0.413)	0.468 (0.577)	0.640 (0.730)	0.7€60 (0.846)
(0,0,0,1.1,0)	170.52	88.24 (44.05)	109.18 (20.15)	2076.6	490.1	0.017 (0.132)	0.050 (0.264)	0.087 (0.364)	0.127 (0.445)	0.159 (0.496)
(0,0,0,1.2,0)	116.24	95.91 (12.17)	101.42 (5.51)	164.8	32.4	0.070 (0.287)	0.202 (0.492)	0.304 (0.626)	0.419 (0.723)	0.514 (0.797)
(0,0,0,1.5,0)	101.29	96.08 (2.59)	96.04 (3.31)	22.1	26.6	0.012 (0.016)	0.065 (0.071)	0.232 (0.240)	0.461 (0.475)	0.705 (0.729)
(0,0,0,0,1.1)	203.23	100.56 (62.66)	119.71 (36.92)	3922.7	1750.2	0.017 (0.075)	0.037 (0.174)	0.063 (0.248)	0.084 (0.308)	0.100 (0.350)
(0,0,0,0,1.2)	136.76	90.65 (25.51)	104.22 (12.51)	737.4	174.3	0.029 (0.208)	0.097 (0.398)	0.152 (0.524)	0.218 (0.609)	0.272 (0.671)
(0,0,0,0,1.5)	102.88	97.10 (3.09)	97.19 (3.96)	18.2	23.6	0.095 (0.134)	0.244 (0.302)	0.422 (0.512)	0.631 (0.674)	0.810 (0.823)
(1.1,1.1,1.1, 1.1,1.1)	139.07	92.24 (27.44)	104.47 (11.88)	812.7	161.1	0.048 (0.202)	0.125 (0.367)	0.194 (0.507)	0.252 (0.593)	0.302 (0.664)
(1.2,1.2,1.2, 1.2,1.2)	106.03	97.25 (5.39)	98.81 (3.54)	36.6	14.1	0.092 (0.243)	0.272 (0.448)	0.438 (0.595)	0.605 (0.728)	0.743 (0.844)
(1.5,1.5,1.5, 1.5,1.5)	101.01	96.10 (2.42)	96.31 (1.79)	21.1	16.8	0.005 (0.008)	0.040 (0.040)	0.212 (0.237)	0.430 (0.478)	0.713 (0.757)

Experiment has a step-change disturbance when $\delta > 1$; standard deviations are shown in parentheses.

Moreover, $MSE(\hat{\tau}_{LT})$ is smaller than $MSE(\hat{\tau}_{SC})$ in most of the cases.

Finally, according to the results obtained from simulation experiments 3 and 4, one can conclude that when the real change type is a linear-trend disturbance, the change-point estimator designed for linear trend outperforms the change-point estimator designed for step-change disturbances, regardless of the shift magnitudes and the process dimension.

Conclusions

When a control chart generates a signal due to an out-of-control condition, process engineers initiate a search to identify and to eliminate the root causes of process variation. However, the signaling time is different from the first time that a change manifests itself into a process (change point), and in most cases, there is a considerable time lag between them. Therefore, knowing the exact time of a change in a process would simplify the search procedure to improve the quality of the process by eliminating the causes of process variation.

In this paper, a new control chart was initially proposed based on a transformation method to monitor multi-attribute processes. The transformation method was used to eliminate the skewness of multivariate binomial processes. Then, the MLE of change points designed for both step changes and linear-trend disturbances were derived. Using Monte Carlo simulation experiments involving multivariate binomial processes with two and five attributes, the performances of the

Table 4 Estimated change points and precision performances of two different MLEs in simulation experiment 3

β	*E(T)*	$E(\hat{\tau}_{LT})$	$E(\hat{\tau}_{SC})$	MSE$(\hat{\tau}_{LT})$	MSE$(\hat{\tau}_{SC})$	*P0*	*P1*	*P2*	*P3*	*P4*
(0.001,0)	145.66	108.45 (19.97)	119.61 (15.11)	470.24	612.85	0.023 (0.011)	0.075 (0.032)	0.126 (0.059)	0.168 (0.084)	0.200 (0.102)
(0.002,0)	128.97	105.07 (12.86)	111.78 (9.41)	190.87	227.17	0.038 (0.017)	0.100 (0.051)	0.168 (0.098)	0.229 (0.136)	0.296 (0.165)
(0.003,0)	121.88	103.61 (10.19)	108.36 (8.64)	116.80	144.36	0.048 (0.028)	0.125 (0.077)	0.215 (0.119)	0.295 (0.159)	0.373 (0.217)
(0.005,0)	115.37	102.59 (7.83)	105.50 (7.01)	67.93	79.27	0.060 (0.033)	0.174 (0.099)	0.288 (0.164)	0.394 (0.236)	0.470 (0.315)
(0.01,0)	109.42	101.29 (5.64)	102.80 (4.96)	33.45	32.47	0.093 (0.072)	0.277 (0.179)	0.440 (0.309)	0.544 (0.476)	0.660 (0.625)
(0.05,0)	102.91	98.90 (4.66)	98.83 (5.24)	22.89	28.84	0.422 (0.404)	0.813 (0.775)	0.896 (0.875)	0.933 (0.902)	0.950 (0.924)
(0,0.001)	137.37	106.01 (16.52)	115.31 (12.36)	308.91	387.06	0.027 (0.014)	0.093 (0.051)	0.144 (0.072)	0.190 (0.099)	0.222 (0.123)
(0,0.002)	123.21	103.47 (11.22)	109.04 (8.78)	137.92	158.76	0.042 (0.028)	0.120 (0.067)	0.201 (0.107)	0.280 (0.164)	0.352 (0.213)
(0,0.003)	117.82	103.33 (8.29)	106.79 (7.73)	79.76	105.81	0.056 (0.031)	0.163 (0.078)	0.254 (0.124)	0.340 (0.188)	0.421 (0.262)
(0,0.005)	112.17	102.15 (7.12)	104.43 (5.76)	55.23	52.86	0.076 (0.041)	0.198 (0.126)	0.295 (0.214)	0.399 (0.301)	0.528 (0.424)
(0,0.01)	107.29	100.71 (5.18)	101.93 (3.79)	27.37	18.09	0.134 (0.084)	0.305 (0.263)	0.497 (0.457)	0.666 (0.650)	0.792 (0.801)
(0,0.05)	102.29	98.80 (3.52)	98.48 (4.59)	13.82	23.36	0.400 (0.416)	0.810 (0.764)	0.908 (0.866)	0.940 (0.905)	0.957 (0.928)
(0.001,0.001)	134.96	105.96 (14.85)	114.61 (11.40)	255.86	343.34	0.026 (0.016)	0.082 (0.046)	0.144 (0.081)	0.191 (0.111)	0.242 (0.132)
(0.002,0.002)	121.92	103.10 (11.79)	108.24 (8.94)	148.65	147.84	0.050 (0.024)	0.121 (0.060)	0.212 (0.104)	0.298 (0.158)	0.367 (0.212)
(0.003,0.003)	116.38	102.74 (9.33)	106.18 (7.39)	94.53	92.86	0.040 (0.28)	0.131 (0.080)	0.229 (0.149)	0.306 (0.224)	0.398 (0.278)
(0.005,0.005)	111.50	101.52 (7.13)	103.78 (5.27)	53.10	42.09	0.067 (0.051)	0.230 (0.144)	0.350 (0.256)	0.452 (0.366)	0.552 (0.471)
(0.01,0.01)	107.03	100.59 (4.79)	101.40 (5.54)	23.31	32.67	0.112 (0.094)	0.334 (0.277)	0.548 (0.492)	0.713 (0.692)	0.824 (0.821)
(0.05,0.05)	102.18	98.40 (3.67)	97.98 (5.70)	15.99	36.51	0.372 (0.380)	0.762 (0.737)	0.876 (0.839)	0.904 (0.867)	0.933 (0.892)

Experiment has a linear-trend disturbance; standard deviations are shown in parentheses.

change-point estimators were evaluated by comparing them in the presence of real step-change and real linear-trend disturbances. The results showed that the change-point estimator designed for step changes outperforms the change-point estimator designed for linear-trend disturbances when the true change type is a step-change disturbance, regardless of the shift magnitudes and process dimensions. In contrast, the change-point estimator designed for linear-trend disturbances outperforms the change-point estimator designed for step changes when the true change type is a linear-trend disturbance, regardless of the shift magnitudes and process dimensions.

Appendix

Appendix 1

Denoting the MLE of the change-point estimator designed for step changes by $\hat{\tau}_{SC}$, the likelihood function for the transformed vector $\mathbf{Y_{ij}}$ is (Johnson and Wichern 2007)

$$L(\tau_{SC}, \boldsymbol{\mu}_{1,Y}, \mathbf{Cov}_{1,Y}|\mathbf{Y}) = \prod_{i=1}^{\tau_{SC}} \frac{1}{(2\pi)^{q/2}\left|\mathbf{Cov}_{0,Y}\right|^{1/2}} e^{-n/2\left(\overline{\mathbf{Y}}_i-\boldsymbol{\mu}_{0,Y}\right)'\mathbf{Cov}_{0,Y}^{-1}\left(\overline{\mathbf{Y}}_i-\boldsymbol{\mu}_{0,Y}\right)} \times \prod_{i=\tau_{SC}+1}^{S} \frac{1}{(2\pi)^{q/2}\left|\mathbf{Cov}_{1,Y}\right|^{1/2}} e^{-n/2\left(\overline{\mathbf{Y}}_i-\boldsymbol{\mu}_{1,Y}\right)'\mathbf{Cov}_{1,Y}^{-1}\left(\overline{\mathbf{Y}}_i-\boldsymbol{\mu}_{1,Y}\right)}, \tag{9}$$

where $\boldsymbol{\mu}_{0,Y}$, $\boldsymbol{\mu}_{1,Y}$, $\mathbf{Cov}_{0,Y}$, and $\mathbf{Cov}_{1,Y}$ are in-control and out-of-control mean vector and covariance matrix of vector $\mathbf{Y}$, respectively. The logarithm of the likelihood function when the constant term is shown by K is

$$\begin{aligned}\mathrm{Ln}\left(L\left(\tau_{SC}, \boldsymbol{\mu}_{1,Y}, \mathbf{Cov}_{1,Y}|\mathbf{Y}\right)\right) &= K - \frac{\tau_{SC}}{2}\mathrm{Ln}\left|\mathbf{Cov}_{0,Y}\right| \\ &- \frac{n}{2}\sum_{i=1}^{\tau_{SC}}\left[\left(\overline{\mathbf{Y}}_i-\boldsymbol{\mu}_{0,Y}\right)'\mathbf{Cov}_{0,Y}^{-1}\left(\overline{\mathbf{Y}}_i-\boldsymbol{\mu}_{0,Y}\right)\right] \\ &- \frac{(S-\tau_{SC})}{2}\mathrm{Ln}\left|\mathbf{Cov}_{1,Y}\right| \\ &- \frac{n}{2}\sum_{i=\tau_{SC}+1}^{S}\left[\left(\overline{\mathbf{Y}}_i-\boldsymbol{\mu}_{1,Y}\right)'\mathbf{Cov}_{1,Y}^{-1}\left(\overline{\mathbf{Y}}_i-\boldsymbol{\mu}_{1,Y}\right)\right].\end{aligned} \tag{10}$$

Table 5 Estimated change points and precision performances of two different MLEs in simulation experiment 4

β	$E(T)$	$E(\hat{\tau}_{LT})$	$E(\hat{\tau}_{SC})$	$MSE(\hat{\tau}_{LT})$	$MSE(\hat{\tau}_{SC})$	*P0*	*P1*	*P2*	*P3*	*P4*
(0.001,0,0,0,0)	151.89	114.67 (23.62)	127.18 (18.73)	763.64	1089.2	0.011 (0.014)	0.048 (0.036)	0.096 (0.057)	0.133 (0.070)	0.177 (0.083)
(0.003,0,0,0,0)	125.85	108.14 (12.14)	113.12 (8.68)	213.47	247.3	0.037 (0.016)	0.100 (0.040)	0.164 (0.065)	0.214 (0.107)	0.267 (0.148)
(0.005,0,0,0,0)	118.16	106.15 (7.81)	109.08 (6.15)	98.85	120.4	0.037 (0.017)	0.114 (0.052)	0.187 (0.100)	0.248 (0.142)	0.319 (0.182)
(0.01,0,0,0,0)	110.71	102.83 (4.82)	104.14 (3.82)	31.26	31.7	0.054 (0.036)	0.171 (0.120)	0.290 (0.221)	0.413 (0.347)	0.534 (0.486)
(0,0.001,0,0,0)	146.88	114.19 (21.85)	124.01 (16.81)	678.51	859.2	0.019 (0.009)	0.049 (0.034)	0.100 (0.056)	0.145 (0.077)	0.185 (0.101)
(0,0.003,0,0,0)	123.95	107.48 (10.49)	111.79 (8.30)	165.91	207.7	0.034 (0.018)	0.096 (0.051)	0.172 (0.083)	0.234 (0.122)	0.295 (0.155)
(0,0.005,0,0,0)	116.54	105.02 (7.52)	108.11 (5.74)	81.71	98.6	0.041 (0.032)	0.130 (0.081)	0.221 (0.120)	0.296 (0.176)	0.363 (0.229)
(0,0.01,0,0,0)	109.87	102.51 (4.08)	103.54 (4.89)	22.90	36.5	0.068 (0.047)	0.211 (0.143)	0.352 (0.270)	0.485 (0.407)	0.621 (0.558)
(0,0,0.001,0,0)	141.68	112.81 (19.59)	121.18 (14.98)	547.65	672.7	0.018 (0.014)	0.062 (0.030)	0.099 (0.059)	0.146 (0.085)	0.205 (0.114)
(0,0,0.003,0,0)	120.06	106.96 (10.03)	110.08 (7.49)	149.05	157.8	0.041 (0.016)	0.112 (0.047)	0.160 (0.087)	0.225 (0.130)	0.286 (0.174)
(0,0,0.005,0,0)	113.69	104.31 (6.48)	106.10 (5.28)	60.59	65.1	0.041 (0.037)	0.133 (0.084)	0.227 (0.146)	0.308 (0.230)	0.396 (0.319)
(0,0,0.01,0,0)	108.42	101.56 (4.47)	102.38 (3.49)	22.39	17.9	0.101 (0.084)	0.282 (0.243)	0.457 (0.429)	0.642 (0.616)	0.777 (0.751)
(0,0,0,0.001,0)	152.12	113.54 (23.31)	125.78 (18.38)	726.13	1001.9	0.015 (0.014)	0.066 (0.042)	0.106 (0.058)	0.158 (0.080)	0.189 (0.092)
(0,0,0,0.003,0)	124.59	107.36 (11.27)	112.24 (8.61)	180.97	224.0	0.036 (0.017)	0.096 (0.052)	0.156 (0.086)	0.219 (0.124)	0.278 (0.158)
(0,0,0,0.005,0)	117.59	105.96 (7.76)	108.94 (6.30)	95.67	119.6	0.039 (0.0.24)	0.109 (0.071)	0.189 (0.111)	0.249 (0.153)	0.314 (0.200)
(0,0,0,0.01,0)	110.68	103.01 (4.87)	104.22 (4.09)	32.72	34.5	0.065 (0.040)	0.174 (0.119)	0.279 (0.200)	0.412 (0.317)	0.547 (0.476)
(0,0,0,0,0.001)	145.57	114.13 (21.90)	123.88 (15.84)	678.77	820.8	0.032 (0.009)	0.073 (0.030)	0.106 (0.056)	0.152 (0.072)	0.197 (0.099)
(0,0,0,0,0.003)	122.04	107.10 (10.58)	111.42 (7.48)	162.38	186.30	0.032 (0.013)	0.083 (0.034)	0.144 (0.075)	0.213 (0.112)	0.269 (0.150)
(0,0,0,0,0.005)	115.36	105.18 (6.63)	107.15 (5.37)	70.77	84.21	0.038 (0.021)	0.131 (0.079)	0.231 (0.137)	0.302 (0.192)	0.366 (0.241)
(0,0,0,0,0.01)	109.15	102.28 (3.95)	103.13 (3.12)	20.76	19.48	0.060 (0.059)	0.237 (0.200)	0.387 (0.340)	0.551 (0.501)	0.705 (0.637)
(0.001,0.001, 0.001,0.001,0.001)	136.81	110.99 (15.57)	118.66 (12.52)	395.10	504.68	0.020 (0.015)	0.082 (0.039)	0.130 (0.062)	0.173 (0.080)	0.212 (0.105)
(0.003,0.003, 0.003,0.003,0.003)	117.34	105.87 (7.63)	108.65 (5.88)	92.57	109.36	0.050 (0.015)	0.139 (0.058)	0.203 (0.098)	0.273 (0.148)	0.333 (0.198)
(0.005,0.005, 0.005,0.005,0.005)	111.91	103.97 (5.72)	105.09 (4.63)	48.50	47.36	0.041 (0.031)	0.132 (0.100)	0.236 (0.192)	0.345 (0.269)	0.441 (0.378)
(0.01,0.01, 0.01,0.01,0.01)	107.45	101.19 (2.81)	101.85 (2.51)	9.27	9.68	0.106 (0.101)	0.346 (0.318)	0.570 (0.558)	0.765 (0.750)	0.905 (0.877)

Experiment has a linear-trend disturbance; standard deviations are shown in parentheses.

There are three unknown parameters (τ_{sc}, $\boldsymbol{\mu}_{1,Y}$, and $\mathbf{Cov}_{1,Y}$) in Equation 10 that should be estimated. In order to find $\hat{\tau}_{SC}$, one must first estimate $\boldsymbol{\mu}_{1,Y}$ and $\mathbf{Cov}_{1,Y}$ for all possible values of the change point. To do this, the MLE of the out-of-control mean vector and covariance matrix, denoted by $\hat{\boldsymbol{\mu}}_{1,Y}$ and $\hat{\mathbf{Cov}}_{1,Y}$, respectively, are first derived as

$$\hat{\boldsymbol{\mu}}_{1,Y} = \frac{\sum_{i=\tau_{SC}+1}^{S} \overline{\mathbf{Y}_i}}{S-\tau_{SC}}, \tag{11}$$

$$\hat{\mathbf{Cov}}_{1,Y} = \frac{n}{S-\tau_{SC}} \sum_{i=\tau_{SC}+1}^{S} \left(\overline{\mathbf{Y}_i} - \hat{\boldsymbol{\mu}}_{1,Y}\right)' \left(\overline{\mathbf{Y}_i} - \hat{\boldsymbol{\mu}}_{1,Y}\right). \tag{12}$$

Then, the estimated unknown parameters in Equations 11 and 12 for all the potential values of τ_{sc} are obtained and are substituted in Equation 13 to find the MLE of the change point:

$$\hat{\tau}_{SC} = arg \underset{0 \le t < S}{\mathrm{Min}} \left\{ \begin{array}{l} \frac{t}{2}\mathrm{Ln}\left|\mathbf{Cov}_{0,Y}\right| + \frac{n}{2}\sum_{i=1}^{t}\left(\overline{\mathbf{Y}_i} - \boldsymbol{\mu}_{0,Y}\right)' \mathbf{Cov}_{0,Y}^{-1}\left(\overline{\mathbf{Y}_i} - \boldsymbol{\mu}_{0,Y}\right) \\ + \frac{S-t}{2}\mathrm{Ln}\left|\hat{\mathbf{Cov}}_{1,Y}\right| + \frac{n}{2}\sum_{i=t+1}^{S}\left(\overline{\mathbf{Y}_i} - \hat{\boldsymbol{\mu}}_{1,Y}\right)' \hat{\mathbf{Cov}}_{1,Y}^{-1}\left(\overline{\mathbf{Y}_i} - \hat{\boldsymbol{\mu}}_{1,Y}\right) \end{array} \right\} \tag{13}$$

Appendix 2

Denoting the MLE of the change-point estimator designed for linear-trend disturbances by $\hat{\tau}_{LT}$, the likelihood function of the transformed vector $\mathbf{Y}_{ij}$ is

$$\begin{aligned} L\left(\tau_{LT}, \boldsymbol{\mu}_{1,Y}, \mathbf{Cov}_{1,Y} \middle| \mathbf{Y}\right) &= \prod_{i=1}^{\tau_{LT}} \frac{1}{(2\pi)^{q/2}\left|\mathbf{Cov}_{0,Y}\right|^{1/2}} e^{-n/2\left(\overline{\mathbf{Y}_i} - \boldsymbol{\mu}_{0,Y}\right)' \mathbf{Cov}_{0,Y}^{-1}\left(\overline{\mathbf{Y}_i} - \boldsymbol{\mu}_{0,Y}\right)} \\ &\times \prod_{i=\tau_{LT}+1}^{S} \frac{1}{(2\pi)^{q/2}\left|\mathbf{Cov}_{1,Y}\right|^{1/2}} e^{-n/2\left(\overline{\mathbf{Y}_i} - \boldsymbol{\mu}_{0,Y} - \boldsymbol{\beta}(i-\tau_{LT})\right)' \mathbf{Cov}_{1,Y}^{-1}\left(\overline{\mathbf{Y}_i} - \boldsymbol{\mu}_{0,Y} - \boldsymbol{\beta}(i-\tau_{LT})\right)}, \end{aligned} \tag{14}$$

where $\boldsymbol{\mu}_{0,Y}$ and $\mathbf{Cov}_{0,Y}$ are the in-control mean vector and covariance matrix, and $\boldsymbol{\mu}_{1,Y}$ and $\mathbf{Cov}_{1,Y}$ are the out-of-control mean-vector and covariance matrix of vector $\mathbf{Y}$, respectively. As a result, the logarithm of the likelihood function in which the constant term is denoted by K is

$$\begin{aligned} \mathrm{Ln}\left(L\left(\tau_{LT}, \boldsymbol{\mu}_{1,Y}, \mathbf{Cov}_{1,Y} \middle| \mathbf{Y}\right)\right) &= K - \frac{\tau_{LT}}{2}\mathrm{Ln}\left|\mathbf{Cov}_{0,Y}\right| \\ &- \frac{n}{2}\sum_{i=1}^{\tau_{LT}}\left[\left(\overline{\mathbf{Y}_i} - \boldsymbol{\mu}_{0,Y}\right)' \mathbf{Cov}_{0,Y}^{-1}\left(\overline{\mathbf{Y}_i} - \boldsymbol{\mu}_{0,Y}\right)\right] - \frac{(S-\tau_{LT})}{2}\mathrm{Ln}\left|\mathbf{Cov}_{1,Y}\right| \\ &- \frac{n}{2}\sum_{i=\tau_{LT}+1}^{S}\left[\left(\overline{\mathbf{Y}_i} - \boldsymbol{\mu}_{0,Y} - \boldsymbol{\beta}(i-\tau_{LT})\right)' \mathbf{Cov}_{1,Y}^{-1}\left(\overline{\mathbf{Y}_i} - \boldsymbol{\mu}_{0,Y} - \boldsymbol{\beta}(i-\tau_{LT})\right)\right]. \end{aligned} \tag{15}$$

There are three unknown parameters τ_{LT}, $\boldsymbol{\beta}$, and $\mathbf{Cov}_{1,Y}$ in Equation 15. Again, we first calculate the MLEs of the unknown parameters in Equations 16 and 17 for all of the possible values of the change point:

$$\hat{\boldsymbol{\beta}} = \frac{\sum_{i=\tau_{LT}+1}^{S} (i-\tau_{LT})\left(\overline{\mathbf{Y}_i} - \boldsymbol{\mu}_{0,Y}\right)}{\sum_{i=\tau_{LT}+1}^{S} (i-\tau_{LT})^2}, \tag{16}$$

$$\begin{aligned} \hat{\mathbf{Cov}}_{1,Y} &= \frac{n}{S-\tau_{LT}} \sum_{i=\tau_{LT}+1}^{S} \left(\overline{\mathbf{Y}_i} - \boldsymbol{\mu}_{0,Y} - \hat{\boldsymbol{\beta}}(i-\tau_{LT})\right)' \\ &\times \left(\overline{\mathbf{Y}_i} - \boldsymbol{\mu}_{0,Y} - \hat{\boldsymbol{\beta}}(i-\tau_{LT})\right). \end{aligned} \tag{17}$$

Then, the estimated parameters $\hat{\boldsymbol{\beta}}$ and $\hat{\mathbf{Cov}}_{1,Y}$ are substituted in Equation 18 in order to find the MLE of the change point:

$$\hat{\tau}_{LT} = arg \underset{0 \le t \le S}{\mathrm{Min}} \left\{ \begin{array}{l} \frac{t}{2}\mathrm{Ln}\left|\mathbf{Cov}_{0,Y}\right| + \frac{n}{2}\sum_{i=1}^{t}\left(\overline{\mathbf{Y}_i} - \boldsymbol{\mu}_{0,Y}\right)' \mathbf{Cov}_{0,Y}^{-1}\left(\overline{\mathbf{Y}_i} - \boldsymbol{\mu}_{0,Y}\right) \\ + \frac{S-t}{2}\mathrm{Ln}\left|\hat{\mathbf{Cov}}_{1,Y}\right| \\ + \frac{n}{2}\sum_{i=t+1}^{S}\left(\overline{\mathbf{Y}_i} - \boldsymbol{\mu}_{0,Y} - \hat{\boldsymbol{\beta}}(i-t)\right)' \hat{\mathbf{Cov}}_{1,Y}^{-1} \\ \times \left(\overline{\mathbf{Y}_i} - \boldsymbol{\mu}_{0,Y} - \hat{\boldsymbol{\beta}}(i-t)\right) \end{array} \right\}. \tag{18}$$

Competing interests

The authors declare that they have no competing interests.

Authors' contributions

STAN and MK first proposed a new control chart to monitor multi-binomial processes based on a transformation method. Then, they derived the maximum likelihood estimators of change points designed for both step changes and linear-trend disturbances. At the end, they evaluated and compared the performances of the proposed change-point estimators using some Monte Carlo simulation experiments, considering that the real change type presented in a process are of either a step change or a linear-trend disturbance. Based on the results obtained, the change-point estimator designed for step changes outperforms the change-point estimator designed for linear-trend disturbances, when the real change type is a step change. In contrast, the change-point estimator designed for linear-trend disturbances outperforms the change-point estimator designed for step changes, when the real change type is a linear-trend disturbance. All authors read and approved the final manuscript.

Authors' information

STAN is a professor at the Department of Industrial Engineering, Sharif University of Technology. His research interests are in the areas of simulation modeling and analysis, applied statistics, and multivariate quality control. Before joining the Sharif University of Technology faculty, he worked as a systems engineer and quality control manager for Iranian Electric Meters Company. He received his degree in Bachelor of Science in Industrial Engineering from the Sharif University of Technology in 1979, his Master's and his Ph.D. degrees both in Industrial Engineering from the West Virginia University in 1989 and 1992, respectively. He is a member of απμ. MK received his Master's of Science degree in Industrial Engineering from the

Sharif University of Technology in 2012. His research interests are in the areas of applied statistics, quality engineering, and operations research.

Acknowledgements
The authors are thankful for the constructive comments of anonymous reviewers as well as the editorial changes suggested by the language editor that certainly improved the presentation of the paper.

References

Abbasi B, Niaki STA, Seyedan SE (2011) A simple transformation method in skewness reduction. Int J Eng Trans A: Basics 24:169–175

Cario MC, Nelson BL (1997) Modeling and generating random vectors with arbitrary marginal distributions and correlation matrix. Technical Report, Department of Industrial Engineering and Management Sciences. Northwestern University, Evanston

Hotelling H (1947) Multivariate quality control. In: Eisenhart C, Hastay MW, Wallis WA (ed) Techniques of statistical analysis. McGraw-Hill, New York, pp 111–184

Jarque CM, Bera AK (1987) A test for normality of observations and regression residuals. Int Stat Rev 55:163–172

Johnson RA, Wichern DW (2007) Applied multivariate statistical analysis, 6th edition. Prentice Hall, Upper Saddle River

Lowry CA, Montgomery DC (1995) A review of multivariate control charts. IIE Trans 27:800–810

Niaki STA, Abbasi B (2007) Skewness reduction approach in multi-attribute process monitoring. Comm Statist Theory Methods 36:2313–2325

Niaki STA, Abbasi B (2008) Generating correlation matrices for normal random vectors in NORTA algorithm using artificial neural networks. J Uncertain Syst 2:192–201

Niaki STA, Khedmati M (2012a) Detecting and estimating the time of a step-change in multivariate Poisson processes. Sci Iran Tans E 19:862–871

Niaki STA, Khedmati M (2012b) Estimating the change point of the parameter vector of multivariate Poisson processes monitored by a multi-attribute T^2 control chart. Int J Adv Manuf Technol. doi:10.1007/s00170-012-4128-x

Nishina K (1992) A comparison of control charts from the viewpoint of change point estimation. Qual Reliab Eng Int 8:537–541

Page ES (1954) Continuous inspection schemes. Biometrika 41:1–9

Perry MB, Pignatiello JJ (2005) Estimation of the change point of the process fraction nonconforming in SPC applications. Int J Reliab Qual Saf Eng 12:95–110

Perry MB, Pignatiello JJ, Simpson JR (2007) Estimating the change point of the process fraction non-conforming with a monotonic change disturbance in SPC. Qual Reliab Eng Int 23:327–339

Samuel TR, Pignatiello JJ (2001) Identifying the time of a step change in the process fraction nonconforming. Qual Eng 13:357–365

Zandi F, Niaki STA, Nayeri MA, Fathi M (2011) Change-point estimation of the process fraction non-conforming with a linear trend in statistical process control. Int J Comput Integr Manuf 24:939–947

8

Process capability improvement of an engine connecting rod machining process

GVSS Sharma[1*] and P Srinivasa Rao[2]

Abstract

Statistical process control is an excellent quality assurance tool to improve the quality of manufacture and ultimately scores on end-customer satisfaction. SPC uses process monitoring charts to record the key quality characteristics (KQCs) of the component in manufacture. This paper elaborates on one such KQC of the manufacturing of a connecting rod of an internal combustion engine. Here the journey to attain the process potential capability index (C_p) and the process performance capability index (C_{pk}) values greater than 1.33 is elaborated by identifying the root cause through quality control tools like the cause-and-effect diagram and examining each cause one after another. In this paper, the define-measure-analyze-improve-control (DMAIC) approach is employed. The definition phase starts with process mapping and identifying the KQC. The next phase is the measurement phase comprising the cause-and-effect diagram and data collection of KQC measurements. Then follows the analysis phase where the process potential and performance capability indices are calculated, followed by the analysis of variance (ANOVA) of the mean values. Finally, the process monitoring charts are used to control the process and prevent any deviations. By using this DMAIC approach, standard deviation is reduced from 0.48 to 0.048, the C_p values from 0.12 to 1.72, and the C_{pk} values from 0.12 to 1.37, respectively.

Keywords: Key quality characteristic; Cause-and-effect diagram; Statistical process control; Process monitoring charts; Failure modes and effects analysis; Analysis of variance

Introduction

One of the major manufacturing processes in engine manufacturing is that of connecting rod manufacturing. This paper implements the define-measure-analyze-improve-control (DMAIC) approach to improve the capability of connecting rod manufacturing process by reducing the thickness variations from a nominal value.

Process mapping and identifying key quality characteristics (KQCs) is carried out in the 'definition phase', while estimation of process capability indices is carried out in the 'measure phase'. The one-way analysis of variance (ANOVA) method of investigation to test for the differences between the manufacturing data is employed in the 'analysis phase'. Finally, the process monitoring chart (PMC) for the thrust face thickness is employed in the 'improve and control phase'.

Statistical process control studies form the basic tool for obtaining the required process capability confidence levels. The various process capability indices are defined as follows:

$$C_P = \frac{\text{USL–LSL}}{6\sigma} \tag{1}$$

$$C_{PKU} = \frac{\text{USL}-\mu}{3\sigma} \tag{2}$$

$$C_{PKL} = \frac{\mu-\text{LSL}}{3\sigma} \tag{3}$$

$$C_{PK} = \min\left\{\frac{\text{USL}-\mu}{3\sigma}, \frac{\mu-\text{LSL}}{3\sigma}\right\} \tag{4}$$

where USL and LSL are the upper and lower specification limits, μ = process mean, and σ = standard deviation.

The term C_p denotes the process potential capability index, and similarly, the term C_{pk} denotes the process performance capability index. C_p gives an indication of

* Correspondence: sarma.gvss@gmail.com
[1]Department of Mechanical Engineering, GMR Institute of Technology, GMR Nagar, Rajam-532127, Andhra Pradesh State, India
Full list of author information is available at the end of the article

the dispersion of the product dimensional values within the specified tolerance zone during the manufacturing process. Similarly, the index C_{pk} denotes for the centering of the manufacturing process with respect to the mean of the specified dimensional tolerance zone of the product. C_{pk} gives us an idea on whether the manufacturing process is performing at the middle of the tolerance zone or nearer the upper or lower tolerance limits. If the manufacturing process is nearer the lower limit, then the process performance capability index is given by C_{pkl}, and if the manufacturing process is nearer the upper limit, then the process performance capability index is given by C_{pku}. As a measure of precautionary safety, the minimum value between the two is taken as the value of C_{pk}.

Literature survey

The literature survey of process capability improvements using the DMAIC approach is discussed in this section. Schilling (1994) emphasized how process control is better than the traditional sampling techniques. During the same era, Locke (1994), in his paper titled 'Statistical measurement control,' emphasized the importance of process charts, cause-and-effect considerations, and control charting. After primitive studies on statistical quality control, Lin (2004) had shed some light on process capability indices for normal distribution. Tong et al. (2004) suggested that a DMAIC approach is useful for printed circuit board quality improvement. They also proved how design of experiments is one of the core statistical tools for six-sigma improvement. Subsequently, Li et al. (2008) once again proved the importance of DMAIC approach to improve the capability of surface mount technology in a solder printing process. Hwang (2006) discussed the DMAIC phases in detail with application to manufacturing execution system. Gentili et al. (2006) applied the DMAIC process to a mechanical manufacturing process line, which manufactures both professional and simple kitchen knives. Sahay et al. (2011) once again brought the DMAIC approach into use to analyze the manufacturing lines of a brake lever at a Connecticut automotive component manufacturing company. Singh (2011) investigated the process capability of polyjet printing for plastic components. In his observation, he traveled the improvement journey of the process of critical dimensions and their C_{pk} value attainment greater than 1.33, which is considered to be an industrial benchmark. In recent studies conducted by Lin et al. (2013), they focused on turbine engine blade inspection as it is a key aspect of engine quality. They elaborated on the accurate yield assessment of processes with multiple characteristics like the turbine blade manufacturing process. Kumaravadivel and Natarajan (2013) dealt with the application of the six-sigma methodology to the flywheel casting process. The primary problem solving tools used were the process map, cause-and-effect matrix, and failure modes and effects analysis (FMEA). Sharma et al. (2013a), b in their papers adopted the DMAIC approach to solve the bolt hole center distance and crankpin bore honing operations of the connecting rod manufacturing process. Chen et al. (2013) discussed the application of ANOVA methodology to find significant parameters that affect the part's quality indices with respect to plastic injection molding.

A careful study from the above literature reveals that the DMAIC approach is the best methodology for problem solving tools to improve the manufacturing process capability levels. Hence, this paper focuses on the application of DMAIC approach for process capability improvement of the crankpin bore honing operation of an engine connecting rod manufacturing process.

Definition phase

The definition phase starts with the correct mapping of the machining process flow of the connecting rod.

Process mapping

The process flow chart for the machining line of the connecting rod machining cell consisted of the following machining operation sequence shown in Figure 1. Table 1 describes the machining operations of the connecting rod manufacturing cell.

Identifying KQCs

The acceptable thrust face thickness variation of the connecting rod forgings was limited from 0.5 mm to a maximum of 1 mm. The forgings from these tolerance limits resulted in a thrust face 'unclean' problem and were subsequently rejected. This caused costly repair and rework. Hence, in this regard, this research aims at improving the connecting rod manufacturing process by reducing the thrust face thickness variations early in the stages of the machining process line so that these variations are not carried to the subsequent machining operations down the manufacturing line. Hence, thrust face thickness was of the main concern and identified as a KQC, whose value is equal to $27.000^{(+0.500/0.000)}$ mm. Figure 2 depicts a diagrammatic view of this key process characteristic (KQC).

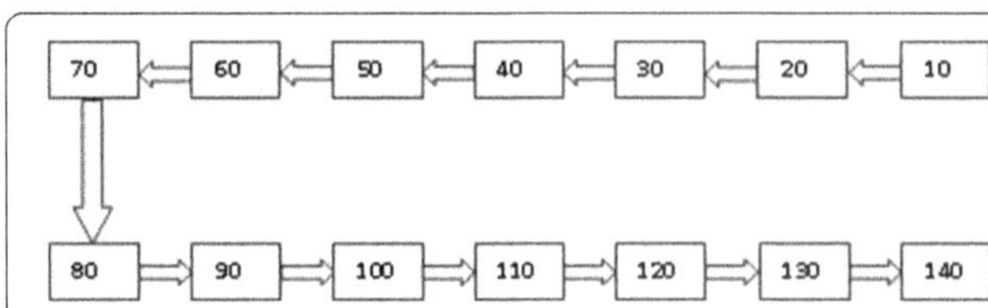

Figure 1 Process flow chart. Refer to Table 1 for the corresponding descriptions of the operations.

Table 1 Machining operations of the connecting rod manufacturing cell

Machining operation number	Description
10	Thrust face rough grinding
20	Gudgeon pin rough boring
30	Crankpin rough boring
40	Side face broaching
50	Finish grinding
60	Bolt hole drilling
70	Key way milling
80	Rod and cap assembly
90	Finish grinding of assembly
100	Finish boring of gudgeon pin
110	Finish boring of crankpin
120	Crankpin bore honing
130	Magnetic crack detection
140	Final quality check, set making, and dispatch to engine assembly line

Measurement phase

In this phase, the thrust face thickness data of 30 nominal consecutive readings is collected and plotted on the process monitoring chart. This data collection was performed in four iterations. In each iteration, the data set of thrust face width measurement readings is taken and analyzed for C_p and C_{pk} values and followed by the suitable corrective action. The corrective action was decided based on the cause-and-effect diagram. After the corrective action was implemented, the next iteration was performed. This procedure was continued until the C_p and C_{pk} values are greater than or equal to 1.33, i.e., up to 4σ quality level as decided by the management of the engine manufacturing plant.

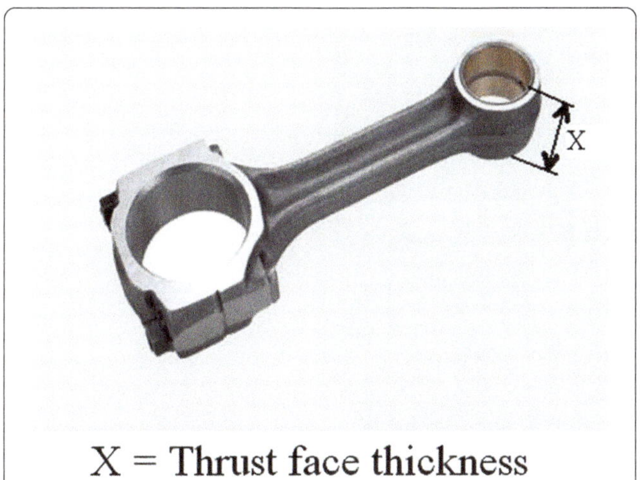

Figure 2 Thrust face thickness.

Cause-and-effect diagram

The key process characteristic identified was the thrust face thickness which is equal to $27.000^{(+0.500/0.000)}$ mm whose machining tolerance zone is equal to 0.500 mm. The C_p value, i.e., the process potential capability index, $\{C_p = (USL - LSL)/6\sigma\}$, was nominally equal to 0.2, which was far below the acceptance level limit greater than 1.33 for the above KQC. The first part of the measurement phase investigation was to track down and differentiate the common causes and special causes involved. For doing so, the cause-and-effect diagram was employed, as shown in Figure 3.

Machine setup

The machine employed was a vertical grinding machine which consisted of in-process sensors which sensed the amount of material removed during the grinding process. Figure 4 shows a picture of the machine loading and unloading platform.

Process failure modes and effects analysis

A FMEA sheet for thrust face thickness is shown in Table 2. From the causes enumerated in the cause-and-effect diagram, the failure modes and effects analysis was performed keeping in view the KPC under study. The risk priority numbers which were above 100 (as per the decision of the management) were considered to be the criteria for implementing the corrective action. It can be noticed that the highest risk priority number (RPN) is for in-process sensing gauge malfunction, followed by improper grinding wheel dressing and worn-out fixture rest pads. Hence, to mitigate these causes, a necessary action plan was devised. After performing the necessary action, the potential process FMEA was once again performed, and the risk priority numbers were recalculated until they attained a RPN below the 100 level mark.

Data collection

Data collection of the key process characteristic was performed for 32 consecutive machined components. Data collection was performed in four iterations spanning a period of 3 weeks, i.e., about 2,500 consecutive components. The data is tabulated in Table 3. The data in Table 3 was plotted on the process monitoring chart with the number of components in the x-axis and component dimension on the y-axis and is shown in Figure 5.

Analysis phase

The analysis phase comprises performing the calculations for the C_p and C_{pk} values across each iteration. This was followed by one-way ANOVA method of investigation to test for the differences between the four iterations of the data sets.

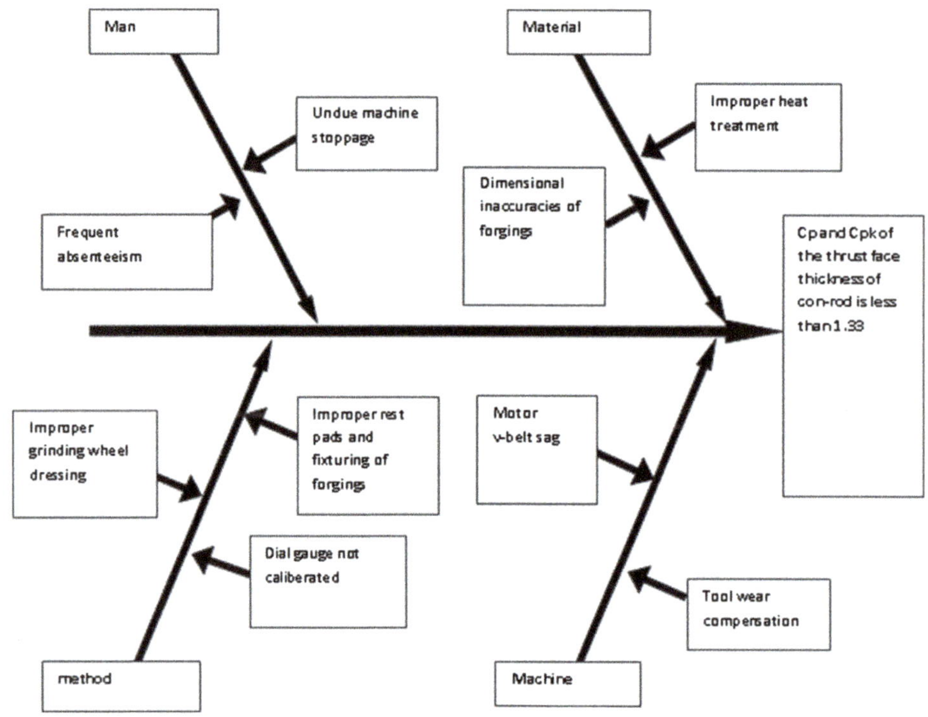

Figure 3 Cause-and-effect diagram showing the variables affecting the thrust face thickness.

Calculations of C_p and C_{pk}

The calculations of C_p and C_{pk} are tabulated in Table 4. From the process monitoring charts and the calculations in Table 4, the following sections are the analysis done for the data set of each iteration.

Iteration 1

The first set of the statistical process capability study is composed of the raw data of the KQC, which depicted the transparent picture of the state of the existing problem. Continuous set of readings of the connecting rod after grinding operation no. 10 were captured with the help of a dial gauge on a metrological surface plate. Hence, it is seen here in the first iteration of statistical process control (SPC) studies that the process is not capable, and the C_p and C_{pk} values of the characteristic under study are 0.20 and 0.12, which are far less than that for process to be capable, i.e., 1.33. Hence, the next

Figure 4 Machine loading and unloading platforms.

Table 2 Process failure modes and effects analysis sheet

Process name	Potential failure	Potential effect	Severity	Potential cause	Occurrence	Current controls	Detection	Risk priority number
Rough grinding of thrust face thickness (operation no. 10)	Thrust face thickness more	Costly rework of the component for attaining correct thrust face thickness	9	In-process sensing gauge malfunction	7	Gauge R&R quality check point	3	189
				Worn-out rest pads of the fixture	6	Fixture calibration procedure document	2	108
		Incorrect location of the component into the gudgeon pin boring fixture in the next operation (no. 20)	9	Grinding wheel pads of the fixture	8	Grinding wheel dressing quality check after production shift changeover	2	144
				Grinding wheel setting procedure not standardized	5	Tool setup control in the check list	2	90
	Thrust face thickness less	Component rejection	6	Improper zero setting and referencing of the grinding wheel positioning	6	Tool setup during shift changeover procedure document	1	36
				Dial gauge not calibrated	5	Gauge measurement system analysis document	1	30
	Thrust face unclean	Component rejection	3	Dimensional inaccuracies of forgings	6	Visual checkpoint	1	18

Table 3 Measured dimensions in tabular form

Series number	Iteration 1	Iteration 2	Iteration 3	Iteration 4
01	26.000	26.500	27.250	27.250
02	27.500	27.250	27.350	27.250
03	27.250	27.250	27.350	27.250
04	27.750	27.500	27.400	27.250
05	27.400	27.400	27.200	27.300
06	26.750	27.750	27.300	27.300
07	27.000	27.000	27.250	27.300
08	27.240	27.240	27.250	27.300
09	27.300	27.300	27.300	27.300
10	27.750	27.000	27.300	27.350
11	26.250	26.750	27.350	27.350
12	27.000	27.000	27.350	27.350
13	27.000	26.900	27.450	27.350
14	27.250	27.250	27.250	27.350
15	27.250	26.900	27.200	27.400
16	27.500	27.500	27.250	27.250
17	27.500	27.500	27.250	27.250
18	27.250	27.250	27.300	27.250
19	27.750	27.750	27.250	27.250
20	27.000	27.000	27.350	27.300
21	26.750	26.750	27.350	27.300
22	27.250	27.250	27.400	27.300
23	27.250	27.250	27.200	27.300
24	27.000	27.000	27.200	27.350
25	27.500	27.500	27.300	27.350
26	27.250	27.000	27.250	27.350
27	27.220	26.900	27.300	27.350
28	27.220	27.250	27.300	27.400
29	27.250	26.900	27.350	27.250
30	27.200	27.500	27.350	27.250
31	26.500	27.500	27.400	27.250
32	26.500	27.250	27.200	27.250

set of data is captured after performing measurement system analysis (MSA) studies in iteration 2 of SPC studies for the KPC under study.

Iteration 2

In this iteration, the data is collected after gauge repeatability and reproducibility (GR&R) was performed for the dial gauge and calibration of the dial gauge as a part of the measurement system analysis procedure. From the set of data in Table 3, it is seen that there is a marginal increase of C_p from 0.20 to 0.28 and C_{pk} from 0.12 to 0.21. This marginal increase is a positive sign, but still, the process is not capable as both C_p and C_{pk} are far less than the desired value of 1.33. This calls for another iteration.

Iteration 3

In this iteration, data is collected after machine preventive maintenance schedule completion and replacement of grinding shoes of the machine as well as rest pads of the fixture. From the set of data in Table 4, it is seen that there is a noticeable increase of C_p from 0.120 to 1.23 and C_{pk} from 0.21 to 0.99. This increase is a positive sign, but still, the process is not capable as both C_p and C_{pk} are far less than the desired value of 1.33. This calls for another iteration, i.e., iteration 4.

Iteration 4

This iteration tackles the 'tool wear compensation' cause. It is a common phenomenon that as any machining operation progresses, there is a calculated wear of the cutting tool responsible for the machining operation. The grinding operation here is no such exception. Hence, here the data was collected after presetting the value for the tool wear compensation knob at 15 μm on the control panel of the machine. This means that after every wear out of 15 μm of the grinding shoes, the grinding wheel surface is lowered by 15 μm in order to compensate for the tool wear so that the dimensions of the product being machined will remain unchanged.

One-way ANOVA method

The one-way ANOVA method of investigation is adopted to test for the differences between the four iterations of data collected.

Procedure describing one-way ANOVA

In general, one-way ANOVA technique is used to study the effect of $k(>2)$ levels of a single factor. A factor is a characteristic under consideration thought to influence the measured observations, and level is a value of the factor.

To determine if different levels of the factor affect measured observations differently, the following hypotheses are to be tested:

$$H_0 : \mu_i = \mu \text{ all } i = 1, 2, 3, 4$$

H_1: $\mu_i \neq \mu$ for some $i = 1, 2, 3, 4$, where μ_i is the population mean for level i and μ is the overall grand mean of all levels.

Here we have four levels (i.e., four iterations), and each level consisted of 32 measurement readings of thrust face thickness of the connecting rod. The sum, sum of squares, mean, and variance for each iteration are tabulated in Table 5.

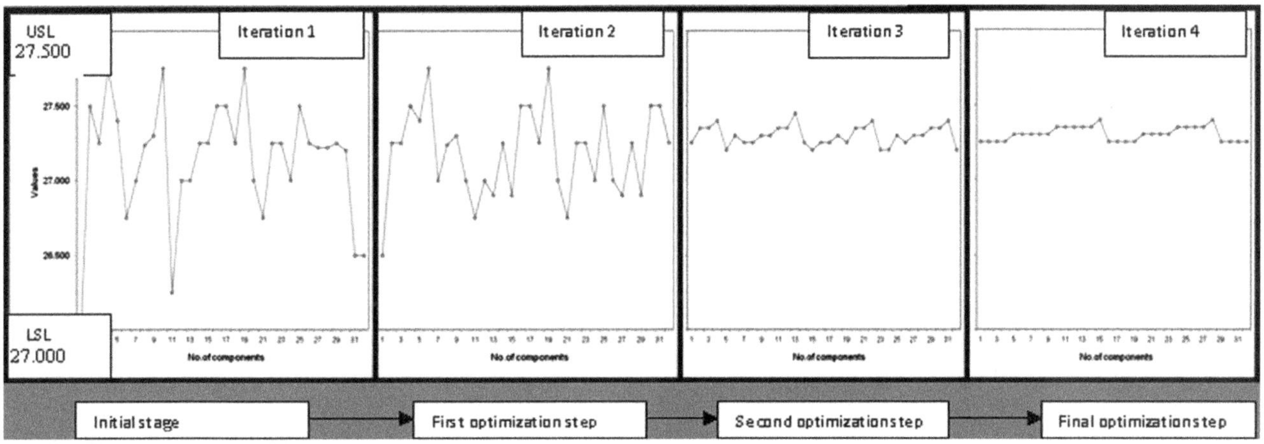

Figure 5 Process monitoring charts.

If x_{ij} denotes the data from the *i*th level and *j*th observation, then the overall or grand mean is given by

$$\mu = \sum_{i=1}^{4}\sum_{j=1}^{32}\frac{x_{ij}}{N} \tag{5}$$

where N is the total sample size of all four iterations, i.e., 32 × 4 = 128. Hence, from Equation 5, we get $\mu = 27.233$.

The sum of squared deviations about the grand mean across all N observations is given by

$$\mathrm{SST}_T = \sum_{i=1}^{4}\sum_{j=1}^{32}(x_{ij} - \mu)^2 \tag{6}$$

The sum of squared deviations for each level mean about the grand mean is given by

$$\mathrm{SST}_L = \sum_{i=1}^{4} 4 \times (\mu_i - \mu)^2 \tag{7}$$

The sum of squared deviations for all observations within each level from that level mean summed across all levels is given by

$$\mathrm{SST}_E = \sum_{i=1}^{4}\sum_{j=1}^{32}(x_{ij} - \mu_i)^2 \tag{8}$$

From Equations 6, 7, and 8, the values of SST_T, SST_G, and SST_E obtained are 8.694, 0.55289, and 8.084, respectively.

On dividing SST_T, SST_L, and SST_E by their associated degrees of freedom (*df*), we get the mean of squared deviations, respectively. Hence, the mean of squared deviations between levels is given by

$$\mathrm{MST}_L = \frac{\mathrm{SST}_L}{df_L} = \frac{0.55289}{(4-1)} = 0.184 \tag{9}$$

The mean of squared deviations within levels is given by

$$\mathrm{MST}_E = \frac{\mathrm{SST}_E}{df_E} = \frac{2.94}{(128 - 4)} = 0.065 \tag{10}$$

Finally, the F statistic is given by the following formula:

$$F_{\text{statistic}} = \frac{\mathrm{MST}_L}{\mathrm{MST}_E} = \frac{0.98}{0.065} = 15.07 \tag{11}$$

On summarizing all the above values in tabular form, the ANOVA table is obtained as shown in Table 6.

An α value of 0.05 is typically used, corresponding to 95% confidence levels. If α is defined to be equal to 0.05, then the critical value for the rejection region is $F_{\text{critical}\,(\alpha,\,K-1,\,N-K)}$ and is obtained to be 2.677. Thus,

$$F_{\text{critical}} = 2.677 \tag{12}$$

From Equations 10 and 11, it is seen that

$$F_{\text{statistic}} > F_{\text{critical}} \tag{13}$$

Therefore, the decision will be to reject the null hypothesis. If the decision from the one-way analysis of variance is to reject the null hypothesis, then it indicates

Table 4 Calculations of C_p and C_{pk}

Formula	Iteration 1	Iteration 2	Iteration 3	Iteration 4
USL	27.500	27.500	27.500	27.500
LSL	27.000	27.000	27.000	27.000
σ	0.408	0.296	0.068	0.048
$C_p = \frac{(USL-LSL)}{6\sigma}$	0.20	0.28	1.23	1.72
$C_{pku} = \frac{(USL\text{-}mean)}{3\sigma}$	0.29	0.35	0.99	1.37
$C_{pkl} = \frac{(mean-LSL)}{3\sigma}$	0.12	0.21	1.47	2.08
$C_{pk} = \min(C_{pku}, C_{pkl})$	0.12	0.21	0.99	1.37

Table 5 Mean and variance of all four iterations

Formula	Iteration 1	Iteration 2	Iteration 3	Iteration 4
Sample size	32	32	32	32
Sum	868.580	870.040	873.550	873.650
Sum of squares	23,581.139	23,658.01	23,846.69	23,852.08
Mean (μ_i)	27.143	27.189	27.298	27.302
Variance (σ^2)	0.166577	0.08734	0.004594	0.002336

that at least one of the means (μ_i) is different from the other remaining means. In order to figure out where this difference lies, a *post hoc* ANOVA test is required.

Post hoc ANOVA test

Since here the sample sizes are the same, we go for Tukey's test for conducting the *post hoc* ANOVA test. In Tukey's test, the honestly significant difference (HSD) is calculated as

$$\mathrm{HSD} = q\sqrt{\frac{\mathrm{MST}_E}{n}} = 3.63\sqrt{\frac{0.065}{32}} = 0.16 \quad (14)$$

where q is the studentized range statistic which is equal to a value of 3.63, for a degree of freedom of 124 and $k = 4$. The difference between the individual mean values of the four iteration levels can be summarized in a tabular form as shown in Table 7.

In Table 7, the absolute difference is of concern, and so the negative signs are to be ignored. Also in Table 7, it is seen that the difference of $\mu_1 - \mu_4 = 0.16$, which is equal with that of the HSD in Equation 14. Hence, it is concluded that the mean set of data between iteration 1 and iteration 4 is statistically significant when compared to the rest. Thus, it is concluded that among all the different causes enumerated in the cause-and-effect diagram, the most influencing cause is the tool wear compensation correction for the measurement data of iteration 4.

Improve and control phase

In this phase, the process monitoring charts are regularly employed for monitoring of the thrust face thickness of the connecting rod. In addition, gauge calibration is done periodically as a part of measurement system analysis, and properly calibrated gauges are used at the work place.

Table 6 ANOVA table

Source of variation	*df*	Sum of squares	Mean of squares	*F*
Level	3	0.55289	0.184	2.83
Within/error	124	8.084	0.065	
Total	127	8.694		

Result and discussion

As part of standardizing the process, the following activities were carried out:

1. The tool wear compensation knob which does not have any graduations was calibrated, and graduations were engraved at a 90° interval
2. Proper fixture maintenance and machine maintenance schedule were established, and regular checks were included in the check lists
3. After every 12,000 components, the grinding shoes need to be replaced
4. Coolant recirculation pressure was set at a value of around 15 kg of force

After incorporating the above actions in the manufacturing control plan, the same procedure can be horizontally deployed for solving of pragmatic problems of similar nature.

Conclusion

SPC studies were found to be useful for eliminating the special cause of errors, streamlining the process, and making the process a capable manufacturing process by improving the C_p and C_{pk} values of the key quality characteristic under study. The cause-and-effect diagram formed an important scientific tool for enlisting the causes behind the poor performance of the process. On adapting the DMAIC approach, the estimated standard deviation 'σ' of the thrust face thickness is reduced from 0.408 to 0.048, while the process performance capability index C_{pk} is enhanced from 0.12 to 1.37.

The C_p/C_{pk} values after performing the few iterations of data collection were greater than 1.33, and hence the process was declared as a capable process. After performing the root cause analysis, the major root cause, confirmed by the one-way ANOVA technique, was the improper setting of the tool wear compensation knob followed by the replacement of worn-out fixture rest pads for the thrust face grinding KPC. Hence, the one-way ANOVA technique was employed successfully for identification of the root cause liable for the low process capability.

Table 7 Differences of means between any two iterations

Difference	Computation	Numerical value
$\mu_1 - \mu_2 =$	27.143 − 27.189	−0.046
$\mu_1 - \mu_3 =$	27.143 − 27.298	−0.155
$\mu_1 - \mu_4 =$	27.143 − 27.302	−0.16
$\mu_2 - \mu_3 =$	27.189 − 27.298	−0.109
$\mu_2 - \mu_4 =$	27.189 − 27.302	−0.113
$\mu_3 - \mu_4 =$	27.298 − 27.302	−0.004

Competing interests

The authors declare that they have no competing interests.

Authors' contributions

GVSSS collected the required data of the Key Quality Characteristic, participated in the Failure Modes and Effects Analysis and performed the statistical analysis. PSR conceived of the study and participated in its analysis, design and coordination. All authors read and approved the final manuscript.

Acknowledgments

The authors are thankful to M/s AVTEC Limited of pithampur Industrial area, Madhya Pradesh State, India, for facilitating in the data collection for the present process capability improvement study.

Author details

[1]Department of Mechanical Engineering, GMR Institute of Technology, GMR Nagar, Rajam-532127, Andhra Pradesh State, India. [2]Department of Industrial Engineering, GITAM Institute of Technology, GITAM University, Visakhapatnam - 530045 Andhra Pradesh State, India.

References

Chen W-L, Huang C-Y, Huang C-Y (2013) Finding efficient frontier of process parameters for plastic injection molding. J Ind Eng Int 9:25, doi:10.1186/2251-712X-9-25

Gentili E, Aggogeri F, Mazzola M (2006) The improvement of a manufacturing stream using the DMAIC method. University of Brescia Paper No. IMECE2006-14469. ASME, New York, pp 127–133, doi: 10.1115/IMECE2006-14469

Hwang Y-D (2006) The practices of integrating manufacturing execution system and six sigma methodology. Int J Adv Manuf Technol 30:761–768, doi:10.1007/s00170-005-0090-1

Kumaravadivel A, Natarajan U (2013) Application of six-sigma DMAIC methodology to sand-casting process with response surface methodology. Int J Adv Manuf Technol, doi:10.1007/s00170-013-5119-2

Li M-HC, Al-Refaie A, Yang C-Y (2008) DMAIC approach to improve the capability of SMT solder printing process. IEEE Trans Electron Packag Manuf 31:126

Lin H-C (2004) The measurement of a process capability for folded normal process data. Int J Adv Manuf Technol 24:223–228, doi:10.1007/s00170-003-1615-0

Lin SJ, Yang DL, Cheng FT, Wu MF (2013) Aircraft turbine engine manufacturing with multiple specifications. J Test Eval 41(1):1–7, doi:10.1520/JTE20120022

Locke JW (1994) Statistical measurement control, quality and statistics. In: Milton Kowalewski J Jr (ed) Total quality management, ASTM STP 1209. American Society for Testing and Materials, Philadelphia

Sahay C, Ghosh S, Bheemarthi PK (2011) Process improvement of brake lever production using DMAIC. University of Hartford, Paper No. IMECE2011-63813. ASME, New York, pp 801–826, doi:10.1115/IMECE2001-63813

Schilling EG (1994) The transition from sampling to SPC, quality and statistics. In: Milton Kowalewski J Jr (ed) Total quality management, ASTM STP1209. American Society for Testing and Materials, Philadelphia

Sharma GVSS, Rao PS, Jagadeesh V, Vishwakarma A (2013a) Process capability improvement – a case study of an engine connecting rod manufacturing process. Int J Mech Eng Technol (IJMET) 4(5):116–129

Sharma GVSS, Rao PS, Jagadeesh V, Vishwakarma A (2013b) Process capability improvement – a case study of crank-pin bore honing operation of an engine connecting rod manufacturing process. Int J Adv Res Eng Technol (IJARET) 4(6):84–97

Singh R (2011) Process capability study of polyjet printing for plastic components. J Mech Sci Tech 25(4):1011–1015

Tong JPC, Tsung F, Yen BPC (2004) A DMAIC approach to printed circuit board quality improvement. Int J Adv Manuf Technol 23:523–531, doi:10.1007/s00170-003-1721-z

Solving an one-dimensional cutting stock problem by simulated annealing and tabu search

Meghdad HMA Jahromi[1*], Reza Tavakkoli-Moghaddam[2], Ahmad Makui[3] and Abbas Shamsi[1]

Abstract

A cutting stock problem is one of the main and classical problems in operations research that is modeled as LP problem. Because of its NP-hard nature, finding an optimal solution in reasonable time is extremely difficult and at least non-economical. In this paper, two meta-heuristic algorithms, namely simulated annealing (SA) and tabu search (TS), are proposed and developed for this type of the complex and large-sized problem. To evaluate the efficiency of these proposed approaches, several problems are solved using SA and TS, and then the related results are compared. The results show that the proposed SA gives good results in terms of objective function values rather than TS.

Keywords: One-dimensional cutting stock problem, Mathematical model, Simulated annealing, Tabu search

Background

A one-dimensional cutting stock problem (1D-CSP) is one of the famous combinatorial optimization problems, which has many applications in industries. In this problem, the amount of residual pieces of processed stock rolls, called trim loss (i.e., wastage) is a significant objective in most studies that should be minimized. A standard one-dimensional cutting stock problem (S1D-CSP) as a kind of the above problem is known as an NP-complete one (Gradisar et al. 2002). Finding solutions for this problem especially for large-sized problems is extremely difficult. In fact, due to the NP-hardness of this problem, it is computationally intractable to obtain its optimum solution in a reasonable time for medium to large-sized problems, and hence researchers have tried to develop meta-heuristic algorithms for solving such a problem. In fact, applying these algorithms is not easy to reliably state that which of these algorithms is best for the given problem. However, it is possible to identify methods that consistently produce better results compared to others for a certain problem. The assumptions of the S1D-CSP are as follows.

- In this problem, all used stock lengths should be cut to the end in as much as it is possible.
- All stock lengths are identical and there is no difference between the lengths of them.

In a standard cutting problem, each cutting stock is consumed totally, because the remaining is as cutting waste. While in a general one-dimensional stock cutting problem (G1D-CSP), as another kind of the 1D-CSP, stock cutting residuals can be used later. Also, in a GlD-CSP, the storage stock has different lengths. Reduction of cutting wastes is one of the main goals in the cutting process and also one of the basic purposes in the 1D-CSP. A standard one dimensional cutting stock problem (S1D-CSP) and different procedures have been reviewed by many researchers. (Varela et al. 2009) worked on a practical cutting stock problem from a production plant of plastic rolls as a variant of the well-known one dimensional cutting stock. They proposed a heuristic solution based on a GRASP algorithm in their work.

(Cui 2005) proposed an algorithm to produce T-shape cutting patterns applied in the manufacturing of circular and sectional blanks for stators and rotors. The proposed algorithm uses the knapsack algorithm and an enumeration method to determine the optimal combination of blank rows in the strips, the strip numbers and directions in the pattern. A heuristic algorithm for the one-dimensional cutting stock problem with usable leftover is presented by (Cui & Yang 2010).

* Correspondence: jahromi@iaukhomein.ac.ir
[1]Department of Industrial Engineering, Khomein Branch, Islamic Azad University, Khomein, Iran
Full list of author information is available at the end of the article

Their algorithm consists of two procedures, namely linear programming and sequential heuristic procedures that fulfill the major and remaining portions of the item demands, respectively. This algorithm is able to balance the cost of the consumed bars, the profit from leftovers and the profit from a shorter stocks reduction. (Armbruster 2002) presented a solution approach for the capacitated one-dimensional cutting stock problem under technological constraints by concentrating on solving the pattern sequencing problem occurring from the technological constraints. (Dikili et al. 2007) developed a novel approach for one-dimensional cutting stock problems in ship production via the cutting patterns obtained by the analytical methods at the mathematical modeling stage. The proposed method is able to capture the ideal solution of analytical methods. The main advantage of their proposed method is to guarantee in obtaining an integer solution. Also, the method behaves so that the trim loss is minimized and the stock usage is maximized.

(Liang et al. 2002a) proposed an evolutionary algorithm based on evolutionary programming for cutting stock problems considering contiguity. Two new mutation operators are proposed in their work. Results found by the evolutionary programming are significantly better than or comparable to those found by the genetic algorithm (GA). It is worth to note that evolutionary programming uses mutation as the primary search operator and does not employ any crossover.

(Poldi & Arenales 2009) proposed heuristics for the one-dimensional cutting stock problem with limited multiple stock lengths. They reviewed some existing heuristics and proposed others for solving the integer one-dimensional cutting stock problem with multiple stock lengths. Their study deals with the classical one-dimensional integer cutting stock problem. The given objective function is to minimize the waste (i.e., trim loss).

(Umetani et al. 2003) considered the 1D-CSP, in which the number of different cutting patterns is constrained within a given bound. They proposed a meta-heuristic algorithm incorporated with an adaptive pattern generation technique. The proposed meta-heuristic searches a solution with small deviations from the given demands while using the fixed number of different cutting patterns. (Reinertsen & Vossen 2010) considered the problem of cutting stock problems with due date while addressing common cutting considerations, such as aggregation of orders, multiple stock lengths and cutting different types of material on the same machine. In their work, meeting due dates is more important than reducing the scrap. Also, in condition that there is no feasible production plan, their models find a cutting plan with the minimum weighted tardiness. (Liang et al. 2002b) applied the newest perfect algorithm, which is similar to genetic algorithm, with a mobility function for a cutting stock problem. (Weng et al. 2003) applied the GA to the manufacturing rule oriented cutting of a sectional steel problem for the shipbuilding industry.

(Sung et al. 2004) applied tabu search (TS) to develop an optimization procedure based on the cutting processes in the shipbuilding industry. (Yang et al. 2006) developed the efficient TS method with a mixed objective function for the one dimensional cutting stock problem under a small amount of various cutting conditions. Their proposed TS method consists of a mixed objective function that helps the algorithm to discover the best solution during the one-by-one neighborhood searching processes. (Gradisar et al. 2002) presented an experimental study of various methods aimed at minimizing trim-loss, in their work they mentioned that the LP-based method is possible only when the stock is of the same length or few groups of standard lengths. When all stock lengths are different frequencies, an approach can be used that treat by each item individually.

In addition, (Chen et al. 1996) proposed a simulated annealing (SA) method for the one-dimensional cutting stock problem. A meta-heuristic method based on ACO was presented to solve a one-dimensional cutting stock problem by (Eshghi & Javanshir 2005). In their work, based on probabilistic laws that designed, ants do select various cuts and then select the best patterns. Also, (Dyckhoff 1990) developed a classification scheme for cutting stock problems through a large variety of applications reported in the literature so that classified problems using four characteristics; namely, dimensionality, kind of assignment, assortment of large objects and assortment of small items. He classified the cutting stock problems solutions into two groups: 1) item-oriented approach that treats each item to be cut separately and 2) pattern-oriented approach. So that in this approach at the beginning order lengths are combined into cutting patterns, for which the frequencies that are necessary to satisfy the demands, are determined in a succeeding step. In the second approach, the classic LP-based namely "delayed pattern generation" proposed (Gilmore & Gomory 1961; Gilmore & Gomory 1963) or any other LP-based method is mostly used. (Javanshir & Shadalooee 3) developed a mathematical model for 1D-CSP by defining the virtual cost for the trim loss of each stock. They solved their developed model with SA algorithm. Also, 1D-CSP considered in their work is taken into account as item-oriented.

It is not easy to define exactly which meta-heuristic algorithm is best for which type of problem. However, it is possible, for a given problem, to identify algorithms that consistently produce better results compared with those results produced by other algorithms. For this reason, in this paper, we compare the efficiency and effectiveness of two well-known meta-heuristic, namely simulated annealing (SA) and tabu search (TS), in producing the cutting plan with the lowest trill loss. In addition, we present a mixed-integer linear programming (MILP) mathematical model for a standard one-dimensional cutting stock problem that minimizes the trim losses of the cutting orders. Because of the NP-hard nature of this problem, the presented mathematical model can find an optimal solution by a standard OR software (e.g., Lingo) for small-sized problems.

This paper is organized as follows. Section 2 presents the mathematical programming model for the S1D-CSP. Simulated annealing (SA) and tabu search (TS) algorithms for the considered problem are proposed in Sections 3 and 4, respectively. The computational experience is represented in Section 5. Finally, conclusion is made in Section 6.

Mathematical model of S1D-CSP

This section presents a mixed-integer linear programming (MILP) model for a standard one-dimensional cutting stock problem (S1D-CSP). Minimizing the sum of cutting wastes or cutting remaining unusable in satisfying of needs (demands) is the objective function of the presented model.

Notations

To formulate the S1D-CSP, the following notations are used.

- i order number ($i = 1,\ldots,n$).
- j stock number ($j = 1,\ldots,m$).
- d_j stock length.
- tl_j stock wastes.
- s_i order lengths.
- n_i required number of orders with s_i length.
- T sum of the cutting wastes in the cutting program.

Decision variables

The following variables are used in the presented MILP model.

- $x_{i,j}$ integer variable, number of orders with s_i length that are cut from stock j.
- y_j zero–one variable that equals to one if the stock j is applied in the cutting plan otherwise, equals to zero.

Mathematical model

The MILP model for a standard one-dimensional cutting stock problem can be defined as follows.

$$\min \quad T = \sum_{j=1}^{m} tl_j \tag{1}$$

s.t.

$$\sum_{j=1}^{m} x_{ij} = n_i \ \forall i \tag{2}$$

$$\sum_{i=1}^{n} x_{ij}.s_i + tl_j = d_j.y_j \ \forall j \tag{3}$$

$$y_j \in (0,1)$$

$$x_{i,j} \in \text{integer}$$

Minimizing the total trim loss is the objective function (1) of the model. Constraint (2) guarantees cutting needs of orders with regards to the storage stock. Constraint (3) calculates the cutting waste of each stock in the cutting process.

Due to NP-hard nature of this problem, the computational time increases exponentially and finding optimal solution for large-size problem is extremely difficult. This model is solved by the Lingo software using a branch-and-bound method, and the computational times and the state of the method for 20 randomly generated problems with different sizes are shown in Table 1. Results show finding optimal solution for large-scale problems (15, 16, 17 and 18), in a reasonable time and gradually with increasing the size of the problems (19 and 20), even finding feasible area of solutions, is so hard.

Since finding the optimal solution in a reasonable time for medium to large-sized problems is intractable, a number of researchers have considered developing meta-heuristics to compute near-optimal solutions. It is noticeable that identifying methods consistently producing better results, as compared to others, for a certain problem is very important and interesting. Among the meta-heuristics, simulated annealing (SA) and tabu search (TS) are the ones whose applications have been widely explored for different combinatorial optimization problems. In this work, effectiveness and efficiency of these meta-heuristics for the given problem are compared.

Simulated annealing for S1D-CSP

(Kirkpatrick et al. 1983) introduced simulated annealing (SA) as a meta-heuristic algorithm in 1983. It draws its inspiration from physical annealing of solids. In this

Table 1 NP-hard nature of the S1D-CSP

Problems	Requested orders	State	CPU time (Sec.)
1	9	Global optimum found	1
2	15		1
3	32		2
4	45		4
5	59		7
6	72		10
7	85		28
8	98		59
9	110		607
10	190		2570
11	305		6077
12	950		15185
13	1503		58106
14	1704		79202
15	1850	Cannot find any global optimum after a 24-hours run	
16	1990		
17	2500		
18	3645		
19	4985	Cannot find any feasible area of solutions after a 24-hours run	
20	5750		

algorithm, it may become stuck in a local optimum. To prevent this from happening and trapping in a local optimum, the algorithm occasionally makes a random change to the solution even through the change does not immediately improve the result. This may allow the program to break free of a local optimum and find a better solution later. If the change does not lead to a better solution, the algorithm will probably undo the change in a short time anyway. SA consists of internal and external loops. Task of the internal loop controls the attaining balance in each temperature and updating temperature.

The Metropolis decision-making loop (i.e., possibility of acceptance of non-improvement) is inserted in the internal loop. According to the decision-making loop, if the current solution $S_{current}$ cannot improve the objective function comparing with the new neighboring solution S_{new}; then, if $y < e^{\frac{-\Delta}{T}}$ (i.e., $\Delta = E(S_{new}) - E(S_{current}) > 0$ and $y \in [0,1]$), the new solution will be accepted. In fact, the acceptance possibility of a neighboring solution is dependent on changing the amount of the objective function, and current temperature. The value of T starts at initial temperature T_0 and is slowly reduce according to the cooling ratio $a < 1$ (i.e., $T_{k+1} = a.T_k$).

Also, the external loop controls the stopping condition of SA. A given SA may iterate for a number of times at each temperature, called the Markov chain length. The neighborhood search structure generates a new solution from the current candidate solution by slightly changing it. How move from one solution to another one in the solution space is dependent on the structure of designing operators. Attaining to the undiscovered area of exploration space pertain to design efficient operators. The generic flow of SA is given below.

Select an initial temperature $T_0 > 0$;
Generate an initial solution, S_0, and make it the current solution, $S_{current}$, and the current best solution, S_{best};
repeat

set repetition counter $n = 1$;

repeat

Generates the solution S_{new} in the neighborhood of $S_{current}$;
Calculate $\Delta = f(S_{current}) - f(S_{new})$;
If $(\Delta \le 0)$ *then* $S_{current} = S_{new}$;
else $S_{current} = S_{new}$ with the probability $P(\Delta, T)$;
if $(S_{new} < S_{best})$ *then* $S_{best} = S_{new}$;
$n = n + 1$;

until $n >$ Markov chain length;

Reduce the temperature T;
until the stopping criterion is met.

Solution representation

The first step in developing of a meta-heuristic procedure is to design the solution representation. In this paper, with regards to the problem nature, the initial solution is created randomly. This solution is constructed as a $(1 \times m)$*-array*, so that m equal to the number of orders. Each order placed in a cell randomly as illustrated in Figure 1.

Neighborhood generation (perturbation)

To search solution space, attaining to undiscovered areas and finding the best solutions, using suitable neighborhood search operators is required. As stated before, designing this operator is dependent on the solution representation. Mechanism of the neighborhood generation should be able to construct the entire solution space. In this paper, in order to generate the neighboring solution, the algorithm randomly changes the place of two orders in the current solution in its array, as shown in Figure 2. This strategy is very simple and guarantees searching the whole of the solution space.

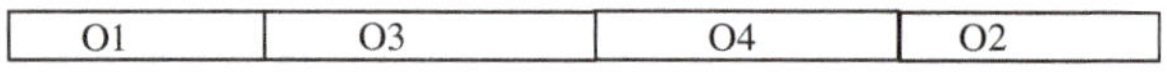

Figure 1 Representation of a solution with four orders.

Fitness function

The objective function considered in this paper minimizes the total trim loss. It is assumed that the stock length is standard with the equal length (e.g., 12 meters considered in this paper). By considering the position of each order in the array from the first (i.e., leftmost) to the end (i.e., rightmost), the total waste is computed. So that all the orders, which the summation of their length is lesser than (or equal to) the length of stock, are assigned to it for cutting and the difference between them considered as the trill loss of that stock and so on, till all orders are assigned.

Temperature decreasing function

Temperature is used to compute the acceptance probability of a solution that is worse than the previous one. In the proposed SA algorithm, a geometric function is used as follows.

$$T_k = \alpha T_{k-1} \tag{4}$$

where, α is a constant less than 1, but it is very close to 1 (i.e., a typical value for α is chosen between 0.85 and o.95). k is the temperature transformations counter. The high amount of α is as the slow temperature reduction that leads to the best search of neighbors; but, the solution time increases. Adversely, less amount of α is an indicator of the quick temperature reduction and leads to quicker search in neighbors and sooner coverage in the algorithm.

Markov chain length

One of the most important parameters in determining the quality of resultant solutions of SA is the number of searched points in solution space at each temperature, called Markov chain length. In our proposed SA, this counter is denoted by *MarChain_Length* and used in the inner loop of SA as a counter. The Markov chain length considered in our proposed SA depends on the problem size (i.e., number of orders).

Outer counter variable

The outer counter variable is increased by one, whenever a solution is not selected by probability and shows the number of rejection of perturbed solutions. In fact, if this variable reaches to a pre-determined fixed value as input parameter, the search procedure is stopped or started again depending on the other criteria. Because there is no superior solution exists in the neighborhood, and the search is reached to a near optimal/optimal solution.

Stopping criterion

In our proposed SA, the stopping criterion is a number of consecutive times in the cooling process that does not lead to improvement of the objective function as input data.

Tabu search for S1D-CSP

(Glover 1990) first proposed tabu search (TS) as a higher-level heuristic algorithm used for solving optimization problems. It avoids entrapment in cycles by forbidding or penalizing moves that take the solution, in the next iteration, to points in the solution space previously visited by using the list of prohibited neighboring solutions known as tabu list. Aspiration defined in the tabu search restricts the search from being trapped into a solution surrounded by tabu neighbors. Thus, it helps in the checking condition for the acceptance of solutions, known as a main feature of TS. It also allows search to override the tabu status of the solution and provides backtracking of the recent solutions as they lead to a new path towards a better solution. The generic structure of the TS algorithm is given below.

Step 1. (initialization)

(A). Generate initial solution $S_{initial}$

(B). $S_{current} = S_{initial}$

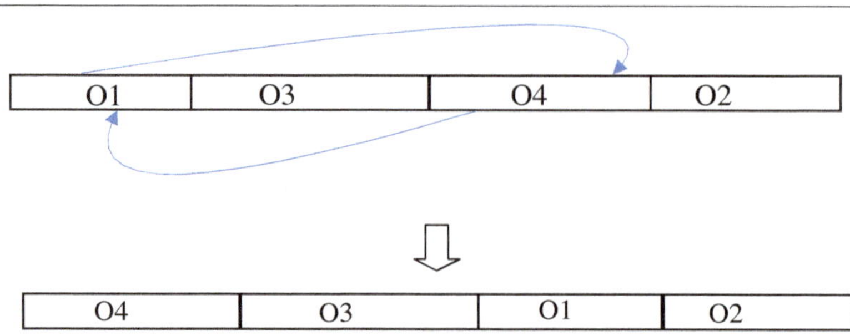

Figure 2 Neighborhood generation.

(C). Record the current best known solution by setting $S_{best} = S_{current}$, define $best\_cost = f(S_{best})$

(D). Set tabu list (H) to empty.

Step 2. (choice and termination)

Determine $Candidate\text{-}N(S_{current})$ as subset of $N(H, S_{current})$.

Select S_{new} from $Candidate\text{-}N(S_{current})$.

(S_{new} is called a highest evaluation element of $Candidate\text{-}N(S_{current})$.

Terminate by a chosen iteration cut-off rule.

Step 3. (update)

Re-set $S_{current} = S_{new}$.

If $f(S_{current}) < best\_cost$ or $f(S_{current}) <$ Aspiration perform Step 1(B)

Update the tabu list (H) and Aspiration conditions.

Return to Step 2.

Tabu list

A tabu list is a set of recently visited solutions and keeps tracking the previously visited solutions in order to restrict the algorithm to revisit the pre-visited solutions. In fact, at every step, the algorithm checks for verifying whether the newly generated solution is lately visited. The tabu list task helps the algorithm to visit more new neighboring solutions.

Aspiration

Aspiration is a tabu search variable whose function is to restrict the search from being trapped at a solution surrounded by tabu neighbors. If a solution has a neighborhood of only tabu solutions, the one with objective function value better than aspiration will be chosen to explore further.

Computational experience

As mentioned earlier, it is not easy to reliably determine which meta-heuristic method is the best for which problem; but, it is possible to identify methods consistently producing better results, as compared to others for a certain problem. In this section, effectiveness and efficiency of the SA and TS algorithms for the 1D-CSP are considered.

The proposed SA and TS algorithms are coded in Microsoft C#.Net and executed on a Pentium processor running at 2.5 GHz and 2 GB of RAM. The results of solving the 14 small and medium-size test problems solved by branch-and-bound (B&B), SA and TS algorithms are shown in Table 2. This table shows the objective function value (OFV) and the gap between solutions obtained from SA and TS with optimal solutions of the B&B method. The results show that SA and TS algorithms are suitable and both of them even are capable to obtain an optimal solution for small-sized problems. Also, these algorithms can find good results for the medium-sized problems so that the gap between them and optimal solution is lower that 20% (the first 14 randomly generated problems are in the small to medium range of size).

Table 2 Result obtained from SA and TS and branch-and-bound methods

Problems	B&B method	SA		TS	
	OFV*	OFV	GAP	OFV	GAP
1	4.5	4.5	0.0	4.5	0.0
2	7.5	7.5	0.0	7.5	0.0
3	16	16	0.0	16	0.0
4	22.5	22.5	0.0	22.5	0.0
5	29.5	29.5	0.0	29.5	0.0
6	36	38	5.6	40	11.1
7	34	37	8.8	39	14.7
8	39.2	43	9.7	44	12.2
9	44	45	2.3	50	13.6
10	76	79	3.9	85	11.8
11	122	125.4	2.8	141	15.6
12	285	302	6.0	336	17.9
13	450.9	511	13.3	540	19.8
14	511.2	599.5	17.3	612	19.7

* Objective function value (meter).

To compare the efficiency of these algorithms with each other, the six large-scale problems (15, 16, 17, 18, 19 and 20) are solved and the total stocks used in each solution, the total trim losses and total trim loss ratio of each algorithm are represented in Table 3. The total trim loss ratio can be defined by:

$$\text{Total trim loss ratio} = \frac{\text{Sum of trim loss of all used stocks}}{\text{Stocks used } \times \text{ standard length}} \times 100\% \quad (5)$$

It is obvious that SA outperforms TS to get a better minimum number of stocks used or less total trim loss (i.e., total trim loss ratio) so that the mean of the total trim loss ratio of SA for the problems is 6.49 percent while it is 7.45 percent in TS. Figure 3 shows CPU times of two algorithms for the twenty mentioned problems. It is noticeable that more CPU times of the SA algorithm in comparative with TS is due to the internal and external loops and extremely depends on the Markov chain length.

Table 3 Comparison between SA and TS algorithms

Problems	Requested orders	SA			TS		
		Stocks used	Trim loss	Total trim loss ratio (%)	Stocks used	Trim loss	Total trim loss ratio (%)
1	9	5	4.5	7.50	5	4.5	7.50
2	15	9	7.5	6.94	9	7.5	6.94
3	32	15	16	8.89	15	16	8.89
4	45	25	22.5	7.50	25	22.5	7.50
5	59	34	29.5	7.23	34	29.5	7.23
6	72	42	38	7.54	44	40	7.58
7	85	51	37	6.05	56	39	5.80
8	98	55	43	6.52	61	44	6.01
9	110	60	45	6.25	69	50	6.04
10	190	111	79	5.93	128	85	5.53
11	305	149	125.4	7.01	173	141	6.79
12	950	501	302	5.02	554	336	5.05
13	1503	850	511	5.01	865	540	5.20
14	1704	951	599.5	5.25	1097	612	4.65
15	1850	1150	751	5.44	1165	1170	8.37
16	1990	1110	897	6.73	1222	1412.4	9.63
17	2500	1350	993	6.13	1480	1955.6	11.01
18	3645	1901	1287	5.64	1984	2192.4	9.21
19	4985	2713	2004.2	6.16	2671	3198	9.98
20	5750	3112	2614	7.00	3221	3930.8	10.17
Average				**6.49**			**7.45**

Conclusions

In this paper, a mixed-integer linear programming (MILP) model for a standard one-dimensional cutting stock problem (S1D-CSP) has been presented. Due to its NP-hardness, finding feasible solution in a reasonable time, especially for medium to large size problems, is impossible. To solve such a hard problem, two well-known meta-heuristics based on simulated annealing (SA) and tabu search (TS) have been proposed and developed. To compare the efficiency of these algorithms, a number of test problems have been solved, and the associated results in terms of the objective function (i.e., trim loss) and CPU time have been compared. The results show the efficiency of our proposed SA in finding

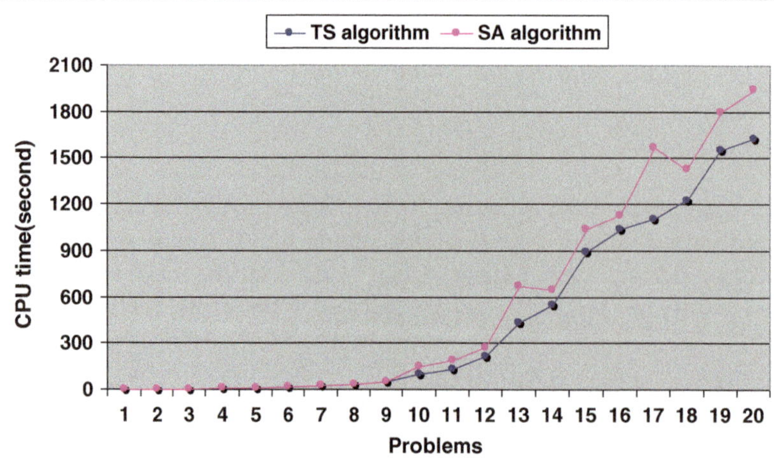

Figure 3 CPU times of the proposed SA and TS algorithms.

cutting plan as its output with lesser trim loss rather than TS. In addition, the results show that TS outperforms SA in terms of the CPU time.

Competing interests
The authors declare that they have no competing interests.

Authors' contributions
MHMAJ studied the associated literature review and presented the mathematical model. RTM managed the study and participated in the problem solving approaches. AM approved and validate the model and was responsible for revising the manuscript. AS managed and analyzed the required data and conducted the experimental analysis. All authors read and approved the final manuscript.

Author details
[1]Department of Industrial Engineering, Khomein Branch, Islamic Azad University, Khomein, Iran. [2]Department of Industrial Engineering, College of Engineering, University of Tehran, Tehran, Iran. [3]Department of Industrial Engineering, Iran University of Science and Technology, Tehran, Iran.

References
Armbruster M (2002) A solution procedure for a pattern sequencing problem as part of a one-dimensional cutting stock problem in the steel industry. Eur J Oper Res 141(2):328–340
Chen CLS, Hart SM, Tham WM (1996) A simulated annealing heuristic for the one-dimensional cutting stock problem. Eur J Oper Res 93(3):522–535
Cui Y (2005) A cutting stock problem and its solution in the manufacturing industry of large electric generators. Comput Oper Res 32(7):1709–1721
Cui Y, Yang Y (2010) A heuristic for the one-dimensional cutting stock problem with usable leftover. Eur J Oper Res 204(2):245–250
Dikili AC, Sarlöz E, Pek NA (2007) A successive elimination method for one-dimensional stock cutting problems in ship production. Ocean Eng 34(13):1841–1849
Dyckhoff H (1990) A typology of cutting and packing problems. Eur J Oper Res 44(3):145–159
Eshghi K, Javanshir H (2005) An ACO algorithm for one-dimensional cutting stock problem. J Ind Eng Int 1(1):10–19
Gilmore PC, Gomory RE (1961) A linear programming approach to the cutting stock problem. Oper Res 9(3):849–859
Gilmore PC, Gomory RE (1963) A linear programming approach to the cutting stock problem, PartII. Oper Res 11(3):863–888
Glover F (1990) Tabu search: A tutorial. Interfaces 20(4):74–94
Gradisar M, Resinovic G, Kljajic M (2002) Evaluation of algorithms for one-dimensional cutting. Comput Oper Res 29(9):1207–1220
Javanshir H, Shadalooee M (2007) The trim loss concentration in one-dimensional cutting stock problem (1D-CSP) by defining a virtual cost. J Ind Eng Int 3(4):10–19
Kirkpatrick S, Gelatt CD, Vecchi MP (1983) Optimization by simulated annealing. Science 220:671–680
Liang KH, Yao X, Newton C, Hoffman D (2002a) A new evolutionary approach to cutting stock problems with and without contiguity. Comput Oper Res 29(12):1641–1659
Liang KH, Yao X, Newton Y, Hoffman D (2002b) A new evolutionary approach to cutting stock problems with and without contiguity. Comput Oper Res 29(12):1641–59
Poldi KC, Arenales MN (2009) Heuristics for the one-dimensional cutting stock problem with limited multiple stock lengths. Comput Oper Res 36(6):2074–2081
Reinertsen H, Vossen TWM (2010) The one-dimensional cutting stock problem with due dates. Eur J Oper Res 201(3):701–711
Sung TC, Weng WC, Yang CT (2004) Optimization of piece arrangement for one-size source cutting problem by tabu search for shipyard. In: Proceeding of the 16th National Conference on Naval Architecture and Marine Engineering. Naval Architecture and Marine Engineering Department, National Cheng-Kung University, Tainan, Taiwan, 5–6
Umetani S, Yagiura M, Ibaraki T (2003) One-dimensional cutting stock problem to minimize the number of different patterns. Eur J Oper Res 146(2):388–402
Varela R, Vela CR, Puente J, Sierra M, González-Rodríguez I (2009) An effective solution for a real cutting stock problem in manufacturing plastic rolls. Ann Oper Res 166(1):125–146
Weng WC, Yang CT, Hung CF (2003) The optimization of section steel arrangement for ship construction associated with cutting rule by genetic algorithm. In: Proceeding of the 17th Asian-Pacific Technical Exchange and Advisory Meeting on Marine Structure. National Cheng Kung University, Tainan, Taiwan, 237–246
Yang CT, Sung TC, Weng WC (2006) An improved tabu search approach with mixed objective function for one-dimensional cutting stock problems. Adv Eng Softw 37(8):502–513

10

An L1-norm method for generating all of efficient solutions of multi-objective integer linear programming problem

Ghasem Tohidi[1*] and Shabnam Razavyan[2]

Abstract

This paper extends the proposed method by Jahanshahloo et al. (2004) (a method for generating all the efficient solutions of a 0–1 multi-objective linear programming problem, Asia-Pacific Journal of Operational Research). This paper considers the recession direction for a multi-objective integer linear programming (MOILP) problem and presents necessary and sufficient conditions to have unbounded feasible region and infinite optimal values for objective functions of MOILP problems. If the number of efficient solution is finite, the proposed method finds all of them without generating all feasible solutions of MOILP or concluding that there is no efficient solution. In any iteration of the proposed algorithm, a single objective integer linear programming problem, constrained problem, is solved. We will show that the optimal solutions of these single objective integer linear programming problems are efficient solutions of an MOILP problem. The algorithm can also give subsets of efficient solutions that can be useful for designing interactive procedures for large, real-life problems. The applicability of the proposed method is illustrated by using some numerical examples.

Keywords: Multi-objective integer linear programming, Single objective integer linear programming, Recession direction, Efficient solution, L1-norm

Background

Multiple criteria decision making suggests, including multiple (two or more) objective functions, a mathematical programming framework. Since most real-life problems include conflicting objectives, multiple objective optimization provides a means for obtaining more realistic models. Multi-objective integer linear programming (MOILP) problem is an important research area as many practical situations require discrete representations by integer variables, and many decision makers have to deal with several objectives (Ulungu and Teghem 1994). Some note-worthy practical environments, where the MOILP problems find their applications, are supply chain design, logistics planning, scheduling, and financial planning. The MOILP problems are theoretically challenging as well, as most of them, even their single objective versions, fall into the class of computationally intractable problems.

Numerous algorithms have been designed to solve an MOILP (Climaco et al. 1997; Rasmussen 1986; Teghem and Kunsch 1986; Ulungu and Teghem 1994) and multiple objective mixed integer linear programs (Mavrotas and Diakoulaki 1998; Sylva and Crema 2007). Klamroth et al. (2004) and Ehrgott (2006) study the general MOIP problem. Klamroth (2004) defined composite functions to obtain upper bounds on the objective function values of the efficient solutions and discussed the use of the upper bounds in generating the efficient set. To form the composite functions, they proposed some classical optimization methods such as cutting plane method and branch and bound algorithm. However, the MOILP have not received the algorithmic attention that continuous problems have. The literature available on this topic is limited.

Using a straightforward theoretical approach, Sylva and Crema's (2004) algorithm enumerates all efficient solutions of MOILP models with bounded feasible regions. Sylva and Crema's (2004) approach solves the problem using a sequence of progressively more

* Correspondence: ghatohidi@yahoo.com
[1]Assistant professor Department of Mathematics, Islamic Azad University, Central Tehran Branch, Tehran, Iran
Full list of author information is available at the end of the article

constrained integer linear programs and generates a new solution at each step. Starting from an initial efficient solution, Sylva and Crema's (2007) algorithm finds at each iteration a new one that maximizes the infinity-norm distance from the set dominated by the previously found solutions. When all variables are integer and feasible region is bounded, their approach generates the whole set of efficient solutions; and it would be expensive using the method for enumerating the whole set of efficient solutions except for small problems.

However, in some cases, the feasible region of an MOILP problem is unbounded. Therefore, an MOILP problem can have infinite objective values. These cases have not been considered in Sylva and Crema (2004) and (2007). This paper considers the recession direction to the MOILP problem and provides necessary and sufficient conditions to have unbounded feasible region and infinite values of MOILP problem and then extends Jahanshahloo et al. (2004) method to solve an MOILP problem. When the number of efficient solutions of an MOILP problem is finite, the proposed algorithm finds all of the efficient solutions without generating all of the feasible solutions of an MOILP. Using a straightforward theoretical approach, the efficient solutions are found using a sequence of progressively more constrained integer linear programs generating new efficient solutions in any iteration. We prove that all of the optimal solutions of this single objective integer linear programming problem are efficient solutions of an MOILP problem. Therefore, the proposed method reduces the number of constrained problems which are solved.

The paper is organized as follows: Section 2 presents a brief background about MOILP problem. Section 3 introduces modified algorithm for finding all of the efficient solutions of an MOILP problem with bounded or unbounded feasible region. Illustration with some numerical examples is given in Section 4. Finally, the concluding results are presented.

An MOILP problem is a special case of multi-objective program. An MOILP problem with s-objectives is defined as:

$$\begin{aligned} &\max\{C_1W, C_2W, \cdots, C_sW\} \\ &\text{s.t.} \quad A_iW \le b_i, i = 1, 2, \cdots, m \\ &\qquad W \in Z_n^+ \end{aligned} \tag{1}$$

where, $C_r = (c_{r1}, c_{r2}, \cdots, c_{rn})(r = 1, 2, \cdots, s)$, $\cdot A_i = (a_{i1}, a_{i2}, \cdots, a_{in})(i = 1, 2, \cdots, m)$, $\cdot Z_n^+ = \{(e_1, \cdots, e_n)|e_j \in Z^+ = \{0, 1, 2, \cdots\}, j = 1, \cdots, n\}$ and $W = (w_1, w_2, \cdots, w_n)^T$

The set X, which is defined as follows:

$$X = \{W|A_iW \le b_i, i = 1, 2, \ldots, m, W \in Z_n^+\} \tag{2}$$

is called the set of feasible solutions of the problem (1). Corresponding to each $W \in X$, the vector Y is defined as follows (Jahanshahloo et al. 2004):

$$Y = (y_1, \cdots, y_s)^T = (C_1W, C_2W, \cdots, C_sW)^T. \tag{3}$$

Definition The *vector* $Y = (y_1, \cdots, y_s)^T$ dominates the vector $Y^o = (y_1^o, \cdots, y_s^o)^T$ if for each r $(r = 1, 2, \ldots, s)$, $y_r \ge y_r^o$, and there is at least one l such that $y_l > y_l^o$.

Definition Let
$F = \{Y|Y=(C_1W, C_2W, \cdots, C_sW)^T,\ A_iW \le b_i,\ i = 1, \cdots, m,\ W \in Z_n^+\}$. F is called the values space of objective functions in problem (1).

Let $g_r = C_rW_r^*(r = 1, \cdots, s)$, where W_r^* is the optimal solution of rth problem from the following problems (Jahanshahloo et al. 2004):

$$\begin{aligned} g_r = &\max C_rW \\ \text{s.t.} \quad &A_iW \le b_i, i = 1, 2, \cdots, m \\ &W \in Z_n^+. \end{aligned} \tag{4}$$

Definition Let $g = (g_1, g_2, \cdots, g_s)^T = (C_1W_1^*, C_2W_2^*, \cdots, C_sW_s^*)^T$. The g is called the ideal vector of problem (1) (Jahanshahloo et al. (2004)).

Efficient solutions of MOILP problem

Feasible region of a 0–1 MOILP is bounded. However, an MOILP problem has bounded or unbounded feasible region, which are discussed as follows.

Methods

MOILP problem with unbounded feasible region

In some cases, feasible region of an MOILP problem is unbounded. For instance, consider the following MOILP problem.

$$\begin{aligned} &max\, w_1 + w_2 \\ &\max 2w_1 + w_2 \\ &\text{s.t.} -5w_1 + 4w_2 \le 20 \\ &\quad -6w_1 + 7w_2 \le 42 \\ &\quad w_1, w_2 \in Z^+. \end{aligned}$$

To explain the unbounded case, we define recession direction for MOILP problems similar to recession direction for linear programming problem (Bazaraa et al. 2007).

Definition: Let $d \ne 0$ and $d \in Z_n^+$, then d is a recession direction of the MOILP problem if and only if for all $W \in X$, and for all $\lambda \in Z^+$, we have $W + \lambda d \in X$.

Theorem *Let* $d(\ne 0)$, *then* d *is a recession direction of problem (1) if and only if* $\lambda \in Z^+$ *and* $W + \lambda d \in X$.

Proof The proof is evident.

Theorem *If there is $d(\neq 0)$ such that*

$$\begin{cases} A_i d \leq 0, i = 1, \cdots, m, \\ C_r d \geq 0, r = 1, \cdots, s \\ d \in Z_n^+ \end{cases}$$

with at least one $p(p \in \{1, \cdots, s\})$ such that $C_r d > 0$ and $d \in Z_n^+$, then the optimal values of the objective functions is infinite, i.e., problem (1) has no efficient solution.

Proof The proof is evident.

Theorem *Suppose that for each recession direction of model (1), say $\tilde{d}$, there is $p(p \in \{1, \ldots, s\})$ such that $C_p \tilde{d} < 0$, then model (1) has an efficient solution.*

Proof The proof is evident.

MOILP problem with bounded feasible region

This section extends the proposed method in (Jahanshahloo et al. 2004) to find all efficient solutions of MOILP with bounded feasible region. To specify some efficient solutions of model (1), we consider the following theorem for an MOILP problem, which is a modification of Theorem 2.1 in (Jahanshahloo et al. 2004)

Theorem *Let $O_l = \left\{W_{1l}^*, W_{2l}^*, \cdots, W_{fl}^*\right\}$ be the set of optimal solutions of lth problem from problems (4), then at least one of these solutions is an efficient solution of problem (1).*

Proof The proof is similar to that of Theorem 2.1 in (Jahanshahloo et al. 2004) and is omitted.

Theorem *For each $W \in X$ as a feasible solution of an MOILP problem with bounded feasible region, the vector $g = (g_1, g_2, \cdots, g_s)^T$ dominates the vector $Y = (C_1 W, C_2 W, \cdots, C_s W)^T \neq g$.*

Proof: See the proof of the Theorem 2.2 in (Jahanshahloo et al. 2004).

As noted in (Jahanshahloo et al. 2004) about a 0–1 MOLP, to find the other efficient solutions of problem (1) with bounded feasible region, we can specify a feasible solution, say $W \in X$, such that $g - Y = (g_1 - C_1 W, g_2 - C_2 W, \cdots, g_s - C_s W)^T$ is minimized. Therefore, we can solve the following MOILP problem.

$$\begin{aligned} &\min \{g_1 - C_1 W, g_2 - C_2 W, \cdots, g_s - C_s W\} \\ &\text{s.t.} \quad A_i W \leq b_i, \quad i = 1, 2, \cdots, m \\ &\qquad W \in Z_n^+ \end{aligned} \tag{5}$$

where X is a bounded feasible solution of an MOILP problem. Since $\forall W \in X,\ g_r \geq C_r W, (r = 1, 2, \cdots, s)$ using L1-norm we have:

$$\begin{aligned} &\min_{W \in X} \sum_{r=1}^{s} |g_r - C_r W| \\ &= \min_{W \in X} \sum_{r=1}^{s} (g_r - C_r W) \\ &= \sum_{r=1}^{s} g_r + \min_{W \in X} \sum_{r=1}^{s} (-C_r W) \\ &= \sum_{r=1}^{s} g_r - \max_{W \in X} \sum_{r=1}^{s} C_r W. \end{aligned}$$

To find some other efficient solutions of the MOILP problem, we solve the following linear integer programming problem:

$$\begin{aligned} &\max \sum_{r=1}^{s} C_r W \\ &\text{s.t.} \quad A_i W \leq b_i, i = 1, 2, \ldots, m \\ &\qquad W \in Z_n^+. \end{aligned} \tag{6}$$

Theorem *Each optimal solution of problem (6) is an efficient solution for an MOILP problem.*

Proof The proof is similar to that of Theorem 2.3 in (Jahanshahloo et al. 2004) and is omitted. □

As noted about a 0–1 MOILP in (Jahanshahloo et al. 2004), let for each $q \in L_o = \{i_1, i_2, \cdots, i_\alpha\} \subseteq \{1, 2, \cdots, s\}$, the following problem has a unique optimal solution,

$$\begin{aligned} &g_q = \max C_q W \\ &\text{s.t.} \quad A_i W \leq b_i, \quad i = 1, 2, \cdots, m \\ &\qquad W \in Z_n^+. \end{aligned} \tag{7}$$

Suppose that $G_o = \left\{W_{i_1}^*, W_{i_2}^*, \cdots, W_{i_\alpha}^*\right\}$ is the set of optimal solutions of problem (7). When G_o is empty, we solve problem (6). In this case, let $G_o = \left\{W_{i_1}^*, W_{i_2}^*, \cdots, W_{i_\beta}^*\right\}$ be the set of optimal solutions of problems (6), and $L_o = \{i_1, i_2, \cdots, i_\beta\}$. To find the other efficient solutions of the MOILP problem, for each $q \in L_o$, we add the constraints

$$\begin{aligned} &C_r W > C_r W_q^* - M t_{rq}, \quad r = 1, 2, \cdots, s \\ &t_{1q} + t_{2q} + \cdots + t_{sq} \leq s - 1 \\ &t_{rq} \in \{0, 1\}, \quad r = 1, 2, \cdots, s \end{aligned} \tag{8}$$

to problem (6), where M is a large positive integer number. Therefore, we have

$$
\begin{aligned}
&\max \sum_{r=1}^{s} C_r W \\
&\text{s.t. } A_i W \le b_i, i = 1, 2, \cdots, m \\
&\quad C_r W > C_r W_q^* - M t_{rq}, r = 1, 2, \cdots, s \ , \ q \in L_o \\
&\quad \sum_{r=1}^{s} t_{rq} \le s - 1, q \in L_o \\
&\quad W \in Z_n^+, \\
&\quad t_{rq} \in \{0, 1\}, r = 1, 2, \cdots, s \ , \ q \in L_o.
\end{aligned} \tag{9}
$$

As can be seen, if $t_{rq} = 1$, then constraint $C_r W > C_r W_q^* - M t_{rq}$ is redundant. The constraint $\sum_{r=1}^{s} t_{rq} \le s - 1,\ q \in L_o$ implies that at least one of the constraints $C_r W > C_r W_q^* - M t_{rq},\ \ r = 1, 2, \cdots, s$ is not redundant (Jahanshahloo et al. 2004). To solve problem (9), the inequalities

$$
C_r W > C_r W_q^* - M t_{rq}, \quad r = 1, \ldots, s,\ q \in L_o \tag{10}
$$

should be converted to ≥ form. For all $W \in Z_n^+$, the values of $C_r W$, $C_r W_q^*$ and $M t_{rq}$ are integers. Therefore, $C_r W - \left(C_r W_q^* + M t_{rq}\right)$ is an integer, and we can add the positive continuous variable, say δ_{rq}, to the right hand side of the constraint (10). By introducing positive variables $\delta_{rq}(r = 1, \ldots, s, q \in L_o)$, problem (9) is converted to the following problem:

$$
\begin{aligned}
&\max \sum_{r=1}^{s} C_r W \\
&\text{s.t. } A_i W \le b_i, i = 1, 2, \cdots, m \\
&\quad C_r W \ge \delta_{rq} + C_r W_q^* - M t_{rq}, r = 1, 2, \cdots, s, q \in L_o \\
&\quad \sum_{r=1}^{s} t_{rq} \le s - 1, q \in L_o \\
&\quad W \in Z_n^+, \\
&\quad \delta_{rq} \ge \epsilon, t_{rq} \in \{0, 1\}, r = 1, 2, \cdots, s \ , \ q \in L_o
\end{aligned} \tag{11}
$$

where ϵ is a small positive real number.

Theorem *Models (9) and (11) are equivalent.*

Proof For each $j, r(j = 1, \ldots, n, r = 1, \ldots, s) w_j \in Z^+$ and c_{rj} is an integer number. Therefore, for each $W \in X, \sum_{r=1}^{s} C_r W = \sum_{r=1}^{s} \sum_{j=1}^{n} c_{jr} w_j$ is an integer number. Therefore, difference of the sides of the strictly constraints are integer numbers. Thus, adding δ_{rq} (with $\delta_{rq} \ge \epsilon, r = 1, \ldots, s, q \in L_o$) to the lower sides of the strictly constraints does not alter the feasible region of model (6). Therefore, models (9) and (11) are equivalent.

Let $A = \left\{W_{i_{k+1}}^*, W_{i_{k+2}}^*, \cdots, W_{i_{k+l}}^*\right\}$ be the set of optimal solutions of problem (11) where $k = \alpha$ or $k = \beta$. We set $G_1 = G_o \cup A$ (Jahanshahloo et al. 2004). To find the other efficient solution of problem (1), for each $W_q^* \in A$, we add the following constraints to problem (11):

$$
\begin{aligned}
&C_r W \ge \delta_{rq} + C_r W_q^* - M t_{rq}, \quad r = 1, 2, \cdots, s, q \in L_o \\
&\sum_{r=1}^{s} t_{rq} \le s - 1 \\
&\delta_{rq} \ge \epsilon, \quad r = 1, 2, \cdots, s, q \in L_o.
\end{aligned}
$$

Finally, problem (11) can be written as follows,

$$
\begin{aligned}
&\max \sum_{r=1}^{s} C_r W \\
&\text{s.t.} A_i W \le b_i, i = 1, 2, \cdots, m \\
&\quad C_r W \ge \delta_{rq} + C_r W_q^* - M t_{rq}, r = 1, 2, \cdots, s, q = i_1, i_2 \cdots, \\
&\quad i_k, i_{k+1}, \cdots, i_{k+l} \\
&\quad \sum_{r=1}^{s} t_{rq} \le s - 1, q = i_1, i_2 \cdots, i_k, i_{k+1}, \cdots, i_{k+l} \\
&\quad W \in Z_n^+, t_{rq} \in \{0, 1\}, r = 1, 2, \cdots, s, q = i_1, i_2 \cdots, \\
&\quad i_k, i_{k+1}, \cdots, i_{k+l} \\
&\quad \delta_{rq} \ge \epsilon, r = 1, 2, \cdots, s, q = i_1, i_2 \cdots, i_k, i_{k+1}, \cdots, i_{k+l}.
\end{aligned} \tag{12}
$$

This process is continued until problem (12) became infeasible. As noted in (Jahanshahloo et al. (2004)), if problem (9) or (11) or (12) has an alternative optimal solutions, we have to determine all of them. All of them are efficient solutions of problem (1).

Theorem *Each optimal solution of problem (12) is an efficient solution for MOILP problem.*

Proof The proof is similar to that of Theorem 2.4 in (Jahanshahloo et al. 2004) and is omitted.

Using the discussions of the previous sections, in the following cases, an MOILP problem has efficient an solution.

1. X is nonempty and bounded.
2. X is unbounded, and there is no $d \neq 0$ such that $A_i d \le 0, i = 1, \cdots, m, C_r d \ge 0, r = 1, \cdots, s$ with at least one $p(p \in \{1, \cdots, s\})$ such that $C_p d > 0$ and $d \in Z_n^+$.

We can modify the proposed algorithm to solve a 0–1 MOILP in Jahanshahloo et al. (2004) to find all of the efficient solutions of MOILP problems with bounded and unbounded feasible regions as follows. When the numbers of efficient solutions are infinite, we can find their subset (for instance see example 3).

The modified algorithm

Stage 0: Solve the system $A_i d \leq 0, i = 1, \ldots, m, C_r d \geq 0$, $r = 1, \ldots, s$, with at least one $p(p \in \{1, 2, \ldots, s\})$ such that $C_p d > 0$ and $d \in Z_n^+$. If this system has a solution, then there is no efficient solution for problem (1) and go to stage 3. Otherwise, go to stage 1.

Stage 1:

Step 1.1 Set $k = 0$ and solve problem (4) and specify $G_o = \left\{ W_{i_1}^*, W_{i_2}^*, \cdots, W_{i_\alpha}^* \right\}$. If G_o is empty go to step 1.2, otherwise go to step 1.3,

Step 1.2 Determine all optimal solutions of problem (6) and set $G_o = \left\{ W_{i_1}^*, W_{i_2}^*, \cdots, W_{i_\beta}^* \right\}$,

Step 1.3 Determine all optimal solutions of problem (11) and set $A = \left\{ W_{i_{k+1}}^*, W_{i_{k+2}}^*, \cdots, W_{i_{k+l}}^* \right\}$,

Step 1.4 If A is not empty, set $G_1 = G_o \cup A$ and go to stage 2. Otherwise, stop; the set G_o is all of the efficient solutions of problem (3),

Stage 2:

Step 2.1 Determine all optimal solutions of problem (12), and suppose B is the set of optimal solutions of problem (12),

Step 1.4 If A is not empty, set $G_1 = G_o \cup A$ and go to stage 2. Otherwise, stop; the set G_o is all of the efficient solutions of problem (3),

Stage 2:

Step 2.1 Determine all optimal solutions of problem (12), and suppose B is the set of optimal solutions of problem (12),

Step 2.2 If B is not empty, set $G_{k+1} = G_k \cup B$ and go to stage 2.

Otherwise, stop; the set G_k is all of the efficient solutions of problem (1),

Stage 3: End.

Theorem*: there is no recession directionLet the number of efficient solutions of an MOILP problem be finite. Then, the modified algorithm generates all of them with bounded or unbounded feasible region.*

Proof The feasible region of MOILP problem is bounded or unbounded. Therefore, we can consider the following cases:

1. If there is a d such that $A_i d \leq 0$, $i = 1, \ldots, m$, $C_r d \geq 0, r = 1, \ldots, s$, with at least one $p(p \in \{1, 2, \ldots, s\})$ such that $C_p d > 0$ and $d \in Z_n^+$, then feasible region of the MOILP is unbounded, and objective function values can become infinite together. In this case, the algorithm is stopped.
2. If X is nonempty and bounded. The proof is similar to that of Theorem 2.5 in (Jahanshahloo et al. 2004).
3. If feasible region is unbounded and there is no d such that $A_i d \leq 0$, $i = 1, \ldots, m$, $C_r d \geq 0, r = 1, \ldots, s$, with at least one $p(p \in \{1, 2, \ldots, s\})$ such that $C_p d > 0$ and $d \in Z_n^+$, then objective functions values of the MOILP cannot become infinite together. In this case, in order to prove the theorem by adding the constraint $w_1 + w_2 + \cdots + w_n \leq M$ to problem (11), its feasible region is converted to a bounded feasible region where M is a large positive number. Also, the rest of the proof for this case is similar to that of Theorem 2.5 in (Jahanshahloo et al. 2004) and is omitted.

Theorem *Let the number of efficient solutions be finite. Then, the modified algorithm is convergent.*

Proof Let the number of efficient solutions be finite. In any iteration, at least one efficient solution is found, and the found efficient solutions are not the same. Therefore, the modified algorithm is convergent.

Examples This section illustrates the proposed algorithm for the four MOILP problems with bounded or unbounded feasible regions.

Example 1 Consider the following MOILP problem with two objective functions.

$$\begin{aligned} \max \quad & w_1 + w_2 \\ \max \quad & 4w_1 + 3w_2 \\ \text{s.t.} \quad & -3w_1 + 2w_2 \leq 6 \\ & -6w_1 + 10w_2 \leq 60 \\ & w_1, \; w_2 \in Z^+. \end{aligned}$$

As can be seen, there is $d = \begin{pmatrix} d_1 \\ d_2 \end{pmatrix}$ such that $A_i d \leq 0$, $i = 1, 2$, $C_r d \geq 0, r = 1, 2$ and $d \in Z_2^+$, where $A_1 = (-3, 2)$, $A_2 = (-6, 10)$, $C_1 = (1, 1)$ and $C_2 = (4, 3)$. That is, feasible region is unbounded and objective functions can become infinite together. Therefore, there is no any efficient solution for this problem.

Example 2 The feasible region of the following MOILP problem is unbounded.

$$\begin{aligned} \max \quad & w_1 + w_2 \\ \max \quad & -2w_2 \\ \text{s.t.} \quad & -w_1 + w_2 \le 1 \\ & -4w_1 + 6w_2 \le 24 \\ & w_1,\ w_2 \in Z^+. \end{aligned}$$

As can be seen, $d = \begin{pmatrix} d_1 \\ d_2 \end{pmatrix} = \begin{pmatrix} 1 \\ 0 \end{pmatrix}$ is a solution of the following system,

$$A_i d \le 0,\ i = 1,2,\ C_1 d > 0,\ C_2 d \ge 0,\ d \in Z_2^+$$

where $A_1 = (-1,1), A_2 = (-4,6),\ C_1 = (1,1)$ and $C_2 = (0,-2)$. Similar to Example 1, there is no efficient solution for this problem.

Example 3 Consider the following MOILP problem

$$\begin{aligned} \max \quad & -2w_1 + w_2 \\ \max \quad & w_1 - 3w_2 \\ \text{s.t.} \quad & -4w_1 + w_2 \le 4 \\ & -9w_1 + 5w_2 \le 45 \\ & w_1,\ w_2 \in Z^+. \end{aligned} \tag{13}$$

It is evident that there is d say $d = \begin{pmatrix} d_1 \\ d_2 \end{pmatrix} = \begin{pmatrix} 1 \\ 1 \end{pmatrix}$, such that $A_i d \le 0,\ i = 1,2,\ d \in Z_2^+$ where $A_1 = (-4,1)$, $A_2 = (-9,5)$. That is, the feasible region of this problem is unbounded. But, there is no recession direction such that,

$$C_r d \le 0,\ r = 1,2, \exists p \in \{1,2\}, \mathrm{C_p d} \ge 0, \mathrm{d} \in \mathrm{Z}_2^+,$$

where $A_1 = (-1,1),\ \ A_2 = (-4,6),\ \ C_1 = (-2,1)$, and $C_2 = (1,-3)$. Therefore, this problem has an efficient solution.

Stage 1, Step 1.1: Consider the following single objective integer programming problems:

$$\begin{aligned} \max \quad & -2w_1 + w_2 \\ \text{s.t.} \quad & -4w_1 + w_2 \le 4 \\ & -9w_1 + 5w_2 \le 45 \\ & w_1,\ w_2 \in Z^+ \end{aligned} \tag{14}$$

and

$$\begin{aligned} \max \quad & w_1 - 3w_2 \\ \text{s.t.} \quad & -4w_1 + w_2 \le 4 \\ & -9w_1 + 5w_2 \le 45 \\ & w_1,\ w_2 \in Z^+. \end{aligned} \tag{15}$$

$W_1^* = (5,18)$ is the optimal solution of problem (14), and $Y^1 = (8,-49)^T$ is its objective value vector. However, the optimal value of problem (15) is infinite, and this problem does not have an optimal solution. Therefore, $G_o = \{1\}$.

Step 1.2: The corresponding problem of $G_o = \{1\}$ is as follows:

$$\begin{aligned} \max \quad & -w_1 - 2w_2 \\ \text{s.t.} \quad & -4w_1 + w_2 \le 4 \\ & -9w_1 + 5w_2 \le 45 \\ & -2w_1 + w_2 - \delta_1 + 100t_{11} \ge 8 \\ & w_1 - 3w_2 - \delta_2 + 100t_{21} \ge -49 \\ & t_{11} + t_{12} \le 1 \\ & t_{11},\ t_{12} \in \{0,1\} \\ & w_1,\ w_2 \in Z^+ \\ & \delta_1, \delta_2 \ge \epsilon. \end{aligned} \tag{16}$$

$W_2^* = (0,0)$ is the optimal solution of problem (16), and $Y^2 = (0,0)^T$ is its objective value vector. Therefore, $A = \{(0,0)\}$.

Step 1.3: $G_1 = A \cup G_o = \{(5,18),(0,0)\}$

Stage 2, *Iteration 1*

Step 2.1: The corresponding problem of G_1 is as follows:

$$\begin{aligned} \max \quad & -w_1 - 2w_2 \\ \text{s.t.} \quad & -4w_1 + w_2 \le 4 \\ & -9w_1 + 5w_2 \le 45 \\ & -2w_1 + w_2 - \delta_1 + 100t_{11} \ge 8 \\ & w_1 - 3w_2 - \delta_2 + 100t_{21} \ge -49 \\ & t_{11} + t_{21} \le 1 \\ & -2w_1 + w_2 - \delta_3 + 100t_{12} \ge 0 \\ & w_1 - 3w_2 - \delta_4 + 100t_{22} \ge 0 \\ & t_{21} + t_{22} \le 1 \\ & t_{11},\ t_{21}, t_{12}, t_{22}, \in \{0,1\} \\ & w_1,\ w_2 \in Z^+ \\ & \delta_1, \delta_2, \delta_3, \delta_4 \ge \epsilon. \end{aligned} \tag{17}$$

$W_2^* = (0,0)$ is the optimal solution of problem (17), and $Y^3 = (-2,1)^T$ is its objective value vector. Thus, $B = \left\{(1,0)^T\right\}$.

Iteration 2

Step 2.1: The corresponding problem of G_1 is as follows:

$$\begin{aligned} \max \quad & -w_1 - 2w_2 \\ \text{s.t.} \quad & -4w_1 + w_2 \le 4 \\ & -9w_1 + 5w_2 \le 45 \\ & -2w_1 + w_2 - \delta_1 + 100t_{11} \ge 8 \\ & w_1 - 3w_2 - \delta_2 + 100t_{21} \ge -49 \\ & t_{11} + t_{21} \le 1 \\ & -2w_1 + w_2 - \delta_3 + 100t_{12} \ge 0 \\ & w_1 - 3w_2 - \delta_4 + 100t_{22} \ge 0 \\ & t_{12} + t_{22} \le 1 \\ & -2w_1 + w_2 - \delta_5 + 100t_{13} \ge 0 \\ & w_1 - 3w_2 - \delta_6 + 100t_{23} \ge 0 \\ & t_{13} + t_{23} \le 1 \\ & t_{11},\ t_{21}, t_{12}, t_{22}, t_{13}, t_{23} \in \{0,1\} \\ & w_1,\ w_2 \in Z^+ \\ & \delta_1, \delta_2, \delta_3, \delta_4, \delta_5, \delta_6 \ge \epsilon. \end{aligned} \tag{18}$$

$W_4^* = (0,1)^T$ and $W_5^* = (2,0)^T$ are optimal solutions of problem (18). Therefore, $Y^4 = (1,-3)^T$ and $Y^5 = (-4,2)^T$ are its objective value vectors, respectively. Hence, $B = \{(0,1),(2,0)\}$.

Using the other single objective integer problems, we find that for each $n \in Z^+$, $W^* = (0,n)$ is an efficient solution of problem (13). Hence, the number of the efficient solution of this problem is infinite, and the proposed approach finds at least one of them in any iteration.

Example 4 Consider the following MOILP problem with $s = m = 2$ and $n = 3$:

$$\begin{array}{ll} \max & w_1 + 2w_2 - w_3 \\ \max & 2w_1 + w_2 + 3w_3 \\ \text{s.t.} & w_1 + w_2 + w_3 \le 3 \\ & 2w_1 + w_2 - w_3 \le 2 \\ & w_1,\ w_2,\ w_3 \in Z^+. \end{array}$$

As can be seen, there is no $d = (d_1, d_2, d_3)^T$ such that $A_i d \le 0$, $i = 1,2$ and $d \in Z_3^+$, where $A_1 = (1,1,1)$ and $A_2 = (2,1,-1)$. Therefore, the feasible region is bounded, and the problem has an efficient solution.

Step 1.1: By solving the following problems, the members of G_o are specified.

$$\begin{array}{ll} \max & w_1 + 2w_2 - w_3 \\ \text{s.t.} & w_1 + w_2 + w_3 \le 3 \\ & 2w_1 + w_2 - w_3 \le 2 \\ & w_1,\ w_2,\ w_3 \in Z^+ \end{array}$$

and

$$\begin{array}{ll} \max & 2w_1 + w_2 + 3w_3 \\ \text{s.t.} & w_1 + w_2 + w_3 \le 3 \\ & 2w_1 + w_2 - w_3 \le 2 \\ & w_1,\ w_2,\ w_3 \in Z^+. \end{array}$$

$W_1^* = (0,2,0)^T$ and $W_2^* = (0,0,3)^T$ are the optimal solutions of the above single objective integer problems, respectively, and their objective value vectors are $Y^1 = (4,2)^T$ and $Y^2 = (-3,9)^T$ (Figure 1). Hence, $G_o = \left\{(0,2,0)^T, (0,0,3)^T\right\}$.

Step 1.2: The corresponding problem of G_o is as follows:

$$\begin{array}{ll} \max & 3w_1 + 3w_2 + 2w_3 \\ \text{s.t.} & w_1 + w_2 + w_3 \le 2w_1 + w_2 - w_3 \le 2 \\ & w_1 + 2w_2 - w_3 - \delta_1 + 100t_{11} \ge 5 \\ & 2w_1 + w_2 + 3w_3 - \delta_2 + 100t_{21} \ge 3 \\ & w_1 + 2w_2 - w_3 - \delta_3 + 100t_{12} \ge -3 \\ & 2w_1 + w_2 + 3w_3 - \delta_4 + 100t_{22} \ge 10 \\ & t_{11} + t_{21} \le 1 \\ & t_{12} + t_{22} \le 1 \\ & w_1,\ w_2,\ w_3 \in Z^+ \\ & \delta_1, \delta_2, \delta_3, \delta_4 \ge \epsilon \\ & t_{11},\ t_{12},\ t_{21},\ t_{22} \in \{0,1\}. \end{array}$$

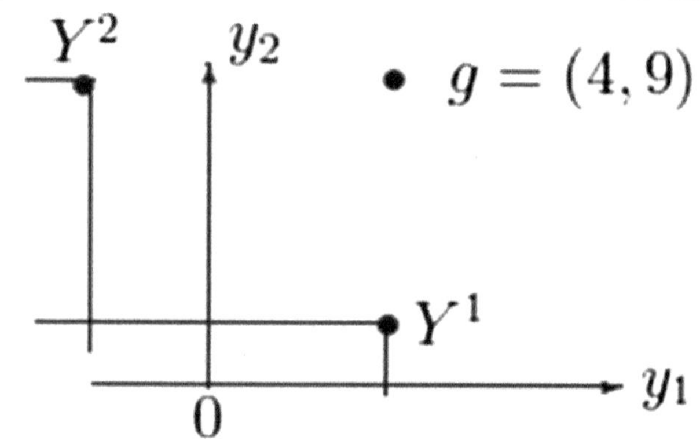

Figure 1 The first two efficient solutions of the MOLP example.

By solving the above problem, we have $W_3^* = (0,2,1)^T$, $W_4^* = (1,1,1)^T$ and $A = \left\{(0,2,1)^T, (1,1,1)^T\right\}$. Therefore, the corresponding objective value vectors are $Y^3 = (3,5)^T$ and $Y^4 = (2,6)^T$ (Figure 2).

Step 1.3: $G_1 = A \cup G_o = \left\{(0,2,0)^T, (0,0,3)^T, (0,2,1)^T, (1,1,1)^T\right\}$

Stage 2, ***Iteration 1***

Step 2.1: The corresponding problem of G_1 is as follows:

$$\begin{array}{ll} \max & 3w_1 + 3w_2 + 2w_3 \\ \text{s.t.} & w_1 + w_2 + w_3 \le 5 \\ & 2w_1 + w_2 - w_3 \le 4 \\ & w_1 + 2w_2 - w_3 - \delta_1 + 100t_{11} \ge 5 \\ & 2w_1 + w_2 + 3w_3 - \delta_2 + 100t_{21} \ge 3 \\ & w_1 + 2w_2 - w_3 - \delta_3 + 100t_{12} \ge -3 \\ & 2w_1 + w_2 + 3w_3 - \delta_4 + 100t_{22} \ge 10 \\ & w_1 + 2w_2 - w_3 - \delta_5 + 100t_{13} \ge 4 \end{array}$$

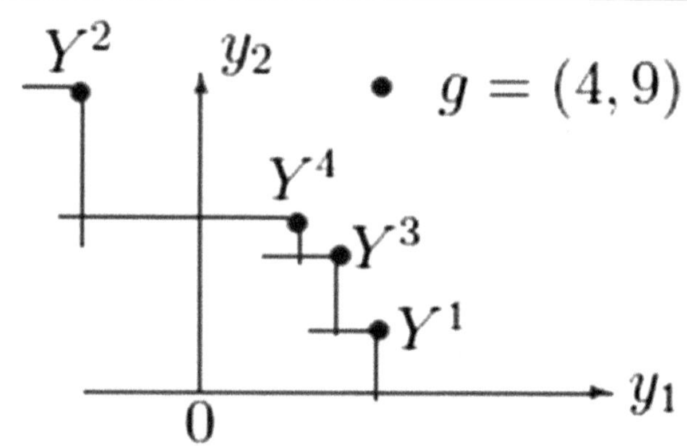

Figure 2 The first two efficient solutions of the MOLP example.

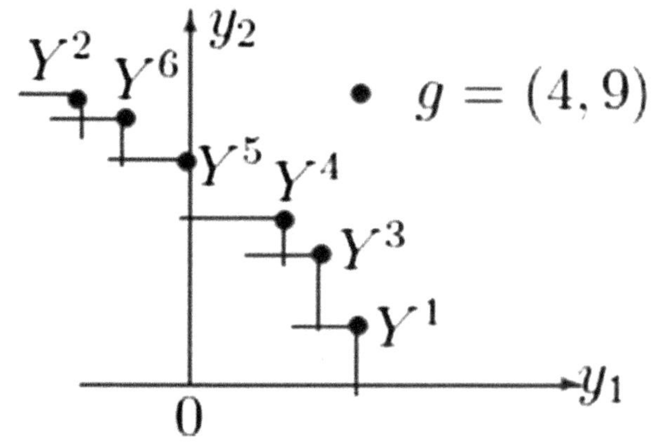

Figure 3 The all of efficient solutions of the MOLP example.

$$2w_1 + w_2 + 3w_3 - \delta_6 + 100t_{23} \geq 6$$
$$w_1 + 2w_2 - w_3 - \delta_7 + 100t_{14} \geq 3$$
$$2w_1 + w_2 + 3w_3 - \delta_8 + 100t_{24} \geq 7$$
$$t_{11} + t_{21} \leq 1 t_{12} + t_{22} \leq 1$$
$$t_{13} + t_{23} \leq 1$$
$$t_{14} + t_{24} \leq 1$$
$$w_1,\ w_2,\ w_3 \in Z^+$$
$$\delta_p \geq \epsilon, \quad p = 1, \ldots, 8$$
$$t_{11},\ t_{12},\ t_{21},\ t_{22},\ t_{13},\ t_{23},\ t_{14},\ t_{24} \in \{0, 1\}.$$

By solving the above problem, we will have $W_5^* = (0,1,2)^T$, $W_6^* = (1,0,2)^T$, $Y^5 = (0,7)^T$, $Y^6 = (-1,8)^T$ (Figure 3), and $B = \{(0,1,2),(1,0,2)\}$.

Step2.2: $G_2 = B \cup G_1 = \left\{(0,2,0)^T, (0,0,3)^T, (0,2,1)^T, (1,1,1)^T, (0,1,2)^T, (1,0,2)^T\right\}$.

Iteration 2

Step 1.1: The corresponding problem of G_2 is infeasible. Hence, the members of G_2 are efficient solutions for the MOILP example.

Conclusions

This paper considered MOILP problems with bounded and unbounded feasible regions. Then, the proposed method in (Jahanshahloo et al. 2004) has been modified to find all of the efficient solutions of an MOILP problem when the number of efficient solutions is finite. In any iteration of the modified algorithm, at least one efficient solution of an MOILP problem is found. If W_p^* and W_{p+1}^* are two efficient solutions of an MOILP which have been obtained in p^{th} and $(p+1)^{th}$ iterations, respectively, then the distance of Y_p is less than Y_{p+1} from g using the $L1$ norm, and the rank of W_p^* is higher than W_{p+1}^*. Therefore, the efficient solutions of the MOILP problem are ranked using the proposed algorithm.

Competing interests
The authors declare that they have no competing interests.

Authors' contributions
The final manuscript is approved by Ghasem Tohidi and Shabnam Razavyan.

Authors' information
Ghasem Tohidi is currently an Assistance Professor in the Department of Mathematics of Islamic Azad University, Central Tehran Branch. His research interests include Data Envelopment Analysis (DEA), Fuzzy and Multi Objective Programming (MOP). He has published many articles in a number of peer reviewed academic journals and conferences.
Shabnam Razavyan is currently an Assistance Professor in the Department of Mathematics of Islamic Azad University, South Tehran Branch. Her research interests include Data Envelopment Analysis (DEA) and Multi Objective Programming (MOP). She has published many articles in a number of peer reviewed academic journals and conferences. She is a member of the Iranian Operations Research Society.

Acknowledgment
Financial support of this research, in a project, is from the Islamic Azad University, Central Tehran Branch is acknowledged.

Author details
[1]Assistant professor Department of Mathematics, Islamic Azad University, Central Tehran Branch, Tehran, Iran. [2]Assistant professor Department of Mathematics, Islamic Azad University, South Tehran Branch, Tehran, Iran.

References

Bazaraa MS, Jarvis JJ, Sherali HD (2007) Linear programming and network flows. John Wiley and Sons Inc., New York

Climaco J, Ferreira C, Captivo ME (1997) Multicriteria integer programming. In: Climaco J (ed) Multicriteria analysis. An overview of different algorithmic approaches, Springer, Berlin, pp 248–258

Ehrgott M (2006) A discussion of scalarization techniques for multiple objective integer programming. Annals of Operational Research 147(1):343–360

Jahanshahloo GR, Hosseinzadeh Lotfi F, Shoja N, Tohidi G (2004) A method for generating all the efficient solutions of a 0–1 multi-objective linear programming problem. Asia-Pacific Journal of Operational Research 21 (1):127–139

Klamroth K, Tind J, Zust S (2004) Integer programming duality in multiple objective programming. Journal of Global Optimization 29:1–18

Mavrotas G, Diakoulaki D (1998) A branch and bound algorithm for mixed zero one multiple objective linear programming. European Journal of Operational Research 107:530–541

Rasmussen LM (1986) Zero one programming with multiple criteria. European Journal of Operational Research 26:83–95

Sylva J, Crema A (2004) A method for finding the set of nondominated vectors for multiple objective integer linear programs. European Journal of Operational Research 158(1):46–55

Sylva J, Crema A (2007) A method for finding well-dispersed subsets of non-dominated vectors for multiple objective mixed integer linear programs. European Journal of Operational Research 180:1011–1027

Teghem J, Kunsch PLJ (1986) A survey of techniques for finding efficient solutions to multi-objective integer linear programming. Asia- Pacific Journal of Operational Research 3:95–108

Ulungu EL, Teghem J (1994) Multi-objective combinatorial optimization problems: a survey. Journal of Multi-Criteria Decision Analysis 3:83–104

Phase II monitoring of auto-correlated linear profiles using linear mixed model

A Narvand, P Soleimani and Sadigh Raissi*

Abstract

In many circumstances, the quality of a process or product is best characterized by a given mathematical function between a response variable and one or more explanatory variables that is typically referred to as profile. There are some investigations to monitor auto-correlated linear and nonlinear profiles in recent years. In the present paper, we use the linear mixed models to account autocorrelation within observations which is gathered on phase II of the monitoring process. We undertake that the structure of correlated linear profiles simultaneously has both random and fixed effects. The work enhanced a Hotelling's T^2 statistic, a multivariate exponential weighted moving average (MEWMA), and a multivariate cumulative sum (MCUSUM) control charts to monitor process. We also compared their performances, in terms of average run length criterion, and designated that the proposed control charts schemes could effectively act in detecting shifts in process parameters. Finally, the results are applied on a real case study in an agricultural field.

Keywords: Profile monitoring; linear mixed model; MCUSUM, MEWMA; Hotelling's T^2; autocorrelation; average run length

Introduction

Control charts are used to detect anomalies in the processes. They are most often used to monitor production-related processes. In many business-related processes, the quality of a process or product can be characterized by a relationship between a response variable and one or more explanatory variables which is referred to as profile. The purpose of the analyzing of profile in phase I is to determine the stability of the process and estimate parameters, however, in phase II, analyzers are interested in rapidly detecting the significant shifts in the process parameters. Phase I analysis of simple linear profiles has been investigated by a number of authors such as Stover and Brill (1998), Kang and Albin (2000), Kim et al. (2003) and Mahmoud et al. (Mahmoud and Woodall 2004, 2007). Many authors including Kang and Albin (2000), Kim et al. (2003), Noorossana et al. (2004), Gupta et al. (2006), Zou et al. (2006), Saghaei et al. (2009), and Mahmoud et al. (2009) have investigated phase II monitoring of simple linear profiles. Noorossana et al. (2010a, b) investigated monitoring of multivariate simple linear profiles on phase II. Zou et al. (2007) and Kazemzadeh et al. (2009a, b) considered cases when the profiles can be characterized by multiple and polynomial regression models respectively. Mahmoud (2008) considered phase I monitoring of multiple linear profiles, and Kazemzadeh et al. (2008) proposed three methods for monitoring the *k*th-order polynomial profile in phase I. Ding et al. (2006), Moguerza et al. (2007), Williams et al. (2007), and Vaghefi et al. (2009) investigated nonlinear profiles. In these studies, it is implicitly assumed that the error terms within or between profiles is independently and identically normally distributed; however in some cases, these assumptions can be violated. Noorossana et al. (2010a, b) analyzed the effects of non-normality on the monitoring of simple linear profiles. Noorossana et al. (2008) and Kazemzadeh et al. (2009a, b) investigated autocorrelation between successive simple linear and polynomial profiles respectively. Soleimani et al. (2009) proposed a transformation to eliminate the autocorrelation between observations within a simple linear profile in phase II. Jensen et al. (2008) proposed two T^2 control charts based on linear mixed model (LMM) to account for the autocorrelation within linear profiles in phase I. They concluded that the linear mixed

* Correspondence: Raissi@azad.ac.ir
School of Industrial Engineering, South Tehran Branch, Islamic Azad University, Tehran, Iran

model is superior to the least square approach for unbalanced or missing data, especially when the number of observation within a profile is small and the correlation is weak. Jensen and Birch (2009) used nonlinear mixed model to account correlation within nonlinear profiles. Qie et al. (2010) investigated nonparametric profile monitoring with arbitrary design using mixed models. They proposed a control chart that combines the exponentially weighted moving average control chart based on local linear kernel smoothing and a nonparametric regression test under the assumption that observations within and between individual profiles are independent of each other.

The present study acts as an extension of the work of Jensen et al. (2008) in applying a linear mixed model on the presence of autocorrelation within linear profiles on phase I control chart applications; conversely, our focus is on phase II of profile monitoring in which one could use the proposed control charts to detect any departures from the given profile parameters.

The remainder of the paper is organized as follows. In 'Linear mixed model' Section, the LMM is mathematically presented. In the 'Proposed methods' Section, our methods including three modified multivariate control charts namely Hotelling T^2, multivariate exponential weighted moving average (MEWMA) and a multivariate cumulative sum control charts (MCUSUM) are illustrated. In 'Simulation studies' Section, the results of simulation study to evaluate the performance of the methods are presented. In addition, a case study from an agriculture field is investigated on the Section 'Case study'. The final section closes with concluding remarks.

Linear mixed model

Linear mixed models (Laird and Ware 1982) are popular for analysis of longitudinal data. A linear mixed model contains fixed and random effects and is linear in these effects. This model allows us to account autocorrelation within profiles. In matrix notation, a mixed model can be represented as

$$\mathbf{y} = \mathbf{X}\boldsymbol{\beta} + \mathbf{Z}\mathbf{b} + \epsilon, \tag{1}$$

where $\mathbf{y}$ is a vector of observations, with mean $E(\mathbf{y}) = \mathbf{X}\boldsymbol{\beta}$, $\boldsymbol{\beta}$ is a vector of fixed effects, $\mathbf{b}$ is a vector of independent and identically distributed (IID) random effects with mean $E(\mathbf{b}) = \mathbf{0}$ and variance-covariance matrix $\mathrm{Var}(\mathbf{b}) = \mathbf{D}$, ϵ is a vector of IID random error terms with mean $E(\epsilon) = \mathbf{0}$ and variance $\mathrm{Var}(\epsilon) = \mathbf{R}$, and $\mathbf{X}$ and $\mathbf{Z}$ are matrices of regressors relating the observations $\mathbf{y}$ to $\boldsymbol{\beta}$ and $\mathbf{b}$.

In the 1950s, Charles Roy Henderson provided the best linear unbiased estimate (BLUE) of fixed effects and best linear unbiased predictions (BLUP) of random effects. Subsequently, mixed modeling has become a major area of statistical research, including work on the computation of maximum likelihood estimates, nonlinear mixed effect models, missing data in mixed effects models, and Bayesian estimation of mixed effects models (West et al. 2007).

Henderson's 'mixed model equations' (MME) are (Robinson 1991) as follows:

$$\begin{pmatrix} \mathbf{X}'\mathbf{R}^{-1}\mathbf{X} & \mathbf{X}'\mathbf{R}^{-1}\mathbf{Z} \\ \mathbf{Z}'\mathbf{R}^{-1}\mathbf{X} & \mathbf{Z}'\mathbf{R}^{-1}\mathbf{Z} + \mathbf{D}^{-1} \end{pmatrix} \begin{pmatrix} \hat{\boldsymbol{\beta}} \\ \hat{\mathbf{b}} \end{pmatrix} = \begin{pmatrix} \mathbf{X}'\mathbf{R}^{-1}\mathbf{y} \\ \mathbf{Z}'\mathbf{R}^{-1}\mathbf{y} \end{pmatrix} \tag{2}$$

The solutions to the MME, $\hat{\boldsymbol{\beta}}$, and $\hat{\mathbf{b}}$ are BLUEs and BLUPs for $\boldsymbol{\beta}$ and $\mathbf{b}$, respectively.

Mixed models require somewhat sophisticated computing algorithms to fit. Solutions to the MME are obtained by methods similar to those used for linear least squares. For complicated models and large datasets, iterative methods may be needed.

In profile monitoring, one could suppose that the jth response follows a LMM; therefore,

$$\mathbf{y}_j = \mathbf{X}_j\boldsymbol{\beta} + \mathbf{Z}_j\mathbf{b}_j + \boldsymbol{\varepsilon}_j; j = 1, 2, \ldots, m, \tag{3}$$

where $\mathbf{X}_i$ is a $(n_j \times p)$ matrix of regressors, and $\mathbf{Z}_j$ is a $(n_j \times q)$ matrix associated with random effects. $\boldsymbol{\beta}$ is a $(p \times 1)$ vector of fixed effects, and $\mathbf{y}_j$ is the $(n_j \times 1)$ response vector for the jth profile. The coefficient vector of the random effect terms is $\mathbf{b}_j \sim MN\ (0,\mathbf{D})$, and $\mathbf{D}$ is assumed to be a diagonal matrix; thus, the random effects are assumed not to be correlated with each other. In addition, it is assumed that $\mathrm{Cov}(\boldsymbol{\varepsilon}_j, \mathbf{b}_j) = 0$ and $\boldsymbol{\varepsilon}_j$ is $(n_j \times 1)$ vector of errors where $\boldsymbol{\varepsilon}_j \sim MN\ (0,\mathbf{R}_j)$. If the errors are assumed to be independent, $\mathbf{R}_j = \sigma^2\mathbf{I}$, but correlated, the functional structure for the error terms may be used.

As noted before, it is considered that $\hat{\boldsymbol{\beta}}$ is an estimator of $\boldsymbol{\beta}$, and $\hat{\mathbf{b}}_j$ is a predictor of $\mathbf{b}_j$, then $\hat{\mathbf{y}}_j = \mathbf{X}_j\hat{\boldsymbol{\beta}}$ is the population average, and $\hat{\mathbf{y}}_j = \mathbf{X}_j\hat{\boldsymbol{\beta}} + \mathbf{Z}_j\hat{\mathbf{b}}_j$ is the profile specific prediction; so if $\mathbf{D}$ and $\mathbf{R}_j$ are known, then it can be shown as follows:

$$\hat{\boldsymbol{\beta}} = \left(\sum\nolimits_{j=1}^{m}\mathbf{X}_j'\mathbf{V}_j^{-1}\mathbf{X}_j\right)^{-1}\left(\sum\nolimits_{j=1}^{m}\mathbf{X}_j'\mathbf{V}_j^{-1}\mathbf{y}_j\right), \tag{4}$$

and the BLUP of $\mathbf{b}$ is

$$\hat{\mathbf{b}}_j = \mathbf{D}\mathbf{Z}_j'\mathbf{V}_j^{-1}\left(\mathbf{y}_j - \mathbf{X}_j\hat{\boldsymbol{\beta}}\right), \tag{5}$$

where $\mathbf{V}_j = \mathbf{Z}_j\mathbf{D}\mathbf{Z}_j' + \mathbf{R}_j$ is the overall estimated variance covariance matrix (Schabenberger and Pierce 2002).

Proposed methods

In this paper, we propose a linear mixed model approach for accounting the correlation within linear profiles in phase II. It is assumed that profiles are correlated based on first-order autoregressive (AR(1)) structure.

If the errors follow an auto-correlated structure such as an AR(1) structure, then $\mathbf{R}_j$ by Schabenberger and Pierce (2002) is given as Equation 6. More details on the types of correlated errors structures could be acquired in this reference as

$$\mathbf{R}j = \sigma^2 \begin{bmatrix} 1 & \rho & \rho^2 & \cdots & \rho^{nj-1} \\ \rho & 1 & \rho & \cdots & \rho^{nj-2} \\ \rho^2 & \rho & 1 & \cdots & \rho^{nj-3} \\ \vdots & \vdots & \vdots & \ddots & \vdots \\ \rho^{nj-1} & \rho^{nj-2} & \rho^{nj-3} & \cdots & 1 \end{bmatrix}. \quad (6)$$

It is assumed that for the jth sample collected over time, our observations are $(\mathbf{X}_i, \mathbf{y}_{ij})$, $i = 1,2,\ldots,n$ and $j = 1,2,\ldots,m$. We considered the case that all the fixed effects have a corresponding random effect, $(\mathbf{X}_j = \mathbf{Z}_j)$. If the process is in control, the problem can be formulated as follows:

$$y_{ij} = (\beta_0 + b_{0j}) + (\beta_1 + b_{1j})x_{i1} + \ldots + \left(\beta_{p-1} + b_{p-1j}\right)x_{p-1i} + \varepsilon_{ij} \quad (7)$$

and

$$\varepsilon_{ij} = \rho\varepsilon_{ij-1} + a_{ij}, \quad (8)$$

where ε_{ij} are the correlated error terms and a_{ij} are white noises as $a_{ij} \sim N(0,\sigma^2)$. The $\beta_0, \beta_1,\ldots,\beta_{p-1}$ are fixed effects that are the same for all profiles. The $b_{0j}, b_{1j},\ldots,b_{p-1j}$ are random effects for the jth profile and they are normal random variables with zero mean and variance of $\sigma_0^2, \sigma_1^2, \ldots, \sigma_{p-1}^2$, respectively, which are not to be correlated with each other and also not to be correlated with the errors. The x values are fixed and constant from profile to profile. In this article we especially focused on phase II of the monitoring process, so all profile's parameters, process variance, and correlation coefficient are known in phase I. Accordingly, we utilized the modified Jensen et al. (2008) approach to monitor autocorrelation on phase II.

The Hotelling's T^2 statistic control chart

As a first proposed control chart, we use T^2 statistic to monitor the fixed effects for each sample. This statistic is given by

$$T_j^2 = \left(\hat{\boldsymbol{\beta}}_j - \boldsymbol{\beta}_0\right)' \Sigma^{-1} \left(\hat{\boldsymbol{\beta}}_j - \boldsymbol{\beta}_0\right), \quad (9)$$

where $\hat{\boldsymbol{\beta}}_j = \left(\mathbf{X}_j^{-1}\mathbf{V}_j^{-1}\mathbf{X}_j\right)^{-1}\left(\mathbf{X}_j^{-1}\mathbf{V}_j^{-1}\mathbf{y}_j\right)$ and $\boldsymbol{\beta}_0$ denote the in-control value of $\boldsymbol{\beta}$.

In Equation 9 the variance covariance matrix of fixed effects is $\sum_{\boldsymbol{\beta}} = \left(\mathbf{X}_j'\mathbf{V}_j^{-1}\mathbf{X}_j\right)^{-1}$.

The upper control limit, UCL, is chosen to achieve a specified in control average run length (ARL).

The MEWMA control chart

Our second proposed control chart is based on MEWMA proposed by Lowry et al. (1992) for monitoring the vector of $\hat{\boldsymbol{\beta}}_j$. Here the MEWMA statistics is as follows:

$$\mathbf{z}_j = \theta\left(\hat{\boldsymbol{\beta}}_j - \boldsymbol{\beta}_0\right)' + (1-\theta)\mathbf{z}_{j-1}, \quad (10)$$

where $\mathbf{z}_0 = 0$ and $\theta (0 < \theta < 1)$ is the smoothing parameter. Therefore, the chart statistic denotes by MEWMA_j is given by

$$\text{MEWMA}_j = \mathbf{z}_j' {\sum}_{\mathbf{Z}}^{-1} \mathbf{z}_j, \quad (11)$$

where $\sum_{\mathbf{z}} = \frac{\theta}{2-\theta}\sum_{\boldsymbol{\beta}} = \frac{\theta}{2-\theta}\left(\mathbf{X}_j'\mathbf{V}_j^{-1}\mathbf{X}_j\right)^{-1}$.

This control chart gives a signal when $\text{EWMA}_j > \text{UCL}$, where $(\text{UCL} > 0)$ is chosen to achieve a specified in control ARL.

The MCUSUM control chart

The third suggested method is based on the MCUSUM control chart which is proposed by Crosier (1988). In this method, the statistic is given by

$$\mathbf{s}_j = \begin{cases} 0 & \text{for } c_j \le k \\ \left(\mathbf{s}_{j-1} + \hat{\boldsymbol{\beta}}_j - \boldsymbol{\beta}_0\right)(1 - k/c_j); & \text{for } c_j > k \end{cases}, \quad (12)$$

where

$$\mathbf{c}_j = \left[\left(\mathbf{s}_{j-1} + \hat{\boldsymbol{\beta}}_j - \boldsymbol{\beta}_0\right)' \sum^{-1} \left(\mathbf{s}_{j-1} + \hat{\boldsymbol{\beta}}_j - \boldsymbol{\beta}_0\right)\right]^{1/2} \text{ and } \mathbf{s}_0 = 0$$

and k is a selected constant.

The estimator of variance covariance matrix is

$$\sum_{\boldsymbol{\beta}} = \left(\mathbf{X}_j'\mathbf{V}_j^{-1}\mathbf{X}_j\right)^{-1}. \quad (13)$$

The chart gives a signal if $\left(\mathbf{s}_j' \sum^{-1} \mathbf{s}_j\right)^{1/2} > \text{UCL}$ where $(\text{UCL} > 0)$ is chosen to achieve the desired in-control ARL.

Simulation studies

To show the performance of the proposed methods, we considered the underlying linear profile as Equation 14:

$$y_{ij} = 3 + b_{0j} + (2 + b_{1j})x_i + \varepsilon_{ij}, \quad (14)$$

where

$$\varepsilon_{ij} = \rho\varepsilon_{i-1j} + a_{ij}, \quad (15)$$

and $a_{ij} \sim N(0,1), b_{0j} \sim N(0,.1), b_{1j} \sim N(0,.1)$. In our simulation investigation, we considered three significant different autocorrelation coefficients: a $\rho = 0.1$ to designate a weak type auto-correlated process, intermediate autocorrelation

Table 1 Simulated UCL for control charts

ρ Value	UCL for control charts		
	T^2	MEWMA	MCUSUM
0.1	10.75	9.8	5.56
0.5	6.68	8.8	4.98
0.9	8.32	7.30	4.15

by $\rho = 0.5$, and strong autocorrelation by $\rho = 0.9$. The in-control ARL is roughly set equal to 200 and the ARL values were evaluated through 10,000 simulation replications under different shifts in intercept, slope, and errors (standard deviation). For MEWMA control chart, the smoothing parameter θ is chosen to be 0.2. As a general rule, to design MCUSUM control chart with the k approach, one chooses k to be half of the delta shift which is the amount of shift in the process that we wish to detect, expressed as a multiple of the standard deviation of the data points. Accordingly, we set k equal to 0.5 (for more detail see Montgomery 2005). UCLs of control charts are designed to achieve a specified in control ARL of 200. The simulated UCLs for each proposed control chart are shown in Table 1.

The three proposed control charts are compared on different scenarios of the example in terms of ARL, and the calculated amounts for the different changes in the intercept is shown in Table 2.

According to the Table 2, under $\lambda\sigma$ shift in the intercept, when autocorrelation is weak ($\rho = 0.1$), the MEWMA method performs relatively similar to MCUSUM control chart, and they also have better performance for detecting the small, moderate, and large shifts than the T^2 control chart. In the intermediate and strong autocorrelation circumstance ($\rho = 0.5$) and ($\rho = 0.9$), MCUSUM performs uniformly better than the other two methods. Moreover, MEWMA uniformly performs better than T^2 control chart. Figure 1 presented the derivative ARL under different shifts in intercept when autocorrelation is different in three levels. Table 3 shows the simulation results under different shifts in slope.

From Table 3, under $\beta\sigma$ shift in slope, while the autocorrelation is weak ($\rho = 0.1$), the proposed MCUSUM method uniformly performs better than MEWMA method. Also, MEWMA performs consistently better than T^2 method. In addition, similar results are obtained when the autocorrelation is intermediate ($\rho = 0.5$). Once the amount of autocorrelation coefficient is high, MCUSUM and MEWMA methods perform uniformly better than the T^2 method and also, MCUSUM method performs relatively similar to MEWMA method. Figure 2 illustrates ARL under different shifts in slope once autocorrelation be changed in the aforementioned levels.

Next comparisons of the proposed three control charts in terms of ARL under $\delta\sigma$ shift in the standard deviation followed. Table 4 shows that the proposed T^2 chart performs significantly better than MEWMA and MCUSUM charts in different amount of correlation coefficients. In addition for strong and intermediate autocorrelation condition, MEWMA and MCUSUM have similar manners and when the autocorrelation is weak, MEWMA relatively achieves better performance. Derivative ARL under different shifts of standard deviation is presented in Figure 3 when autocorrelation is different.

Based on the simulation results, it is evident that the proposed MEWMA and MCUSUM methods act relatively better than the T^2 chart in detecting shift in the parameters of profile; conversely, the proposed T^2 chart performs better than the MEWMA and MCUSUM in detecting shift in the variation.

Case study

Consider the case study carried out by Schabenberger and Pierce (2002). It was a real data set from ten apple trees which 25 apples are randomly chosen on each tree. Their focus was on the analysis of the apples in the largest size, with initial diameters exceeded 2.75 in. Totally there were 80 apples in aspiration size. Diameters of the apples were recorded in every 2 weeks during 12 weeks. Figure 4 shows 16 diameters out of 80 apples in the time domain. In their investigation, functional profile between

Table 2 ARL comparisons under different $\lambda\sigma$ shift in intercept

Method	λ	0.2	0.4	0.6	0.8	1	1.2	1.4	1.6	1.8	2
T^2	$\rho = 0.1$	184.1	147.3	109.3	78	52.6	36.4	25.5	18.1	13.2	9.7
MEWMA		129.2	60.3	31.2	18.2	12.2	9.2	7.1	5.8	4.9	4.3
MCUSUM		126.4	57	29.4	17.1	11.1	9	7.1	6.1	5.3	4.7
T^2	$\rho = 0.5$	192	168.2	135.2	99.4	73.2	53.4	38.8	27.4	20.2	14.7
MEWMA		152.4	79	40.5	23.7	15.4	11.2	8.6	6.9	5.8	5
MCUSUM		133	63.5	33.1	20.4	13.8	10.5	8.2	6.4	5.7	5
T^2	$\rho = 0.9$	198.5	196	192.1	185.4	173.7	162.5	150	135.5	121.1	107.2
MEWMA		189.5	162.8	128.8	93.2	65.2	44.4	29.1	20.9	15.3	11.9
MCUSUM		186.5	158.9	118.9	77.5	45.8	28.4	19.7	14.8	11.9	9.7

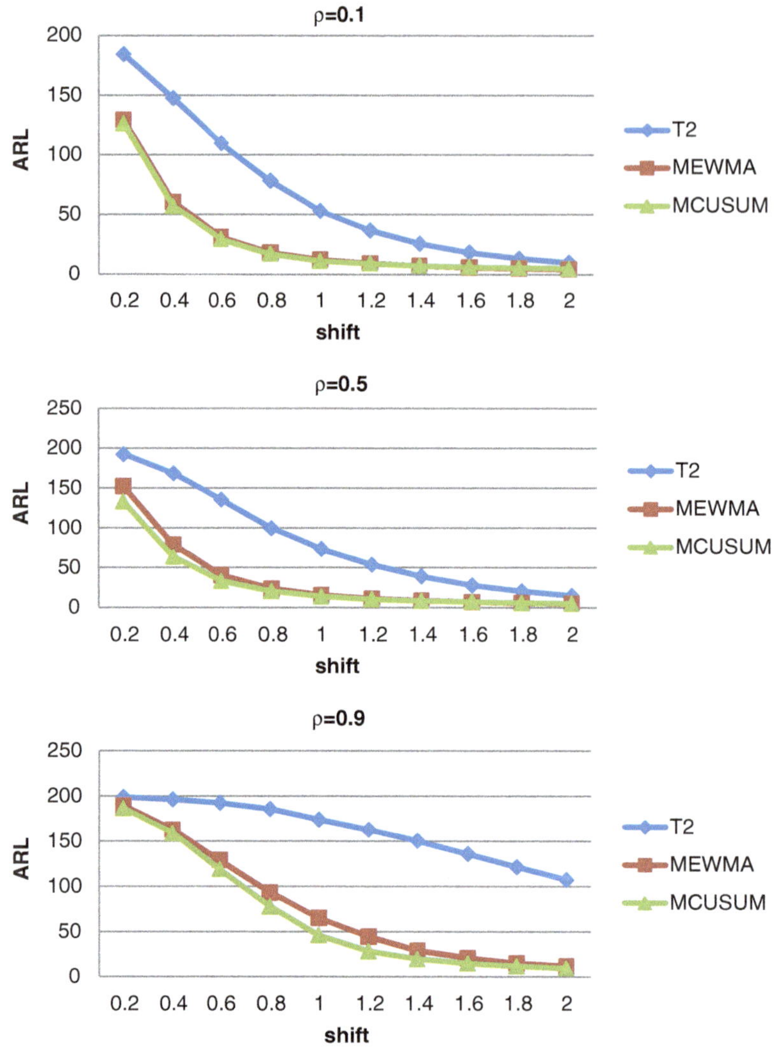

Figure 1 ARL comparisons under $\lambda\sigma$ shift in intercept with $\rho = 0.1$, 0.5, and 0.9.

Table 3 ARL comparisons under different $\beta\sigma$ shift in slope

Method	β	0.025	0.05	0.075	0.1	0.125	0.15	0.175	0.2	0.225	0.25
T^2	$\rho = 0.1$	196.3	194.2	178.3	164	146.6	128	109.8	96.6	82.1	68.3
MEWMA		184.2	144.3	106.3	78.1	56.6	41.5	32.5	25.1	20.2	17.2
MCUSUM		180.4	136.5	95.3	66.6	47.5	33.2	26.2	21.5	16.7	14.9
T^2	$\rho = 0.5$	198.3	195.2	181.4	163.3	143.7	126.5	108.6	93.4	78.5	66.8
MEWMA		188	142.3	107.3	76.4	56.1	41.4	31.7	24.5	19.8	16.6
MCUSUM		178	133.4	92.9	65.5	47.2	35.4	28.2	22	17.8	13.3
T^2	$\rho = 0.9$	197.1	193.6	179.2	162.1	141.7	125.6	107.1	91.9	79.5	67.5
MEWMA		185.9	143.9	105.3	78.2	57.4	43.1	33.6	26.6	21.3	17.9
MCUSUM		184.1	143.6	104.3	75.8	57.2	42.1	32.3	25.4	21.2	17.3

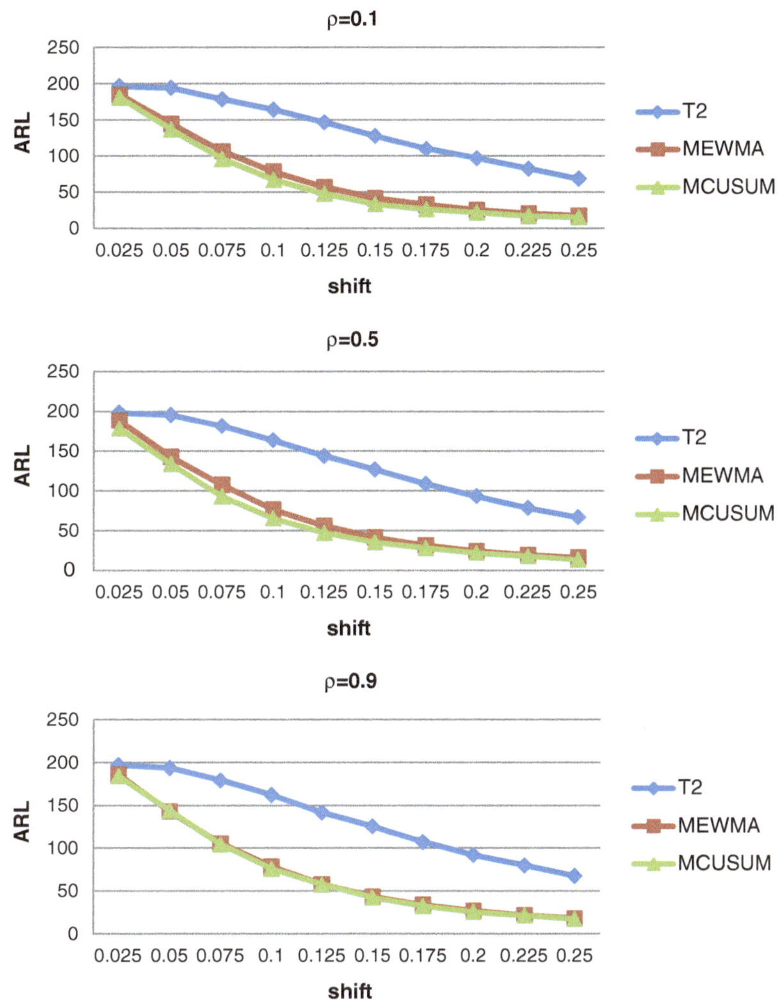

Figure 2 ARL comparisons under βσ shift in slope with $\rho = 0.1$, 0.5, and 0.9.

time and diameter considered as quality characteristic that needs to be monitored over time. Schabenberger and Pierce (2002) and also later Soleimani et al. (2009) modeled such correlation between observations by a first-order autoregressive model of AR(1). Based on the preceding analysis, the following statements hold a linear mixed model equation for the declared case study:

$$y_{ij} = 2.8321 + b_{0j} + (0.02875 + b_{1j})x_i + \varepsilon_{ij}, \qquad (16)$$
$$\varepsilon_{ij} = .3825\varepsilon_{i-1j} + a_{ij},$$

Table 4 ARL comparisons under δσ shift in standard deviation

Method	δ	1.2	1.4	1.6	1.8	2	2.2	2.4	2.6	2.8	3
T^2	$\rho = 0.1$	71.3	32.3	17.6	10.9	7.8	5.9	4.8	4.2	3.5	3.3
MEWMA		87.3	46.3	29.4	20.1	14.6	11.8	9.6	8.3	6.8	5.9
MCUSUM		95.2	53.3	33.7	25	18.4	14.5	11.7	10.2	8.5	7.3
T^2	$\rho = 0.5$	83.2	36.9	20.1	12.4	8.5	6.3	4.9	4	3.5	3.1
MEWMA		95.1	51.5	31.5	21.9	15.9	12.3	10	8.1	6.8	6
MCUSUM		95.3	52.2	32.6	23.3	17.4	13.4	11.3	9.3	8	7.1
T^2	$\rho = 0.9$	89.2	58.2	34.3	20.3	14.3	10.5	7.9	6.3	5.5	4.2
MEWMA		117.7	71.2	48.1	33.1	23.7	18.2	14.4	11.8	9.8	8.2
MCUSUM		117.5	71	46.8	32.4	23.4	17.6	14.1	11.7	9.8	8.2

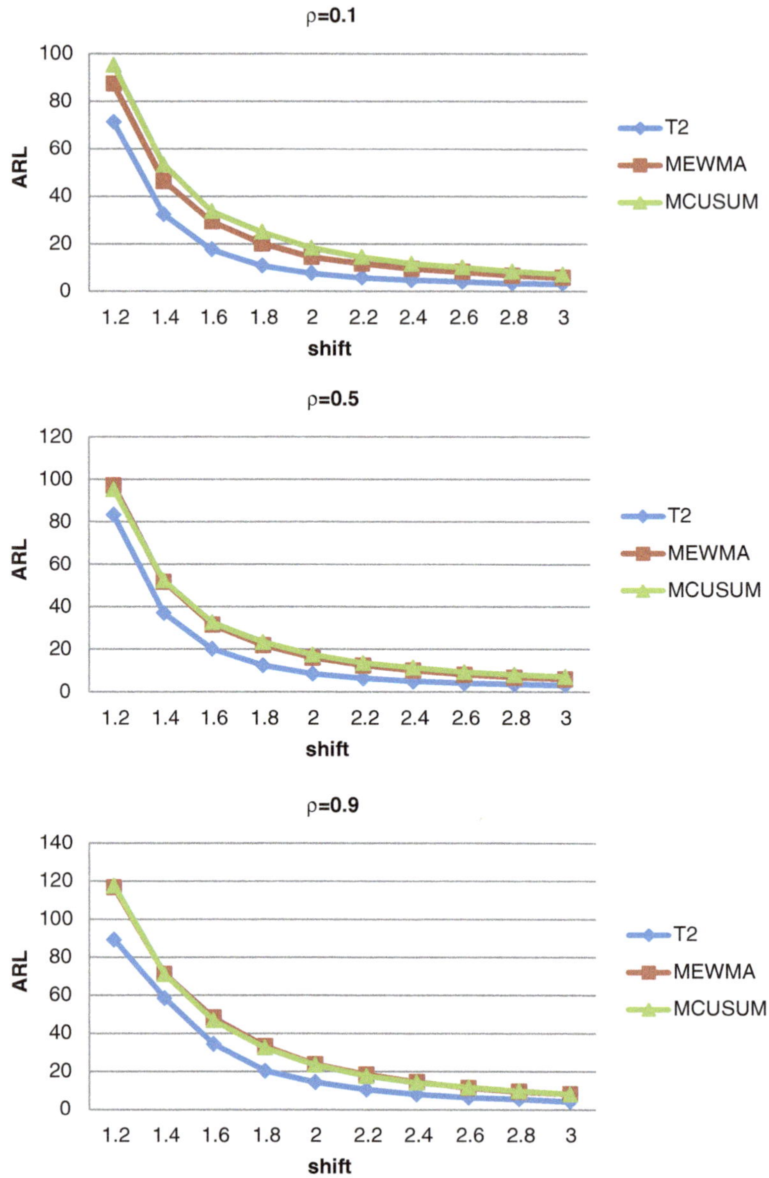

Figure 3 **ARL comparisons under δσ shift of standard deviation with $\rho = 0.1$, 0.5, and 0.9.**

where $b_{0j} \sim N(0,0.008653)$, $b_{1j} \sim N(0,0.00005)$, and $a_{0j} \sim N(0,0.000365)$.

The estimator of variance-covariance matrix of fixed effects gives by

$$\sum\begin{bmatrix} 1.4244 & -.3034 \\ -.3034 & 0.0870 \end{bmatrix}. \tag{17}$$

Consequences of simulation run in the previous section leads us to use MEWMA and MCUSUM which have relatively similar performance on detecting shift in the profile parameters rather than the T^2 method. Hence, the proposed MEWMA control chart was applied in monitoring the linear profile. The smoothing constant (θ) is set equal to 0.2. In order to achieve an in control ARL of 200, the upper control limit is set equal to 7 based on 10,000 simulation runs. In order to examine performance of the control chart, six random samples from the in control simple linear profile are initially generated. Formerly, three random samples are generated to show an out-of-control condition under the intercept shift coefficient of 0.6. Figure 5 illustrates sensitivity of the MEWMA control chart based on our proposed method which temperately depicts quick signal.

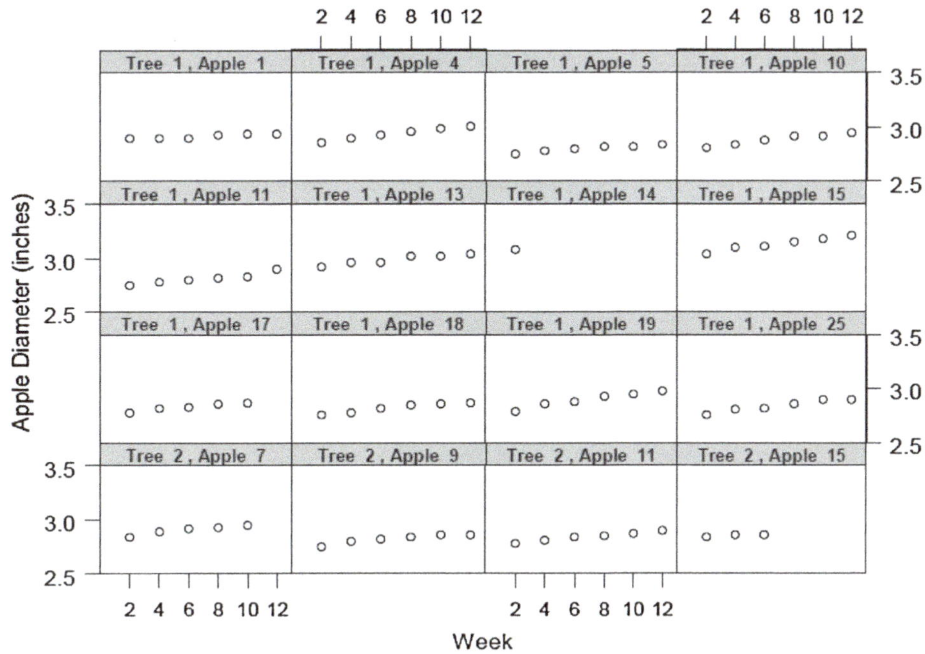

Figure 4 Measured diameters of apples during a period of 12 weeks. (Schabenberger and Pierce 2002).

Concluding remarks

We have studied the sensitivity of three multivariate control charts to detect one-step permanent shift in any parameters of a mixed model linear profile. Our specially designed MCUSUM, MEWMA, and T^2 control charts were also studied as competitors of each other to depict shifts in intercept and slope parameters and also process variation while first-order autoregressive model describes correlations within observations.

The performances of the methods were compared in terms of average run length criteria. Table 5 shows the summarized results.

The following summary recommendations are made:

1 The proposed approach has good performance across the range of possible shifts and it can be used in phase II of linear profile monitoring on the presence of autocorrelation within observations.

2 The anticipated MEWMA and MCUSUM methods almost uniformly perform better efficiency than the T^2 Hotelling control chart under different step shifts in the intercept and slope parameters of linear profile.

3 The Hotelling T^2 control chart has better performance in comparison with the MEWMA and MCUSUM methods under shifts in the process standard deviation.

4 The process circumstance in terms of correlation coefficient has no significant effects on selecting the best choice for monitoring method.

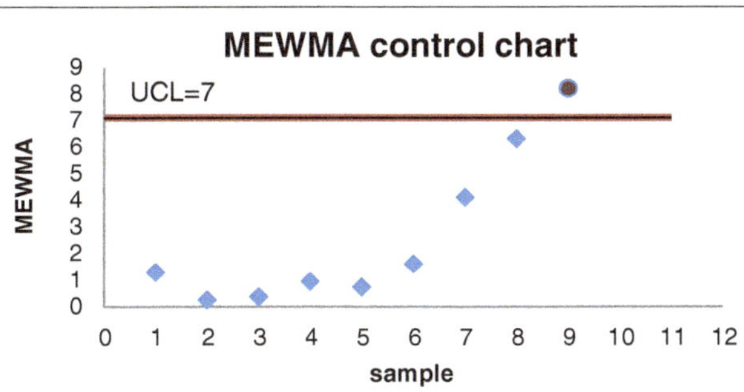

Figure 5 MEWMA control chart under intercept shift coefficient of 0.6.

Table 5 Comparisons on the performance of control charts under different shifts and autocorrelation status

Process circumstance	Shift on	Priority to use
Weakly auto-correlated	Intercept	MEWMA ≈ MCUSUM > T^2
	Slope	MCUSUM > MEWMA > T^2
	Standard deviation	T^2 > MEWMA > MCUSUM
Intermediately auto-correlated	Intercept	MCUSUM > MEWMA > T^2
	Slope	MCUSUM > MEWMA > T^2
	Standard deviation	T^2 > MCUSUM ≈ MEWMA
Strongly auto-correlated	Intercept	MCUSUM > MEWMA > T^2
	Slope	MEWMA ≈ MCUSUM > T^2
	Standard deviation	T^2 > MCUSUM ≈ MEWMA

Competing interests

The authors declare that they have no competing interests.

Authors' contributions

AN and PS studied the associated literature review and presented the mathematical models. SR managed the study and participated in the problem solving approaches. PS and SR approved and validate the model and SR was responsible for revising the manuscript. SR as corresponding author read and approved the final manuscript.

Authors' information

SR (IRAN, Male, 1962). Ph.D., Associate Professor, worked at School of Industrial Engineering in Islamic Azad University, Tehran South Branch (IAU-STB). He obtained his Ph.D., Master and Bachelor degrees in Industrial Engineering. He was engaged in the industrial system engineering technology development and the technical consultant from 1988 up to year. He worked in different management positions both in private sector and university. The last one was deputy of research and planning in Islamic Azad University-Tehran South Branch. During his management period more than 10 scientific journals initiated and research activities facilitated. Currently, Dr. Raissi act as Editor-in-Chief of Journal of Industrial Engineering, International (JIEI) a Springer open journal and also has major responsibility on Ph.D. department of Industrial Engineering in IAU-STB. His main research fields are statistical quality control, quality engineering, system simulation and statistical methods in engineering. PS is an assistant professor of Industrial Engineering at IAU-STB. She especially focused on multivariate profile monitoring techniques under correlated process in her MS and PhD thesis. Her research interests are in the areas of Statistical Quality Control, Time Series, Multivariate Analysis and Metaheuristic algorithms. AN received his Master's of Science degrees in Industrial Engineering from IAU-STB. He especially focused on developing profile monitoring techniques as his academicals task under supervision of PS and SR. His research interests are in the areas of Time Series Analysis and Statistical Quality Control.

Acknowledgments

The authors thank the referee and the grateful editor in chief of *JIEI* for their valuable comments and suggestions. This research was supported by deputy of research in the Modeling and Optimization Research Center of Industrial Engineering Faculty on Islamic Azad University, South Tehran Branch.

Author details

[1]School of Industrial Engineering, South Tehran Branch, Islamic Azad University, Tehran, Iran. [2]Modeling and Optimization Research Centre in Engineering and Science, Sepahbod Gharani St., No. 165, Tehran, Iran.

References

Crosier RB (1988) Multivariate Generalizations of Cumulative Sum Quality Control Schemes. Technometrics 30:291–303

Ding Y, Zeng L, Zhou S (2006) Phase I analysis for monitoring nonlinear profiles in manufacturing processes. J QualTechnol 38(3):199–216

Gupta S, Montgomery DC, Woodall WH (2006) Performance evaluation of two methods for online monitoring of linear calibration profiles. Int J Prod Res 44:1927–1942

Hotelling H (1947) Multivariable Quality Control—Illustrated By The Air Testing Of Sample Bombsights. In: Eisenhart C, Hastay MW, Wallis WA (ed) Techniques of Statistical Analysis. McGraw Hill, New York, pp 111–184

Jensen WA, Birch JB (2009) Profile monitoring via nonlinear mixed model. J QualTechnol 41:18–34

Jensen WA, Birch JB, Woodall WH (2008) Monitoring correlation within linear profiles using mixed models. J QualTechnol 40(2):167–183

Kang L, Albin SL (2000) On-line monitoring when the process yields a linear profile. J QualTechnol 32(4):418–426

Kazemzadeh RB, Noorossana R, Amiri A (2008) Phase I monitoring of polynomial profiles. Commun Stat-Theor M 37(10):1671–1686

Kazemzadeh RB, Noorossana R, Amiri A (2009a) Phase II monitoring of autocorrelated polynomial profiles in AR(1) processes. Int J SciTechnol, Sci Iran 17(1):12–24

Kazemzadeh RB, Noorossana R, Amiri A (2009b) Monitoring polynomial profiles in quality control applications. Int J Adv Manuf Technol 42(7):703–712

Kim K, Mahmoud MA, Woodall WH (2003) On the monitoring of linear profiles. J QualTechnol 35(3):317–328

Laird NM, Ware JH (1982) Random affects models for longitudinal data. Biometrics 38:963–974

Lowry CA, Woodal WH, Champ CH, Rigdon SE (1992) A Multivariate Exponentially Weighted Moving Average Control Chart. Technometrics 34:46–53

Mahmoud MA (2008) Phase I analysis of multiple regression linear profiles. Commun Stat Simul C 37(10):2106–2130

Mahmoud MA, Woodall WH (2004) Phase Ianalysis of linear profiles with applications. Technometrics 46:380–391

Mahmoud MA, Parker PA, Woodall WH, Hawkins DM (2007) A change point method for linear profile data. Qual Reliab Eng Int 23:247–268

Mahmoud MA, Morgan JP, Woodall WH (2009) The monitoring of simple linear regression profiles with two observations per sample. J Appl Stat 37:1249–1263

Moguerza JM, Muñoz A, Psarakis S (2007) Monitoring nonlinear profiles using support vector machines. Lect Notes Comput Sc 4789:574–583

Montgomery DC (2005) Introduction to statistical quality control, 5th edition. John Wiley & Sons, New Jersey, NJ

Noorossana R, Amiri A, Vaghefi A, Roghanian E (2004) Monitoring Quality Characteristics Using Linear Profile. In: Proceedings of the 3rd InternationalIndustrial Engineering Conference. , Tehran, Iran

Noorossana R, Amiri A, Soleimani P (2008) On the monitoring of autocorrelated linear profiles. Commun Stat-Theor M 37(3):425–442

Noorossana R, Eyvazian M, Vaghefi A (2010a) Phase II monitoring of multivariate simple linear profiles. Comput Ind Eng 58:563–570

Noorossana R, Vaghefi A, Dorri M (2010b) Effect of non-normality on the monitoring of simple linear profiles. Qual Reliab Eng Int 27:1015–1021

Qie P, Zhang Y, Wang Z (2010) Nonparametric profile monitoring by mixed effects modeling. Technometrics 52(3):283–285

Robinson GK (1991) That BLUP is a good thing: the estimation of random effects. Stat Sci 6(1):15–32

Saghaei A, Mehrjoo M, Amiri A (2009) A CUSUM-based method for monitoring simple linear profiles. Int J Adv Manuf Technol. doi:10.1007/s00170-009-2063-2

Schabenberger O, Pierce FJ (2002) Contemporary statistical models for the plant and soil sciences. CRCPress, Boca Raton

Soleimani P, Noorossana R, Amiri A (2009) Monitoring simple linear profiles in the presence of autocorrelation within profiles. Comput Ind Eng Int 57(3):1015–1021

Stover FS, Brill RV (1998) Statistical quality control applied to ion chromatography calibration. J Chromatogr A 804:37–43

Vaghefi A, Tajbakhsh SD, Noorossana R (2009) Phase II monitoring ofnonlinear profiles. Commun Stat-Theor M 38:1834–1851

West BT, Welch KB, Galecki AT (2007) Linear mixed models: a practical guide to using statistical software. Chapman & Hall/CRC, New York

Williams JD, Woodall WH, Birch JB (2007) Statistical monitoring of nonlinear product and process quality profiles. Qual Reliab Eng Int 23:925–941

Zou C, Zhang Y, Wang Z (2006) Control chart based on change-point model for monitoring linear profiles. IIE Transactions 38(12):1093–1103

Zou C, Tsung F, Wang Z (2007) Monitoring general linear profiles using multivariate exponentially weighted moving average schemes. Technometrics 49(4):395–408

12

Forecasting S&P 500 index using artificial neural networks and design of experiments

Seyed Taghi Akhavan Niaki* and Saeid Hoseinzade

Abstract

The main objective of this research is to forecast the daily direction of Standard & Poor's 500 (S&P 500) index using an artificial neural network (ANN). In order to select the most influential features (factors) of the proposed ANN that affect the daily direction of S&P 500 (the response), design of experiments are conducted to determine the statistically significant factors among 27 potential financial and economical variables along with a feature defined as the number of nodes of the ANN. The results of employing the proposed methodology show that the ANN that uses the most influential features is able to forecast the daily direction of S&P 500 significantly better than the traditional logit model. Furthermore, experimental results of employing the proposed ANN on the trades in a test period indicate that ANN could significantly improve the trading profit as compared with the buy-and-hold strategy.

Keywords: S&P 500 index, Financial time series, Artificial neural networks, Design of experiments, Analysis of variance, Logit model

Background

Due to the great volume of trading in stock markets, significant profit can be made by improving trading performance using adequate forecasting of financial variables such as stock prices, stock market indices, and prices of financial derivatives; that is why several research works in different fields of study have been performed on this subject so far. Atsalakis and Valavanis (2009) and Vanstone and Finnie (2009) comprehensively surveyed these works and their involved methodologies. Among other financial variables, stock market indices have received significant attention, and many researchers such as Atsalakis and Valavanis (2009), Vanstone and Finnie (2009), and Leung et al. (2000) proposed different methodologies to forecast them.

The efficient market hypothesis states that all of the available information in a market is reflected in the price of a stock at any moment, and therefore, forecasting the future price of a stock using the historical data and consequently gaining a return above the market average is impossible (Peters 1991). However, many empirical research works questioned this hypothesis and claimed that markets often do not follow the efficient market hypothesis completely. Jensen (1978) stated that because of the psychological factors along with varying groups of human decision makers, markets do not react immediately to the newly released information and that this phenomenon makes the markets inefficient. Additionally, Ferson (1989), Fama and French (1989), and Fama and Schwert (1977) showed that financial variables could be forecasted using time series on some other financial and economical variables.

Artificial neural networks (ANNs) are one of the relatively newly developed methods that have been widely used for forecasting purposes in different fields. Zhang et al. (1998) made a comprehensive review of using ANNs as a forecasting tool.

One of the fields of study in which the ANN application received a significant attention is finance. Trippi and Turban (1996) reviewed the application of ANNs in different areas in finance and investment. One of the areas of finance, in which using ANNs gained attention, is forecasting financial time series. Financial time series have some characteristics that make them hard to forecast especially when a traditional statistical method is employed (see for example Eakins and Stansell 2003; Hussain et al. 2007; Lam 2004; Lin et al. 2006; Motiwalla

* Correspondence: Niaki@Sharif.edu
Department of Industrial Engineering, Sharif University of Technology, P.O. Box 11155-9414 Azaadi Ave, Tehran 1458889694, Iran

and Wahab 2000; Thawornwong and Enke 2004; Versace et al. 2004; Yao et al. 1992). These characteristics are as follows:

1. Non-stationarity. It means that because of the different business and economic cycles, the statistical properties of financial data change over time.
2. Nonlinearity. It means that the relationship between the financial and economical independent variables and the desired dependent variable is not linear.
3. Noisiness. It means that there are day-to-day variations in financial time series.

There are two main approaches of univariate and multivariate analyses in forecasting financial time series where ANNs can be used in both (Cao and Tay 2001). In univariate analyses, the features are limited to the variable being forecasted, while in multivariate case, any variable that is thought to be related to the output is considered a feature (an input of the model). The following points justify the use of ANNs to forecast financial time series (Eakins and Stansell 2003; Hornik et al. 1989; Hussain et al. 2007; Lam 2004):

1. ANNs are nonlinear. It means that they can capture nonlinear relations between feature (input or independent) and response (output or dependent) variables.
2. ANNs are data driven. It means that they do not need any explicit assumption on the model between inputs and outputs.
3. ANNs can generalize. It means that after training, they can produce good results even when they face new input patterns.
4. Unlike statistical techniques, ANNs do not need assumptions on the distribution of input data.

Although ANNs have the above-mentioned advantages, the robustness of their outcomes has been questioned (Saad et al. 1998). Besides, the ANNs' limitations are as follows (Hussain et al. 2007: Lam 2004):

1. Determining the optimal combination of the network parameters such as learning rate, momentum, number of hidden layers, and number of hidden nodes in each layer is difficult.
2. Selecting the relevant features of an ANN is not an easy job.
3. Great volume of data is required to train the network to achieve an accurate result.

Designing an optimal ANN with only relevant features is essential in obtaining good forecasting results (Zhu et al. 2008). In fact, using irrelevant features along with not using relevant ones can reduce the performance of an ANN (Thawornwong and Enke 2004). For example, Cao et al. (2005) concluded that extending the model from CAPM to that of the 3-factor model of Fama and French (1989) negatively affects the forecast accuracy in the Shanghai stock exchange. Hence, the key factor to success in forecasting the financial time series is designing an ANN that has the least complexity with only relevant and most influential features (Atsalakis and Valavanis 2009).

There are several techniques in the literature to select relevant features of ANNs. These techniques can be classified into two main categories. The first category uses variable relevance analysis to examine whether a variable has useful information to forecast the output variable. If the answer to this question is yes, then the variable will be used as an input for the ANN. Thawornwong and Enke (2004) used this method to select features of an ANN. The disadvantage of the methods in this category is that even if a variable has some useful information, ANNs may not be able to extract that information and use it to capture the relationship between that variable and the output.

The techniques of the second category compare the performance of different ANNs that contain different features. Zhu et al. (2008), Cao et al. (2005), and O'Connor and Madden (2006) used one of these techniques. The advantage of the techniques of this category is that they not only select relevant features, but they also propose a method to extract information on the features to forecast the output. That is why the techniques of this category are more common in the literature. However, there are several drawbacks in the use of the techniques of this category in previous studies. The main problem is that they lack a scientific design of experiments, which leads to robust results that can be statistically trusted. In other words, most of the available studies did not undergo statistical analysis to prove that their designed ANN would significantly improve the results (Lam 2004).

The main objective of this research is to forecast the daily direction of Standard & Poor's (S&P 500) index using an ANN that uses the most influential financial and economical features. We are also going to use tests of hypothesis to prove statistically that the proposed ANN significantly improves the forecasting ability of the S&P 500 index.

The remainder of this paper is organized as follows. 'The propose ANN' section describes the developed ANN and the designed experiments that will be used to select the most influential features. In the 'Conducting the experiments' section, the experiments are conducted and the results are analyzed. In the 'Examining the financial performances of the proposed ANN' section, the hypothesis that using the forecasts produced by the proposed ANN significantly increases the profit is tested. The 'Forecasting S&P 500 daily direction using regression' section contains the comparison results of employing both the proposed ANN and the traditional logit model approach. Finally, the research is concluded in the 'Conclusions and further research' section.

The proposed ANN

ANNs are flexible computing methods that have the ability to capture the patterns among variables. ANNs have some characteristics (mentioned in the 'Background' section) that make them reasonable to be used for a wide range of problems.

In order to design an appropriate ANN for a particular problem, one should decide on the network topology, number of network layers, number of nodes in each layer, activation function of the nodes, and finally, the learning algorithm.

Based on the topology, ANNs are mainly divided into two groups of feed-forward and recurrent networks. For univariate forecasting analysis, the use of recurrent topology is more common than the feed-forward networks. Hussain et al. (2007), Lin et al. (2006), and Saad et al. (1998) are the ones who employed recurrent topology. However, in the field of financial forecasting, especially in multivariate forecasting analysis, the feed-forward topology has gained much more attention. As a result, the latter topology will be used in this research as well.

The number of network layers depends on the complexity of the problem being modeled. In addition to the input and the output layers that are essential for an ANN design, many feed-forward networks have one or more hidden layers. Several researchers have proposed different methods to determine the optimal number of hidden layers and hidden nodes. However, these methods are very complex and hard to apply (Zhang et al. 1998). Moreover, none of the existing methods can guarantee that the obtained ANNs are optimal. Therefore, the common practice of identifying the proper network design, which is comparing the performances of ANNs with different designs and selecting the network that results in the best performance, is taken in this research (Hosseini et al. 2006). In addition, while different researchers in the literature recommended not using more than two hidden layers, in this research the optimal number of hidden nodes in only one hidden layer is determined using design of experiments (DOE). In other words, the number of hidden nodes is considered a potential influential factor in DOE that affect the forecasting ability of the proposed ANN. The levels of this factor will be discussed later in this section.

The input layer of an ANN consists of the input variables (features) that seem to be influential to the output variable. These influential features are determined using DOE as well. Again, the features of the proposed ANN are the potential factors of DOE. These features along with their levels will be discussed later in this section.

The output layer of an ANN consists of nodes associated with the dependent variables. Since the objective of this research is to forecast the daily direction of S&P 500 index properly and since the direction is either positive or negative, the output layer of the proposed ANN consists of only one node.

As the activation function, the tangent hyperbolic sigmoid (Tansig) function, which is the most common one in the relevant literature, is used for the nodes of all layers. Furthermore, the error back-propagation algorithm is employed to train the designed ANN.

In order to design, train, and simulate the proposed ANN, the neural network toolbox of the MATLAB 7 package software (Mathworks 2004) is used in this research. For a detailed description of the neural network toolbox, the reader is referred to the work of Demuth and Beale (1998) who demonstrated the application of ANNs in this toolbox comprehensively.

Data compilation

The dataset of this research has been compiled containing the daily closing value of S&P 500 as the output and 27 financial and economical variables as the potential features for 3,650 days partially taken on the period starting 01 March 1994 to 30 June 2008. The first 80% of the compiled data (for 2,920 days) is used for the training purpose, the next 10% is used for verification, and finally, the last 10% is used to test the performance of the designed ANN. The verification and the test periods consist of the data for 365 days each.

The main consideration for selecting the potential features is whether they have significant influence on the output variable (the direction of S&P 500) in the next day. While some of the features considered in this study were used in previous studies, some of them are new and are used in this study for the first time. Analyzing the results of the designed experiments will indicate which of these features have a meaningful impact on the response. The list, the description, and the sources of the potential features are given in Table 1.

Selecting the design and choosing the factor levels

In order to determine the relevant and the most influential variables of the forecasting process, a factorial design is employed in this research. The factors include the 27 features given in the Table 1, each with two levels (0 and 1). Level 1 of a factor indicates that the corresponding factor is used as a feature, and level 0 represents the nonexistence of the factor. In addition to the 27 features, the number of hidden nodes is also considered as a potential factor in designing the experiments. Note that experiments with fewer input

Table 1 The 27 potential influential features

1	SPYt-1	The return of the S&P 500 index in day t-1. Source: finance.yahoo.com
2	SPYt-2	The return of the S&P 500 index in day t-2. Source: finance.yahoo.com
3	SPYt-3	The return of the S&P 500 index in day t-3. Source: finance.yahoo.com
4	Oil	Relative change in the price of the crude oil (Cushing, OK WTI Spot Price FOB (dollars per barrel)). Source: Energy Information Administration, available at http://tonto.eia.doe.gov/dnav/pet/hist/rwtcd.htm
5	Gold	Relative change in the gold price. Source: www.usagold.com
6	CTB3M	Change in the market yield on US Treasury securities at 3-month constant maturity, quoted on investment basis. Source: H.15 Release - Federal Reserve Board of Governors
7	CTB6M	Change in the market yield on US Treasury securities at 6-month constant maturity, quoted on investment basis. Source: H.15 Release - Federal Reserve Board of Governors
8	CTB1Y	Change in the market yield on US Treasury securities at 1-year constant maturity, quoted on investment basis. Source: H.15 Release - Federal Reserve Board of Governors
9	CTB5Y	Change in the market yield on US Treasury securities at 5-year constant maturity, quoted on investment basis. Source: H.15 Release - Federal Reserve Board of Governors
10	CTB10Y	Change in the market yield on US Treasury securities at 10-year constant maturity, quoted on investment basis. Source: H.15 Release - Federal Reserve Board of Governors
11	AAA	Change in the Moody's yield on seasoned corporate bonds - all industries, Aaa. Source: H.15 Release - Federal Reserve Board of Governors
12	BBB	Change in the Moody's yield on seasoned corporate bonds - all industries, Baa. Source: H.15 Release - Federal Reserve Board of Governors
13	USD-Y	Relative change in the exchange rate between US dollar and Japanese yen. Source: OANDA.com
14	USD-GBP	Relative change in the exchange rate between US dollar and British pound. Source: OANDA.com
15	USD-CAD	Relative change in the exchange rate between US dollar and Canadian dollar. Source: OANDA.com
16	HIS	Hang Seng index return in day t-1. Source: finance.yahoo.com
17	FCHI	CAC 40 index return in day t-1. Source: finance.yahoo.com
18	FTSE	FTSE 100 index return in day t-1. Source: finance.yahoo.com
19	GDAXI	DAX index return in day t-1. Source: finance.yahoo.com
20	V	Relative change in the trading volume of S&P 500 index. Source: finance.yahoo.com
21	XOM	Exxon Mobil stock return in day t-1. Source: finance.yahoo.com
22	GE	General Electric stock return in day t-1. Source: finance.yahoo.com
23	MSFT	Micro Soft stock return in day t-1. Source: finance.yahoo.com
24	PG	Procter and Gamble stock return in day t-1. Source: finance.yahoo.com
25	JNJ	Johnson and Johnson stock return in day t-1. Source: finance.yahoo.com
26	DJI	Dow Jones Industrial Average index return in day t-1. Source: finance.yahoo.com
27	IXIC	NASDAQ composite index return in day t-1. Source: finance.yahoo.com

factors need fewer hidden nodes and vice versa. In other words, the number of hidden nodes of the proposed ANN for experiments on a large number of features should be high enough to have the ability of capturing the patterns. Therefore, the number of hidden nodes is not constant in all experiments and is considered a factor in the experiments. Considering the number of other features, four levels of the number of hidden nodes including 5, 15, 30, and 60 are considered in this research. As a result, the total number of combinations is 2^{29} (four for the number of hidden nodes and 2^{27} for the other features). We also note that only one replicate of this design needs 2^{29} experiments, which is impractical. To make the total number of experiments practical, two approaches are used.

In the first approach, in each pair of the features that are highly correlated, only one is kept and the other is excluded. This is based on the fact that if two features are highly correlated, then both are inferred to contain similar information. Hence, using one of them as the input is almost sufficient for an ANN to extract the desired pattern. By 'highly correlated,' we mean that the absolute value of the correlation is at least 0.5. The application of this approach leads to exclude 7 of the 27 initial financial and economical variables. The omitted variables of this approach are DJI, IXIC, CTB6M, CTB1Y, CTB5Y, CTB10Y, and BAA.

The second approach of decreasing the number of features is based on the concept of grouping. In this approach, the remaining features of the first approach with almost similar nature are placed in one group, making them group factors. In this way, the existence and nonexistence of the group factors matters. Zhu et al. (2008) and O'Connor and Madden (2006) used a similar technique for comparing the performance of different ANNs. Employing the second approach makes the remaining 20 features as the six group factors, as shown in Table 2.

Conducting the experiments

Based on the derivations given in the 'Selecting the design and choosing the factor levels' subsection, there are now six group factors, each with two levels along with

Table 2 Grouping features

Group number	Group description	Group title	Included input variables
1	S&P 500 index return in three previous days	G1	SPYt-1
			SPYt-2
			SPYt-3
2	Financial and economical indicators	G2	Oil
			Gold
			CTB3M
			AAA
3	The return of the five biggest companies in S&P 500	G3	XOM
			GE
			MSFT
			PG
			JNJ
4	Exchange rate between USD and three other currencies	G4	USD-Y
			USD-GBP
			USD-CAD
5	The return of the four world major indices	G5	HIS
			FCHI
			FTSE
			GDAXI
6	S&P 500 trading volume	G6	V

the number of hidden nodes with four levels. This decreases the total number of the factor level combination to a practical value of 2^8. However, four of these combinations are not possible. Since one cannot have an ANN without input, the combinations in which the level of all features is zero are not feasible. The number of these infeasible combinations is four. Consequently, the number of possible combinations would be 252. Each combination corresponds to a specific ANN; each observation for a combination uses the specified ANN to produce the response variable, and a total of 10 replicates are considered for each combination, leading to the total number of 2,520 experiments. Experiments are conducted in a way that for each observation of each possible combination, the corresponding ANN is initialized with a random value. The initialized ANN is then trained with the training dataset until the early stopping[a] rule terminates the training process. Next, the trained network is simulated with the test dataset, and the forecasts that are generated by the ANN in the test period are gathered. Since the test period consists of 365 days, each ANN corresponding to an experiment forecasts 365 daily directions of S&P 500 in the test period. The response variable is the number of correct forecasts that the ANN produces in the test period.

Determining the influential factors

After conducting the experiments and obtaining the results, the analysis of variance (ANOVA) approach is used to determine the factors that have statistically significant effects on the response variable. The general linear model procedure of the SAS software package (SAS Institute Inc. 2009) is employed for this purpose. We assume that only the main effects and the second-order mutual (interaction) effects exist, and mutual effects of higher orders can be neglected.

The underlying assumptions of the ANOVA approach are that the error terms are independently and normally distributed with mean zero and a constant variance. By a residual analysis (not shown here), we concluded that the assumptions hold, and hence, the results of ANOVA are valid. The results are summarized in Table 3.

The results in Table 3 show that four main effects, including *G*1, *G*4, *G*6, and hidden nodes, and one mutual effect, *G*2 × *G*3, are statistically significant at 5% significance level. This means that the existence or nonexistence of the other factors including *G*1 and *G*5 does not meaningfully influence the response variable.

In order to determine the levels of the selected factors on which the response has the best value, Duncan's multiple range test is used. Table 4 presents the optimal level of the influential factors.

Table 3 The ANOVA results

Source	Type III sum of squares	*df*	Mean square	*F*
G1	1,070.88	1	1,070.88	30.26*
G2	121.07	1	121.07	3.42
G3	23.34	1	23.34	0.66
G4	5,057.11	1	5,057.11	142.91*
G5	109.34	1	109.34	3.09
G6	258.63	1	258.63	7.31*
Hidden nodes	1,844.41	3	614.8	17.37*
G1 × *G2*	117.66	1	117.66	3.33
G1 × *G3*	22.83	1	22.83	0.65
G1 × *G4*	29.77	1	29.77	0.84
G1 × *G5*	30.62	1	30.62	0.87
G1 × *G6*	0.51	1	0.51	0.01
G1 × hidden nodes	151.75	3	50.58	1.43
G2 × *G3*	154.65	1	154.65	4.37*
G2 × *G4*	6.98	1	6.98	0.2
G2 × *G5*	7.98	1	7.98	0.23
G2 × *G6*	9.84	1	9.84	0.28
G2 × hidden nodes	72.87	3	24.29	0.69
G3 × *G4*	0.59	1	0.59	0.02
G3 × *G5*	0.63	1	0.63	0.02
G3 × *G6*	130.42	1	130.42	3.69
G3 × hidden nodes	48.41	3	16.14	0.46
G4 × *G5*	92.6	1	92.6	2.62
G4 × *G6*	23.59	1	23.59	0.67
G4 × hidden nodes	257.08	3	85.69	2.42
G5 × *G6*	16.4	1	16.4	0.46
G5 × hidden nodes	239.54	3	79.85	2.26
G6 × hidden nodes	121.54	3	40.51	1.14
Error	87,650.55	2,477	35.39	
Total	97,862.62	2,519		

*p value < 0.05.

Table 4 Influential factors and their optimal level

Significant factors	Optimal level	Description
G1	0	This means that the variables of group 1 should not be used as a feature of the optimal ANN.
G4	1	This means that the variables of group 4 should be used as a feature of the optimal ANN.
G6	0	This means that the variables of group 6 should not be used as a feature of the optimal ANN.
Hidden nodes	60	This means that the optimal ANN should have 60 hidden nodes.
G2 × *G3*	*G3* = 1 *G2* = 0	This means that the variables of group 3 should be used, while the variables of group 2 should not be used as features of the optimal ANN.

Examining the financial performances of the proposed ANN

There are many criteria to evaluate the performances of a forecasting method. While the number of correct forecasts in the test period has been used as a criterion so far, the main goal of developing any financial forecasting method is making profit in real markets. Deciding on to what extent the optimal ANN can result in making profit comparing with other trading strategies is the purpose of this section. The common trading strategies that have been used in most of the relevant literature are as follows:

1. Buy-and-hold strategy (passive strategy).
2. Investing on an index that reflects the whole market.

However, since the objective of this research is to properly forecast the direction of S&P 500 index in a day, the buy-and-hold strategy and the investing strategy on S&P 500 are the same.

The following assumptions are made in trading in the S&P 500 market:

1. At the beginning of the test period, there is $100,000 available as initial investment capital.
2. Every trade is made at the end of the day with the closing price of that day.
3. Transaction cost is flat rate and is $8 per trade (O'Connor and Madden 2006).
4. Transaction costs are not deducted directly from the investment capital and are paid from a separate source. At the end of the test period for calculating the final capital, this cost will be considered.

Since the proposed ANN produces different results when initialized with different random values, one should initialize the designed ANN with different random values a number of times and compare the mean value of the final capitals produced by the ANN with the final capitals produced by other trading strategies. This helps to reach a statistically supportable conclusion. It should be noted that some of the previous studies such as those of Leung et al. (2000), Hussain et al. (2007), Yao et al. (1992), and O'Connor and Madden (2006) lacked the consideration of the effect of initialization.

For simulating trades in the real market using the forecasts produced by the proposed ANN, the following five steps are performed 100 times:

1. Initialize the network and train it by the training dataset.
2. Simulate the trained ANN with the test dataset and obtain the signals that the ANN produces in the test period.
3. When the ANN forecasts an increase in S&P 500 in the next day (the output of the ANN is positive), all of it will be used for buying SPDRs[b] if the capital is in the form of cash, and no trade will be performed if the capital is in the form of stocks.
4. When the ANN forecasts a decrease in S&P 500 in the next day (the output of the ANN is negative), all of it will be liquidated if the capital is in the form of stocks, and no trade will be performed if the capital is in the form of cash.
5. At the end of the test period, all of the capital will be liquidated, and the total transaction cost is deducted. This leads us to the final capital.

In the buy-and-hold strategy, we spend all of the initial capital investment to buy SPDRs at the beginning of the test period and liquidate all of the purchased stocks at the end of the test period. This strategy produces an amount of $89,593.3 as the final capital in the specified test period. For testing whether using the ANN produces more final capital than the buy-and-hold strategy does, one needs to test the following hypothesis:

$$H_0 : \mu \leq 89,593.3,$$

$$H_1 : \mu > 89,593.3,$$

in which μ is the average of 100 final capitals produced by trading in the market using the proposed ANN.

Since the variance of the final capitals is unknown and needs to be estimated by the sample variance and since the sample size is large enough (more than 30), to test the hypotheses, a t test was carried out with a p value of 0.02503. Hence, it is concluded that at the 5% significance level, the null hypothesis is rejected in the favor of the alternative hypothesis. The result of accepting alternative hypothesis is that using the proposed ANN significantly increases the final capital in comparison with the buy-and-hold strategy.

Forecasting S&P 500 daily direction using regression

According to the relevant literature, the relation between financial and economical variables is nonlinear. One of the reasons to employ a nonlinear method (ANN) to forecast the S&P 500 daily direction in this research is the nonlinearity of the relation. However, in order to test whether the nonlinearity assumption is true and to evaluate the performance of the proposed ANN methodology, in this section, we compare the performance of the proposed methodology with the ones of a linear model. Furthermore, since the response variable of interest, the daily direction of S&P 500, which is a binary variable, a logistic regression model (the logit model) is employed. For this model, the first 3,284 data (including the training and the verification data) are used to model the relationship between the independent and the dependent variables. All of the 27 initial financial and economical variables are used as the independent variables of the logit model. In order to eliminate the redundant input variables, the forward selection method of the SAS software package is used. The estimated regression function becomes:

$$z = -0.1394 + 6.5569(SPY_{t-2}),$$

$$P = f(z) = \frac{1}{1+e^{-z}},$$

in which SPY_{t-2} is the return of S&P 500 in day t-2 and P is the probability that the S&P 500 index decreases in the next day.

In order to validate the estimated regression function, a residual analysis based on the Box-Pierce method is conducted to see whether the error terms are white noise, i.e., they are independent standard normal random variables. The p value of the test statistic becomes 0.1145; indicating the null hypothesis cannot be rejected at 5% confidence level. Thus, the regression model is valid for forecasting. Applying the obtained regression equation to the test data, the logit model produces 189 correct forecasts in the test period.

For comparing the performance of the logit model with the ones of the proposed ANN, a statistical test of hypothesis is conducted as follows.

$$H_0 : \mu_{NN} \leq 189,$$

$$H_1 : \mu_{NN} > 189,$$

in which μ_{NN} is the mean of the number of correct forecasts that the optimal ANN produces with 50 random initializations. Once more, a t test may be employed for the test of hypothesis for which the p value is obtained as 0.001. Therefore, at 5% significance level, the null hypothesis is rejected in favor of the alternative hypothesis. This means that the proposed ANN as a nonlinear model can significantly outperform the logit model. Thus, it can be inferred that the relation between the independent variables and the dependent variable is nonlinear and that using a nonlinear model such as ANN is absolutely reasonable.

Conclusions and further research

In this research, the financial and economical features that have a significant influence on the daily direction of S&P 500 index were determined using DOE. Furthermore, an artificial neural network as a forecasting method was employed for which the influential features were used to forecast the daily direction of the S&P 500 index. The main findings of this research are as follows:

1. A factor that significantly influences the daily direction of S&P 500 is the exchange rates between the US dollar and three main currencies including the British pound, Canadian dollar, and Japanese yen.
2. Some of the variables that were found influential in previous studies on monthly forecasts do not have a significant influence on the daily forecast.
3. Using some of the variables including return of S&P 500 in previous days and relative change in the trading volume of S&P 500 not only is unhelpful, but also confuses the ANN and decreases the performance of ANNs for forecasting the daily direction of S&P 500.
4. Unlike some of the previous studies that did not consider the effect of network initialization, this paper does consider this effect.
5. The results obtained in this research are statistically supportable.
6. The factorial design of experiments helped to examine not only the main effects of the factors, but also the mutual effects between them. This led us to find out that the combination of the network features is also important for an ANN to produce good results.
7. By statistical analysis, it was proven that the designed ANN outperforms the logit model regarding the number of correct forecasts and outperforms the buy-and-hold strategy in terms of the obtained profit.

It should be mentioned that the conclusions are based on the use of only the feed-forward networks, and the scope extends to see if better results are achievable through the use of recurrent networks and other feed-forward varieties such as radial basis, adaptive neuro-fuzzy, and some other architectures. Future research may also include making the proposed ANN more reliable. As the results of the designed experiments showed, the mean value of the profit produced using the proposed ANN was more than the profit produced by the buy-and-hold strategy. However, the outputs of the proposed ANN are still not quite reliable because of two reasons. First, ANNs with different initializations produce different signals for a certain feature. Second, the proposed ANN with a certain initialization produces incorrect signals in some cases. Therefore, the first step in making the optimal ANN more reliable is initializing the optimal ANN in a way to produce as more correct signals as possible. Also, the second step identifies market situations in which using the proposed ANN with the best initialization is suitable and more likely to produce correct signals. Hence, it is recommended to use the signals produced by the proposed ANN in only certain market conditions. For example, signals produced by the proposed ANN may be useful in only a bull or bear market. As a result, by examining the relation between market conditions and signals produced by the proposed ANN, one can discover the market conditions in which the signals produced are more reliable.

Endnotes

[a]Early stopping rule is a method to avoid over-fitting, but it improves the generalization ability of ANNs. This method halts the training process when the performance with validation data stops improving.

[b]A short form of Standard & Poor's depositary receipt, an exchange-traded fund managed by the State Street Global Advisors that tracks the S&P 500 index. Each share of spider contains one-tenth of the S&P index and trades at roughly one-tenth of the dollar-value level of the S&P 500.

Competing interests
The authors declare that they have no competing interests.

Authors' contributions
STAN and SH forecasted the daily direction of S&P 500 index using ANN. In order to select the most influential features of the proposed ANN that affect the daily direction of S&P 500, design of experiments were conducted to determine the statistically significant factors among 27 potential financial and economical variables along with a feature defined as the number of nodes of ANN. The results of employing the proposed methodology showed that ANN which uses the most influential features is able to forecast the daily direction of S&P 500 significantly better than the traditional logit model. Furthermore, experimental results of employing the proposed ANN on the trades in a test period indicated that ANN could significantly improve the trading profit as compared with the buy-and-hold strategy. All authors read and approved the final manuscript.

Authors' information
STAN is a professor of Industrial Engineering at Sharif University of Technology. His research interests are in the areas of Simulation Modeling and Analysis, Applied Statistics, and Multivariate Quality Control. Before joining Sharif University of Technology, he worked as a systems engineer and quality control manager for Iranian Electric Meters Company. He received his Bachelor of Science in Industrial Engineering from Sharif University of Technology in 1979, his Master's and PhD degrees both in Industrial Engineering from West Virginia University in 1989 and 1992, respectively. He is a member of απμ. SH received his Bachelor of Science and Master's of Science degrees in Industrial Engineering from Sharif University of Technology in 2007 and 2009, respectively. His research interests are in the areas of Time Series Analysis and Finance.

Acknowledgments
The authors are thankful for the constructive comments of the anonymous reviewers as well as the editorial changes suggested by the language editor that certainly improved the presentation of the paper.

References
Atsalakis GS, Valavanis KP (2009) Surveying stock market forecasting techniques – Part II: Soft computing methods. Expert Syst Appl 36:5941–5950
Cao L, Tay F (2001) Financial forecasting using vector machines. Neural Comput Appl 10:184–192
Cao Q, Leggio KB, Schniederjans MJ (2005) A comparison between Fama and French's model and artificial neural networks in predicting the Chinese stock market. Comput Oper Res 32:2499–2512
Demuth H, Beale M (1998) Neural network toolbox: for use with MATLAB, 5th edn. The Math Works, Inc., Natick, MA
Eakins SG, Stansell SR (2003) Can value-based stock selection criteria yield superior risk-adjusted returns: an application of neural networks. Int Rev Financ Anal 12:83–97
Fama EF, French KR (1989) Business conditions and expected returns on stocks and bonds. J Financ Econ 25:23–49
Fama EF, Schwert WG (1977) Asset returns and inflation. J Financ Econ 5:115–146
Ferson W (1989) Changes in expected security returns, risk, and the level of interest rates. J Finance 44:1191–1217
Hornik K, Stinchcombe M, White H (1989) Multilayer feed forward networks are universal approximators. Neural Netw 2:359–366
Hosseini H, Luo D, Reynolds KJ (2006) The comparison of different feed forward neural network architectures for ECG signal diagnosis. Medical Eng Phys 28:372–378
Hussain AJ, Knowles A, Lisboa PJG, El-Deredy W (2007) Financial time series prediction using polynomial pipelined neural networks. Expert Syst Appl 35:1186–1199
Jensen M (1978) Some anomalous evidence regarding market efficiency. J Financ Econ 6:95–101
Lam M (2004) Neural network techniques for financial performance prediction: integrating fundamental and technical analysis. Decis Support Syst 37:567–581
Leung MT, Daouk H, Chen A (2000) Forecasting stock indices: a comparison of classification and level estimation models. Int J Forecast 16:173–190
Lin C, Huang J, Gen M, Tzeng G (2006) Recurrent neural network for dynamic portfolio selection. Appl Math Comput 175:1139–1146
Mathworks (2004) MATLAB, version 7.0.0.19920 (R14). Mathworks, Inc, Natick, MA
Motiwalla L, Wahab M (2000) Predictable variation and profitable trading of US equities: a trading simulation using neural networks. Comput Oper Res 27:1111–1129
O'Connor N, Madden MG (2006) A neural network approach to predicting stock exchange movements using external factors. Knowl Base Syst 19:371–378
Peters EE (1991) Chaos and order in the capital markets: a new view of cycles, prices, and market volatility. Wiley, New York
Saad EW, Prokhorov DV, Wunsch DC (1998) Comparative study of stock trend prediction using time delay, recurrent and probabilistic neural networks. IEEE Trans Neural Netw 9:1456–1470
SAS Institute Inc (2009) SAS 9.1 TS Level 1M0. SAS Institute Inc, Cary, NC
Thawornwong S, Enke D (2004) The adaptive selection of financial and economic variables for use with artificial neural networks. Neurocomputing 56:205–232
Trippi RR, Turban E (1996) Neural networks in finance and investing: using artificial intelligence to improve real-world performance. Irwin Professional Pub, Burr Ridge
Vanstone B, Finnie G (2009) An empirical methodology for developing stock market trading systems using artificial neural networks. Expert Syst Appl 36:6668–6680
Versace M, Bhatt R, Hinds O, Shiffer M (2004) Predicting the exchange traded fund DIA with a combination of genetic algorithms and neural networks. Expert Syst Appl 27:417–425
Yao J, Tan CL, Poh H (1992) Neural networks for technical analysis: a case study on KLCI. Int J Theor Appl Finance 2:221–241
Zhang G, Patuwo BE, Hu MY (1998) Forecasting with artificial neural networks: the state of the art. Int J Forecast 14:35–62
Zhu X, Wang H, Xu L, Li L (2008) Predicting stock index increments by neural networks: the role of trading volume under different horizons. Expert Syst Appl 34:3043–3054

13

Behavioral study of piston manufacturing plant through stochastic models

Arvind K Lal[1], Manwinder Kaur[1*] and Sneh Lata[2]

Abstract

Piston plays a vital role in almost all types of vehicles. The present study discusses the behavioral study of a piston manufacturing plant. Manufacturing plants are complex repairable systems and therefore, it is difficult to evaluate the performance of a piston manufacturing plant using stochastic models. The stochastic model is an efficient performance evaluator for repairable systems. In this paper, two stochastic models and computation algorithm of the piston manufacturing plant are illustrated using the state-space transition diagram and availability parameter is used for its behavioral study. Finally, the conclusion is discussed based on the resulting computations.

Keywords: Piston manufacturing plant; Behavioral study; Stochastic model; Time-dependent availability

Introduction

Piston is a cylindrical shaped component tightly fitted within another cylinder. Piston plays an important role in all vehicles and thus, the performance of the piston manufacturing process is equally important as that of piston quality and needs to be evaluated. The present study discusses the behavioral study of a piston manufacturing plant.

Manufacturing plants are complex repairable systems, and therefore, it is difficult to evaluate their performance metric including reliability and availability. The stochastic process is an efficient performance evaluator model for repairable systems. Several authors have considered such models to analyze the reliability and availability of manufacturing plants such as plastic manufacturing plant (Gupta et al. 2007), cement industries (Gupta et al. 2005b), butter oil manufacturing plant (Gupta et al. 2005c), and thermal power plant (Gupta and Tiwari 2009) for performance evaluation. Gupta et al. (2005a) discussed the performance of the polymer powder production process using the supplementary variable method.

In some of above studies, analytical approaches like Laplace transforms (Gupta 2003), matrix method (Gupta et al. 2007), and Lagrange's method (Mahajan and Singh 1999) are used to discuss transient state of the resulting mathematical expression in order to study the performance metrics including reliability and time-dependent availability. However, the computational part and analysis have not been effectively performed using these analytic methods and thus these were not used in the analysis of steady state behavior of the process in these studies. Further, various other techniques such as genetic algorithm (Kumar et al. 2010), GABLT (Sharma and Kumar 2010), and stochastic reward petri nets (Sachdeva et al. 2009) have also been used to analyze the steady state behavior of the systems. However, these analytic methods are not useful to obtain the solution of the transient state of complex manufacturing system. Therefore, we have used a numerical method to solve such a complex mathematical problem relating to the piston manufacturing plant.

In this paper, two stochastic models and computation algorithm of the piston manufacturing plant are illustrated using the state transition diagram. The objective of this study is to help the industry management by developing a mathematical model for piston manufacturing plant to evaluate its performance results based on availability. In the subsequent section, we will discuss the piston manufacturing process and its further division into subsystems in order to develop the stochastic models.

* Correspondence: manwinder.k.a@gmail.com
[1]School of Mathematics and Computer Applications, Thapar University, Patiala, PB 147004, India
Full list of author information is available at the end of the article

The process and its systems and subsystems

In the piston manufacturing plant, a gravity die casting process is carried out in the foundry to make the piston with the help of semiautomatic die machines without external pressures. The resultant pistons are sent for cleaning up the surface of the piston from the risers and runners and later, the strength, hardness, and other foundry defects are evaluated. Then a bunch of pistons is placed in the piston blank for its machining operations after which a fixture seat machine is used. Then, two rough grooves are made on the piston by using a carbide tool tip to fit a pin in them. Subsequently, another hole is drilled on the piston for oil passage and then finishing is given to the rough grooves using a diamond tool. Later, the piston is shaped into an oval to ensure its smooth movement within the cylinder using the carbide tool. In the next process, finish crown and cavity machines are used to put a rough cavity first and then a finish cavity on the piston. Further, valves are made on the piston using the diamond tool followed by a rounding-off operation on the corners of piston. Then, circlip grooves are made on the piston in the absence of any coolant and air pressure. After that, deburring, cleaning, and surface treatment operations are performed. Final inspections are carried out on the manufactured piston under some operational tests before packing.

In order to reduce the complexity for the effective behavioral (or performance) analysis, the piston manufacturing process is categorized into two systems namely, S1 and S2. The system which formed the piston into oval shape is denoted by S1 and the system which produces the final product as piston is denoted by S2. The flow chart of the piston manufacturing process is given in Figure 1.

Further, system S1 is divided into six subsystems, namely, A, B, C, D, E, and F. The brief description of these subsystems is as follows:

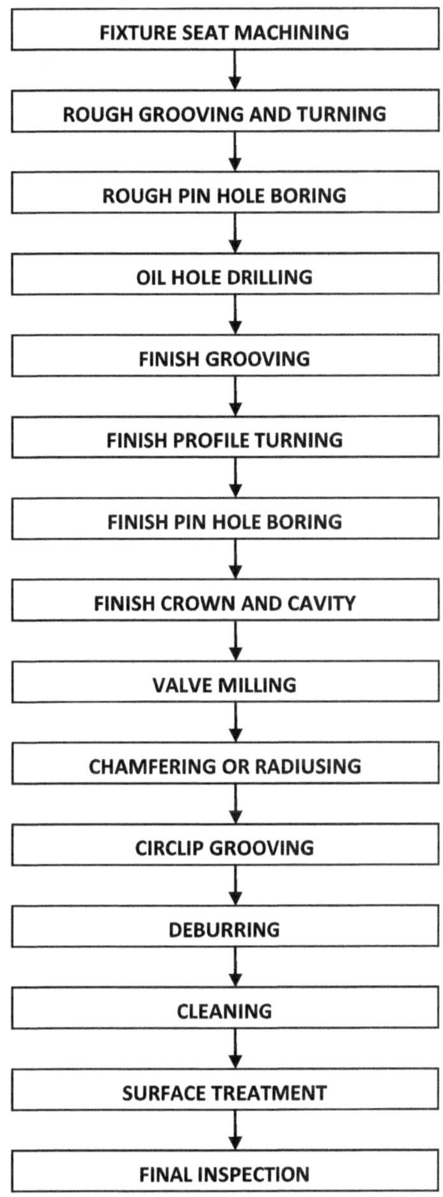

Figure 1 Flow chart of piston manufacturing plant.

1. Subsystem A is a fixture seat machining operation which is performed for clamping of pistons.
2. Subsystem B is a machining operation of rough grooving and turning operation.
3. Subsystem C is rough pin hole boring machine used to make pin holes on the piston to give proper size to the holes, which are used to fit the pin by which a rod is connected to crankshaft.
4. An oil hole is drilled on the piston for oil passage and the operating machine; oil hole drilling is considered as sub-system D.
5. Subsystem E is the finishing grooving machine used to give finishing to the rough grooves.
6. Subsystem F is the finish profile turning operation in which the piston is formed into oval shape in order to overcome expansion problems during working of the piston at high temperatures.

Similarly, system S2 is divided into nine subsystems namely G, H, I, J, K, L, M, N, and O which are described as follows:

1. Subsystem G is a finish pin hole boring machine.
2. Finish crown and cavity is subsystem H, which is operated to give finishing to the crown of piston, i.e., the upper part of the piston.

3. Subsystem I is a valve milling machine to make the valve recession of the piston.
4. Subsystem J is the chamfering or radiusing machine used to round off the corners of the piston to ensure smooth run in the cylinder.
5. Circlip grooves are made on the piston through subsystem K, a circlip grooving machine.
6. Subsystem L is the deburring machine. The deburring or brushing operation is performed on the piston using this machine.
7. Subsystem M is the cleaning machine which helps to clean the inside and outside of the piston.
8. Subsystem N is the surface treatment operation to coat the piston with some mixture.
9. Subsystem O is the final inspection of the manufactured product though operational tests before being packed.

In addition to above discussion, we have considered two kinds of failures for each subsystem, which are major and minor. Major failures are those in which the system shows complete breakdown while the minor failures are those which can be repaired during working conditions (reduced state). Now, we proceed with each system description and its subsystem. The subsystems A, B, C, D, F, F, I, J, and M are subject to major failure while C, E, G, and K are subject to both major and minor failures. Subsystems L, N, and O are considered to have no failure. Next, the state space of stochastic process for both systems is described in a diagrammatic form, known as transition diagram.

Transition diagrams

Transition diagrams for system S1 and S2 are shown in Figures 1 and 2, respectively. In the transition diagrams, we consider system S1 or system S2 in good state when all of its subsystems show good conditions, in reduced state when any of its subsystem is in reduced state, and in failed state when any of its subsystem fails. The corresponding subsystem state notations are described in the Section 'Notations'.

Notations

In this section, the notations are presented as follows:

1. The good condition of each subsystem is represented by capital letters, as A, B, C, D, E, and F for subsystems in S1 and G, H, I, J, K, and M for subsystems in S2, respectively.
2. Lowercase letters, including a, b, c, d, e, and f and g, h, i, j, k, and m, represent the failed state of subsystems A, B, C, D, E, and F, and G, H, I, J, K, and M, respectively.
3. The reduced working state of subsystems C, E, and G is indicated by $\bar{C}$, $\bar{E}$, and $\bar{G}$, respectively.
4. The constant failure rates of subsystems A, B, D, F, $\bar{C}$, $\bar{E}$, C, and E are represented by α_i (i = 1 to 8), respectively, in Figure 2. This parameter α_i (i = 1 to 7) also represents the respective constant failure rates of subsystems H, I, J, K, M, $\bar{G}$, and G in Figure 3.
5. The parameter β_i represents the respective constant repair rates of subsystems A, B, D, F, $\bar{C}$, $\bar{E}$, C, and E, for i = 1 to 8 in Figure 2. In Figure 3, it

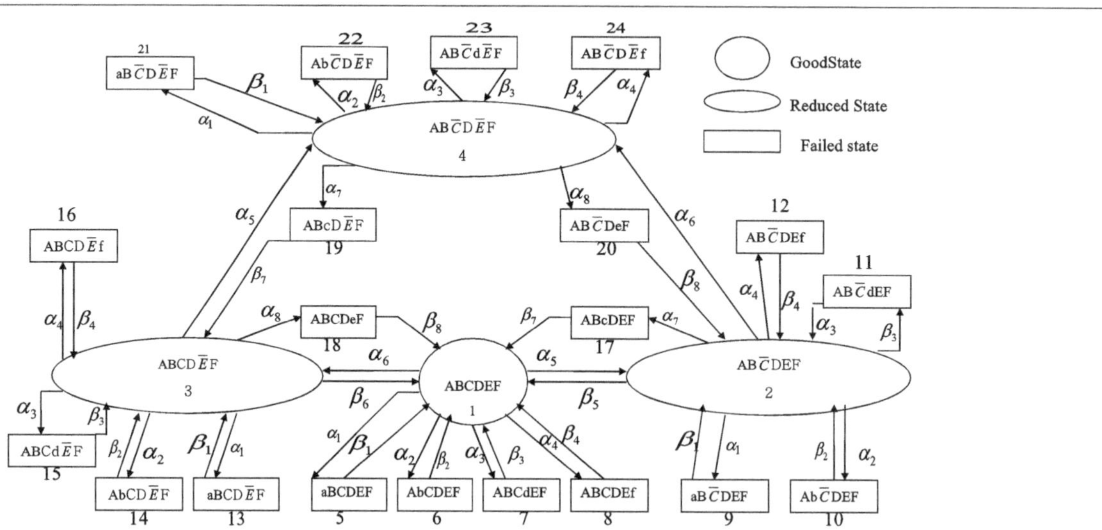

Figure 2 State transition diagram for system S_1. The constant failure rates of subsystems A, B, D, F, $\bar{C}$, $\bar{E}$, C, and E are represented by α_i (i = 1 to 8), respectively. The parameter β_i represents the respective constant repair rates of subsystems A, B, D, F, $\bar{C}$, $\bar{E}$, C, and E, for i = 1 to 8. Lowercase letters, including a, b, c, d, e, and f represent the failed state of subsystems A, B, C, D, E, and F respectively.

represents respective constant repair rates of subsystems H, I, J, K, M, $\bar{G}$, and G for i = 1 to 7.

6. Symbol $P_j(t)$ represents the probability that the system is in jth state at time t (j = 1 to 24 for system S1 and j = 1 to 13 for system S2). Symbol $P_j'(t)$ represent the rate of change of probability of jth state with respect to time t. (j = 1 to 24 for system S1 and j = 1 to 13 for system S2).

Assumptions

The following assumptions are used to model the performance analysis of the process for both systems S1 and S2:

1. Failure and repair rates are independent with each other and their unit is per hour.
2. There are no simultaneous failures among the subsystems.
3. Subsystems C, E, and G fail only through reduced states.
4. Repair of C, E, and G in the reduced state does not make these subsystems completely functional.
5. Switching to standby components is perfect.
6. Subsystems L, N, and O are considered as subsystems that never failed.

Stochastic models and computation algorithm of piston manufacturing plant

Systems S1 and S2 have been formulated by following the mnemonic rule discussed in the work of (Gupta et al. 2005a, b) using transition diagrams. The resulting Chapman-Kolmogorov first-order differential-difference equations for both systems under probabilistic considerations of each state discussed in state transition diagrams (Figures 2 and 3) are obtained as shown in the succeeding sections.

Mathematical formulation of system S_1

In this section, the mathematical formulation of system S1 is performed, with the help of transition diagram shown in Figure 2, as follows:

$$P_1'(t) + \lambda_1 P_1(t) = \sum_{i=1}^{4} \beta_i P_{4+i}(t) + \sum_{i=2}^{3} \beta_{3+i} P_i(t) + \sum_{i=7}^{8} \beta_i P_{10+i}(t), \quad (1)$$

$$P_2'(t) + \lambda_2 P_2(t) = \sum_{i=1}^{4} \beta_i P_{8+i}(t) + \beta_8 P_{20}(t) + \alpha_5 P_1(t), \quad (2)$$

$$P_3'(t) + \lambda_3 P_3(t) = \sum_{i=1}^{4} \beta_i P_{12+i}(t) + \beta_7 P_{19}(t) + \alpha_6 P_1(t), \quad (3)$$

$$P_4'(t) + \lambda_4 P_4(t) = \sum_{i=1}^{4} \beta_i P_{20+i}(t) + \alpha_5 P_3(t) + \alpha_6 P_2(t), \quad (4)$$

$$P_{4+i}'(t) + \beta_i P_{4+i}(t) = \alpha_i P_1(t), i = 1, 2, 3, 4 \quad (5)$$

$$P_{8+i}'(t) + \beta_i P_{8+i}(t) = \alpha_i P_2(t) \quad (6)$$

$$P_{12+i}'(t) + \beta_i P_{12+i}(t) = \alpha_i P_3(t) \quad (7)$$

$$P_{16+i}'(t) + \beta_7 P_{16+i}(t) = \alpha_7 P_j(t), i = 1, 2; j = 2, 4 \quad (8)$$

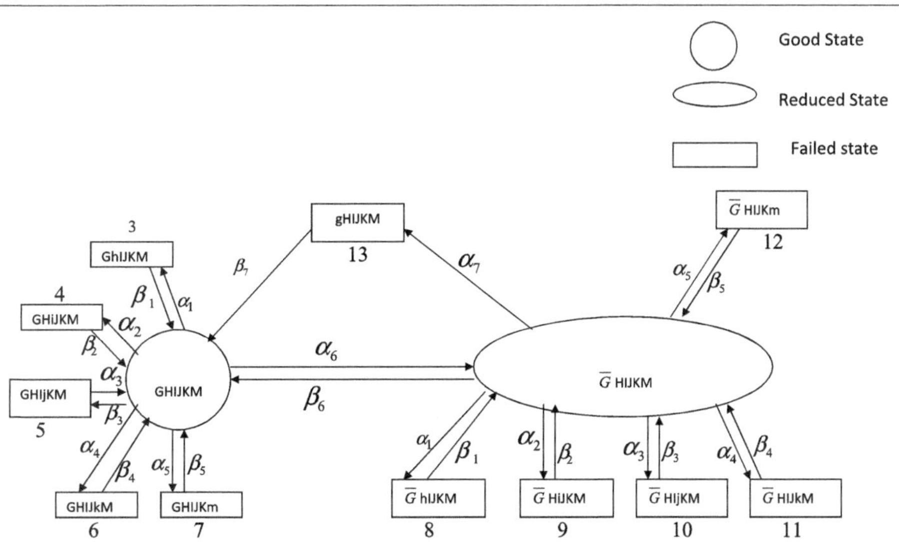

Figure 3 State transition diagram for system S_2. Parameter α_i (i = 1 to 7) represents the respective constant failure rates of subsystems H, I, J, K, M, $\bar{G}$, and G. Parameter β_i represents the respective constant repair rates of subsystems H, I, J, K, M, $\bar{G}$, and G for i = 1 to 7. Lowercase letters, including g, h, i, j, k, and m, represent the failed state of subsystems G, H, I, J, K, and M, respectively.

$$P'_{18+i}(t) + \beta_8 P_{18+i}(t) = \alpha_8 P_j(t), i = 1, 2 j = 3, 4 \quad (9)$$

$$P'_{20+i}(t) + \beta_i P_{20+i}(t) = \alpha_i P_4(t), i = 1, 2 3, 4, \quad (10)$$

where

$$\lambda_1 = \alpha_1 + \alpha_2 + \alpha_3 + \alpha_4 + \alpha_5 + \alpha_6 \quad (11)$$

$$\lambda_2 = \alpha_1 + \alpha_2 + \alpha_3 + \alpha_4 + \alpha_6 + \alpha_7 + \beta_5 \quad (12)$$

$$\lambda_3 = \alpha_1 + \alpha_2 + \alpha_3 + \alpha_4 + \alpha_5 + \alpha_8 + \beta_6 \quad (13)$$

$$\lambda_4 = \alpha_1 + \alpha_2 + \alpha_3 + \alpha_4 + \alpha_7 + \alpha_8 \quad (14)$$

with initial conditions $P_1(0) = 1$ and 0, otherwise.

The time-dependent availability $A_1(t)$ of the system (S1) is

$$A_1(t) = \sum_{i=1}^{4} P_i(t). \quad (15)$$

Mathematical formulation of system S2

In this section, the mathematical formulation of system S2 is performed, with the help of transition diagram shown in Figure 3, as follows:

$$P'_1(t) + \lambda_1 P_1(t) = \sum_{i=1}^{5} \beta_i P_{2+i}(t) + \beta_6 P_2(t) + \beta_7 P_{13}(t) \quad (16)$$

$$P'_2(t) + \lambda_2 P_2(t) = \sum_{i=1}^{5} \beta_i P_{7+i}(t) + \alpha_6 P_1(t) \quad (17)$$

$$P'_{2+i}(t) + \beta_i P_{2+i}(t) = \alpha_i P_j(t), \quad (18)$$

where

$$j = \begin{cases} 1 \text{ for } i = 1, 2, ..., 5 \\ 2 \text{ for } i = 6, 7, ..., 13 \end{cases}, \lambda_1 = \alpha_1 + \alpha_2 + \alpha_3 + \alpha_4 + \alpha_5 + \alpha_6 \quad (19)$$

$$\lambda_2 = \alpha_1 + \alpha_2 + \alpha_3 + \alpha_4 + \alpha_5 + \alpha_7 + \beta_6 \quad (20)$$

with initial conditions

$$P_1(0) = 1 \text{ and } P_i(0) = 0, \quad i = 2\text{–}15. \quad (21)$$

The time-dependent availability $A_2(t)$ of the system (S2) is

$$A_2(t) = \sum_{i=1}^{2} P_i(t). \quad (22)$$

System S1 completes its target only when system S2 starts, with the output of the previous system for further operation. Thus, S1 and S2 are in series which helps us to compute the time-dependent availability of the process industry as

$$A_P(t) = A_1(t) \times A_2(t). \quad (23)$$

As the system of differential Equations (1, 2, 3, 4, 5, 6, 7, 8, 9, and 10 and 16, 17, and 18) is very complex, it is difficult to find its analytical solution, particularly using Laplace transformation method. Therefore, a numerical procedure is used to obtain the computation for the availability of the process, which is discussed next.

Computational algorithm

In order to obtain the availability parameter of the process industry, the results are approximated for the following transient probabilities:

$$p_{1i}(t) = P[X(t) = i | X(0) = 1] \quad (24)$$

using initial distribution

$$P_1(0) = 1, P_j(0) = 0, \text{for } j = 2, ..., 24 \quad (25)$$

to study the effect of failure and repair rate variations on the plant performance. Also, as the transition matrix changes with respect to the change in transition rates, the complete transition matrix data for this case study was, thus, not evaluated for both the systems under different failure and repair rate variations. The algorithm for solving the system of equations (1, 2, 3, 4, 5, 6, 7, 8, 9, and 10) that corresponds to system S1 is as follows:

1. Input initial time, step size, transition rates (i.e., failure and repair rates) of each state.
2. Set initial distribution $P_1(0) = 1, P_j(0) = 0$, for $j = 2, ..., 24$.
3. Apply Runge–Kutta fourth-order method on differential equations (1, 2, 3, 4, 5, 6, 7, 8, 9, and 10) to obtain probabilities $p_{1i}(t), p_{12}(t), ..., p_{1n}(t), n = 24$.
4. Compute time-dependent availability using Equation 15.

Further, in order to compute the second row of transition matrix, replace step 2 with initial distribution, $P_2(0) = 1, P_j(0) = 0$, for $j = 1, 3, ..., 24$ for computing the following transition probabilities:

$$p_{2i}(t) = P[X(t) = i | X(0) = 2] \quad (26)$$

and proceed this way to analyze the results to compute each row of transition matrix. Similarly, results for S2 systems were computed.

We have performed 72,000 iterations on the system of differential-difference equations (1, 2, 3, 4, 5, 6, 7, 8, 9, and 10 and 16, 17, and 18) together with conditions (Equations 14 and 21) to solve them using the above algorithm, taking step size $t = 0.005$ with the assumption

Table 1 Time-dependent availability of the systems S_1 and S_2 and process industry

	Value of parameters described in first column of this table											
Time (h)	30	60	90	120	150	180	210	240	270	300	330	360
$A_1(t)$	0.9777	0.9705	0.9663	0.9634	0.9613	0.9596	0.9582	0.9572	0.9564	0.9557	0.9552	0.9548
$A_2(t)$	0.9914	0.9893	0.9887	0.9885	0.9884	0.9883	0.9883	0.9883	0.9883	0.9883	0.9883	0.9883
$A_p(t)$	0.9692	0.9601	0.9554	0.9523	0.9501	0.9484	0.9471	0.9460	0.9452	0.9446	0.9440	0.9436

that it equals to 1 h, considering the fact discussed (Gupta et al. 2007). This leads us to computations of time-dependent availability from 30 to 360 h. The resulting computations are shown in Table 1. The availability results are also simulated for various fluctuations of repair as well as failure rates of each subsystem (Tables 2,3,4,5,6,7,8,9,10,11,12,13). The simulated data for various fluctuating environment of failure and repair rates can be referred from Tables 2,3,4,5,6,7 for system S1 and from Tables 8,9,10,11,12,13 for system S2. Based on these results, the behavior study of both the system is discussed next.

Behavior study of both systems

Results are obtained for the overall time-dependent availability of the process, using Equation 23, with empirical data collected from industry. The data has been taken from industry in the form of operating hours, number of failures, number of repairs, and corresponding time for each repair for all the subsystems of the process industry. This data is composed into parameterized form of exponentially distributed failure rate and repair rates and given as $\alpha_1 = 0.0014$, $\alpha_2 = 0.0006$, $\alpha_3 = 0.0009$, $\alpha_4 = 0.0009$, $\alpha_5 = 0.0208$, $\alpha_6 = 0.0208$, $\alpha_7 = 0.0009$, and $\alpha_8 = 0.0039$ and $\beta_1 = 1.08$, $\beta_2 = 0.043$, $\beta_3 = 0.5$, $\beta_4 = 0.286$, $\beta_5 = 0.154$, $\beta_6 = 0.25$, $\beta_7 = 0.059$ and $\beta_8 = 0.087$ for system S1, respectively; and $\alpha_1 = 0.0014$, $\alpha_2 = 0.0003$, $\alpha_3 = 0.0001$, $\alpha_4 = 0.0003$, $\alpha_5 = 0.0001$, $\alpha_6 = 0.0208$, and $\alpha_7 = 0.004$ and $\beta_1 = 0.33$, $\beta_2 = 0.5$, $\beta_3 = 0.67$, $\beta_4 = 0.035$, $\beta_5 = 3.03$, $\beta_6 = 0.222$, and $\beta_7 = 0.125$ for system S2, respectively.

The computations of overall time-dependent availability of the process industry are shown in Table 1. The tabular form of computations helps us to conclude that the process is 96% reliable for operation at 30 h.

Further, we found that the time-dependent availability of system S1 is more affected by fixture seat machining (A) in comparison to the effect of other systems as observed from the simulated results shown in Tables 2,3,4,5,6,7. The time-dependent availability of the system S1 decreases by 1.02% from the empirical results (shown in Table 1), while it decreases by approximately 0.4% with the increase in time from 30 to 360 h. However, other subsystems' failure rate affects the system availability by 0.2% (for rough grooving cum turning and oil hole drilling), 0.06% (rough pin hole boring), 0.01% (for finish grooving), and 0.03% (finish profile turning) in comparison to fixture seat

Table 2 Fixture seat machining's failure (α_1) and repair rate (β_1) effect on time-dependent availability of S_1

Time in hours	α_1 value				β_1 value			
	0.0054	0.0104	0.154	0.0204	1.08	1.1800	1.2800	1.3800
30	0.9742	0.9698	0.9655	0.9612	0.9777	0.9778	0.9779	0.9780
60	0.9670	0.9627	0.9585	0.9543	0.9705	0.9706	0.9707	0.9707
90	0.9629	0.9587	0.9545	0.9503	0.9663	0.9664	0.9665	0.9666
120	0.9600	0.9558	0.9517	0.9475	0.9634	0.9635	0.9636	0.9637
150	0.9579	0.9537	0.9495	0.9454	0.9613	0.9614	0.9615	0.9615
180	0.9562	0.9520	0.9479	0.9438	0.9596	0.9597	0.9598	0.9598
210	0.9549	0.9507	0.9466	0.9425	0.9582	0.9583	0.9584	0.9585
240	0.9538	0.9497	0.9456	0.9415	0.9572	0.9573	0.9574	0.9575
270	0.9530	0.9489	0.9447	0.9406	0.9564	0.9565	0.9566	0.9566
300	0.9524	0.9482	0.9441	0.9400	0.9557	0.9558	0.9559	0.9560
330	0.9518	0.9477	0.9436	0.9395	0.9552	0.9553	0.9554	0.9554
360	0.9514	0.9473	0.9432	0.9391	0.9548	0.9549	0.9550	0.9550

Values of other parameters taken, in respect to varying values of α_1 of S1 are as follows: $\alpha_2 = 0.0006$, $\alpha_3 = 0.0009$, $\alpha_4 = 0.0009$, $\alpha_5 = 0.0208$, $\alpha_6 = 0.0208$, $\alpha_7 = 0.0009$, $\alpha_8 = 0.0039$; $\beta_1 = 1.08$, $>\beta_2 = 0.043$, $\beta_3 = 0.5$, $\beta_4 = 0.286$, $\beta_5 = 0.154$, $\beta_6 = 0.25$, $\beta_7 = 0.059$, $\beta_8 = 0.087$, respectively. Values of other parameters taken, in respect to varying values of β_1 of S1 are as follows: $\alpha_1 = 0.0014$, $\alpha_2 = 0.0006$, $\alpha_3 = 0.0009$, $\alpha_4 = 0.0009$, $\alpha_5 = 0.0208$, $\alpha_6 = 0.0208$, $\alpha_7 = 0.0009$, $\alpha_8 = 0.0039$; $\beta_2 = 0.043$, $\beta_3 = 0.5$, $\beta_4 = 0.286$, $\beta_5 = 0.154$, $\beta_6 = 0.25$, $\beta_7 = 0.059$, $\beta_8 = 0.087$, respectively.

Table 3 Rough grooving and turning's failure (α_2) and repair rate (β_2) effect on time-dependent availability of S1

Time in hours	α_2 value				β_2 value			
	0.0007	0.0008	0.0009	0.0010	0.044	0.045	0.046	0.047
30	0.9761	0.9744	0.9728	0.9712	0.9778	0.9779	0.9780	0.9781
60	0.9684	0.9664	0.9644	0.9624	0.9707	0.9709	0.9711	0.9713
90	0.9642	0.9621	0.9600	0.9579	0.9666	0.9669	0.9671	0.9673
120	0.9613	0.9592	0.9571	0.9550	0.9637	0.9640	0.9643	0.9645
150	0.9591	0.9570	0.9549	0.9528	0.9616	0.9618	0.9621	0.9624
180	0.9575	0.9553	0.9532	0.9511	0.9599	0.9601	0.9604	0.9607
210	0.9561	0.9540	0.9519	0.9498	0.9585	0.9588	0.9591	0.9593
240	0.9551	0.9530	0.9509	0.9488	0.9575	0.9578	0.9580	0.9583
270	0.9543	0.9522	0.9501	0.9480	0.9567	0.9569	0.9572	0.9574
300	0.9536	0.9515	0.9494	0.9473	0.9560	0.9563	0.9565	0.9568
330	0.9531	0.9510	0.9489	0.9468	0.9555	0.9558	0.9560	0.9563
360	0.9527	0.9506	0.9485	0.9464	0.9551	0.9553	0.9556	0.9559

Values of other parameters taken, in respect to varying values of α_2 of S1 are as follows: $\alpha_1 = 0.0014$, $\alpha_3 = 0.0009$, $\alpha_4 = 0.0009$, $\alpha_5 = 0.0208$, $\alpha_6 = 0.0208$, $\alpha_7 = 0.0009$, $\alpha_8 = 0.0039$; $\beta_1 = 1.08$, $\beta_2 = 0.043$, $\beta_3 = 0.5$, $\beta_4 = 0.286$, $\beta_5 = 0.154$, $\beta_6 = 0.25$, $\beta_7 = 0.059$, $\beta_8 = 0.087$, respectively. Values of other parameters taken, in respect to varying values of β_2 of S1 are as follows: $\alpha_1 = 0.0014$, $\alpha_2 = 0.0006$, $\alpha_3 = 0.0009$, $\alpha_4 = 0.0009$, $\alpha_5 = 0.0208$, $\alpha_6 = 0.0208$, $\alpha_7 = 0.0009$, $\alpha_8 = 0.0039$; $\beta_1 = 1.08$, $\beta_3 = 0.5$, $\beta_4 = 0.286$, $\beta_5 = 0.154$, $\beta_6 = 0.25$, $\beta_7 = 0.059$, $\beta_8 = 0.087$, respectively.

machining when compared to empirical results of system S1 availability, A_1 (Table 1).

Similarly, repair rate fluctuations have been studied for the effect of maintenance on time-dependent availability of the system. We found that the time-dependent availability of system S1 shows 0.001% increase for fixture seat machining and rough pin hole boring, and for other subsystems, time-dependent availability shows 0.01% (for rough grooving cum tuning and finish profile turning), 0.02% (for rough oil drilling), and 0.025%

Table 4 Rough pin hole boring failure rate (α_7) and repair rate (β_7) effect on time-dependent availability of S1

Time in hours	α_7 value				β_7 value			
	0.0010	0.0011	0.0012	0.0013	0.060	0.061	0.062	0.063
30	0.9775	0.9773	0.9771	0.9770	0.9777	0.9777	0.9777	0.9777
60	0.9702	0.9699	0.9696	0.9693	0.9705	0.9705	0.9706	0.9706
90	0.9659	0.9656	0.9652	0.9648	0.9664	0.9664	0.9665	0.9665
120	0.9630	0.9626	0.9621	0.9617	0.9635	0.9636	0.9636	0.9637
150	0.9608	0.9603	0.9599	0.9594	0.9613	0.9614	0.9615	0.9616
180	0.9591	0.9586	0.9581	0.9576	0.9597	0.9597	0.9598	0.9599
210	0.9577	0.9572	0.9567	0.9562	0.9583	0.9584	0.9585	0.9586
240	0.9567	0.9562	0.9557	0.9552	0.9573	0.9574	0.9575	0.9576
270	0.9558	0.9553	0.9548	0.9543	0.9565	0.9566	0.9567	0.9568
300	0.9552	0.9547	0.9542	0.9536	0.9558	0.9559	0.9560	0.9561
330	0.9547	0.9541	0.9536	0.9531	0.9553	0.9554	0.9555	0.9556
360	0.9543	0.9537	0.9532	0.9527	0.9549	0.9550	0.9551	0.9552

Values of other parameters taken, in respect to varying values of α_7 of S1 are as follows: $\alpha_1 = 0.0014$, $\alpha_2 = 0.0006$, $\alpha_3 = 0.0009$, $\alpha_4 = 0.0009$, $\alpha_5 = 0.0208$, $\alpha_6 = 0.0208$, $\alpha_8 = 0.0039$; $\beta_1 = 1.08$, $\beta_2 = 0.043$, $\beta_3 = 0.5$, $\beta_4 = 0.286$, $\beta_5 = 0.154$, $\beta_6 = 0.25$, $\beta_7 = 0.059$, $\beta_8 = 0.087$, respectively. Values of other parameters taken, in respect to varying values of β_7 of S1 are as follows: $\alpha_1 = 0.0014$, $\alpha_2 = 0.0006$, $\alpha_3 = 0.0009$, $\alpha_4 = 0.0009$, $\alpha_5 = 0.0208$, $\alpha_6 = 0.0208$, $\alpha_7 = 0.0009$, $\alpha_8 = 0.0039$; $\beta_1 = 1.08$, $\beta_2 = 0.043$, $\beta_3 = 0.5$, $\beta_4 = 0.286$, $\beta_5 = 0.154$, $\beta_6 = 0.25$, $\beta_8 = 0.087$, respectively.

Table 5 Oil hole drilling's failure (α_3) and repair (β_3) rate on time-dependent availability of S1

Time in hours	α_3 value				β_3 value			
	0.0010	0.0011	0.0012	0.0013	0.6	0.7	0.8	0.9
30	0.9775	0.9773	0.9771	0.9769	0.9780	0.9782	0.9783	0.9784
60	0.9703	0.9701	0.9699	0.9697	0.9708	0.9710	0.9711	0.9712
90	0.9661	0.9660	0.9658	0.9656	0.9666	0.9668	0.9670	0.9671
120	0.9633	0.9631	0.9629	0.9627	0.9637	0.9639	0.9641	0.9642
150	0.9611	0.9609	0.9607	0.9605	0.9615	0.9617	0.9619	0.9620
180	0.9594	0.9592	0.9590	0.9588	0.9599	0.9600	0.9602	0.9603
210	0.9581	0.9579	0.9577	0.9575	0.9585	0.9587	0.9589	0.9590
240	0.9570	0.9568	0.9567	0.9565	0.9575	0.9577	0.9578	0.9579
270	0.9562	0.9560	0.9558	0.9556	0.9566	0.9568	0.9570	0.9571
300	0.9555	0.9553	0.9552	0.9550	0.9560	0.9562	0.9563	0.9564
330	0.9550	0.9548	0.9546	0.9545	0.9555	0.9557	0.9558	0.9559
360	0.9546	0.9544	0.9542	0.9541	0.9551	0.9552	0.9554	0.9555

Values of other parameters taken, in respect to varying values of α_3 of S1 are as follows: $\alpha_1 = 0.0014$, $\alpha_2 = 0.0006$, $\alpha_4 = 0.0009$, $\alpha_5 = 0.0208$, $\alpha_6 = 0.0208$, $\alpha_7 = 0.0009$ $\alpha_8 = 0.0039$; $\beta_1 = 1.08$, $\beta_2 = 0.043$, $\beta_3 = 0.5$, $\beta_4 = 0.286$, $\beta_5 = 0.154$, $\beta_6 = 0.25$, $\beta_7 = 0.059$, $\beta_8 = 0.087$, respectively. Values of other parameters taken, in respect to varying values of β_3 of S1 are as follows: $\alpha_1 = 0.0014$, $\alpha_2 = 0.0006$, $\alpha_3 = 0.0009$, $\alpha_4 = 0.0009$, $\alpha_5 = 0.0208$, $\alpha_6 = 0.0208$, $\alpha_7 = 0.0009$ $\alpha_8 = 0.0039$; $\beta_1 = 1.08$, $\beta_2 = 0.043$, $\beta_4 = 0.286$, $\beta_5 = 0.154$, $\beta_6 = 0.25$, $\beta_7 = 0.059$, $\beta_8 = 0.087$, respectively.

Table 6 Finish grooving's failure (α_8) and repair rate (β_8) effect on time-dependent availability of S1

Time in hours	α_8 value				β_8 value			
	0.0040	0.0041	0.0042	0.0043	0.097	0.107	0.117	0.127
30	0.9776	0.9775	0.9773	0.9772	0.9780	0.9783	0.9785	0.9788
60	0.9703	0.9701	0.9699	0.9697	0.9712	0.9717	0.9722	0.9727
90	0.9661	0.9659	0.9656	0.9654	0.9673	0.9681	0.9688	0.9693
120	0.9632	0.9629	0.9627	0.9624	0.9646	0.9656	0.9664	0.9671
150	0.9610	0.9607	0.9605	0.9602	0.9626	0.9637	0.9647	0.9655
180	0.9593	0.9590	0.9587	0.9585	0.9611	0.9623	0.9633	0.9642
210	0.9580	0.9577	0.9574	0.9571	0.9598	0.9612	0.9622	0.9632
240	0.9569	0.9566	0.9564	0.9561	0.9589	0.9603	0.9614	0.9624
270	0.9561	0.9558	0.9555	0.9553	0.9581	0.9595	0.9607	0.9617
300	0.9554	0.9552	0.9549	0.9546	0.9575	0.9590	0.9602	0.9612
330	0.9549	0.9546	0.9544	0.9541	0.9570	0.9585	0.9598	0.9608
360	0.9545	0.9542	0.9540	0.9537	0.9567	0.9582	0.9595	0.9605

Values of other parameters taken, in respect to varying values of α_8 of S1 are as follows: $\alpha_1 = 0.0014$, $\alpha_2 = 0.0006$, $\alpha_3 = 0.0009$, $\alpha_4 = 0.0009$, $\alpha_5 = 0.0208$, $\alpha_6 = 0.0208$, $\alpha_7 = 0.0009$; $\beta_1 = 1.08$, $\beta_2 = 0.043$, $\beta_3 = 0.5$, $\beta_4 = 0.286$, $\beta_5 = 0.154$, $\beta_6 = 0.25$, $\beta_7 = 0.059$, $\beta_8 = 0.087$, respectively. Values of other parameters taken, in respect to varying values of β_8 of S1 are as follows: $\alpha_1 = 0.0014$, $\alpha_2 = 0.0006$, $\alpha_3 = 0.0009$, $\alpha_4 = 0.0009$, $\alpha_5 = 0.0208$, $\alpha_6 = 0.0208$, $\alpha_7 = 0.0009$, $\alpha_8 = 0.0039$; $\beta_1 = 1.08$, $\beta_2 = 0.043$, $\beta_3 = 0.5$, $\beta_4 = 0.286$, $\beta_5 = 0.154$, $\beta_6 = 0.25$, $\beta_7 = 0.059$.

Table 7 Finish profile turning's failure (α_4) and repair (β_4) rate effect on time-dependent availability of S1

Time in hours	α_4 value				β_4 value			
	0.0010	0.0011	0.0012	0.0013	0.296	0.306	0.316	0.326
30	0.9774	0.9770	0.9767	0.9764	0.9778	0.9779	0.9780	0.9781
60	0.9701	0.9698	0.9695	0.9692	0.9706	0.9707	0.9708	0.9708
90	0.9660	0.9657	0.9654	0.9650	0.9664	0.9665	0.9666	0.9667
120	0.9631	0.9628	0.9625	0.9622	0.9635	0.9636	0.9637	0.9638
150	0.9609	0.9606	0.9603	0.9600	0.9614	0.9615	0.9615	0.9616
180	0.9593	0.9589	0.9586	0.9583	0.9597	0.9598	0.9599	0.9599
210	0.9579	0.9576	0.9573	0.9570	0.9583	0.9584	0.9585	0.9586
240	0.9569	0.9566	0.9562	0.9559	0.9573	0.9574	0.9575	0.9575
270	0.9561	0.9557	0.9554	0.9551	0.9565	0.9566	0.9566	0.9567
300	0.9554	0.9551	0.9548	0.9544	0.9558	0.9559	0.9560	0.9561
330	0.9549	0.9546	0.9542	0.9539	0.9553	0.9554	0.9555	0.9555
360	0.9545	0.9541	0.9538	0.9535	0.9549	0.9550	0.9551	0.9551

Values of other parameters taken, in respect to varying values of α_4 of S1 are as follows: $\alpha_1 = 0.0014$, $\alpha_2 = 0.0006$, $\alpha_3 = 0.0009$, $\alpha_5 = 0.0208$, $\alpha_6 = 0.0208$, $\alpha_7 = 0.0009$, $\alpha_8 = 0.0039$; $\beta_1 = 1.08$, $\beta_2 = 0.043$, $\beta_3 = 0.5$, $\beta_4 = 0.286$, $\beta_5 = 0.154$, $\beta_6 = 0.25$, $\beta_7 = 0.059$, $\beta_8 = 0.087$, respectively. Values of other parameters taken, in respect to varying values of β_4 of S1 are as follows: $\alpha_1 = 0.0014$, $\alpha_2 = 0.0006$, $\alpha_3 = 0.0009$, $\alpha_4 = 0.0009$, $\alpha_5 = 0.0208$, $\alpha_6 = 0.0208$, $\alpha_7 = 0.0009$, $\alpha_8 = 0.0039$; $\beta_1 = 1.08$, $\beta_2 = 0.043$, $\beta_3 = 0.5$, $\beta_5 = 0.154$, $\beta_6 = 0.25$, $\beta_7 = 0.059$, $\beta_8 = 0.087$, respectively.

increase (for finish grooving) with corresponding increase in repair rates as observed from Tables 2,3,4,5,6,7. So the weak subsystem was found to be the fixture seat machining which is due to the effect maintenance practice as well.

Table 8 Effect of failure rate (α_1) and repair rate (β_1) of finish crown and cavity on time-dependent availability of S2

Time in hours	α_1 value				β_1 value			
	0.0054	0.0104	0.154	0.0204	0.34	0.35	0.36	0.37
30	0.9905	0.9894	0.9883	0.9873	0.9914	0.9914	0.9914	0.9914
60	0.9885	0.9874	0.9864	0.9854	0.9894	0.9894	0.9894	0.9894
90	0.9878	0.9868	0.9858	0.9848	0.9887	0.9887	0.9887	0.9887
120	0.9876	0.9866	0.9855	0.9846	0.9885	0.9885	0.9885	0.9885
150	0.9875	0.9865	0.9855	0.9845	0.9884	0.9884	0.9884	0.9884
180	0.9875	0.9864	0.9854	0.9845	0.9884	0.9884	0.9884	0.9884
210	0.9875	0.9864	0.9854	0.9844	0.9883	0.9884	0.9884	0.9884
240	0.9875	0.9864	0.9854	0.9844	0.9883	0.9884	0.9884	0.9884
270	0.9875	0.9864	0.9854	0.9844	0.9883	0.9883	0.9884	0.9884
300	0.9875	0.9864	0.9854	0.9844	0.9883	0.9883	0.9884	0.9884
330	0.9875	0.9864	0.9854	0.9844	0.9883	0.9883	0.9884	0.9884
360	0.9875	0.9864	0.9854	0.9844	0.9883	0.9883	0.9884	0.9884

Values of other parameters taken, in respect to varying values of α_1 of S2 are as follows: $\alpha_2 = 0.0003$, $\alpha_3 = 0.0001$, $\alpha_4 = 0.0003$, $\alpha_5 = 0.0001$, $\alpha_6 = 0.0208$, $\alpha_7 = 0.004$; $\beta_1 = 0.33$, $\beta_2 = 0.5$, $\beta_3 = 0.67$, $\beta_4 = 0.035$, $\beta_5 = 3.03$, $\beta_6 = 0.222$, $\beta_7 = 0.125$, respectively. Values of other parameters taken, in respect to varying values of β_1 of S2 are as follows: $\alpha_1 = 0.0014$, $\alpha_2 = 0.0003$, $\alpha_3 = 0.0001$, $\alpha_4 = 0.0003$, $\alpha_5 = 0.0001$, $\alpha_6 = 0.0208$, $\alpha_7 = 0.004$, $\beta_2 = 0.5$, $\beta_3 = 0.67$, $\beta_4 = 0.035$, $\beta_5 = 3.03$, $\beta_6 = 0.222$, $\beta_7 = 0.125$, respectively.

Table 9 Effect of failure rate of valve milling (α_2) on time-dependent availability of the system S2

Time in hours	α_2 value				β_2 value			
	0.0004	0.0005	0.0006	0.0007	0.6	0.7	0.8	0.9
30	0.9913	0.9913	0.9913	0.9913	0.9914	0.9914	0.9914	0.9914
60	0.9893	0.9893	0.9893	0.9893	0.9893	0.9894	0.9894	0.9894
90	0.9887	0.9887	0.9886	0.9886	0.9887	0.9887	0.9887	0.9887
120	0.9884	0.9884	0.9884	0.9884	0.9885	0.9885	0.9885	0.9885
150	0.9884	0.9883	0.9883	0.9883	0.9884	0.9884	0.9884	0.9884
180	0.9883	0.9883	0.9883	0.9883	0.9884	0.9884	0.9884	0.9884
210	0.9883	0.9883	0.9883	0.9883	0.9883	0.9883	0.9884	0.9884
240	0.9883	0.9883	0.9883	0.9883	0.9883	0.9883	0.9883	0.9884
270	0.9883	0.9883	0.9883	0.9883	0.9883	0.9883	0.9883	0.9884
300	0.9883	0.9883	0.9883	0.9883	0.9883	0.9883	0.9883	0.9884
330	0.9883	0.9883	0.9883	0.9883	0.9883	0.9883	0.9883	0.9884
360	0.9883	0.9883	0.9883	0.9883	0.9883	0.9883	0.9883	0.9884

Values of other parameters taken, in respect to varying values of α_2 of S2 are as follows: $\alpha_1 = 0.0014$, $\alpha_2 = 0.0003$, $\alpha_3 = 0.0001$, $\alpha_4 = 0.0003$, $\alpha_5 = 0.0001$, $\alpha_6 = 0.0208$, $\alpha_7 = 0.004$, $\beta_1 = 0.33$, $\beta_2 = 0.5$, $\beta_3 = 0.67$, $\beta_4 = 0.035$, $\beta_5 = 3.03$, $\beta_6 = 0.222$, $\beta_7 = 0.125$, respectively. Values of other parameters taken, in respect to varying values of β_2 of S2 are as follows: $\alpha_1 = 0.0014$, $\alpha_2 = 0.0003$, $\alpha_3 = 0.0001$, $\alpha_4 = 0.0003$, $\alpha_5 = 0.0001$, $\alpha_6 = 0.0208$, $\alpha_7 = 0.004$; $\beta_1 = 0.33$, $\beta_3 = 0.67$, $\beta_4 = 0.035$, $\beta_5 = 3.03$, $\beta_6 = 0.222$, $\beta_7 = 0.125$, respectively.

Likewise, the results were simulated for systems in S2 and can be referred from Tables 8,9,10,11,12,13. We observed that the time-dependent availability of this system is more affected by circlip grooving (K), in comparison to other subsystems. It is observed that the

Table 10 Effect of failure rate (α_3) and repair rate (β_3) of chamfering or radiusing on time-dependent availability of S2

Time in hours	α_3 value				β_3 value			
	0.0002	0.0003	0.0004	0.0005	0.68	0.69	0.70	0.71
30	0.9912	0.9911	0.9909	0.9908	0.9914	0.9914	0.9914	0.9914
60	0.9892	0.9891	0.9889	0.9888	0.9893	0.9893	0.9893	0.9894
90	0.9885	0.9884	0.9882	0.9881	0.9887	0.9887	0.9887	0.9887
120	0.9883	0.9882	0.9880	0.9879	0.9885	0.9885	0.9885	0.9885
150	0.9882	0.9881	0.9879	0.9878	0.9884	0.9884	0.9884	0.9884
180	0.9882	0.9881	0.9879	0.9878	0.9883	0.9884	0.9884	0.9884
210	0.9882	0.9880	0.9879	0.9878	0.9883	0.9883	0.9883	0.9883
240	0.9882	0.9880	0.9879	0.9878	0.9883	0.9883	0.9883	0.9883
270	0.9882	0.9880	0.9879	0.9878	0.9883	0.9883	0.9883	0.9883
300	0.9882	0.9880	0.9879	0.9878	0.9883	0.9883	0.9883	0.9883
330	0.9882	0.9880	0.9879	0.9878	0.9883	0.9883	0.9883	0.9883
360	0.9882	0.9880	0.9879	0.9878	0.9883	0.9883	0.9883	0.9883

Values of other parameters taken, in respect to varying values of α_3 of S2 are as follows: $\alpha_1 = 0.0014$, $\alpha_2 = 0.0003$, $\alpha_4 = 0.0003$, $\alpha_5 = 0.0001$, $\alpha_6 = 0.0208$, $\alpha_7 = 0.004$; $\beta_1 = 0.33$, $\beta_2 = 0.5$, $\beta_3 = 0.67$, $\beta_4 = 0.035$, $\beta_5 = 3.03$, $\beta_6 = 0.222$, $\beta_7 = 0.125$, respectively. Values of other parameters taken, in respect to varying values of β_3 of S2 are as follows: $\alpha_1 = 0.0014$, $\alpha_2 = 0.0003$, $\alpha_3 = 0.0001$, $\alpha_4 = 0.0003$, $\alpha_5 = 0.0001$, $\alpha_6 = 0.0208$, $\alpha_7 = 0.004$, $\beta_1 = 0.33$, $\beta_2 = 0.5$, $\beta_4 = 0.035$, $\beta_5 = 3.03$, $\beta_6 = 0.222$, $\beta_7 = 0.125$, respectively.

Table 11 Effect of failure rate (α_4) and repair rate (β_4) of circlip grooving on time-dependent availability of the system S2

Time in hours	α_4 value				β_4 value			
	0.0004	0.0005	0.0006	0.0007	0.045	0.055	0.065	0.075
30	0.9895	0.9877	0.9859	0.9841	0.9920	0.9925	0.9930	0.9933
60	0.9869	0.9845	0.9820	0.9796	0.9906	0.9916	0.9923	0.9928
90	0.9860	0.9834	0.9807	0.9781	0.9903	0.9914	0.9922	0.9928
120	0.9857	0.9830	0.9803	0.9776	0.9902	0.9914	0.9922	0.9928
150	0.9856	0.9829	0.9801	0.9774	0.9902	0.9914	0.9922	0.9928
180	0.9856	0.9828	0.9801	0.9774	0.9902	0.9914	0.9922	0.9928
210	0.9856	0.9828	0.9801	0.9774	0.9902	0.9914	0.9922	0.9928
240	0.9856	0.9828	0.9801	0.9773	0.9902	0.9914	0.9922	0.9928
270	0.9856	0.9828	0.9801	0.9773	0.9902	0.9914	0.9922	0.9928
300	0.9856	0.9828	0.9801	0.9773	0.9902	0.9914	0.9922	0.9928
330	0.9856	0.9828	0.9801	0.9773	0.9902	0.9914	0.9922	0.9928
360	0.9856	0.9828	0.9801	0.9773	0.9902	0.9914	0.9922	0.9928

Values of other parameters taken, in respect to varying values of α_4 of S2 are as follows: $\alpha_1 = 0.0014$, $\alpha_2 = 0.0003$, $\alpha_3 = 0.0001$, $\alpha_5 = 0.0001$, $\alpha_6 = 0.0208$, $\alpha_7 = 0.004$; $\beta_1 = 0.33$, $\beta_2 = 0.5$, $\beta_3 = 0.67$, $\beta_4 = 0.035$, $\beta_5 = 3.03$, $\beta_6 = 0.222$, $\beta_7 = 0.125$, respectively. Values of other parameters taken, in respect to varying values of β_4 of S2 are as follows: $\alpha_1 = 0.0014$, $\alpha_2 = 0.0003$, $\alpha_3 = 0.0001$, $\alpha_4 = 0.0003$, $\alpha_5 = 0.0001$, $\alpha_6 = 0.0208$, $\alpha_7 = 0.004$; $\beta_1 = 0.33$, $\beta_2 = 0.5$, $\beta_3 = 0.67$, $\beta_5 = 3.03$, $\beta_6 = 0.222$, $\beta_7 = 0.125$, respectively.

availability of system S2 decreases by 0.18% when compared with the empirical values of A_2; the availability of system S2 is presented in Table 1. This decreases by 0.34% with the increase in time from 30 to 360 h. However, the

Table 12 Effect of failure rate (α_5) and repair rate (β_5) of cleaning machine on time-dependent availability of the system S_2

Time in hours	α_5 value				β_5 value			
	0.0002	0.0003	0.0004	0.0005	3.13	3.23	3.33	3.43
30	0.9913	0.9913	0.9913	0.9912	0.9914	0.9914	0.9914	0.9914
60	0.9893	0.9893	0.9892	0.9892	0.9893	0.9893	0.9893	0.9893
90	0.9886	0.9886	0.9886	0.9886	0.9887	0.9887	0.9887	0.9887
120	0.9884	0.9884	0.9884	0.9883	0.9885	0.9885	0.9885	0.9885
150	0.9883	0.9883	0.9883	0.9882	0.9884	0.9884	0.9884	0.9884
180	0.9883	0.9883	0.9882	0.9882	0.9883	0.9883	0.9883	0.9883
210	0.9883	0.9883	0.9882	0.9882	0.9883	0.9883	0.9883	0.9883
240	0.9883	0.9883	0.9882	0.9882	0.9883	0.9883	0.9883	0.9883
270	0.9883	0.9883	0.9882	0.9882	0.9883	0.9883	0.9883	0.9883
300	0.9883	0.9883	0.9882	0.9882	0.9883	0.9883	0.9883	0.9883
330	0.9883	0.9883	0.9882	0.9882	0.9883	0.9883	0.9883	0.9883
360	0.9883	0.9883	0.9882	0.9882	0.9883	0.9883	0.9883	0.9883

Values of other parameters taken, in respect to varying values of α_5 of S2 are as follows: $\alpha_1 = 0.0014$, $\alpha_2 = 0.0003$, $\alpha_3 = 0.0001$, $\alpha_4 = 0.0003$, $\alpha_6 = 0.0208$, $\alpha_7 = 0.004$; $\beta_1 = 0.33$, $\beta_2 = 0.5$, $\beta_3 = 0.67$, $\beta_4 = 0.035$, $\beta_5 = 3.03$, $\beta_6 = 0.222$, $\beta_7 = 0.125$, respectively. Values of other parameters taken, in respect to varying values of β_5 of S2 are as follows: $\alpha_1 = 0.0014$, $\alpha_2 = 0.0003$, $\alpha_3 = 0.0001$, $\alpha_4 = 0.0003$, $\alpha_5 = 0.0001$, $\alpha_6 = 0.0208$, $\alpha_7 = 0.004$, $\beta_1 = 0.33$, $\beta_2 = 0.5$, $\beta_3 = 0.67$, $\beta_4 = 0.035$, $\beta_6 = 0.222$, $\beta_7 = 0.125$, respectively.

Table 13 Effect of failure rate (α_7) and repair rate (β_7) of finish pin hole boring on time-dependent availability of the system S2

Time in hours	α_7 value				β_7 value			
	0.005	0.006	0.007	0.008	0.135	0.145	0.155	0.165
30	0.9907	0.9901	0.9895	0.9889	0.9915	0.9917	0.9918	0.9919
60	0.9887	0.9881	0.9874	0.9868	0.9895	0.9897	0.9898	0.9900
90	0.9880	0.9874	0.9868	0.9862	0.9889	0.9890	0.9892	0.9893
120	0.9878	0.9872	0.9865	0.9859	0.9886	0.9888	0.9890	0.9891
150	0.9877	0.9871	0.9865	0.9858	0.9886	0.9887	0.9889	0.9890
180	0.9877	0.9871	0.9864	0.9858	0.9885	0.9887	0.9889	0.9890
210	0.9877	0.9871	0.9864	0.9858	0.9885	0.9887	0.9888	0.9890
240	0.9877	0.9871	0.9864	0.9858	0.9885	0.9887	0.9888	0.9890
270	0.9877	0.9871	0.9864	0.9858	0.9885	0.9887	0.9888	0.9890
300	0.9877	0.9871	0.9864	0.9858	0.9885	0.9887	0.9888	0.9890
330	0.9877	0.9871	0.9864	0.9858	0.9885	0.9887	0.9888	0.9890
360	0.9877	0.9871	0.9864	0.9858	0.9885	0.9887	0.9888	0.9890

Values of other parameters taken, in respect to varying values of α_7 of S2 are as follows: $\alpha_1 = 0.0014$, $\alpha_2 = 0.0003$, $\alpha_3 = 0.0001$, $\alpha_4 = 0.0003$, $\alpha_5 = 0.0001$,$\alpha_6 = 0.0208$; $\beta_1 = 0.33$, $\beta_2 = 0.5$, $\beta_3 = 0.67$, $\beta_4 = 0.035$, $\beta_5 = 3.03$, $\beta_6 = 0.222$, $\beta_7 = 0.125$, respectively. Values of other parameters taken, in respect to varying values of β_7 of S2 are as follows: $\alpha_1 = 0.0014$, $\alpha_2 = 0.0003$, $\alpha_3 = 0.0001$, $\alpha_4 = 0.0003$, $\alpha_5 = 0.0001$, $\alpha_6 = 0.0208$, $\alpha_7 = 0.004$, $\beta_1 = 0.33$, $\beta_2 = 0.5$, $\beta_3 = 0.67$, $\beta_4 = 0.035$, $\beta_5 = 3.03$, $\beta_6 = 0.222$, $\beta_7 = 0.125$, respectively.

decrease in percentage effect of the other subsystem's failure rate on system time-dependent availability is observed as 0.06% for finish pin hole boring, 0.02% for chamfering, 0.01% for finish crown and cavity, 0.003% for cleaning machine, and 0.002% decrease for valve milling in comparison to circlip grooving, as compared to empirical results of system S2 time-dependent availability (Table 1) which can be referred from Tables 8,9,10,11,12,13.

Similarly, the repair rate fluctuations have been studied for the effect of maintenance on time-dependent availability in the case of system S2 as well. We found that time-dependent availability of the system S2 shows 0.5% increase for circlip grooving with decrease in value of its repair rate parameter. In case of finish pin hole boring, the time-dependent availability of the system improved by 0.2% when its repair rates decrease. The effect on time-dependent availability was found to be negligible in case of other subsystems and can be referred from Tables 8,9,10,11,12,13.

Conclusion

From the comparative study of this paper, we have observed that the time-dependent availability of the system S1 and system S2 is affected by fixture seat machining (A) and circlip grooving (K), respectively. Therefore, it is recommended that management should pay more attention to fixture seat machining (A) and circlip grooving (K) in order to increase the time-dependent availability of the piston manufacturing plant. The conclusion drawn from the analysis is consistent with the actual performance of the manufacturing plant and will help the industry to improve its performance. The main advantage of this work is that we can examine the time-dependent analysis.

Competing interests
The authors declare that they have no competing interests.

Authors' contributions
The industrial based research work for data collection was carried by SL. The modeling work has been done by SL and MK. The computation based work done by SL with the help of AKL and MK. The paper drafting, revision was carried in collaboration by all authors. All authors read and approved the final manuscript.

Authors' information
AKL, Associate Professor in the School of Mathematics and Computer Application, Thapar University Patiala. His research fields are computational Astrophysics and Reliability Engineering.
MK is a Ph.D student in the School of Mathematics and Computer Applications, Thapar University, Patiala. She worked as JRF and SRF in UGC sponsored major research project entitled "Reliability, Availability and Maintainability of industrial process" under the supervision of AKL and Prof. S.S. Bhatia. Her research interests are in the field of Reliability Engineering and Stochastic Models.
SL was a Ph.D student in School of Mathematics and Computer Applications, Thapar University. She, presently, works as an assistant professor of University institute of Engineering, Chandigarh University. Her research interests are in the field of Reliability Engineering and Fuzzy Theory.

Acknowledgements
Authors wish to thank all the reviewers for their valuable comments in improving this manuscript. Authors also like to thank Islamic Azad University for providing financial support to journal authors and readers.

Author details
[1]School of Mathematics and Computer Applications, Thapar University, Patiala, PB 147004, India. [2]University Institute of Engineering, Chandigarh University, Gharuan-Mohali, PB 140413, India.

References
Gupta P (2003) Mathematical analysis of reliability and availability of some process industries. Thapar University, India, Dissertation

Gupta S, Tiwari PC (2009) Simulation modeling and analysis of a complex system of a thermal power plant. J Indu Eng Manage 2:387–406

Gupta P, Singh J, Singh IP (2005a) Mission reliability and availability prediction of flexible polymer powder production. Oper Res Soc India 42:152–167

Gupta P, Lal AK, Sharma RK, Singh J (2005b) Behavioral study of cement manufacturing plant: a numerical approach. J Math Sys Sci 1:50–70

Gupta P, Lal AK, Sharma RK, Singh J (2005c) Numerical analysis of reliability and availability of the serial processes in butter-oil processing plant. Int J Qual Reliab Manage 22:303–316

Gupta P, Lal AK, Sharma RK, Singh J (2007) Analysis of reliability and availability of serial processes of plastic-pipe manufacturing plant: a case study. Int J Qual Reliab Manage 24:404–419

Kumar S, Tewari PC, Kumar S, Gupta M (2010) Availability optimization of CO-shift conversion system of a fertilizer plant using genetic algorithm technique. Bangladesh J Sci Ind Res 45:133–140

Mahajan P, Singh J (1999) Reliability of utensils manufacturing plant. OPSEARCH 36:260–269

Sachdeva A, Kumar P, Kumar D (2009) Behavioral and performance analysis of feeding system using stochastic reward petri nets. Int J Adv Manuf Technol 45:156–169

Sharma SP, Kumar D (2010) Stochastic behavior and performance analysis of an industrial system using GABLT technique. Int J Indu Sys Eng 2:1–23

14

Simultaneous robust estimation of multi-response surfaces in the presence of outliers

Mahdi Bashiri[1*] and Amir Moslemi[1,2]

Abstract

A robust approach should be considered when estimating regression coefficients in multi-response problems. Many models are derived from the least squares method. Because the presence of outlier data is unavoidable in most real cases and because the least squares method is sensitive to these types of points, robust regression approaches appear to be a more reliable and suitable method for addressing this problem. Additionally, in many problems, more than one response must be analyzed; thus, multi-response problems have more applications. The robust regression approach used in this paper is based on M-estimator methods. One of the most widely used weighting functions used in regression estimation is Huber's function. In multi-response surfaces, an individual estimation of each response can cause a problem in future deductions because of separate outlier detection schemes. To address this obstacle, a simultaneous independent multi-response iterative reweighting (SIMIR) approach is suggested. By presenting a coincident outlier index (COI) criterion while considering a realistic number of outliers in a multi-response problem, the performance of the proposed method is illustrated. Two well-known cases are presented as numerical examples from the literature. The results show that the proposed approach performs better than the classic estimation, and the proposed index shows efficiency of the proposed approach.

Keywords: Multi-response problem, Robust regression, Outliers, M-estimator

Introduction

A common method of explaining and analyzing the results of experiments is response surface modeling. This term is used for a regression equation that shows the whole behavior of the control variables, the nuisance factors, and the response or responses. We can use the estimated function to predict the response in each value of specific controllable factors. After gathering experimental data, a relationship between the factors (input data) and the response or responses (output results) should be defined to complete the analysis procedure. If we cannot construct a suitable model to define the precise relation between the input variables and the response or the responses' consequents, then the interpretations will not be reliable. After determining an experimental design and performing experiments, the next steps include the statistical analysis and the selection of the optimal input variables.

* Correspondence: bashiri@shahed.ac.ir
[1]Department of Industrial Engineering, Faculty of Engineering Shahed University, Tehran, Iran
Full list of author information is available at the end of the article

One of the most common approaches of regression coefficient estimation is the Least Squares (LS) method. A solution given by LS determines the coefficient values that minimize the sum of the squares of the residuals, in other words, the sum of the square differences between the experimental response values and those calculated by the fitted equation.

The quality of a manufactured product is often evaluated by several performance measures, which are called quality characteristics, each of which is described by a response variable. The values of these response variables are affected by one or more process parameters, which are the input variables. Often, processes with two or more response variables operate in a conflicting way. A group of responses often characterizes the performance of a manufactured product. These responses are usually correlated and measured by different measurement scales. Therefore, a decision-maker must resolve the parameter selection problem to optimize each response. This problem is considered to be a multi-response optimization problem, which is subject to different response requirements.

It is usually difficult to realize an optimal level of the input variables that can result in values close to the ideal or target values for all of the response variables. The main goal of multi-response optimization is, therefore, to find the settings of the input variables that achieve an optimal compromise in the response variables.

In many cases, especially in experimental results, some of the data should be treated outliers. These points, which may occur because of operator reading faults or other similar factors, may have a confusing effect on the total interpretation of the results. These points are called outliers. A data observation or a group of data points that are well separated from the majority of the whole pattern of observation, in other words, data that deviate from the general pattern, are called outliers. However, they are avoidable during the processing to some extent. The main concept in robustness is the presence of outliers and, more precisely, the changes in the distribution of the data. A common way to address outliers and to find them when using LS is to identify the bad observations. To detect the outliers, some graphical procedures such as normal probability plots and numerical regression diagnostics have been proposed. These procedures are defined in Weisberg (1985). Wisnowskia et al. (2001) studied the analysis of multiple outlier detection procedures for a linear regression model. Monte Carlo simulation is used to compare different approaches, and the performances and limitations of each method are discussed. Outlier detection in multivariate problems is not simple to understand; to describe this problem, simple visual methods can be applied. Fernandez Pierna et al. (2002) compared this type of method, called the convex hull method, with classical techniques and robust methods.

The concept of outlier data is qualitative in the sense that it is not the same as incorrect data but rather refers to data that are different from the majority. Often, the presence of outlier data illustrates the existence of an unexpected phenomenon at the start of experimentation but that can be explained, possibly from experimental causes. A problem that we often encountered in the application of regression is the presence of an outlier or outliers in the data. Outliers can be generated by a simple operational mistake, a small sample size, or other factors. Even one outlying observation can destroy an LS estimation, resulting in parameter estimates that do not provide useful information for the majority of the data. Robust regression analysis was developed to improve LS estimation in the presence of outliers and to provide additional information about valid observations. The primary purpose of robust regression analysis is to fit a model that represents the information that is in the majority of the data.

To address this obstacle, some robust approaches were proposed by different authors. Robust regression methods were introduced to address the above-mentioned problems. Ample (Fernandez Pierna et al. 2002) introduced robustness and computational approaches that include Huber (1981) robust statistics and different estimation algorithms. One common robust estimation approach is the M-estimator, which is based on a maximum likelihood estimation (MLE). LS was derived by this type of estimation and considers a special residuals function. The main idea of M-

Table 1 A brief review of single and multi-response robust regression in the literature

	Robust single response (using M-estimators)	Robust multi-response		
		Independent responses		Dependent responses (robust multivariate)
		Individuals	Simultaneous	
Hampel (1971)	✓			
Huber (1981)	✓			
Cummins and Andrews (1995)	✓			
Morgenthaler and Schumacher (1999)	✓			
Hund et al. (2002)	✓			
Wiens and Wu (2010)	✓			
Koksoy (2006)		✓		
Koksoy (2008)		✓		
Quesada and Del Castillo (2004)				✓
Daszykowski et al. (2007)				✓
Rousseeuw et al. (2004)				✓
Our research			✓	

Table 2 Experimental data of the tire tread compound problem

Experiment number	x_1	x_2	x_3	y_1	y_2	y_3	y_4
1	−1	−1	+1	102	900	470	67.5
2	+1	−1	−1	120	860	410	65
3	−1	+1	−1	117	800	570	77.5
4	+1	+1	+1	198	2,294	240	74.5
5	−1	−1	−1	103	490	640	62.5
6	+1	−1	+1	132	1,289	270	67
7	−1	+1	+1	132	1,270	410	78
8	+1	+1	−1	139	1,090	380	70
9	−1.633	0	0	102	770	590	76
10	+1.633	0	0	154	1,690	260	70
11	0	−1.633	0	96	700	520	63
12	0	+1.633	0	163	1,540	380	75
13	0	0	−1.633	116	2,184	520	65
14	0	0	+1.633	153	1,784	290	71
15	0	0	0	133	1,300	380	70
16	0	0	0	133	1,300	380	68.5
17	0	0	0	140	1,145	430	68
18	0	0	0	142	1,090	430	68
19	0	0	0	145	1,260	390	69
20	0	0	0	142	1,344	390	70

estimators is to replace the squared residuals by another function. The M-estimator works by an iterative procedure. As a consequence, several authors (e.g., Cummins and Andrews 1995) have called this estimator iteratively reweighted least squares, or the IRLS method. Additionally in our case, to estimate the regression coefficients, the iterative weighting method can be applied to estimate robust coefficients. One M-estimator function is from Huber (1981), which has become increasingly popular. Since then, more robust approaches have been discussed by investigators. However since M-estimators are simple to understand and considering that recent methods are sometimes so sensitive that they wrongly identify good points as outliers (Hund et al. 2002), we chose to use these type of estimators. Maronna et al. (2006) explained the most recent robust regression algorithms.

Table 3 Summary of the least squares regression coefficients for each response in the first example

Coefficients	$\hat{y}_1$	$\hat{y}_2$	$\hat{y}_3$	$\hat{y}_4$
x_1	16.49	268.15	−99.67	−1.41
x_2	17.88	246.5	−31.4	4.32
x_3	10.91	139.48	−73.92	1.63
x_1^2	−4.01	−83.55	7.93	1.56
x_2^2	−3.45	−124.79	17.31	0.06
x_3^2	−1.57	199.17	0.43	−0.32
x_1x_2	5.13	69.38	8.75	−1.63
x_1x_3	7.13	94.13	6.25	0.13
x_2x_3	7.88	104.38	1.25	−0.25
Intercept	139.12	1261.11	400.38	68.91

Morgenthaler and Schumacher (1999) discussed robust response surfaces in chemistry based on the design of experiments. Hund et al. (2002) presented various methods of outlier detection and evaluated robustness tests with different experimental designs. Robust regression methods and reconstruction experimental design methods have been compared. Wiens and Wu (2010) proposed a comparative study of M-estimators and presented a design that is more optimal compared with possible regression models.

In multi-response problems, the first step is the accurate determination of the regression coefficient because contamination and outlier data can have a negative effect on the models. The robustness concept in multi-responses has been presented by different authors; however, robust design was developed by Taguchi (1986, 1987). This approach is often used in process improvement project, to redesign processes for the purpose of increasing customer satisfaction by improving operational performances. Usually, the model parameters are estimated by LS in robust design. This methodology specifically utilizes both experimentation and optimization methods to determine the system's optimum operating conditions. Koksoy (2008, 2006) presented MSE as a robust design criterion in multi-response problems. Additionally, genetic algorithms and generalized reduced gradients method were used in their solution stage. In the mentioned studies, the general framework for multivariate problems in which data are collected from a combined array has been presented. For example, Quesada and Del Castillo (2004) proposed a dual response approach to multivariate robust parameter designs.

There are also several papers that consider correlations between responses, allowing the variance-covariance structure of the multiple responses to be accounted for. In this case, some multivariate techniques can be applied to these problems. Daszykowski et al. (2007) reviewed robust models and both univariate and multivariate outliers, and the effects of data analysis have been studied. One of the most efficient and useful robust multivariate regressions is the minimum covariance determinant, which was proposed by Rousseeuw et al. (2004).

In multi-response problems, robust regression approaches can be used to decrease the effects of contaminations and to focus outliers. In this paper, it is assumed that there is no correlation between responses;

Table 4 Scaled residuals of responses in the first iteration for the tire tread compound problem

Experiment number	R1	R2	R3	R4	Sum of absolute value of residuals
1	1.284007	0.604262	−0.97866	0.050079	2.917006
2	**1.608107**	−0.85258	−0.53883	0.674586	3.674095
3	0.563234	−0.83752	0.372694	0.891048	2.664497
4	0.612035	0.476129	0.181098	1.11425	2.383513
5	−0.47289	**−1.4899**	−0.49364	−1.05129	3.507725
6	−0.42409	−0.17625	−0.68524	−0.82808	2.113666
7	**−1.46897**	−0.1612	0.226282	−0.61162	2.468066
8	−1.14487	**−1.61804**	0.666114	0.012886	3.441902
9	0.123999	0.67264	0.386427	0.471895	1.654961
10	−0.33271	0.848008	0.082396	−0.56633	1.829443
11	−1.1557	0.691078	**1.502795**	0.737001	**4.086571**
12	0.946989	0.82957	−1.03398	−0.83144	3.641976
13	−0.27294	**2.456887**	−0.15226	−0.29297	3.175049
14	0.064229	−0.93624	0.621073	0.198524	1.820065
15	**−1.4959**	0.154146	−1.38046	0.811253	3.841755
16	**−1.4959**	0.154146	−1.38046	−0.30476	3.335263
17	0.215309	−0.46058	**2.005573**	−0.67677	3.35823
18	0.704225	−0.67871	**2.005573**	−0.67677	**4.065276**
19	**1.4376**	−0.00449	−0.70325	0.067244	2.212588
20	0.704225	0.32865	−0.70325	0.811253	2.547379

a similar assumption is made in other studies, such as Koksoy (2008). However, by considering each response and using univariate M-estimators to estimate the coefficients, a problem could occur, and outlier detection is required to consider all of the responses simultaneously. An outlier appearance in only one response cannot be considered a wrong observation while the other responses are considered to contain normal behavior. If we consider the responses individually, an experiment could treat an observation as outlier data, while in an iterative procedure, that point would be down-weighted more than it is appropriate. Considering all of the responses simultaneously leads to calculating the real number of outliers in a multi-response problem. In this paper, we suppose that the outlier data will occur because of a mistake in the experimentation, but not in the recording of the data. Hence considering one response as an outlier while considering others not to be

Table 5 Actual, individual robust, and SIMIR regression coefficient estimations for the tire tread compound problem

	$\hat{y}_1$			$\hat{y}_2$			$\hat{y}_3$			$\hat{y}_4$		
	Actual	Robust individual	SIMIR	Actual	Robust individual	SIMIR	Actual	Robust individual	SIMIR	Actual	Robust individual	SIMIR
x_1	14.77	17.17	16.75	280.27	302.80	276.86	−62.32	−89.92	−89.69	−1.41	−1.35	−1.57
x_2	19.59	17.79	17.13	258.6	248.54	242.62	−23.21	−31.37	−33.70	4.32	3.97	3.82
x_3	12.62	12.25	12.60	227.3	314.15	291.86	−45.45	−74.16	−75.05	1.63	1.68	1.42
x_1^2	−5	−4.52	−4.25	−10.95	−27.36	−23.64	5.24	12.05	7.18	1.56	1.54	1.64
x_2^2	−4.43	−3.85	−3.63	−52.2	−62.86	−53.89	7.33	17.42	16.68	0.06	0.11	0.16
x_3^2	−2.56	−2.07	−2.40	51.02	0.025	45.04	2.43	4.52	0.99	−0.32	−0.33	−0.13
x_1x_2	6.16	5.18	4.90	163.26	62.65	21.04	4.07	11.44	10.20	−1.63	−0.79	−0.57
x_1x_3	8.16	7.22	6.83	73.93	85.13	128.98	2.34	7.91	8.97	0.13	−0.21	0.14
x_2x_3	6.83	6.81	7.35	84.18	52.89	62.07	0.58	3.94	2.27	−0.25	0.58	0.82
Intercept	141.33	139.16	139.32	1241.7	1265.49	1225.18	396.04	391.21	403.17	68.91	68.85	68.74

Table 6 SE of the estimation of each regression coefficient in the least squares, individual robust and SIMIR approaches

	$\hat{y}_1$			$\hat{y}_2$			$\hat{y}_3$			$\hat{y}_4$		
	LS	**Robust individual**	**SIMIR**	**LS**	**Robust individual**	**SIMIR**	**LS**	**Robust individual**	**SIMIR**	**LS**	**Robust individual**	**SIMIR**
x_1	2.9584	5.79	3.93	146.89	507.94	11.61	1,395.02	762.00	749.18	0	0.0036	0.025
x_2	2.9241	3.20	6.05	146.41	101.17	255.27	67.07	66.62	110.22	0	0.12	0.25
x_3	2.9241	0.13	0.00	7,712.35	7,544.56	4,168.97	810.54	824.36	876.42	0	0.002	0.044
x_1^2	0.9801	0.22	0.55	5,270.76	269.58	161.11	7.23	46.46	3.79	0	0.0004	0.0064
x_2^2	0.9604	0.33	0.63	5,269.30	113.79	2.88	99.60	101.91	87.56	0	0.0025	0.01
x_3^2	0.9801	0.23	0.02	21,948.42	2,600.46	35.72	4	4.40	2.05	0	0.0001	0.036
x_1x_2	1.0609	0.95	1.57	8,813.45	10,122.05	20,225.64	21.90	54.37	37.59	0	0.70	1.12
x_1x_3	1.0609	0.88	1.75	408.04	125.47	3,030.72	15.28	31.05	43.96	0	0.115	0.0001
x_2x_3	1.1025	0.00	0.27	408.04	978.48	488.47	0.44	11.31	2.87	0	0.688	1.14
Intercept	4.8841	4.67	4.03	376.74	566.38	272.60	18.83	23.30	50.97	0	0.003	0.028
SSE	19.8356	16.43	18.83	50,500.43	22,929.92	28,653.03	2,439.95	1,925.83	1,964.64	0	1.645	2.66

SSE, sum of squared error.

contaminated is not rational. A brief review on the literature is given in Table 1, as follows:

To the best of our knowledge, there are a few studies on multi-response robust regression, and this paper focuses on the multi-response robust regression that considers the response residuals simultaneously. To estimate the regression coefficients in the multi-response problem, we propose a procedure in which we apply a simultaneous independent multi-response iterative reweighting procedure and change the M-estimator weighting function; a more precise estimation of each response can be obtained. By considering this procedure, a new criterion is proposed named the coincident outlier index (*COI*), and the performance of this procedure is analyzed.

This paper is organized as follows. 'Using M-estimators for robust estimation of regression coefficients' section presents the robust M-estimator procedure and the modification of the response surface by an iterative weighting procedure. The proposed method for the multi-response problem is defined in 'Robust simultaneous estimation of multi-response problem' section. To illustrate the proposed method, a numerical example is presented before the 'Conclusions' section. Finally, the last section provides the conclusions of this paper.

Using M-estimators for robust estimation of regression coefficients

The M-estimator proposed by Huber (1981) is the generalized form of the (MLEs). This part is extracted from

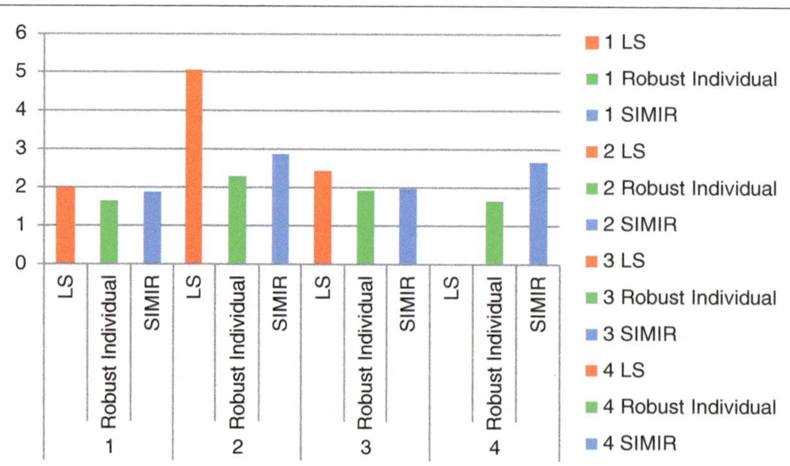

Figure 1 Evaluation of the sum of squared errors of proposed approach and classical approaches for the first example.

Maronna et al. (2006). The M-estimator is the solution of Equation (1), as follows:

$$\hat{\theta} = \arg \min_{\theta} \sum_{i=1}^{n} \rho(x_i, \theta) \quad (1)$$

where ρ is a function with specific properties. Supposed that f is a density function and $\rho = -\log f$, where f is a density function, then $\hat{\theta}$ will be introduced as the MLE of the parameter. There are several ρ functions. One common ρ function is Huber (1981). This function is well-defined in Equation (2):

$$\rho_k(x) = \begin{cases} x^2 & if \quad |x| \le k \\ 2k|x| - k^2 & if \quad |x| > k \end{cases} \quad (2)$$

By considering this function and by defining $\psi = \rho'$ and a weighting function by Equation (3), the iterative algorithm to estimate the unknown parameter can be defined.

$$W(x) = \begin{cases} \psi(x)/x & if \quad x \neq 0 \\ \psi'(x) & if \quad x = 0 \end{cases} \quad (3)$$

M-estimates for estimating regression coefficients are developed in the same way as defined in previous part. Equation (4) should be considered, and the coefficients can be obtained by solving following equation:

$$\sum_{i=1}^{n} \psi\left(\frac{r_i(\hat{\beta})}{\hat{\sigma}}\right) \mathrm{x_i} = 0 \quad (4)$$

The $\hat{\sigma}$ in Equation (4) can also be estimated individually by Equation (5), as follows, or can be solved simultaneously in Equation (4). r_i is a residual of the response.

$$\hat{\sigma} = \frac{1}{0.675} \boldsymbol{Med}_i(|\boldsymbol{r}_i| \boldsymbol{r}_i \neq 0) \quad (5)$$

To make the estimation procedure invariant with respect to the scale of the residuals, the r_is are divided by 's'. The value of 's' is often taken to be equal to 1.4826 MAD, where MAD is the median of the absolute deviations of the residuals from their median, and 1.4826 is a

Table 7 Experimental data of the elastic element of the force transducer problem

Experiment number	x_1	x_2	x_3	z_1	z_2	y_1	y_2
1	−1	−1	−1	−1	1	1.810	1.10
2	−1	−1	−1	1	−1	1.690	1.11
3	−1	−1	1	−1	−1	1.900	1.07
4	−1	−1	1	1	1	1.780	1.07
5	−1	1	−1	−1	−1	1.800	1.47
6	−1	1	−1	1	1	1.630	1.18
7	−1	1	1	−1	1	1.920	1.41
8	−1	1	1	1	−1	1.780	1.58
9	1	−1	−1	−1	−1	1.360	1.57
10	1	−1	−1	1	1	1.220	2.03
11	1	−1	1	−1	1	1.480	1.38
12	1	−1	1	1	−1	1.440	1.68
13	1	1	−1	−1	1	0.693	3.37
14	1	1	−1	1	−1	0.616	3.75
15	1	1	1	−1	−1	0.950	2.81
16	1	1	1	1	1	0.817	2.83
17	−1	0	0	0	0	1.790	1.24
18	1	0	0	0	0	1.030	2.46
19	0	−1	0	0	0	1.530	1.23
20	0	1	0	0	0	1.220	1.73
21	0	0	−1	0	0	1.300	1.63
22	0	0	1	0	0	1.440	1.67
23	0	0	0	0	0	1.380	1.73
24	0	0	0	0	0	1.390	1.74
25	0	0	0	0	0	1.400	1.74

Table 8 Summary of the least squares regression coefficient for each response in the second example

Coefficients	$\hat{y}_1$	$\hat{y}_2$
x_1	−0.36	0.59
x_2	−0.15	0.43
x_3	0.07	−0.09
z_1	−0.05	0.06
z_2	−0.01	−0.04
$\boldsymbol{x_1^2}$	0.02	0.11
$\boldsymbol{x_2^2}$	0.2	−0.25
$\boldsymbol{x_3^2}$	0	−0.08
$\boldsymbol{z_1^2}$	0	0.33
x_1x_2	−0.14	0.3
x_1x_3	0.02	−0.14
x_1z_1	0.01	0.08
x_1z_2	0	0.01
x_2x_3	0.01	−0.033
x_2z_1	0	−0.03
x_2z_2	0	−0.06
x_3z_1	0	0
x_3z_2	0	−0.01
z_1z_2	0	−0.08
Intercept	1.39	1.73

bias adjustment for the standard deviation under the normal distribution.

An iterative reweighting method can be defined as follows: First, compute an initial estimate β_0 and compute $\widehat{\sigma}$ from Equation (5). After that, for k = 0, 1, 2, ... :

First, Compute an initial estimate β_0 and compute $\widehat{\sigma}$ from Equation (5). After that, For k = 0, 1, 2, ... :

(a) Given $\widehat{\beta}_k$, for i = 1, . . . , n, compute $r_{i,k} = y_i - X'_i \widehat{\beta}_k$ and $w_{i,k} = W\left(r_{i,k}/\widehat{\sigma}\right)$.
(b) Compute $\widehat{\beta}_{k+1}$ by solving the following:

$$\sum_{i=1}^{n} w_{i,k} X_i \left(y_i - X'_i \widehat{\beta}_k\right) = 0 \tag{6}$$

Finally, the algorithm stops when $\max\left(\left|r_{i,k} - r_{i,k+1}\right|\right)/\widehat{\sigma} < \varepsilon$

This algorithm converges if $W(x)$ is non-increasing for x >0 (Maronna et al. 2006). If ψ is monotone, because the solution is essentially unique, the choice of the starting point influences the number of iterations but not the final result. This procedure is called (IRWLS).

The procedure is as follows: compute the first coefficients of the regression model, then compute the residuals and weights, and finally compute the new coefficients using Equation (6). This procedure can be repeated because the values of the coefficients and the values of the residuals and weights are different; as a result, this procedure can be repeated until a good solution is obtained. The procedure terminates when the change in the estimation from one iteration to the next is sufficiently small. The estimators of coefficients based on the LS method are basically unbiased; the robust estimators like LS methods are basically unbiased too.

Robust simultaneous estimation of multi-response problem

In a multi-response problem, similar to a single response problem, robust estimation of the regression model is an important issue. A simple approach to estimating the regression models in multi-response problems is to consider the responses individually and to estimate the solution to each problem by the robust M-estimator approach. However, this approach could cause some problems. Assume in one experiment that a specific response residual appears to be an outlier. However, other responses do not show any signs of being unacceptable data for that specific experiment. The outlier data could occur because of a fault in the experimentation. It is not rational to say, then, whether one experiment's result is an outlier for one response because it may not be unacceptable data for the other responses. From this type of deduction, a large amount of the experiment's results could become outliers, and for each response, some points will be assumed to be outliers by mistake. Thus, assuming independence between responses, a simultaneous independent multi-response iterative reweighting (SIMIR) approach is proposed to solve this problem. In this approach, based on M-estimators, some changes are applied in the procedure of weighting functions to estimate the coefficients of the model. The weighting function proposed in this method, down-weights the residuals by considering each response of the multi-response problem in each iteration, simultaneously. We have j responses in this problem and i experiments. The variable $r(i)_j$ defines the residual for the ith replicate of the jth response. The proposed weighting function is given in Equation (7):

$$\begin{cases} w_i = 1 & \text{if} \quad \forall \quad \left|r(i)_j\right| < c \\ w_i = \dfrac{c}{\left|\sum_{j=1}^{l} r(i)_j\right|} & \text{if} \quad \exists \quad \left|r(i)_j\right| > c \end{cases}, \tag{7}$$

Table 9 Scaled residuals of two responses in the first iteration for proposed example

Experiment number	R1	R2	Sum of absolute value of residuals
1	−0.72	−0.172	0.894
2	−0.72	−0.172	0.894
3	−0.33	−0.892	1.229
4	−0.33	−0.892	1.229
5	−0.70	1.006	1.711
6	−0.70	1.006	1.711
7	−0.319	0.286	0.605
8	−0.319	0.286	0.605
9	0.319	−0.286	0.605
10	0.319	−0.286	0.605
11	0.704	−1.006	1.711
12	0.704	−1.006	1.711
13	0.337	0.892	1.229
14	0.337	0.892	1.229
15	0.722	0.172	0.894
16	0.722	0.172	0.894
17	**4.165**	−0.459	**4.624**
18	**−4.165**	0.459	**4.624**
19	0.072	**4.716**	**4.788**
20	−0.072	**−4.716**	**4.788**
21	**1.540**	**−2.880**	**4.420**
22	**−1.540**	**2.880**	**4.420**
23	**−2.166**	−0.166	2.333
24	0	0.083	0.083
25	**2.166**	0.083	2.250

The proposed pseudo code is as follows:

1. *Compute the actual values of the responses in each experiment by performing all of the experiments*
2. *Estimate the regression coefficients of the initial regression model by applying the proper method. While* $\left(\left|r_{i,k}-r_{i,k+1}\right|\right)/\widehat{\sigma} < \varepsilon$ *Do* (ε *is determined by the analyzer).*
3. *Calculate the residuals of each response in all of the experiments*
4. *Compute the* $\widehat{\sigma}$ *by Equation (5).*
5. *If all* $r(i)_j$ *are smaller than the threshold determined in the Huber M-estimator method, then dedicate the weight to be equal to 1 for the residuals in this iteration and go to step (7); else, go to step (6).*
6. *Down weight the residuals by considering the values of the residuals in all of the responses by a function in Equation (7).*
7. *Estimate the regression coefficients by solving Equation (6).*

The performance of the proposed method is presented by a numerical example in the next section. By applying the SIMIR procedure, the squared errors (SE) of the estimated parameters, which are regression coefficients, are reduced compared to those of the least squares estimation of each response; however, to some extent, this strategy is not as precise as the robust individual estimation. One important problem in multi-response problems is the number of real outliers.

The SE criterion is computed in Equation (8):

$$SE = \left(\theta - \hat{\theta}\right)^2 \tag{8}$$

Individually, residuals computed for one response could not be outliers in the whole multi-response problem. To detect the outliers in a multi-response problem, it is not correct to mention the outliers in each individual response. We present the COI to detect the real number of outliers in a robust estimation of the regression coefficients in a multi-response problem. This index can be computed by this procedure, for which we define the threshold by considering the suggested C (defined in Equation (7)) in the Huber procedure and by considering scaled residuals. If we consider the number of responses as n, then the proposed threshold is defined as $T = \left(\left[\frac{n}{2}\right] + 1\right) \times C$. If the sum of the residuals is greater than this threshold, then that experiment is treated as an outlier. Thus, the COI is equal to the number of points

Table 10 Actual, individual robust, and SIMIR approach for regression coefficient estimation for the proposed problem

	$\hat{y}_1$				$\hat{y}_2$			
	LS	**Actual**	**Robust individual**	**SIMIR**	**LS**	**Actual**	**robust Individual**	**SIMIR**
x_1	−0.36	−0.36	−0.35	−0.35	0.59	0.58	0.585	0.585
x_2	−0.15	−0.15	−0.15	−0.15	0.43	0.43	0.43	0.43
x_3	0.07	0.07	0.07	0.07	−0.09	−0.1	−0.09	−0.09
z_1	−0.05	−0.05	−0.05	−0.05	0.06	0.06	0.06	0.06
z_2	−0.01	−0.01	−0.01	−0.01	−0.04	−0.04	−0.04	−0.04
$\boldsymbol{x}_1^2$	0.02	0.02	0.02	0.02	0.11	0.35	0.21	0.21
$\boldsymbol{x}_2^2$	0.2	−0.01	0.09	0.09	−0.25	−0.2	−0.22	−0.22
$\boldsymbol{x}_3^2$	0	−0.02	−0.01	−0.01	−0.08	0	−0.07	−0.07
$\boldsymbol{z}_1^2$	0	0.05	0.02	0.02	0.33	0	0.15	0.15
x_1x_2	−0.14	−0.14	−0.14	−0.14	0.3	0.3	0.3	0.3
x_1x_3	0.02	0.02	0.02	0.02	−0.14	−0.14	−0.14	−0.14
x_1z_1	0.01	0.01	0.01	0.01	0.08	0.07	0.08	0.08
x_1z_2	0	0	0	0	0.01	0.01	0.01	0.01
x_2x_3	0.01	0.01	0.01	0.01	−0.033	−0.033	−0.033	−0.033
x_2z_1	0	0	0	0	−0.03	−0.03	−0.03	−0.03
x_2z_2	0	0	0	0	−0.06	−0.06	−0.06	−0.06
x_3z_1	0	0	0	0	0	0	0	0
x_3z_2	0	0	0	0	−0.01	0	0	0
z_1z_2	0	0	0	0	−0.08	0	0	0
Intercept	1.39	1.39	1.39	1.39	1.73	1.74	1.73	1.73

that are greater than T. Equation (9) defines the proposed COI, as follows:

$$COI = \sum_{i=1}^{N} z_{ij} \quad for \quad each \quad j = 1, \ldots, l$$

$$z_{ij} = \begin{cases} 1 & if \quad \sum_{j=1}^{4} r_{ij} > T \quad for \quad each \quad i \\ 0 & if \quad \sum_{j=1}^{4} r_{ij} < T \quad for \quad each \quad i \end{cases} \tag{9}$$

where N is the number of experiments.

Numerical example

In this section, the efficiency of our proposed approach compared with existing approaches is illustrated for two cases. In case one, the number of coinciding outliers differs from the number of outliers that were detected by an individual procedure. In the second case, the previous experiments contain most of the outliers because of a true fault by the experimenter. The number of coinciding outliers and the outliers that are detected individually does not differ significantly in this case.

Case 1. tire tread compound problem

In this section, we illustrate the proposed method using the well-known problem 'tire tread compound problem', which was originally presented by Derringer and Suich (1980). In this model, three main chemical materials, such as silica (x_1), silane (x_2), and sulfur (x_3), and four responses are assumed. The experimental data results are given in Table 2.

As a first step, we attempt to find a primary regression model with four responses. A central composite design (CCD) with six center points is applied to describe the model. All of the controllable variables are $-1.63 \le x_i \le 1.63, i = 1, 2, 3$. The regression coefficients that are obtained by the least squares estimation method and according to the CCD are given in Table 3, as follows:

The scaled residuals of this multi-response problem are reported in Table 4, as follows:

For the first response, the residuals obtained by the experiments numbered 2, 7, 15, 16, and 19 appear to be outliers, and for the second response residuals, those numbered 5, 8, and 13 appear to be outliers. For the third

Table 11 SE of the estimation of regression coefficients in LS, robust individual, and SIMIR methods

	$\hat{y}_1$			$\hat{y}_2$		
	LS	Robust individual	SIMIR	LS	robust Individual	SIMIR
x_1	0.0001	0	0	0.0001	0.000025	0.000025
x_2	0	0	0	0	0	0
x_3	0	0	0	0.0001	0.0001	0.0001
z_1	0	0	0	0	0	0
z_2	0	0	0	0	0	0
x_1^2	0	0	0	0.057	0.019	0.019
x_2^2	0.0441	0.01	0.01	0.0025	0.0004	0.0004
x_3^2	0.0004	0.0001	0.0001	0.006	0.0049	0.004
z_1^2	0.0025	0.0009	0.0009	0.108	0.022	0.022
x_1x_2	0	0	0	0	0	0
x_1x_3	0	0	0	0	0	0
x_1z_1	0	0	0	0	0.0001	0.0001
x_1z_2	0	0	0	0	0	0
x_2x_3	0	0	0	0	0	0
x_2z_1	0	0	0	0	0	0
x_2z_2	0	0	0	0	0	0
x_3z_1	0	0	0	0	0	0
x_3z_2	0	0	0	0.0001	0	0
z_1z_2	0	0	0	0.0064	0	0
Intercept	0	0	0	0.0001	0.0001	0.0001
SSE	0.0471	0.011	0.011	0.1823	0.047	0.047

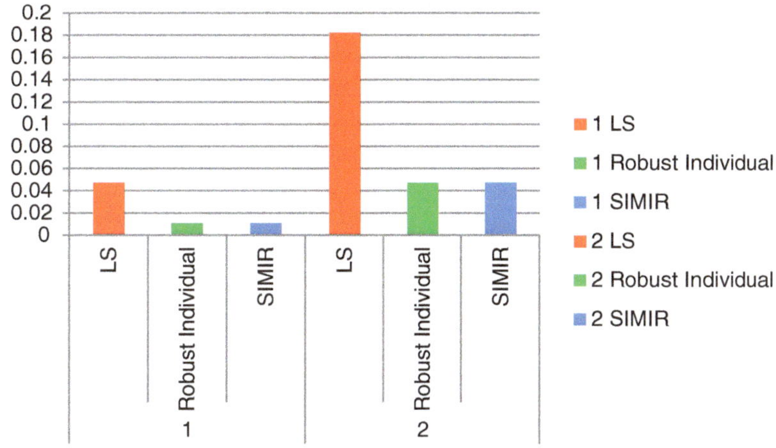

Figure 2 Evaluation of the sum of squared errors of proposed and classical approaches for the second example.

response residuals, those numbered 11, 17, and 18 appear to be outliers. The fourth model does not contain a residual that implies an outlier. By omitting the values of the implied outliers, actual regression models can be obtained by a least squares method because least square is the most efficient method in the absence of outlier data. This method is considered to be the actual result in Table 5. In addition, to obtain the robust regression models by two different approaches, the robust individual regression approach and the SIMIR approach are applied and can be compared by considering the SE criteria.

First, we consider each response individually and apply the M-estimator procedure to each response. The constant C in this example is assumed to be 1.37. Finally, the SIMIR procedure is applied to the data. Coefficients obtained by actual, robust individual and SIMIR procedures are given in Table 5.

Then, to evaluate the procedures mentioned (actual, individual robust, and SIMIR approaches), the SE of these estimated parameters are computed using Equation (8), and the results are reported in Table 6. In Equation (8), it is assumed that the value of θ is the regression coefficient in the actual method, and $\hat{\theta}$ is the regression coefficient in the considered method (actual, individual robust, and SIMIR approaches).

Additionally, the evaluation is given in Figure 1, as follows:

A comparison among the three approaches by considering the sum of squared errors (SSE) is computed in Table 6 and illustrated in Figure 1. In this figure, for the first three responses, the scaled values of the SSE are given.

The results show that the robust individual regression estimation is more precise than that of the least squares estimation or the proposed SIMIR approach, but the coincident outlier index that was presented in the previous section is more reliable and realistic. Moreover, the results state that in the case with an absence of outliers, LS performs better than both the robust individual and the SIMIR procedures. In this example, if we want to count the outliers as a multi-response problem individually, the results would be 11 experiments. However, by the proposed COI index, the real number of the multi-response problem outliers is 2, and it would be more rational that in 20 experiments, almost 10% of the experiments result in outliers.

Case 2. elastic element of a force transducer problem

We provide another example to illustrate the efficiency of the proposed method. The following example was presented as a case study in Romano et al. (2004), in which the problem was about the elastic element of a force transducer. This example involves a combined array design with three control (x) and two noise (z) variables. The control factors are the three parameters that describe the element configuration, namely the lozenge angle (x1), the bore diameter (x2), and the half-length of the vertical segment (x3). Noise factors are the deviation of the lozenge angle from its nominal value (z1) and the

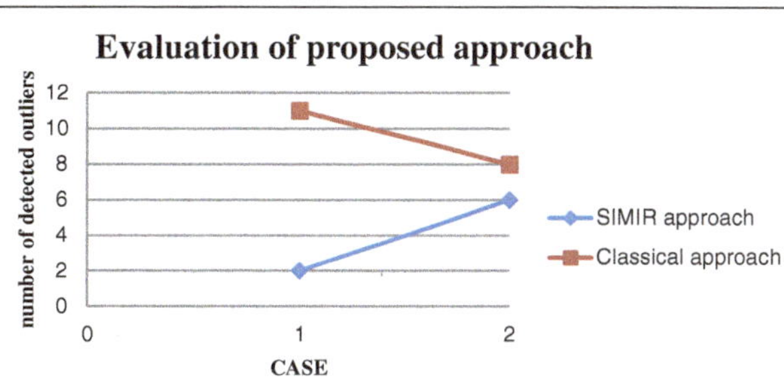

Figure 3 Evaluating the performance of the proposed method and the existing method for both cases, by considering the COI index.

deviation of the bore diameter from its nominal value (z2). These internal noise factors are undeniably independent. The two indicators, namely the non-linearity (y1) and the hysteresis (y2), define the responses. Table 7 displays the data from this experiment.

First, we need to find a primary regression model for the two responses. All of the controllable variables are $-1 \le x_i \le 1, i = 1, 2$. The regression coefficients that were obtained by the least squares estimation method are given in Table 8, as follows:

The scaled residuals of this multi-response problem are reported in Table 9, as follows:

Consequently, for the first response, outliers appear with the residuals obtained by the experiments, numbered 17, 18, 21, 22, 23 and 25; for second response residuals, they are numbered 19, 20, 21 and 22. By omitting these values, the actual regression models can be obtained. To obtain the robust regression models, two different approaches are applied, and these two approaches are compared by considering the SE criteria.

First, we consider each response individually and apply the M-estimator procedure to each response. The constant C in this example is assumed to be 1.37. Thus, three groups of coefficients are reported in Table 10.

To evaluate the three procedures, the proposed SE criterion are calculated, and the results are given in Table 11 as follows:

In addition, to provide more illustration, Figure 2 is given as follows:

Similar to the previous example, our results showed that the robust individual regression estimation performs better, but not significantly better than the least squares method; the SIMIR approach was also considered, but the COI is more realistic and accurate. In this example, if we want to count the outliers in a multi-response problem individually, the results would consist of eight experiments. However, by the proposed COI index, the real number of outliers in the multi-response problem is six.

Therefore, by considering these two examples, the number of detected outliers calculated by both the classical method and the SIMIR proposed approach is shown in Figure 3, as follows:

Conclusions

As mentioned in the previous sections, a robust simultaneous estimation of regression coefficients in the multi-response problem in the case in which contaminated data exists was presented in this paper. In addition, the results showed that the proposed approach would consider a number of points to be real outliers in the multi-response problem, although individual robust regression shows some other points as outliers. Thus, an aggregative approach in the weighting function was proposed, in which all of the responses were surveyed. The SIMIR approach performed better than the classic method for detecting outliers and estimating regression coefficients. Additionally, our results show that the proposed approach would provide a better COI index than the classical approach for outlier detection. For future research, other robust regression approaches can be studied. In addition, considering a problem with correlated responses can be another aspect of related future research.

Competing interests
The authors declare that they have no competing interests.

Authors' contributions
MB has worked on the literature and modeling; he also proposed the response surface methodology. AM has performed the robust concept and numerical example. Both authors read and approved the final manuscript.

Authors' information
MB is an associate professor of the Department of Industrial Engineering, Shahed University, and his research interests are Multiple Response Optimization and Facilities Location Problem. AM is a Ph.D. student of the Department of Industrial Engineering at Amirkabir University of Technology (Tehran Polytechnic).

Acknowledgments
Authors are thankful to the reviewers for their valuable comments.

Author details
[1]Department of Industrial Engineering, Faculty of Engineering Shahed University, Tehran, Iran. [2]Department of Industrial Engineering, Amirkabir University of Technology (Tehran Polytechnic), Hafez, Tehran, Iran.

References
Cummins DJ, Andrews CW (1995) Iteratively reweighted partial least squares: a performance analysis by Monte Carlo simulation. J Chemometr 9:489–507
Daszykowski M, Kaczmarek K, Vander Heyden Y, Walczak B (2007) Robust statistics in data analysis–a review basic concepts. Chemometr Intell Lab 85:203–219
Derringer G, Suich R (1980) Simultaneous optimization of several response variables. J Qual Technol 12:214–219
Fernandez Pierna JA, Wahl F, De Noord OE, Massart DL (2002) Methods for outlier detection in prediction. Chemometr Intell Lab 63:27–39
Hampel FR (1971) A general definition of qualitative robustness. Ann Math Statist 42:1887–1896
Hund E, Massart DL, Smeyers-Verbeke J (2002) Robust regression and outlier detection in the evaluation of robustness tests with different experimental designs. Anal Chim Acta 463:53–73
Huber PJ (1981) Robust statistics. Wiley, New York
Maronna RA, Martin RD, Yohai VJ (2006) Robust statistics: theory and methods. Wiley, New York
Morgenthaler S, Schumacher MM (1999) Robust analysis of a response surface design. Chemometr Intell Lab 47:127–141
Koksoy O (2008) A nonlinear programming solution to robust multiresponse quality problem. Appl Math Comput 196:603–612
Koksoy O (2006) Multiresponse robust design: mean square error (MSE) criterion. Appl Math Comput 175:1716–1729
Quesada GM, Del Castillo E (2004) A dual-response approach to the multivariate robust parameter design problem. Technometrics 46:176–187
Romano D, Varetto M, Vicario G (2004) Multiresponse robust design: a general framework based on combined array. J Qual Technol 36:27–37
Rousseeuw P, Van Aelst S, Van Driessen K, Agull J (2004) Robust Multivariate Regression. Technometrics 46:293–305

Taguchi G (1986) Introduction to quality engineering. Kraus International Publications, White Plains, NJ
Taguchi G (1987) System of Experimental Design: Engineering Methods to Optimize Quality and Minimize Cost. Quality Resources, White Plains, NJ
Weisberg S (1985) Applied linear regression, 2nd edition. Wiley, New York
Wiens D, Wu EKH (2010) A comparative study of robust designs for M-estimated regression models. Comput Stat Data An 54:1683–1695
Wisnowskia JW, Montgomery DC, Simpson JR (2001) A comparative analysis of multiple outlier detection procedures in the linear regression model. Comput Stat Data An 36:351–382

15

The analysis of residuals variation and outliers to obtain robust response surface

Mahdi Bashiri* and Amir Moslemi

Abstract

In this paper, the main idea is to compute the robust regression model, derived by experimentation, in order to achieve a model with minimum effects of outliers and fixed variation among different experimental runs. Both outliers and nonequality of residual variation can affect the response surface parameter estimation. The common way to estimate the regression model coefficients is the ordinary least squares method. The weakness of this method is its sensitivity to outliers and specific residual behavior, so we pursue the modified robust method to solve this problem. Many papers have proposed different robust methods to decrease the effect of outliers, but trends in residual behaviors pose another important issue that should be taken into account. The trends in residuals can cause faulty estimations and thus faulty future decisions and outcomes, so in this paper, an iterative weighting method is used to modify both the outliers and the residuals that follow abnormal trends in variation, like descending or ascending trends, so they will have less effect on the coefficient estimation. Finally, a numerical example illustrates the proposed approach.

Keywords: Coefficient estimation, Response surface, Ordinary least squares, Outliers, Robust model

Introduction

In many cases, especially in experimental results, some of the data are wrong and should be treated as outliers. These points, which may occur because of operator reading faults and the like, may have a confusing effect on the total interpretation of the results. A common method of explaining and analyzing the results of experiments is by response surface design. This term is used for a regression equation that shows the whole behavior of the control variables, the nuisance factors, and the response or responses. We can use the estimated function to predict the response to changes in the values of specific controllable factors. After determining an experimental design and performing experiments, the next steps are generally statistical analysis and then the selection of values for the input variables so as to optimize the output. This can be done by fitting a regression model between the controllable factors and the response variables. Future interpretations are based on this regression model, so the exact model is very important and may affect the optimization stage. This model is generally constructed by the ordinary least squares (OLS) method. But basic OLS is very sensitive to outliers, and they may have an inordinate effect on the ultimate conclusion. So a robust method or a modified OLS should be used for decreasing the outliers' sensitivity.

Our goal in this study is to decrease the destructive effect of outliers. In order to do so, at the first stage, the robust regression or modified regression model should be computed. Then the trend of residuals in response surface design is another aspect which should be considered. The trend behaviors among residuals, both descending and ascending trends, can cause faulty interpretations. However, there are some assumptions in estimating the regression coefficients that should not be violated. It seems that by decreasing the effects of some residuals in the coefficient estimation stage, the initial assumptions can be satisfied, and moreover, because this decreases the overall variability, the robustness of the model will be increased. The main purpose of this paper is to decrease the effects of outliers that violate the variance equality test of residuals. An example of such trends and abnormal behavior is shown in Figure 1.

As mentioned before, the OLS method is very sensitive to outliers. To diminish the effect of these points, some alternative methods of model fitting, such as least absolute

* Correspondence: bashiri@shahed.ac.ir
Department of Industrial Engineering, Faculty of Engineering, Shahed University, Tehran, Iran

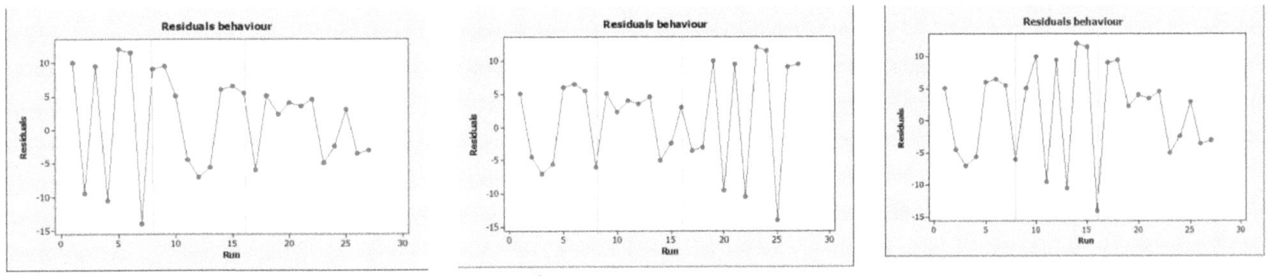

a. Descending trend b. Ascending trend c. oscillation trend

Figure 1 Trends in residual behavior. (a) Descending trend. **(b)** Ascending trend. **(c)** Oscillation trend.

deviations and other robust approaches that simplify the task of outlier identification by weighting the large residuals, are used instead of OLS. Response surfaces have been studied by many researchers, and many approaches have been proposed either to obtain efficient response surfaces or optimize the response surface by different models. Hejazi et al. (2010) proposed a novel approach based on goal programming to find the best combination of factors to optimize multi-response-multi-covariate surfaces by considering location and dispersion effects. Kazemzadeh et al. (2008) proposed a method to optimize multi-response surfaces based on a goal programming method. Robust regression approaches have also been surveyed by many researchers. Huber (Bertsimas and Shioda 2007) proposed M-estimator methods to obtain robust regression. Morgenthaler and Schumacher (1999) discussed robust response surfaces in chemistry based on design of experiment. Because of the weakness of the previous methods in compensating for outliers, the redescending M-estimators (also named GM-estimators) were proposed by Andrews et al. (1972), which are able to reject extreme outliers entirely. Hund et al. (2002) presented various methods of outlier detection and evaluated robustness tests with different experimental designs. Bickela and Frühwirthb (2006) compared different robust estimators with their applications. The M- and GM-estimators work by an iterative procedure. As a consequence, several authors (e.g., Cummins and Andrews 1995) have called these estimators as iteratively reweighted least squares, or IRLS methods. Ortiz et al. (2006) discussed some of the robust methods used for robust regression in analytical chemistry. To obtain a more efficient and yet robust method, Siegel (1982) proposed the repeated median estimator. Also another useful robust method is the least median squared (LMS) method proposed by Rousseeuw (1984). Massart et al. (1986) showed the advantages of its use in chemical analysis. The other useful method is the least trimmed squared (LTS) that was proposed by Rousseeuw and Leroy (1987). Nguyena and Welsch (2010) studied outlier detection and proposed a new least trimmed squares approximation. Both the LMS and the LTS are defined by minimizing a robust measure of the scatter of the residuals. Generalizing this, Rousseeuw and Yohai (1984) introduced S-estimators which are significantly more efficient than the previous estimators. A more recent suggestion is the constrained M-estimates, or CM, proposed by Mendes and Tyler (1995), which combines the good local properties of the M-estimates and the good global robustness properties of the S-estimates. A 'partial' version of the M-estimator based on the 'fair' ψ function and an appropriate weighting scheme was recently proposed by Serneels et al. (2005). The authors believed that the partial robust M-regression outperforms existing methods for robust partial least square regression. Bertsimas and Shioda (2007) presented mixed integer programming or MIP models for the classification and robust regression problems. Zioutas and Avramidis (2005) examined the effect of deleting outliers in the regression model obtained by mixed integer programming and compared the performance of this model with that of least squares, or LS, and LMS. Another new method in robust regression is the mixed linear model surveyed by Dornheim and Brazauskas (2011). (Pop and Sârbu 1996) proposed a new fuzzy regression algorithm to obtain robust models. Maronna et al. (2006) proposed many M-estimators using robust regression methods in both single response and multiple responses. Shahriari et al. (2011) proposed a novel two-step robust estimation of the process mean method based on M-estimator and their method is less sensitive to the presence of outliers. For better illustration of proposed method, the literature review has been classified in Table 1.

In this paper, a novel robust approach considering both outlier data and trends in residuals variations which do not violate the normality assumption is discussed.

This paper is organized as follows. The section 'Robust estimation of the coefficients by iterative weighting methods' presents the robust modification of the response surface by an iterative weighting procedure. The proposed method is defined in section 'Robust estimation of coefficients by testing equality of variations in specified intervals'. To illustrate the proposed method, a numerical

Table 1 Summary of literature review

Author (year)	Characteristic				
	Iterative method	Considering equality of variation of residuals	Ordinary least square	Different M-estimators	Other regression models
Huber (1981)	✓			✓	
Siegel (1982)	✓			✓	
Rousseeuw (1984)	✓			✓	
Massart et al. (1986)	✓			✓	
Rousseeuw and Leroy (1987)	✓			✓	
Cummins and Andrews (1995)	✓		✓		
Pop and Sârbu (1996)					✓
Morgenthaler and Schumacher (1999)	✓		✓	✓	
Hund et al. (2002)			✓		
Zioutas and Avramidis (2005)					✓
Serneels et al. (2005)	✓			✓	
Bickela and Frühwirthb (2006)	✓			✓	
Ortiz et al. (2006)	✓		✓	✓	
Bertsimas and Shioda (2007)					✓
Nguyena and Welsch (2010)	✓			✓	
Dornheim and Brazauskas (2011)					✓
Proposed REVIM	✓	✓	✓		

example is presented in section 'Numerical example'. Finally, the last section is the 'Conclusion' of this paper.

Robust estimation of the coefficients by iterative weighting methods

To compensate for the effects of the outlier values, we can either remove the outlier data or modifying them by weighting the residuals. The first approach is not rational, so we choose to modify them in order to decrease the effect of outliers in the coefficient estimation stage. The proposed idea is as follows:

$$E(y_i) = \mu_i\left(\beta_1, ...\beta_p\right). \tag{1}$$

In this equation, μ_i is a function defined by unknown coefficients (β_i). For example, if $\mu_1 = \beta_1 + \beta_2 x_1$ and x_i are constants, the response y_i can be obtained from the experimental results, and the regression model describes the relation between the variables and the expected values of the y_i.

If all the measurements are good, then the OLS method provides a reasonable model and the coefficients are estimated by minimizing the following equation:

$$\left(y_1 - \mu_1\left(\hat{\beta}_1, ..., \hat{\beta}_p\right)\right)^2 + ... + \left(y_n - \mu_n\left(\hat{\beta}_1, ..., \hat{\beta}_p\right)\right)^2. \tag{2}$$

However, if the results appear abnormal, which may be a consequence of residual behavior in the experiments, the coefficients are determined by minimizing the following equation. The abnormality occurs when a residual behaves like an outlier:

$$w_1\left(y_1 - \mu_1\left(\hat{\beta}_1, ..., \hat{\beta}_p\right)\right)^2 + ... + w_n\left(y_n - \mu_n\left(\hat{\beta}_1, ..., \hat{\beta}_p\right)\right)^2 \tag{3}$$

The weights are not pre-assigned values because the quality of each y_i is not known in advance. The reasonable values for the weights are based on the residuals defined by the following equation:

$$r_i = y_i - \mu_i\left(\hat{\beta}_1, ..., \hat{\beta}_p\right). \tag{4}$$

To make the estimator invariant with respect to the scale of the residuals, the r_i is divided by 's,' which is a robust estimation of the scale. The value of 's' is often taken to be equal to 1.4826 MAD, where MAD is the median of the absolute deviations of the residuals from their median and 1.4826 is a bias adjustment for the standard deviation under the normal distribution.

The weights should be inversely proportional to the value of the residuals, $w_i = \frac{c}{|r_i|}$. In other words, the residuals with large values are weighted less, and this method produces better coefficient estimates. These

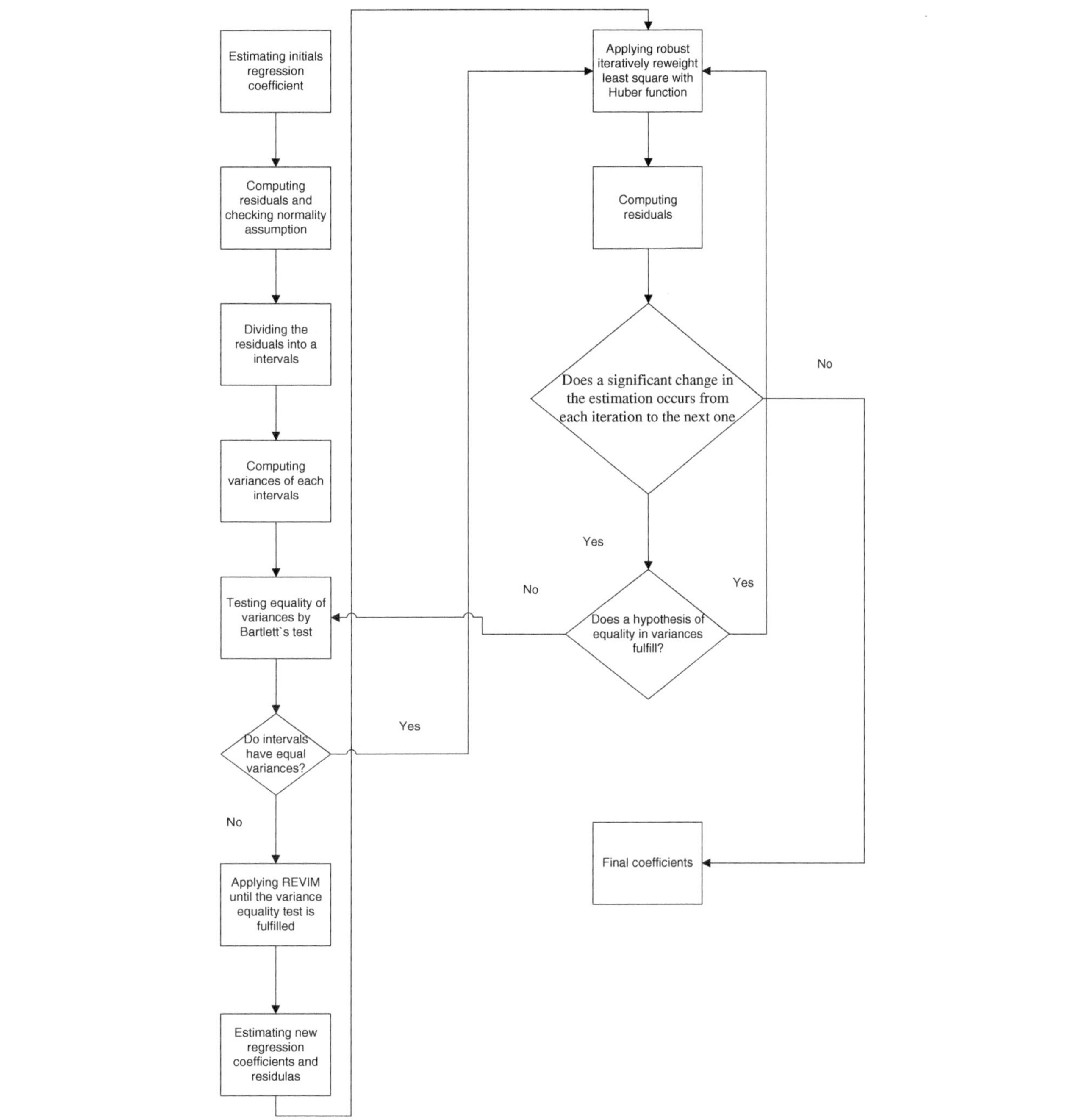

Figure 2 Flowchart of REVIM method.

weights can be chosen by a function such as the Huber weight function:

$$\begin{cases} w_i = 1 & if|r_i| < c \\ w_i = \dfrac{c}{|r_i|} & f|r_i| > c \end{cases}, \tag{5}$$

where c is a constant. The procedure is as follows: compute the first coefficients of the regression model, compute the residuals and weights, and then compute the new coefficients by the equation. This procedure can be repeated until a good solution is obtained, because the values of the coefficients and the values of the residuals and weights are different. This procedure is known as iterative weighting OLS. The procedure terminates when the change in the estimation from one iteration to the next is sufficiently small.

This iterative method is good for modifying outliers, but the trends in residual behavior are not considered. As illustrated before, another approach, in addition to taking outliers into account, is the equality of variation

Table 2 A hypothetical data created according to Box-Behnken design

Material 1	Material 2	Material 3	Material 4	Block	*Y* value	x_1	x_2	x_3	x_4
0.82	1	−55	0	1	96.49	−0.1	−0.1	0	0
0.82	1	−45	−25	1	93.22	−0.1	−0.1	1	−1
0.82	1	−65	−25	1	87.43	−0.1	−0.1	−1	−1
0.91	0.91	−55	0	1	77.20	0.8	−1	0	0
0.91	1.09	−55	0	1	82.83	0.8	0.8	0	0
0.73	0.91	−55	0	1	94.87	−1	−1	0	0
0.73	1.09	−55	0	1	63.46	−1	0.8	0	0
0.82	1	−65	25	1	91.88	−0.1	−0.1	−1	1
0.82	1	−45	25	1	91.88	−0.1	−0.1	1	1
0.83	1.02	−55	0	2	100.28	0	0.1	0	0
0.92	1.02	−55	−25	2	90.44	0.9	0.1	0	−1
0.74	1.02	−55	−25	2	92.53	−0.9	0.1	0	−1
0.83	0.93	−65	0	2	90.32	0	−0.8	−1	0
0.83	1.11	−65	0	2	91.45	0	1	−1	0
0.83	1.11	−45	0	2	90.85	0	1	1	0
0.83	0.93	−45	0	2	69.08	0	−0.8	1	0
0.92	1.02	−55	25	2	88.55	0.9	0.1	0	1
0.74	1.02	−55	25	2	91.55	−0.9	0.1	0	1
0.83	1.02	−55	0	3	90.96	0	0.1	0	0
0.83	0.92	−55	−25	3	85.67	0	−0.9	0	−1
0.83	1.11	−55	−25	3	80.21	0	1	0	−1
0.74	1.02	−45	0	3	84.74	−0.9	0.1	1	0
0.93	1.02	−65	0	3	93.70	1	0.1	−1	0
0.74	1.02	−65	0	3	92.24	−0.9	0.1	−1	0
0.93	1.02	−45	0	3	95.60	1	0.1	1	0
0.83	1.11	−55	25	3	87.83	0	1	0	1
0.83	0.92	−55	25	3	83.21	0	−0.9	0	1

between residuals which is the main idea of the rest of this paper.

Robust estimation of coefficients by testing equality of variations in specified intervals

First, normality assumption is checked. If the normality assumption is violated, this robust approach based on Huber function cannot be applied. The method proposed in this part begins by dividing the experiment runs into a intervals to examine the hypothesis of equality in variations in these intervals. The equality test used in this paper is Bartlett's test (Anderson and McLean 1974). The number of points in each interval can be chosen in the analysis stage. This stage satisfies one of the OLS hypotheses. If this parameter is small, the variation between points might be large and if the number of the points is large, the equality test of the variances may not be reliable, so this value should be determined rationally.

Bartlett's test

Although residual plots are frequently used to diagnose inequality of variance, statistical tests have also been proposed. One widely used procedure is Bartlett's test. The procedure involves computing a statistic whose sampling distribution is closely approximated by the chi-square distribution with *a-1* degrees of freedom, where the a random samples are from independent normal populations. This statistic is defined as

$$x^2 = 2.3026\frac{q}{c} \tag{6}$$

where

$$q = (N{-}a)\log_{10}^{s_p^2} - \sum_{i=1}^{a}(n_i{-}1)\log_{10}^{s_i^2}, \tag{7}$$

$$c = 1 + \frac{1}{3(a{-}1)}\left(\sum_{i=1}^{a}(n_i{-}1)^{-1} - (N{-}a)^{-1}\right), \tag{8}$$

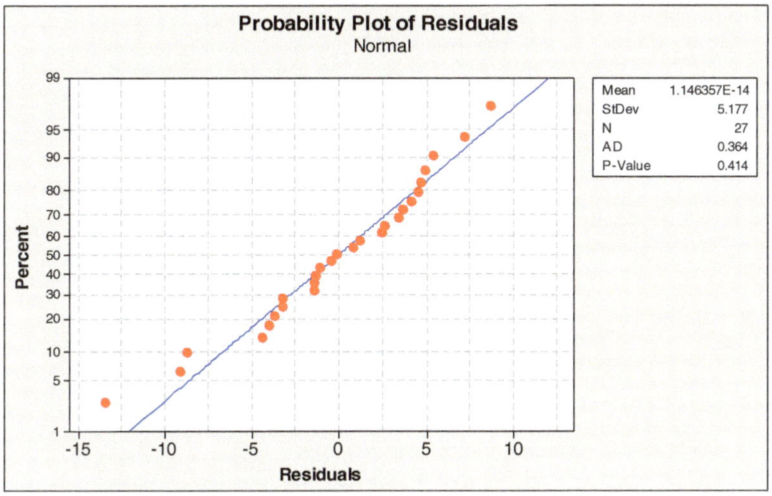

Figure 3 Normality assumption and adequacy checking of residuals.

$$s_p^2 = \frac{\sum_{i=1}^{a}(n_i-1)s_i^2}{N-a}, \qquad (9)$$

and s_i^2 is the sample variance of the ith population.

The hypothesis of equality of variances is rejected if $x_0^2 > x_{\alpha,a-1}^2$, where $x_{\alpha,a-1}^2$ is the upper α percentiles of the chi-square distribution with a-1 degrees of freedom.

Table 3 The ANOVA results

Term	Coefficient	Standard error	*T* value	*P* value
Constant	93.6635	4.732	19.794	0
Block 1	−1.6534	2.305	−0.717	0.49
Block 2	1.1817	2.284	0.517	0.616
x_1	0.329	2.636	0.125	0.903
x_2	0.408	2.636	0.155	0.88
x_3	−1.9282	2.416	−0.798	0.443
x_4	0.3978	2.416	0.165	0.872
$x_1 * x_1$	−3.898	4.215	−0.925	0.377
$x_2 * x_2$	−9.5556	4.173	−2.29	0.045
$x_3 * x_3$	−0.1949	3.571	−0.055	0.958
$x_4 * x_4$	−0.9236	3.578	−0.258	0.802
$x_1 * x_2$	12.3039	5.054	2.434	0.035
$x_1 * x_3$	2.0222	4.366	0.463	0.653
$x_1 * x_4$	−0.3262	4.619	−0.071	0.945
$x_2 * x_3$	4.733	4.564	1.037	0.324
$x_2 * x_4$	2.4782	4.342	0.571	0.581
$x_3 * x_4$	−1.4475	4.174	−0.347	0.736

Proposed robust approach (robust equal variances iterative method)

The following steps are proposed as an iterative method to decrease the effect of trends in the residuals and improve the robust estimation of coefficients. The proposed model is based on OLS method which should be modified.

First of all, our goal is that the residuals that violate the hypothesis of equality in variances should have less effect on the estimates of the coefficients of the regression model, so we should consider modifying these points to have equal variances. Therefore, the residuals derived by experiments are divided into a intervals, and then the variances of each interval are calculated and denoted by s_i^2. The next step is to test the equality of variances with Bartlett's test; if the result shows that the variances in a intervals do not have significant differences, this part of the procedure is stopped, whereas if the result shows that the

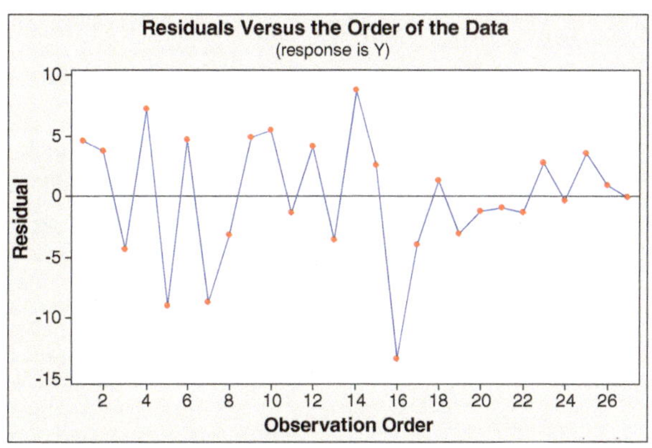

Figure 4 The least squares residuals of the model in order of the runs.

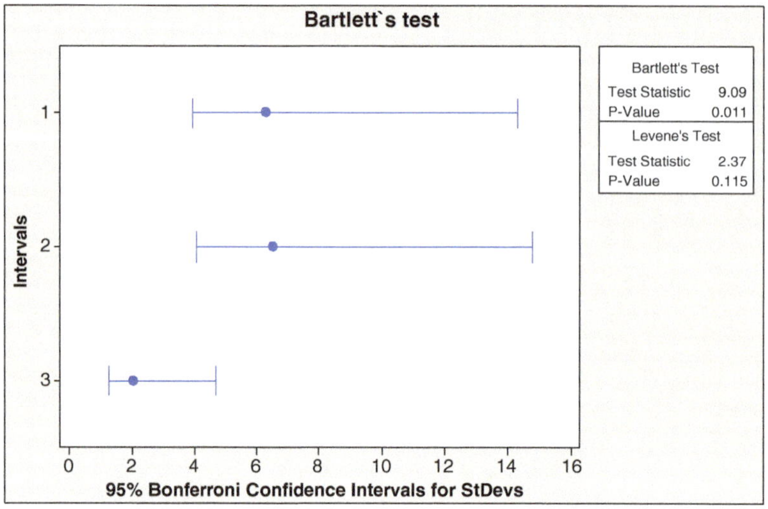

Figure 5 The results of equality test of variances.

variances in the a intervals differ significantly, the iterative weighting procedure is used to modify. As the next step, the critical q statistic in Bartlett's test, for which the hypothesis will be rejected, is computed. The critical q for Bartlett's test is denoted by q_c, and in our case, $q > q_c$. The $s^2_{(1)}$ is the ith rank-sorted variances of the intervals, in descending order. The $s^{2(j)}_{(i)}$ is the variances of the points in ith ranked interval in the jth. In the proposed approach, in higher and higher iterations approach, the residual variances approach equality. The maximum feasible variance of each intervals for which Bartlett's test is fulfilled is denoted as s^2_i max.

Because the q value is greater than q_c after its computation, it should be decreased iteratively until the both values are equal. To do this, we select the largest variance of all intervals. This value is decreased in each iteration. If the value of $s^2_{(1)}$ equals $s^2_{(2)}$ in this decreasing procedure, both $s^2_{(1)}$ and $s^2_{(2)}$ decrease in the next iteration. If the value q does not equal q_c, the values of the variances decrease again. If the values of $s^2_{(1)}$ and $s^2_{(2)}$ equal $s^2_{(3)}$ in this decreasing procedure, all three variances decrease in the next iteration, and this procedure continues until q is equal to q_c. All maximum feasible variances of intervals that satisfy the hypothesis of Bartlett's test are computed with this iterative procedure. Next, we consider two parallel lines $l_1{:}y = +\omega$ and $l_2{:}y = -\omega$, with slope zero and parallel to the x-axis. We decrease the ω values iteratively and compute the variances of points between these two lines. Until the variances of the points are equal to the maximum variance of intervals derived from the last step, this decreasing procedure continues. After that, the points outside these lines are weighted by function in (10).

Table 4 Bartlett's test results (test statistic = 9.09)

Intervals	Number	Lower	Standard deviation	Upper
1	9	3.94377	6.32629	14.3264
2	9	4.07722	6.54037	14.8112
3	9	1.29441	2.07639	4.7021

The pseudo code of the proposed method illustrates this approach:

1. Divide residuals into a intervals.
2. Compute the variances of outliers in each interval, and denote these variances by s^2_i.
3. Sort the variances of each interval in descending order and denote them by $s^2_{(i)}$.
4. Calculate the critical value of Bartlett's test in terms of confidence level and the number of intervals.
5. Compute the q value by formula in (7).

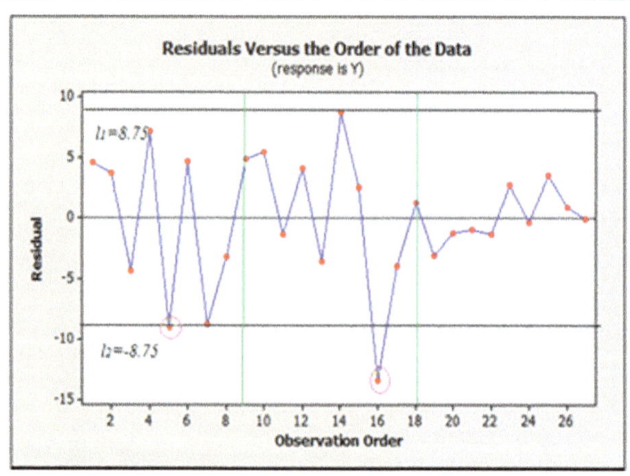

Figure 6 Proposed method.

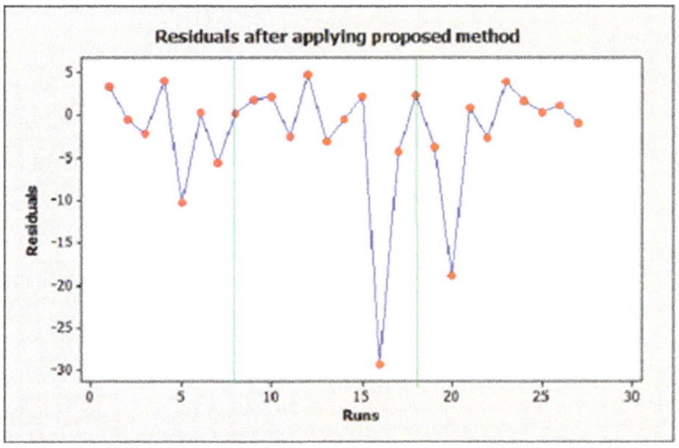

Figure 7 **Proposed method residuals.**

6. Compute the statistic of Bartlett's test by formula in (6).
7. Compute the q_c by considering the critical value of Bartlett's test based on formula (6).
8. Do while $q \geq q_c$,

$$s^{2(j+1)}_{(1)} = s^{2(j)}_{(1)} - \delta \qquad If \quad s^{2(j)}_{(i)} > s^{2(j)}_{(2)}$$
$$s^{2(j+1)}_{(1)} = s^{2(j+1)}_{(1)} = \ldots = s^{2(j+1)}_{(f)} = s^{2(j)}_{(1)} - \delta \quad If \quad s^{2(j)}_{(1)} \leq s^{2(j)}_{(f)} \; f = 2, 3, \ldots$$
$$j = j + 1.$$

9. Determine s_i^2 max as the maximum feasible variances of each intervals to satisfy the hypothesis of equal variances.
10. Consider two parallel lines, l_1, l_2, $l_1{:}y = +\omega$ and $l_2{:}y = -\omega$.
11. Do while $s_k^2 < s_i^2$ max, $l_1{:}y = +m\omega$ and $l_2{:}y = -m\omega$.
12. Determine the points outside lines $l_1$and l_2.

Table 5 Final robust coefficients

Coefficient	Value
Intercept	94.19
x_1	0.63
x_2	1.6
x_3	−1.5
x_4	0.39
x_1x_2	13.29
x_1x_3	2.06
x_1x_4	−0.32
x_2x_3	3.59
x_2x_4	2.46
x_3x_4	−1.44
x_1^2	−4.13
x_2^2	−9.32
x_3^2	−0.19
x_4^2	−1.47
Block 1	−1.36
Block 2	1.28

13. Determine the weight of the outside residuals by function below (formula 10).

$$\begin{cases} w_i = 1 & if\, l_2 < r_i < l_i \\ w_i = \dfrac{1}{|r_i|} & f\, r_i > l_1, r_i < l_2 \end{cases} \tag{10}$$

After this step, by the robust iterative weighting method, the outliers are modified. After each iteration, the test of equality in variances is checked, and if the

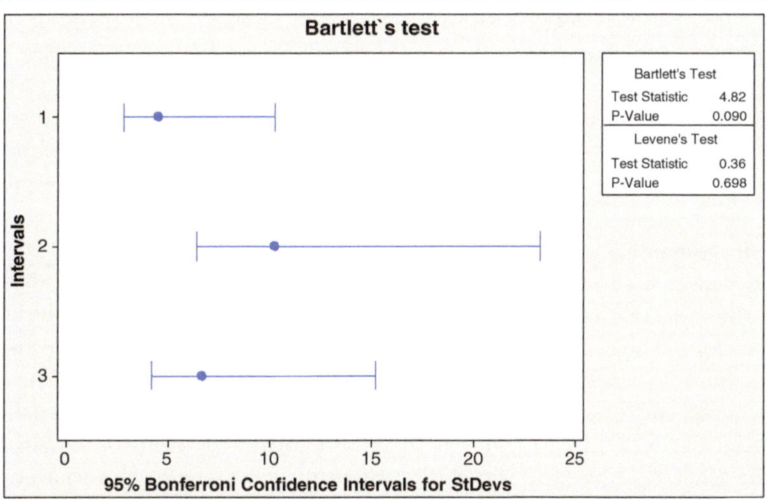

Figure 8 **Bartlett's test results after applying REVIM.**

Table 6 Final residuals of robust model

Run	Residual
1	3.89
2	3.09
3	−3.94
4	8.07
5	−11.6
6	4.44
7	−9.28
8	−2.75
9	4.28
10	4.73
11	−1.69
12	4.51
13	−2.1
14	5.97
15	1.2
16	−14.5
17	−4.28
18	1.65
19	−3.38
20	−13.6
21	−2
22	−1.29
23	2.63
24	0.19
25	2.71
26	−0.13
27	1.23

hypothesis is violated, the above mentioned method is applied. After the equality of variances is satisfied, the iterative weighting method continues, and this procedure continues as long as a minor change in the estimation occurs from each iteration to the next one. This procedure is called the robust equal variances iterative method procedure (REVIM).

The flowchart in Figure 2 illustrates this method.

Numerical example

This is a hypothetical numerical experiment. Suppose that we have an experiment containing one response variable and four explanatory control variables, each of which has three levels, and the objective of the study is to optimize the yield of a product. The data to be used are shown in Table 2. We want to explore the yield response surface by using a second-order regression model. A Box-Behnken design with 27 treatments is used for this experiment. The blocking is used to diminish the effect of nuisance factors, and the blocks are assigned, for example, to 3 days.

Table 7 Comparison between REVIM approach and OLS method

Coefficient	Method		
	Actual	OLS	REVIM
Intercept	95.67	93.66	94.19
x_1	1.25	0.32	0.63
x_2	−1.16	0.40	1.6
x_3	0.79	−1.92	−1.5
x_4	0.39	0.39	0.39
x_1x_2	14.73	12.30	13.29
x_1x_3	2.29	2.02	2.06
x_1x_4	−0.32	−0.32	−0.32
x_2x_3	−2.66	4.73	3.59
x_2x_4	2.47	2.47	2.46
x_3x_4	−1.44	−1.44	−1.44
x_1^2	−5.4	−3.89	−4.13
x_2^2	−7.71	−9.55	−9.32
x_3^2	1.11	−0.19	−0.19
x_4^2	−3.48	−0.92	−1.47
Block 1	−1.99	−1.65	−1.36
Block 2	3.03	1.18	1.28

The primary fitted response regression model is as follows:

$$\begin{aligned}\hat{y} &= 93.65 + 0.32x_1 + 0.4x_2 - 1.92x_3 + 0.39x_4 \\ &- 3.89x_1^2 - 9.55x_2^2 - 0.19x_3^2 - 0.92x_4^2 + 12.3x_1x_2 \\ &+ 2.02x_1x_3 - 0.32x_1x_4 + 4.73x_2x_3 + 2.47x_2x_4 \\ &- 1.44x_3x_4 + \text{block effect.}\end{aligned} \tag{11}$$

The normality assumptions are checked in this example and the results are given in Figure 3. The P value obtained by normality test is 0.414. This value shows that the residuals follow normal distribution.

The analysis of variance (ANOVA) results are shown in Table 3.

Figure 4 shows the residuals of the model in the order of runs. As shown in the figure, the residuals have a rough trend, and if we divide the runs into 3 equal intervals, each containing 9 runs, the second interval has larger variance. This can be proved by Bartlett's test.

The result of Bartlett's test is illustrated in Figure 5 and Table 4.

In this case, we have three intervals, so a-1 = 2. If we consider the significance level 0.95, the critical statistic $x^2_{0.05,2}$ is equal to 5.99 and the test statistic = 9.09 is greater than 5.99, then the hypothesis is rejected. Therefore, we want to compute the maximum standard deviation of each interval that satisfies the hypothesis of Bartlett's test. By the proposed method, the maximum values of these standard variations are ordered as follows: $s_1 = 2.07$, $s_2 = 4.95$, $s_3 = 4.95$. Based on these values, we can compute the limits, $l_1 = 8.7$ and $l_1 = -8.75$. Two points 5 and 16 are outside these limits (circled in the figure), so the weighting procedure is applied and new coefficients are calculated by the robust weighting method. Figure 6 shows the method graphically.

The residuals are computed by this procedure, and the hypothesis of equal variance in three intervals is satisfied. Bartlett's test is applied to the residuals obtained from the proposed method, and the results are given in Figure 7.

The value of Bartlett's test is 4.32, and the hypothesis of equal variances is not rejected. Figure 8 illustrates the result better.

Therefore, the iterative weighting method based on these coefficients to modify the effect of outliers by Huber function with $c = 2$ is applied. This process is applied in each iteration, and if the equality test is not satisfied, the modification is applied. The final robust coefficients and residuals are presented in Tables 5 and 6.

So the residuals in final iteration show that residuals which hinted to be an outlier in first iteration are really outliers, and the model estimation is more accurate and close to the model with no outlier data that we call it *actual model*. These results can be compared with the results obtained by the model with no outlier data. The comparison shows that the proposed model is more precise and accurate than the OLS method in estimation of the regression coefficients by considering unequal variation between residuals. The results are given in Table 7.

Conclusions

A robust estimation of the response surface is the primary goal of this paper. To this end, the proposed method is defined, instead of the common ordinary least squares method of estimating coefficients of the response surface, to decrease the effects of two main causes of the imprecise estimation of coefficients, outliers, and trends in residuals. As the effect of trends in residuals should be taken into account, the proposed method simultaneously modifies the effects of trends and outliers. For each iteration, an equality test of residual variances is performed, and after this hypothesis is satisfied, the outliers are modified. A goal for future research may be to examine the weighting; instead of computing the distance of the residuals from base line after plotting the responses in a normal probability plot (NPP), the weighting function may be proportional to the distance of the response from the NPP regression line.

Competing interests

The authors declare that they have no competing interests.

Authors' contributions

MB has worked on the literature and modeling, he also proposed response surface methodology. AM has formulated the robust concept and numerical example. Both authors read and approved the final manuscript.

Authors' information

MB is an associate professor of industrial engineering and his research interests are multiple response optimization and facilities location problem. AM is a Ph.D. student of industrial engineering at Amirkabir University of Technology (Tehran Polytechnic).

Acknowledgments

Authors are thankful to the reviewers for their valuable comments.

References

Anderson VL, McLean RA (1974) Design of experiments: a realistic approach. Dekker, New York

Andrews DF, Bickel PJ, Hampel FR, Huber PJ, Rogers WH, Tukey JW (1972) Robust estimates of location: survey and advances. Princeton University Press, Princeton

Bertsimas D, Shioda R (2007) Classification and regression via integer optimization. Oper Res 55:252–271

Bickela DR, Frühwirthb R (2006) On a fast, robust estimator of the mode: comparisons to other robust estimators with applications. Comput Stat Data An 50:3500–3530

Cummins DJ, Andrews CW (1995) Iteratively reweighted partial least squares: a performance analysis by Monte Carlo simulation. J Chemometr 9:489–507

Dornheim H, Brazauskas V (2011) Robust-efficient fitting of mixed linear models: methodology and theory. J Stat Plan Infer 141:1422–1435

Hejazi TH, Bashiri M, Noghondarian K, Atkinson AC (2010) Multiresponse optimization with consideration of probabilistic covariates. Qual Reliab Eng Int 27:437–449

Huber PJ (1981) Robust statistics. Wiley, New York

Hund E, Massart DL, Smeyers-Verbeke J (2002) Robust regression and outlier detection in the evaluation of robustness tests with different experimental designs. Anal Chim Acta 463:53–73

Kazemzadeh RB, Bashiri M, Atkinson AC, Noorossana R (2008) A general framework for multiresponse optimization problems based on goal programming. Eur J Oper Res 189:421–429

Maronna RA, Martin RD, Yohai VJ (2006) Robust statistics: theory and methods. Wiley, New York

Massart DL, Kaufman L, Rousseeuw PJ, Leroy A (1986) Least median of squares: a robust method for outlier and model error detection in regression and calibration. Anal Chim Acta 187:171–179

Mendes B, Tyler DE (1995) Robust statistics, data analysis and computer intensive methods. Springer, New York

Morgenthaler S, Schumacher MM (1999) Robust analysis of a response surface design. Chemometr Intell Lab 47:127–141

Nguyena TD, Welsch R (2010) Outlier detection and least trimmed squares approximation using semi-definite programming. Comput Stat Data An 54:3212–3226

Ortiz MC, Sarabia LA, Herrero A (2006) Robust regression techniques. A useful alternative for detection of outlier data. Talanta 70:499–512

Pop HF, Sârbu C (1996) A new fuzzy regression algorithm. Anal Chem 68:771–778

Rousseeuw PJ (1984) Least median of squares regression. J Am Stat Assoc 79:871–880

Rousseeuw PJ, Yohai VJ (1984) Robust and nonlinear time series analysis. Springer, New York

Rousseeuw PJ, Leroy AM (1987) Robust regression and outlier detection. Wiley, New York

Serneels S, Croux C, Filzmoser P, Van Espen PJ (2005) Partial robust M-regression. Chemometr Intell Lab 79:55–64

Shahriari H, Ahmadi O, Shokouhi AH (2011) A two-step robust estimation of the process mean using M-estimator. J Appl Stat 38:1289–1301

Siegel AF (1982) Robust regression using repeated medians. Biometrika 69:242–244

Zioutas G, Avramidis A (2005) Deleting outliers in robust regression with mixed integer programming. Acta Math Appl Sin 21:323–334

16

Effective heuristics and meta-heuristics for the quadratic assignment problem with tuned parameters and analytical comparisons

Mahdi Bashiri* and Hossein Karimi

Abstract

Quadratic assignment problem (QAP) is a well-known problem in the facility location and layout. It belongs to the NP-complete class. There are many heuristic and meta-heuristic methods, which are presented for QAP in the literature. In this paper, we applied 2-opt, greedy 2-opt, 3-opt, greedy 3-opt, and VNZ as heuristic methods and tabu search (TS), simulated annealing, and particle swarm optimization as meta-heuristic methods for the QAP. This research is dedicated to compare the relative percentage deviation of these solution qualities from the best known solution which is introduced in QAPLIB. Furthermore, a tuning method is applied for meta-heuristic parameters. Results indicate that TS is the best in 31% of QAPs, and the IFLS method, which is in the literature, is the best in 58 % of QAPs; these two methods are the same in 11% of test problems. Also, TS has a better computational time among heuristic and meta-heuristic methods.

Keywords: Quadratic assignment problem, Heuristics, Meta-heuristics, Tuning method

Background

Quadratic assignment problem (QAP) is a well-known problem which delineates assigning number of economic facility to the same number of location. The objective function of this problem needs to be minimized. This problem was originally initiated by Koopmans and Beckmann (1957). QAP is used in combinational optimization, participating with traveling salesman problem (TSP) and graph problem. QAP is a kind of location problem and, additionally, a kind of layout problem. A lot of research presented exact solutions for QAP in the literature. Figure 1 shows the number of research, which applies the exact methods for QAP until 2005 (Loiola et al. 2007). These methods include branch and bound, dynamic programming, cutting plan, and branch and cut.

Sahni and Gonzalez (1976) proved that QAP belongs to the NP-complete class. In this class of the problem, computational time for solving problems, which have large instances, is too much and grows exponentially, so heuristic approaches and meta-heuristic methods are recommended for solving this problem. Figure 2 shows the literature about solving QAP with heuristic and meta-heuristic methods. This figure reveals that the meta-heuristic methods have been applied more than the heuristic approaches.

For the first time, a simulated annealing (SA) approach in solving QAP was proposed by Burkard and Rendl (1984). Wilhelm and Ward (1987) then presented the new equilibrium components for solving QAP with SA, and after them, research suggested better SA solution methods. Skorin-Kapov (1990) executed the tabu search (TS) technique for QAP. After that, Taillard (1991) demonstrated robust TS for QAP. TS was also proposed for QAP in later research. Particle swarm optimization (PSO) is a new approach to solve QAP that Kennedy and Eberhart (1997), Shi et al. (2004), and Hongbo et al. presented (Hongbo and Ajith 2007; Hongbo et al. 2007). Figure 3 shows the application of meta-heuristic methods for QAP until 2005 (Loiola et al. 2007).

Figure 3 shows that many research used hybrid algorithm for QAP in recent years. Table 1 shows some of the recent research that applied TS, SA, and PSO for QAP from 2007 to 2009.

This paper is organized as follows: in the next section, QAP formulation is displayed; afterwards, the heuristic

* Correspondence: Bashiri@shahed.ac.ir
Department of Industrial Engineering, Shahed University, Tehran 3319118651, Iran

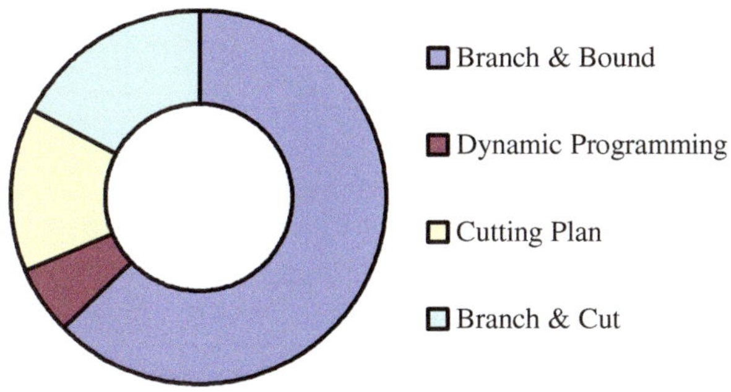

Figure 1 Exact methods for QAP.

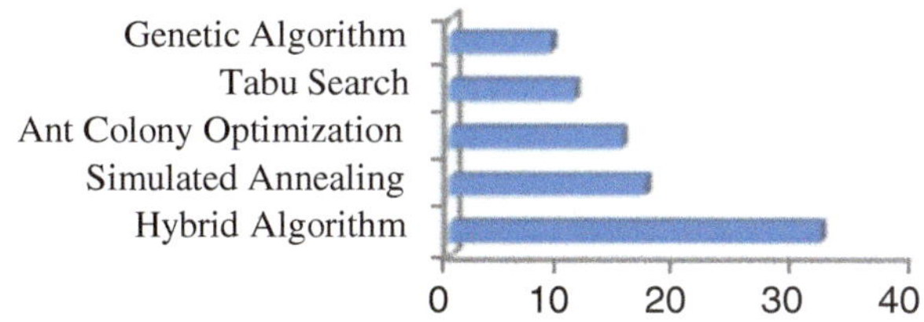

Figure 3 Meta-heuristic approach for QAP.

methods are described. After that, the meta-heuristic methods such as SA, TS, and PSO are presented; the tuning method is described next. Computational analyses and comparison results are mentioned in the 'Results and discussion' section, and the last section is the 'Conclusions' section.

Methods

QAP formulation

This section is dedicated to introduce the classical formulation of QAP considered in this research. This formulation is as follows:

$$\min \sum_{i=1}^{n}\sum_{j=1}^{n}\sum_{k=1}^{n}\sum_{l=1}^{n} c_{ijkl}x_{ik}x_{jl}, \tag{1}$$

$$\sum_{i=1}^{n} x_{ij}\,,\ \forall\, j = 1\ldots n, \tag{2}$$

$$\sum_{i=1}^{n} x_{ij}\,,\ \forall\, i = 1\ldots n, \tag{3}$$

$$x_{ij} = \{0,1\}\ ,\ \forall\, i,j = 1\ldots n, \tag{4}$$

where c_{ijkl} is the cost of assigning facility i in location k and simultaneously facility j in location l, and $x_{ik} = 1$ if location k is assigned to facility i; otherwise, $x_{ik} = 0$. Also, $x_{jl} = 1$ if location l is assigned to facility j; otherwise, $x_{jl} = 0$. The objective function (Equation 1) of this model must be minimized. Each location must be assigned just to one facility, as Equation 2 shows. Equation 3 displays that each facility must be assigned just in one location. The number of facility and location is the same and is equal to n. The variable in this model is binary.

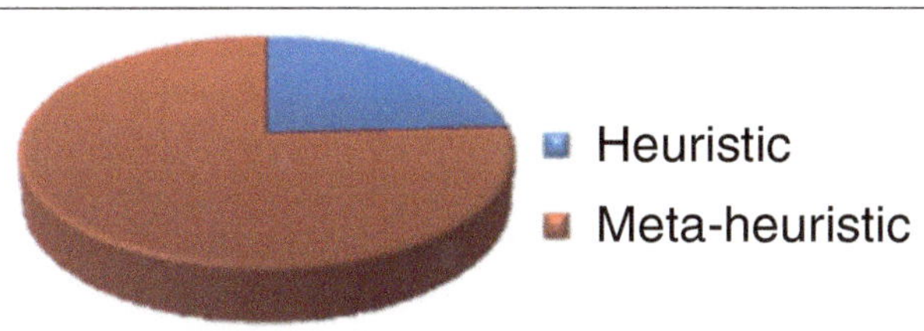

Figure 2 Heuristic and meta-heuristic methods for QAP.

Heuristic methods

Heuristic algorithms do not provide an assurance for optimization of the problem. These methods are an approximation. They have an additional property that worst-case solutions are known. In this section, some heuristic methods as procedures to search the better solution that contains 2-opt, greedy 2-opt, 3-opt, greedy 3-opt, and Vollman, Nugent, Zartler (VNZ) are contemplated.

2-Opt algorithm

Among simple local search algorithms, the most famous ones are 2-opt and 3-opt. The 2-opt algorithm was first proposed by Croes (1958) for TSP. If there are four locations and four facilities, the transposition of facility location in 2-opt method is like that in Figure 4. This figure illustrates that for the first transposition, the facility in location one can be changed with the facility in location two, and for the second transposition, the facility in location one can be changed with the facility in location three, so that if the number of location and facility is shown by n, the number of transposition in each iteration will be $n(n-1)/2$.

Initially, the algorithm considers the transposition of facilities 1 and 2. If the resulting solution's objective function value (OFV) is smaller than that of the initial solution, then it is stored as a candidate for future

Table 1 TS, SA, and PSO approach for QAP from 2007 to 2009

Authors	Meta-heuristic		
	TS	SA	PSO
Hongbo and Ajith (2007); Hongbo et al. (2007)	-	-	·
Singh and Sharma (2008)	-	·	-
Zhu et al. (2009a,b); James et al. (James et al. 2007, James et al. 2009); Fescioglu and Kokar (2008)	·	-	-
This research	·	·	·

Dot (·), the authors used the method; hyphen (–), the authors did not use the method.

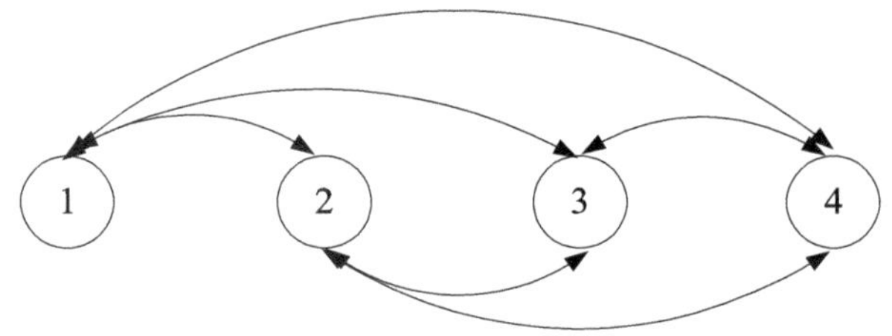

Figure 4 Transposition of facility in the 2-opt method.

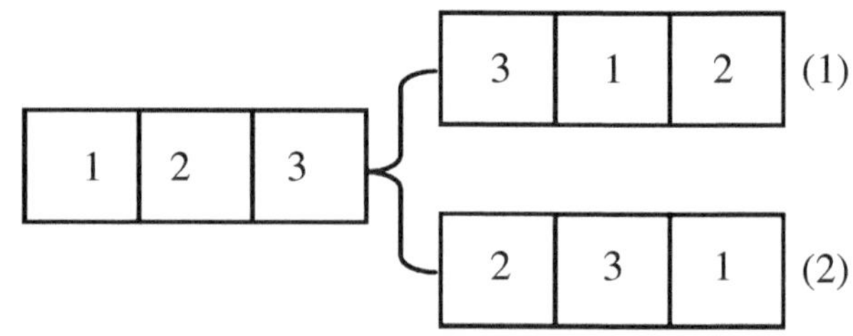

Figure 5 Types of transposition in 3-opt algorithm.

consideration. Otherwise, it is discarded, and the algorithm considers the transposing of facilities 1 and 3. If this exchange generates a better solution, then it is stored as a candidate for future consideration; otherwise, it is discarded, and so on. Thus, whenever a better solution is found, the algorithm discards the previous best solution. This procedure continues until all the pair-wise exchanges are considered.

For n location in the QAP problem, the 2-opt algorithm consists of three steps:

Step 1. Let S be the initial feasible solution and Z its objective function value; then, set $S^* = S$, $Z^* = Z$, $i = 1$ and $j = i + 1 = 2$.

Step 2. Consider the exchange results in a solution S' that has OFV $Z' < Z^*$, set $Z^* = Z'$ and $S^* = S'$. If $j < n$, then repeat step 2; otherwise, set $i = i + 1$ and $j = i + 1$. If $i < n$, repeat step 2; otherwise, proceed to step 3.

Step 3. If $S \neq S^*$, set $S = S^*$, $Z = Z^*$, $i = 1$, $j = i + 1 = 2$ and go to step 2. Otherwise, output S^* is selected as the best solution, and the process is terminated.

Greedy 2-opt algorithm

The greedy 2-opt algorithm is a variant of the 2-opt algorithm. The difference between this method and 2-opt is in selecting the best transposition. This method transposes the facility location if the OFV is better than the known OFV and stabilizes this assignment; it then goes to transpose the facility location from the start. It also consists of three steps. Like the 2-opt algorithm, greedy 2-opt also considers pair-wise exchanges. Initially, it considers transposing facilities 1 and 2. If the resulting OFV is less than the previous one, two facilities are immediately transposed. Otherwise, the algorithm will go on to facility 3 and evaluate the exchange, and so on until improvement is found. If facilities 1 and 2 are transposed, then the algorithm will take it as an initial solution and will repeat the algorithm until it is impossible to improve the solution any further. Greedy 2-opt algorithm makes the exchange permanent whenever an improvement is found and thus consumes less computational time than the 2-opt algorithm. On the other hand, greedy 2-opt algorithm produces slightly worse solutions than the 2-opt algorithm.

3-Opt algorithm

The 3-opt algorithm is similar to the 2-opt algorithm except that it considers transposing two facilities at a time. This algorithm is originally applied for TSP by Bock (1958). For example, if there are three facilities in the same location, two types of transposition can be used with the 3-opt algorithm. These types are shown in Figure 5. Type (2) is applied in this research.

Greedy 3-opt algorithm

Greedy 3-opt algorithm is also similar to the greedy 2-opt algorithm, but it makes the three facility exchange permanent whenever its resulting OFV is better than the current OFV and repeats the algorithm with the new transposition as the initial solution. The transposition in this method is similar to that in 3-opt.

VNZ method

The VNZ method was introduced by Vollman et al. (1968). This method is using less storage space than 2-opt.

There is not any randomization, and also, these methods cannot orient the current solution to the optimum solution in a limited time. However, the meta-heuristic methods contain a good search approach with a reasonable time. These methods are considered in the next subsection.

Meta-heuristic methods

In the original definition, meta-heuristics are solution methods that manage an interaction between the local improvement procedures and higher level strategies to create a process capable of escaping from local optimum solution and performing a good search of solution space. These methods have also come to include any procedures that employ strategies for overcoming the trap of local optimality in complex solution spaces, especially those procedures that take advantage of one or more neighborhood structures as a means of defining acceptable moves to transformation from one solution to another. In this research, TS, SA, and PSO are applied for the QAP, and their comparison has been done for the selected data sets.

Tabu search

Classical methods often face great difficulty when confronted with hard optimization problems present in real situations. Tabu search (TS), for the first time, was proposed by Glover (1989, 1990). This meta-heuristic approach is, in a theatrical manner, changing the ability of solving problems of practical significance. The pseudo code of TS, which is applied in this research, is as follows:

Step 1. Let S be the initial feasible solution and Z its objective function value; then, set $S^* = S$, $Z^* = Z$, max short-term memory (STM) = 5, and max iteration = 1,000; iter = 1. Best O value = O value.
Step 2. Random (i, j) = rand/Long-term memory (LTM) (i, j), $(n1, n2)$ = the indices of maximum value in random.
Step 3. If there is none $(n1, n2)$ in STM matrix, change $n1$ and $n2$ locations; otherwise, repeat step 2.
Step 4. Insert $n1$ and $n2$ in STM and release the last indices from STM (e.g., $m1$, $m2$); and LTM$(m1, m2)$ = LTM$(m1, m2) + 1$.
Step 5. Calculate the objective function value (Z) of the new permutation.
Step 6. If $Z \le Z^*$, then $Z^* = Z$, $S^* = S$, and iter = iter + 1.
Step 7. If iter $\le$ max iteration, then repeat step 2; otherwise, print Z^* and S^*.

Simulated annealing

Simulated annealing is a famous and popular local search meta-heuristic applied to address discrete and continuous optimization problems. This method, like the other meta-heuristic methods, can be escaping from the local solution. Simulated annealing is so named because of its similarity to the process of physical annealing with solids, in which a crystalline solid is heated and then allowed to cool very slowly until it achieves its most regular possible crystal lattice configuration and thus is free of crystal defects. If the cooling schedule is sufficiently slow, the final configuration results in a solid with such superior structural integrity. Equation 5 shows the Metropolis acceptance criteria for each move in the cooling process (Metropolis et al. 1953).

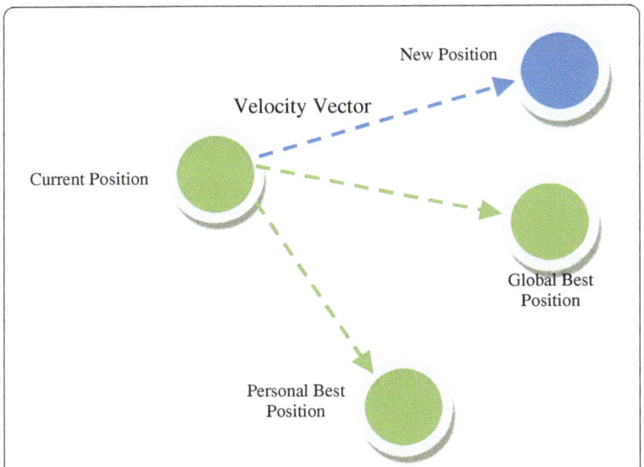

Figure 6 Each particle movement.

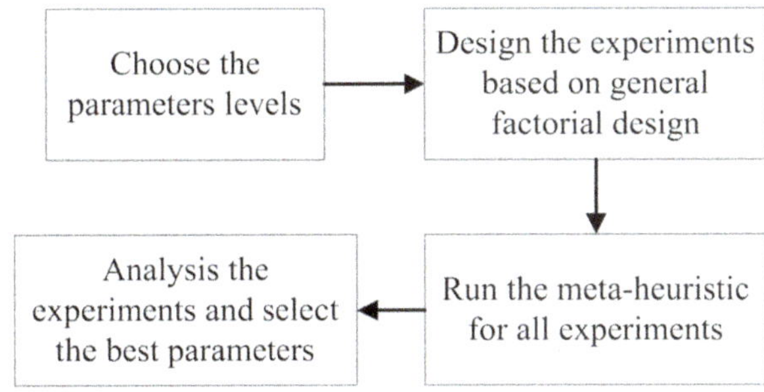

Figure 7 The procedure of tuning the parameters.

$$P = \exp\left(\frac{(\mathrm{OFV}_B - \mathrm{OFV})}{T}\right), \tag{5}$$

where OFV and OFV_B are the objective function values for this iteration and are the best computed one until this iteration. T is the temperature of the algorithm in the iteration, and P is the probability of acceptance for each move in the annealing process. The proposed SA pseudo code for QAP is as follows:

Step 1. Let S be the initial feasible solution and Z its objective function value; then, set $S^* = S$, $Z^* = Z$. $T = 100$, $T_0 = 0.1$, $r = 0.95$, $n$limit max = 5 and $n$over max = 10.
Step 2. $n$limit = 0 and $n$over = 0.
Step 3. Transpose two facilities in the current layout randomly and calculate the objective function value (Z).
Step 4. If $Z \le Z^*$, then accept the transposition; it means that $Z^* = Z$, $S^* = S$, and then $n$limit = $n$limit + 1, $n$over = $n$over + 1; if $n$over = $n$over max or $n$limit =

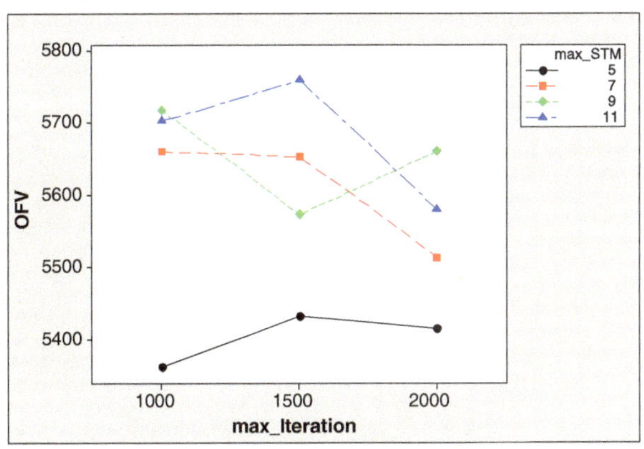

Figure 8 Interaction plot for OFV of 'Nug27'.

*n*limit max, then proceed to step 6.Otherwise, repeat step3; if $Z > Z^*$, then proceed to step 5.

Step 5. Calculate Equation 3; if $P \geq$ rand (0, 1), then $Z^* = Z$, $S^* = S$, and then *n*limit = *n*limit + 1, *n*over = *n*over + 1; if *n*over = *n*over max or *n*limit = *n*limit max, then proceed to step 6; otherwise, repeat step 3.If $P <$ rand (0, 1), then *n*over = *n*over + 1; if *n*over = *n*over max, then proceed to step 6; otherwise, repeat step 3.

Step 6. $T = r \times T$, where r is the rate of cooling. If $T \leq T_0$, then proceed to step 7; otherwise, repeat step 2.

Step 7. Print S^* and Z^*.

Partial swarm optimization

The particle swarm optimization (PSO) is a population-based search algorithm founded on the simulation of the social behavior of bees, birds, or a school of fish. This method is a significant member of the swarm intelligence. It was proposed by Eberhart and Kennedy as an optimization method (Eberhart and Kennedy 1995; Kennedy and Eberhart 1995). Each individual within the swarm is represented by a vector in multidimensional search space. This vector has one assigned vector that determines the next movement of the particle and is called the velocity vector. This technique also determines how to update the velocity of a particle. Each particle updates its velocity based on the current velocity and the best position (*p*best) it has explored so far, as well as based on the global best position (*g*best) explored by a swarm. Movement of each particle is shown in Figure 6, and it is based on Equations 6 and 7. Equation 6 illustrates that the velocity vector is updated by the global best position, personal best position, and current position of each particle. Equation 7 shows that each particle moves by its velocity.

$$v_i(t+1) = w.v_i(t) + b_1.\text{rand}(p\text{best} - x_i(t)) + b_2.\text{rand}(g\text{best} - x_i(t)), \tag{6}$$

$$x_i(t+1) = x_i(t) + v_i(t), \tag{7}$$

where i is the index of the particle, t is the index of an iteration, v_i is the vector of velocity, x_i is the position, w is the weight of current velocity, b_1 is the weight of difference between personal best and current positions, b_2 is the weight of difference between global best and current positions, and rand is used for randomization. The PSO

Table 2 Heuristic computational results for the test problems

Names of instances	n	2-Opt		Greedy 2-opt		3-Opt		Greedy 3-opt		VNZ		Exact
		Best OFV	Time (s)	Best OFV	Time (s)	Best OFV	Time (s)	Best OFV	Time (s)	Best OFV	Time (s)	Best OFV
Nug12	12	630	0.035	720	0.070	612	0.060	662	0.038	630	0.115	578
Nug14	14	1040	0.033	1270	0.031	1076	0.120	1122	0.037	1094	0.079	1014
Nug15	15	1168	0.080	1410	0.031	1194	0.158	1266	0.081	1210	0.094	1150
Nug16a	16	1636	0.036	2032	0.061	1710	0.150	1744	0.092	1708	0.160	1610
Nug16b	16	1312	0.062	1478	0.039	1304	0.175	1342	0.047	1338	0.092	1240
Nug17	17	1764	0.043	2204	0.051	1824	0.183	1900	0.052	1812	0.094	1732
Nug18	18	1988	0.033	2438	0.038	2002	0.325	2102	0.078	2084	0.104	1930
Nug20	20	2676	0.095	3070	0.037	2700	0.300	2780	0.089	2762	0.076	2570
Nug21	21	2560	0.055	3224	0.073	2528	0.482	2612	0.145	2568	0.096	2438
Nug22	22	3836	0.050	4580	0.043	3706	0.702	3850	0.070	4002	0.172	3596
Nug24	24	3670	0.051	4550	0.052	3580	0.889	3776	0.121	3826	0.143	3488
Nug25	25	3816	0.056	4746	0.053	3850	0.999	4014	0.194	3946	0.186	3744
Nug27	27	5582	0.069	6436	0.037	5622	1.580	5932	0.214	5792	0.138	5234
Nug30	30	6180	0.081	7450	0.052	6282	3.476	6670	0.260	6648	0.238	6124
Bur26a	26	5536606	0.104	5686514	0.079	5450887	5.318	5499462	0.390	5642823	0.269	5426670
Bur26b	26	3865109	0.201	4078658	0.082	3825543	6.888	3921950	0.390	3898585	0.340	3817852
Bur26c	26	5559892	0.158	5691003	0.089	5432628	6.066	5488940	0.377	5560520	0.300	5426795
Bur26d	26	3955005	0.160	4032499	0.079	3822905	5.472	3845665	0.408	3999274	0.324	3821225
Bur26e	26	5457341	0.131	5656466	0.082	5387936	6.510	5441312	0.382	5610566	0.351	5386879
Bur26f	26	3882393	0.149	4038741	0.079	3783547	4.828	3818354	0.390	3982126	0.282	3782044
Bur26g	26	10178342	0.148	10668053	0.080	10119845	5.332	10243826	0.389	10405839	0.281	10117172
Bur26h	26	7246369	0.101	7644484	0.085	7100009	5.651	7216120	0.391	7286571	0.281	7098658

Table 3 Solution qualities with CPU time of small-size test problems

Names of instances	*n*	SA		TS		PSO		IFLS with local search (OXPM)		Exact
		RPD	Time (s)	RPD	Time (s)	RPD	Time (s)	RPD	Time (s)	OFV
Nug12	12	**0.00**	0.16	**0.00**	**0.12**	7.61	0.27	1.38	0.70	578
Nug14	14	1.78	0.19	**0.39**	**0.17**	1.38	0.29	1.58	1.28	1,014
Nug15	15	3.30	0.16	0.87	**0.15**	2.26	0.27	**0.17**	1.69	1,150
Nug16a	16	0.75	0.16	1.37	**0.14**	0.75	0.31	**0.00**	2.17	1,610
Nug16b	16	0.65	0.17	**0.00**	**0.14**	**0.00**	0.32	**0.00**	2.14	1,240
Nug17	17	1.04	0.17	**0.69**	**0.15**	3.00	0.32	1.27	2.77	1,732
Nug18	18	2.69	0.18	1.04	**0.14**	3.01	0.34	**0.93**	3.42	1,930
Nug20	20	2.49	0.20	1.56	**0.16**	1.95	0.36	**0.70**	5.28	2,570
Nug21	21	1.72	0.20	**0.25**	**0.17**	2.71	0.40	**0.25**	6.89	2,438
Nug22	22	1.78	0.20	**0.50**	**0.17**	1.33	0.41	1.33	7.77	3,596
Nug24	24	4.59	0.21	**1.15**	**0.18**	3.96	0.45	2.87	10.91	3,488

pseudo code, which is presented for QAP in this research, is as follows:

Step 1. max iteration = 100, number of particle = 15, and $w = n - 1$, $b_1 = n/2$ and $b_2 = (n/2) + 2$, where n is the dimension of the problem. Make 15 permutations as the initial solutions and $Z^* = \min$ (OFV), $S^* = S$, and iter = 1.

Step 2. For $i = 1$ to w, transpose two facility. Do this step for each particle.

Step 3. Calculate objective function value (Z) for each new particle; find the personal best OFV for each particle (*p*best), and find the global best OFV (*g*best).

Step 4. Generate a random discrete number between 0 and b_1, and for $i = 1$ to this random number, simulate each particle to *p*best.

Step 5. Generate a random discrete number between 0 and b_2, and for $i = 1$ to this random number, simulate each particle to *g*best.

Table 4 Solution qualities with CPU time of medium-size test problems

Names of instances	*n*	SA		TS		PSO		IFLS with local search (OXPM)		Exact
		RPD	Time (s)	RPD	Time (s)	RPD	Time (s)	RPD	Time (s)	OFV
Nug25	25	3.69	0.24	1.55	**0.18**	1.92	0.48	**0.21**	12.97	3,744
Bur26a	26	0.10	0.62	**0.09**	**0.53**	0.20	1.32	**0.09**	15.86	5,426,670
Bur26b	26	0.67	0.61	0.19	**0.54**	0.42	1.30	**0.17**	15.56	3,817,852
Bur26c	26	0.06	0.60	0.26	**0.55**	0.30	1.34	**0.00**	15.41	5,426,795
Bur26d	26	0.12	0.60	0.02	**0.54**	0.05	1.30	**0.01**	15.38	3,821,225
Bur26e	26	0.01	0.60	**0.03**	**0.51**	0.05	1.33	0.26	15.16	5,386,879
Bur26f	26	0.26	0.61	0.05	**0.54**	0.33	1.29	**0.00**	15.55	3,782,044
Bur26g	26	0.06	0.59	**0.01**	**0.54**	0.42	1.34	**0.01**	15.28	10,117,172
Bur26h	26	0.03	0.60	0.01	**0.52**	0.35	1.32	**0.00**	14.88	7,098,658
Nug27	27	2.67	0.26	**1.38**	**0.22**	4.43	0.48	2.79	17.42	5,234
Nug30	30	5.94	0.27	2.65	**0.22**	6.01	0.54	**1.32**	22.45	6,124
Tai30a	30	4.70	0.27	4.75	**0.25**	5.90	0.85	**2.46**	18.28	1,818,146
Tai30b	30	3.55	0.26	**1.62**	**0.25**	5.18	0.87	2.33	18.23	637,117,113
Tai40a	40	5.31	0.37	6.12	**0.32**	5.53	1.16	**3.16**	62.66	3,139,370
Tai40b	40	5.00	0.34	**3.07**	**0.29**	3.84	1.07	4.18	60.44	637,250,948
Tai50a	50	5.90	0.47	6.49	**0.39**	7.19	1.36	**2.90**	158.94	4,938,796
Tai50b	50	3.86	0.45	5.15	**0.40**	7.23	1.35	**1.87**	161.27	458,821,517

Table 5 Solution qualities with CPU time of large size test problems

Names of instances	n	SA		TS		PSO		IFLS with local search (OXPM)		Exact
		RPD	Time (s)	RPD	Time (s)	RPD	Time (s)	RPD	Time (s)	OFV
Lipa80a	80	0.25	0.89	**0.15**	**0.79**	40.70	2.39	0.72	1734.11	253,195
Lipa80b	80	23.46	0.94	23.55	**0.80**	23.93	2.38	**20.65**	1889.81	7,763,962
Tai80a	80	6.13	0.94	6.66	**0.79**	7.01	2.37	**2.87**	1390.58	13,527,910
Tai80b	80	2.93	0.96	3.91	**0.82**	4.55	2.48	**0.71**	1122.56	841,223,593
Lipa90a	90	2.01	1.12	**0.49**	**0.93**	1.03	3.07	0.67	2964.41	360,630
Lipa90b	90	13.55	1.17	14.59	**0.96**	14.71	3.35	**0.00**	3200.88	12,490,441
Tai100a	100	4.66	1.69	4.11	**1.20**	4.33	4.82	**2.69**	3560.03	21,090,402
Tai100b	100	3.06	2.15	2.02	**1.73**	4.17	5.21	**0.93**	3595.17	1.186E + 09

Step 6. iter = iter + 1; if iter < max iteration, then repeat step 2;otherwise, proceed to step 7.

Step 7. Print Z^* = *g*best S^* = *g*best permutation (*S*).

Tuning method

The parameters in meta-heuristic methods which are introduced, such as max STM and max iteration, were tuned by carrying out the general factorial design. To achieving this purpose, some levels were defined initially for each parameter. These levels are determined to be focused on the logic of meta-heuristics. For instance, when the number of iteration in PSO is increased, then the time of search is intensified. This is just one characteristic of this meta-heuristic approach. After defining the levels, an experimental design is determined by general factorial design. The meta-heuristic for these experiments is then run, and their results are analyzed. Thus, the best parameter level can be found. Figure 7 illustrates this tuning method.

For example, for 'Nug' instances taken from the Quadratic Assignment Problem Library (QAPLIB), the analysis is done for 'Nug27' for the TS method. The proposed levels for max STM are 5, 7, 9, and 11, and for max iteration, they are 1,000, 1,500, and 2,000. We consider two replicates for each combination of factor levels; hence, for this instance, 24 treatment combinations are arranged. Figure 8 shows the interaction plot for OFV, and it also shows the best combination of these factors. The best max STM is 5, and the best max iteration is 1,000. Hence, we tune the TS parameter based on this experiment. Moreover, the parameters like *n*over max, *n*limit max, *T*, T_0, and others were set by analyzing the general factorial design similar to TS.

Results and discussion

Computational analyses

In this research, an analytical comparison between heuristics and meta-heuristics, which are described as the Nug and 'Bur' instances selected from QAPLIB, are applied. The computer which carried out the experiments is equipped with a 2 53-GHz processor with programming coded in MATLAB7.8 (MathWorks, Natick, MA, USA). Table 2 shows the computational results for heuristic methods. Results show that 3-opt as a heuristic method is the best solution method for QAP among the presented heuristic methods in this research. In this table, sizes of the problems are determined by *n*.

Tables 3, 4, and 5 reveal the computational results for meta-heuristic methods. In this computational analysis, the test problems are classified in small-, medium-, and large-size cases. These groups are taken from the study of Ramkumar et al. (2008). Our experiment results are compared with iterated fast local search (IFLS) when the order crossover with random pair-wise interchange mutation (OXPM) is used (Ramkumar et al 2008). The criterion considered for evaluating the performance is the relative percentage deviation (RPD) of the solution quality from the best known solution. The number of location is *n*.

The superior methods in CPU time and OFV are illustrated in the boldface. The average of gaps for heuristic methods and meta-heuristic methods introduced in this research with the exact solution are shown in Figure 9. These results are related to the Nug and Bur instances.

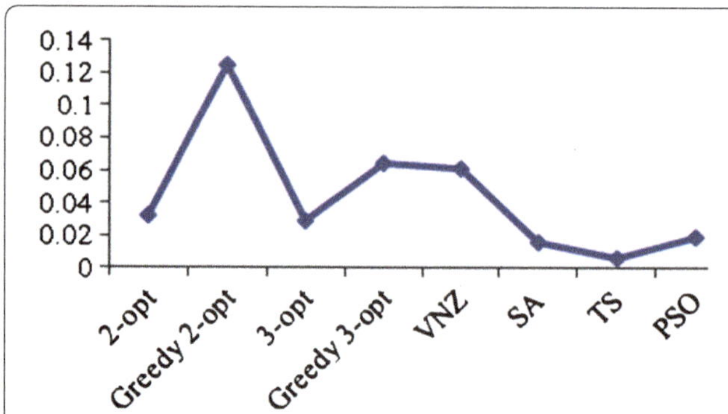

Figure 9 Average gaps for heuristics.

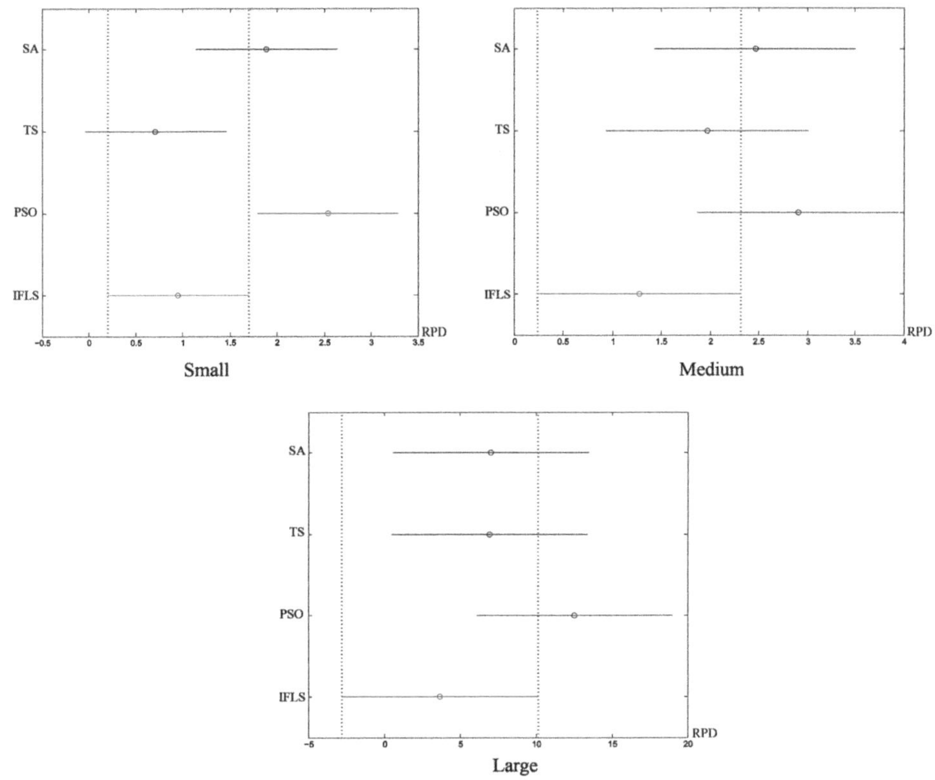

Figure 10 ANOVA results give 95% confidence interval for the difference among meta-heuristics and IFLS.

This figure shows that TS is the best in our methods. In addition, it can be concluded that meta-heuristics have better solution quality; hence, we compare meta-heuristic methods with IFLS. Furthermore, Figure 9 shows that the greedy-2opt is not a good method because the average gap for this method is about 12% in selected data sets.

The performances of our proposed meta-heuristic approaches and IFLS with local search are studied by analysis of variance (ANOVA) test.

Figure 10 summarizes the results of our investigation where the critical value for this analysis is considered as 0.05. As we can observe from Figure 10, in small-size test problems, the means of IFLS and PSO are significantly different, and no meta-heuristics have means significantly different from IFLS in the medium- and large-test problems. It is clear from this figure that TS and IFLS have the best solution quality in small-, medium-, and large-size instances, respectively. Hence, it is not necessary that the p values of these analyses be declared.

Conclusions

This research considers the heuristic and meta-heuristic solution methods for QAP, and the comparison among them has been done. In addition, a tuning method is declared. The comparison has been executed for the selected data set which was extracted from QAPLIB. The results show that 3-opt has better results than the other heuristic methods. Moreover, our meta-heuristic methods are better than the heuristic methods in solution quality. In this paper, some small, medium, and large test problems are used for comparing our meta-heuristic methods to a method from literature (i.e., IFLS with local search). ANOVA test is run for the results, and it showed that our methods are not considerably different from IFLS. TS is the most excellent method in computational time. Comparisons between the selected solution methods for more instances from the QAPLIB and comparisons of these algorithms with other meta-heuristics and hybrid algorithms can be conducted in future research.

Competing interests

The authors declare that they have no competing interests.

Authors' contributions

MB has worked on the literature and modeling; he also proposed the TS, SA, and PSO for his work. HK has performed numerical experiments and carried out sensitivity analysis. Both authors read and approved the final manuscript.

Acknowledgments

The authors would like to thank the anonymous referees for their constructive comments on earlier version of this work.

References

Bock F (1958) An algorithm for solving traveling-salesman and related network optimization. Research report associated with talk presented at the Operations Research Society of America 14th National Meeting, St. Louis
Burkard R, Rendl F (1984) A thermodynamically motivated simulation procedure for combinatorial optimization problems. European Journal of Operational Research 17:169–174
Croes GA (1958) A method for solving traveling salesman problems. Operation Research 6:791–812
Eberhart RC, Kennedy J (1995) A new optimizer using particle swarm theory. Paper presented at the Proceedings of the Sixth International Symposium on Micro Machine and Human Science, Nagoya, Japan
Fescioglu U, Kokar M (2008) Application of Self Controlling Software Approach to Reactive Tabu Search. SASO, Venezia, pp 297–305
Glover F (1989) Tabu search: Part I. ORSA J Comput 1:190–206
Glover F (1990) Tabu search: Part II. ORSA J Comput 2:4–32
Hongbo L, Ajith A (2007) An hybrid fuzzy variable neighborhood particle swarm optimization algorithm for solving quadratic assignment problems. Jucs 13 (9):1309–1331
Hongbo L, Ajith A, Jianying Z (2007) A particle swarm approach to quadratic assignment problems. Soft Comput Ind App 39:213–222
James T, Rego C, Glover F (2007) A cooperative parallel tabu search algorithm for the quadratic assignment problem. Eur J Oper Res 195:810–826
James T, Rego C, Glover F (2009) Multistart tabu search and diversification strategies for the quadratic assignment problem. Systems, Man and Cybernetics, Part A: Systems and Humans 39(3):579–596
Kennedy J, Eberhart RC (1995) Particle swarm optimization. In: Proceedings of IEEE International Conference on Neural Networks, Nov-Dec 1995, 4th edn. IEEE, Piscataway, pp 1942–1948
Kennedy J, Eberhart RC (1997) A discrete binary version of the particle swarm algorithm. In: Proceedings of International Conference on Systems, Man, and Cybernetics, October 1994, 1st edn. IEEE Computer Society Press, New York, pp 4104–4108
Koopmans T, Beckmann M (1957) Assignment problems and the location of economic activities. Econometrica 25:53–76
Loiola E, de Abreo N, Boaventura-Nett P, Hahn P, Querido T (2007) A survey for the quadratic assignment problem. Eur J Oper Res 176:657–690
Metropolis N, Rosenbluth A, Rosenbluth M, Teller A, Teller M (1953) Equation of state calculations for fast computing machines. Journal of Chemical 21: 1087–1092
Ramkumar AS, Ponnambalam SG, Jawahar N, Suresh RK (2008) Iterated fast local search algorithm for solving quadratic assignment problems. Rob Comput-Integr Manuf 24:392–401
Sahni S, Gonzalez T (1976) P-complete approximation problems. J Assoc Comput Mach 23:555–565
Shi X, Xing X, Wang Q (2004) A discrete PSO method for generalized TSP problem. In: Proceedings of International Conference on Machine Learning and Cybernetics, August 2004, 4th edn. IEEE Computer Society Press, New York, pp 2378–2383
Singh SP, Sharma RK (2008) Two-level modified simulated annealing based approach for solving facility layout problem. Int J Prod Res 46(13):3563–3582
Skorin-Kapov J (1990) Tabu search applied to the quadratic assignment problem. ORSA J Comput 2:33–45
Taillard E (1991) Robust taboo search for the quadratic assignment problem. Parallel Comput 17:443–455
Vollman TE, Nugent CE, Zartler RL (1968) A computerized model for office layout. J Ind Eng 19:321–327
Wilhelm MR, Ward TL (1987) Solving quadratic assignment problems by simulated annealing. IEEE Trans 19:107–119
Zhu W, Curry J, Marquez A (2009a) GPU-accelerated simt tabu search for the quadratic assignment problem. NAMRC, Greenville, pp 435–442
Zhu W, Curry J, Marquez A (2009b) SIMD tabu search for the quadratic assignment problem with graphics hardware acceleration. Int J Prod Res 48:1035–1047

Optimisation of shock absorber process parameters using failure mode and effect analysis and genetic algorithm

Arokiasamy Mariajayaprakash[1*], Thiyagarajan Senthilvelan[2] and Krishnapillai Ponnambal Vivekananthan[3]

Abstract

The various process parameters affecting the quality characteristics of the shock absorber during the process were identified using the Ishikawa diagram and by failure mode and effect analysis. The identified process parameters are welding process parameters (squeeze, heat control, wheel speed, and air pressure), damper sealing process parameters (load, hydraulic pressure, air pressure, and fixture height), washing process parameters (total alkalinity, temperature, pH value of rinsing water, and timing), and painting process parameters (flowability, coating thickness, pointage, and temperature). In this paper, the process parameters, namely, painting and washing process parameters, are optimized by Taguchi method. Though the defects are reasonably minimized by Taguchi method, in order to achieve zero defects during the processes, genetic algorithm technique is applied on the optimized parameters obtained by Taguchi method.

Keywords: Ishikawa diagram; FMEA; Taguchi method; Genetic algorithm

Introduction

Shock absorber is one of the major components used in automobiles. It is used to absorb vibrations when the vehicle is moving. Otherwise, those vibrations will affect the vehicle and the rider. The defects which occurred during the process affect the quality of the shock absorber. In order to minimize the defects and improve the quality, the following tools are employed. First, the cause and effect diagram or Ishikawa diagram is used for identifying the parameters that may affect the quality of the shock absorber during the process. Then, failure mode and effect analysis (FMEA) tool is applied to find out the most significant process parameters that may affect the quality of the shock absorber. Finally, Taguchi method is used to optimize the process parameters. Taguchi method is an efficient and powerful tool that can reduce the experimental trials necessary to determine the optimal conditions. It has been widely used in many different fields (Liu et al. 2010).

Tsai et al. (2008) used Taguchi's design of experiments to identify the optimal abrasive jet polishing parameters when applied to the polishing of electrical-discharged-machined SKD 61 mold steel specimens. This paper shows that the analysis of variance (ANOVA) provides an indication of the significance of the individual control factors. Mahapatra and Chaturvedi (2009) proved that Taguchi's experimental design is a simple and systematic way of analyzing a complex process with less experimental design.

In this paper, test parameters are optimized for minimum wear by Taguchi method. Ganapathy et al. (2009) optimized the operating parameters in Jatropha biodiesel engine by Taguchi method.

In this paper, it is proved that signal-to-noise (S/N) ratio is used to optimize the various input parameters of the model. This paper shows that the Taguchi approach-based thermodynamic model has improved the performance parameters slightly. Pishbina et al. (2010) investigated the effects of processing parameters, including suspension concentration, pH, and electric field by Taguchi's design of experiment (DOE) approach in electrophoretic deposition of bioglass suspension. Momeni

* Correspondence: jayaprakashrgcet@gmail.com
[1]Department of Mechanical Engineering, Rajiv Gandhi College of Engineering and Technology, Kirumabakkam, Puducherry 607402, India
Full list of author information is available at the end of the article

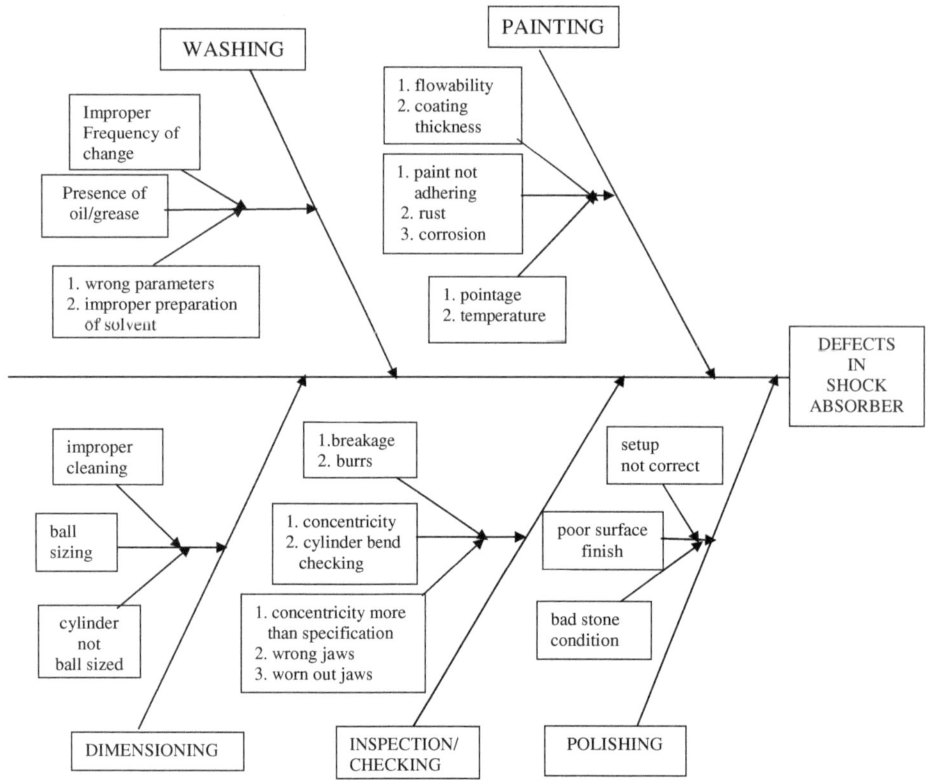

Figure 1 Cause and effect diagram.

et al. (2010) obtained the optimal condition for the degree of sensitization (DOS) measurement based on double-loop EPR technique in H_2SO_4-containing KSCN solution by Taguchi method. In this paper, it is concluded that the higher S/N ratio, the greater is the effect on the DOS measurement. Li et al. (2009) explained how Taguchi method is utilized for optimizing the parameters in thermal flow techniques for sub-35-nm contact-hole fabrication. In this article, the optimal thermal flow parameter settings are determined by signal-to-noise ratio and ANOVA analysis. Gopalsamy et al. (2009) applied Taguchi method to find the optimum process parameters for end milling while hard machining of hardened steel. Results obtained from Taguchi method closely match with ANOVA, and it is found that cutting speed is the most influencing parameter corresponding to quality characteristics.

Hossain et al. (2010) illustrated that the cause and effect diagram provides the solution to reduce the emission of CO_2. Andrejkovic et al. (2011) explained that the Ishikawa diagram quickly identifies the causes of quality problems. Tlale et al. (2008) applied the usage of the cause and effect diagram in cost drivers for manufacturing process. Liu et al. (2011) reported about the usage of FMEA tool and how it is applied to identify the failure modes. Arabian-Hoseynabadi et al. (2010) explained that

Table 1 Process parameters with their ranges and values at three levels

Parameter designation	Process parameters	Range	Level 1	Level 2	Level 3
A	Flowability (s)	15 to 25	15	20	25
B	Coating thickness (μm)	20 to 50	20	40	50
C	Pointage	6 to 12	6	8	12
D	Temperature (°C)	20 to 50	2	35	50

Table 2 L9 orthogonal array

Run	A	B	C	D
1	1	1	1	1
2	1	2	2	2
3	1	3	3	3
4	2	1	2	3
5	2	2	3	1
6	2	3	1	2
7	3	1	3	2
8	3	2	1	3
9	3	3	2	1

Table 3 Experimental L9 array

Trial no.	A Flow ability (s)	B Coating thickness (μm)	C Pointage	D Temperature (°C)
1	15	20	6	20
2	15	40	8	35
3	15	50	12	50
4	20	20	8	50
5	20	40	12	20
6	20	50	6	35
7	25	20	12	35
8	25	40	6	50
9	25	50	8	20

Table 5 Painting defects values and S/N ratios against trial numbers

Trial no.	Defects in experiment (%)				S/N ratio
	1	2	3	Average	
1	4.00	5.00	5.19	4.73	−13.55
2	2.67	2.50	1.30	1.71	−7.00
3	5.33	5.00	5.19	5.18	−14.28
4	1.33	3.75	2.60	2.56	−8.77
5	2.67	2.50	2.60	2.59	−8.26
6	2.67	2.50	2.60	2.59	−8.26
7	2.67	2.50	2.60	2.59	−8.26
8	2.67	3.75	5.19	4.31	−12.06
9	4.00	2.50	2.60	3.03	−9.85

the FMEA tool is applied in wind turbine systems. In this paper, the reliability of the wind turbine system is improved by applying FMEA tool. Boldrin et al. (2009) identified and reduced the risks of failure in ITER NB injector using the FMEA tool.

Cause and effect diagram

Cause and effect diagram is a tool for identifying the root causes of quality problems. It is also called as the Ishikawa diagram or fishbone diagram (Ilie and Ciocoiu 2010). The Ishikawa diagram is constructed, as shown in Figure 1. Using this diagram, the various process parameters that affect the quality of the shock absorber are identified (Sen and Shan 2005). The identified process parameters are listed below:

1. Washing parameters.
2. Welding parameters.
3. Painting parameters.
4. Damper sealing parameters.
5. Assembly parameters.
6. Lubrication parameter.
7. Inspection and checking parameters.
8. Dimensioning parameters.
9. Polishing parameter.
10. Testing parameters.

Failure mode and effect analysis

After constructing the Ishikawa diagram, it is important to perform failure mode and effect analysis. FMEA is a powerful tool which is used to define, identify, and eliminate known potential failures, problems, errors, and so on (Liu et al. 2011). Today, FMEA technique has been applied in many places, such as automobiles, aerospace, military, electricity, and mechanical industries. Basically, FMEA can be classified into two main types: design FMEA and process FMEA. Design FMEA deals with design activities, such as product design, machine, or tooling design. It involves the steps of breaking down the product into smaller parts such as sub-assemblies, sub-systems, or components. Process FMEA is used to solve problems due to manufacturing processes. It starts with a process flow chart that shows each of the manufacturing steps of a product. The potential failure modes and potential causes for each of the process steps are identified; then, the current controls are determined,

Table 4 Defects and the number of defects that occurred during the experiments

Defects				No. of defects		
				Trial 1	Trial 2	Trial 3
Paint not adhering	Rusting	Not drying	Not cleaning	4	4	4
Paint not adhering				1	2	1
Paint not adhering	Manual coating	Powder deposit	Over heating	4	4	4
Powder deposit	Not drying			2	2	2
Rusting	Not cleaning			2	2	2
Paint not coming	Powder deposit			2	2	2
Paint not coming	Rusting	Over heating		2	4	4
Paint not coming	Not drying			3	2	2
Total				20	22	21

Table 6 Average values of painting defects and S/N ratios at different levels

Factors	Level 1		Level 2		Level 3	
	Painting defects	S/N ratios	Painting defects	S/N ratios	Painting defects	S/N ratios
A	4.02	−11.61	2.58	−8.43	3.16	−10.06
B	3.29	−10.19	2.87	−9.10	3.60	−10.80
C	3.73	−11.29	2.58	−8.54	3.45	−10.27
D	3.45	−10.55	2.45	−7.84	3.87	−11.70

followed by the effects of failures on the manufacturing line operators and product end-users. The risks of these effects are then assessed accordingly. FMEA is carried out by a cross-functional team of experts from various departments FMEA is conventionally carried out by a team of engineers. Using their knowledge and past data, risk priority number (RPN) value is assigned for each failure component (Zhang and Chu 2011). RPN is the product of the occurrence (o), severity (s), and detection (d) of a failure. The three risk factors are evaluated using a ten-point scale. Failure modes with higher RPN values are assumed to be more important and are given higher priorities than those with lower RPN values (Wang et al. 2009). In this paper, washing and painting process parameters are having higher RPN values. Hence, these two process parameters are considered as the most significant process parameters.

Taguchi method

Taguchi method is an efficient tool for acquiring optimal process parameters. In this method, the number of experiments is reduced. In this method, two important tools are used: (1) orthogonal array and (2) signal-to-noise ratio (Gologlu and Sakarya 2008). First, experiment parameter factors and their levels are selected. In this study, four painting process parameters are used as control factors, and each parameter is designed to have three levels (Kilickap 2010). The selected process parameters, ranges, and their levels are shown in Table 1.

Selection of orthogonal array

An orthogonal array for process design is applied on the knowledge of control factors and levels. The number of experiments can be reduced using orthogonal array (Yen et al. 2011). The degrees of freedom for the orthogonal array should be greater than or at least equal to those for the process parameters. In this study, L9 orthogonal array having eight degrees of freedom is selected (Anawa and Olabi 2008). This orthogonal array has four columns and nine experiment runs, and it is shown in Table 2.

Conducting the experiment

After selecting the orthogonal array, the factors at different levels are assigned for each trial. The assigned experimental array is shown in Table 3. The experiments are conducted by single randomization technique. Each experiment is repeated three times for the same set off parameters (Ross 1988; Mohamed et al. 2008). In each trial, the defects occurring are recorded, and it is shown in Table 4. In this work, the defects that occur during the welding, washing, painting, and tamper sealing process are considered. The percentage of the defects for each repetition was calculated using the given formula, and then, the average of the defects was determined for each trial condition, and it was shown in Table 5.

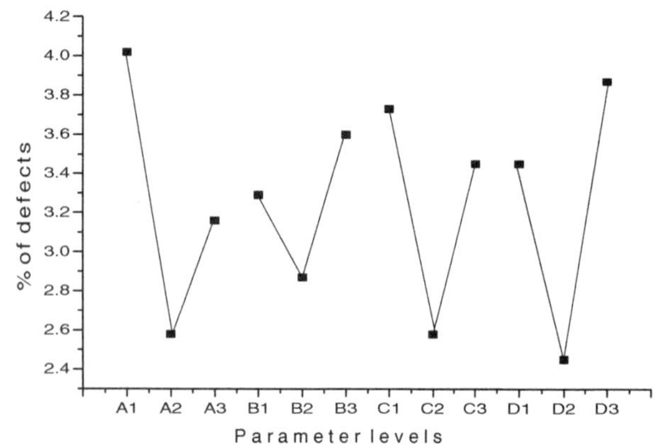

Figure 2 Average values of painting process defects for each parameter at different levels.

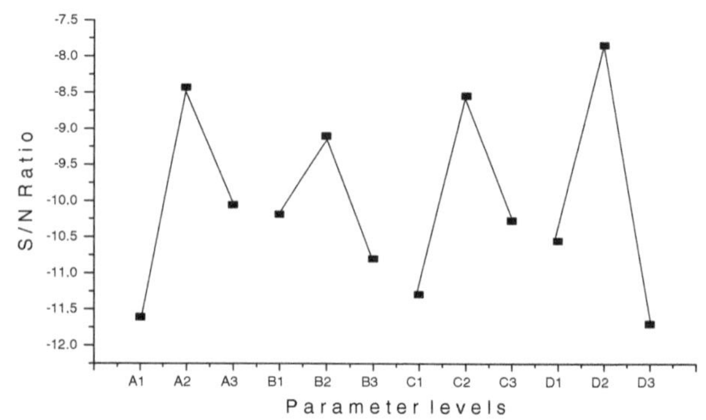

Figure 3 Average values of S/N ratio for each parameter at different levels.

$$\text{Percentage of defects} = \frac{\text{No. of defects occuring due to painting process}}{\text{Total no. of defects in the process (welding, washing, painting, and Damper sealing)}}$$

S/N ratio

Taguchi technique utilizes the signal-to-noise ratio approach to measure the quality characteristic deviating from the desired value. S/N ratio is used as an objective function for optimizing parameters. (Prihandana et al. 2009). Control factors are easily adjustable, and it is set by the manufacturer. These factors are most important in determining the quality characteristics. Noise factors are difficult, impossible, or expensive to control (weather, temperature, humidity, etc.). The S/N ratio is the ratio of mean (signal) to the standard deviation (noise). There are several S/N ratios available depending on the type of characteristics (Mahapatra et al. 2008; Moshat et al. 2010).

(1) smaller is better

$$\eta = -10 \log 1/n \sum_{i=1}^{n} y_{i^2},$$

(2) nominal is the best

$$\eta = 10 \log 1/n \sum_{i=1}^{n} \mu^2/\sigma^2,$$

(3) higher is better.

$$\eta = -10 \log 1/n \sum_{i=1}^{n} 1/y_{i^2},$$

where, η = S/N ratio, y_i = value of quality characteristic at ith setting, μ = mean, n = total number of trial runs at ith setting, and σ = standard deviation.

In this study, in order to minimize the defects, smaller is better S/N ratio is chosen. Smaller is better S/N ratios are computed, and the values are recorded in Table 5. The main objective of shock absorber process is to minimize the defects by determining optimal level of each factor. Since − log is a monotone decreasing function, it implies that we should maximize the S/N ratio (Shiou and Hsu 2008). The average values of the painting defects and S/N ratios for each parameter at different levels are calculated and are recorded in Table 6.

The values given in Table 6 are plotted in Figures 2 and 3. Figures 2 and 3 show the average values of casting defects and S/N ratios for each parameter at different levels. From Figures 2 and 3, it is clear that the casting defects are minimum at the parameter levels A2, B2, C2, and D2, and the S/N ratios are maximum at the same levels of the parameters A2, B2, C2, and D2. Since a higher S/N ratio means a better quality characteristic (minimum defects), the optimal combination of control factor levels is therefore determined as A2 B2 C2 D2.

Analysis of variance

The purpose of ANOVA is to investigate which process parameters significantly affect the quality characteristic. The total variation may be decomposed into many components. In this paper, the total variation present in the process is decomposed to the following components:

1. variation due to factor A.
2. variation due to factor B.
3. variation due to factor C.
4. variation due to factor D.
5. variation due to error.

The total variation is calculated using the values given in Table 4.

$$\text{Total variation } SS_T = SS_A + SS_B + SS_C + SS_D + SS_E.$$

Variation due to error

$$SS_e = SS_T - (SS_A + SS_B + SS_C + SS_D).$$

Table 7 ANOVA for painting defects and S/N ratios

Source	Sum of squares		Degrees of freedom		Variance		*F* ratio	
	Painting defects	S/N ratio	Painting defects	S/N ratio	Painting defects	S/N ratio	Painting defects	S/N ratio
A	9.44	14.97	2	2	4.72	7.49	8.90	24.87
B	2.40	4.20	2	2	1.20	2.10	2.26	6.98
C	6.44	11.39	2	2	3.22	5.70	6.07	18.92
D	9.63	23.41	2	2	4.81	11.70	9.07	38.89
Error	9.56	0.60	18	2	0.53	0.30	1.00	1.00
Total	37.47	54.57	26	8	1.44	6.82		

Table 8 ANOVA for painting defects, including percent contribution

Source	Sum of squares	Degrees of freedom	Variance	*F* ratio	Expected SS'	Percent contribution (*P*)
A	9.44	2	4.72	8.90	6.94	18.52
B	2.40	2	1.20	Pooled (2.6)		
C	6.44	2	3.22	6.07	6.44	17.19
D	9.63	2	4.81	9.07	9.63	25.69
Error (pooled)	9.56	18	0.53	1.00	14.46	38.60
Total	37.47	26	1.44		37.47	100.00

Total degrees of freedom

$$\begin{aligned} v_T &= (v_A + v_B + v_C + v_D + v_e) \\ v_e &= v_T - (v_A + v_B + v_C + v_D) \\ &= 26 - (2 + 2 + 2 + 2) \\ &= 18\ . \end{aligned}$$

The results of ANOVA are shown in Table 7. In Table 7, it is clear that the parameters A, C, and D significantly affect both mean and variation in the painting defects.

Once the experiment has been conducted, the ANOVA is carried out using the results of the experiments. The significant factors and/or their interactions were identified, graphs were plotted for various trial conditions, and the parameters which significantly influence the mean and variation in the shock absorber defects were determined. However, the results obtained so far are not sufficient enough to find the optimum parameters in order to minimize the requirement of some more information to conclude with an optimum parameter set. These sets of information are obtained using the interpretation methods available in the literature (Syrcos 2002).

Percent contribution

Percent contribution is the function of the sum of squares of each significant item. Percent contribution to the total sum of square can be used to evaluate the importance of a change in the process parameter on these quality characteristics (Lakshminarayan and Balasubramanian 2008; Gopalsamy et al. 2009). It is calculated using the formulae given below:

$$\begin{aligned} \text{Percent contribution } (P) &= (SS'_A/SS_T) \times 100 \\ SS'_A &= SS_A - (v_e)\,(v_A) \\ V_A &= V_A' + V_{\text{error}}, \end{aligned}$$

where V_A' is the expected amount of variation due solely to factor A given below:

$$V_A = SS_A/v_A,$$

$$V_A' = SS'_A/v_A.$$

Estimating the mean

From Table 8, it is clear that factor B has the least effect on the quality characteristic. In order to prevent over estimation, factor B is not considered, and the estimation of mean for painting defects is calculated by the following equation (Li et al. 2009):

$$\mu = T + (A2 - T) + (C2 - T) + (D2 - T),$$

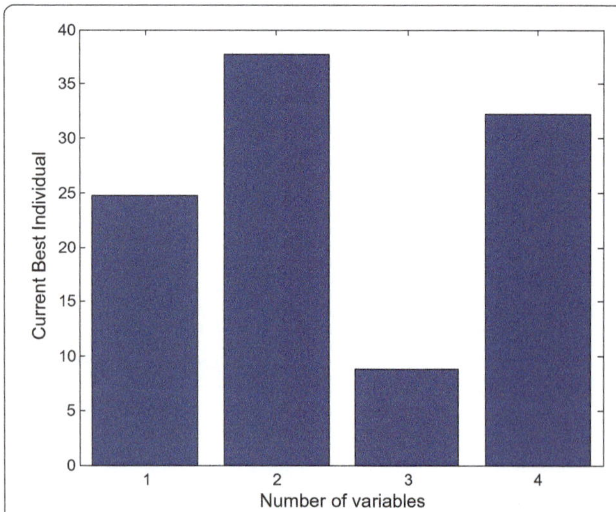

Figure 4 Evaluation of generations for painting process optimization.

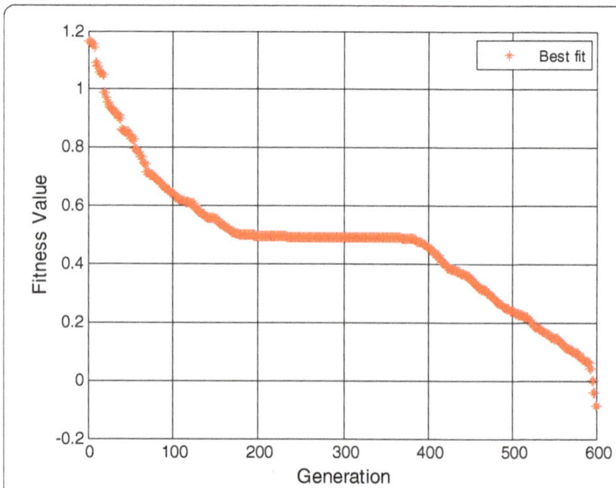

Figure 5 Best individual values obtained by GA for painting process optimisation.

Table 9 Optimisation results obtained by Taguchi method and GA

	Flow ability (s)	Coating thickness (μm)	Pointage	Temperature (°C)	Percentage of defects
Taguchi method	20	40	8	35	0.83
Genetic algorithm	24	40	8	31	0.001

where T is the average values of painting defects at different levels.

The mean for a selected trial condition for parameters at A2, B2, C2, and D2 is 1.11.

Confidence interval around the estimated mean

An important step in Taguchi's optimization technique is to conduct confirmation experiments for validating the predicted values. Thus, a 95% confidence interval (CI) for the predicted mean of optimum quality characteristic on a confirmation test is estimated using the following two equations (Wu and Gu 2010):

$$\text{CI3} = [F(\alpha, 1, v_e) V_e (1/\eta\,\text{eff} + 1/r)]^{1/2} \quad \textit{and}$$

$$\eta\ \text{eff} = N/(1 + \text{total DOF associated in the estimate of mean}),$$

where α is the level of risk, V_e is the error variance, ν_e is the degrees of freedom for the error, η eff is the effective number of replications, and r is number of test trials.

Using the values in Table 5, the CI was calculated as follows:

$$\begin{aligned}\eta\ \text{eff} &= N/1 + (\text{total DOF associated in the estimate of mean})\\ &= 27/(1+2)\\ &= 9,\end{aligned}$$

$$\begin{aligned}\alpha &= 1\text{–confidence limits } (95\%)\\ &= 0.05,\end{aligned}$$

$$\begin{aligned}F\ \text{ratio} &= (1, 0.05, 18)\\ &= 4.41(\text{tabulated}),\end{aligned}$$

Confidence interval CI3 = ±1.02.

The 95% confidence interval of the predicted optimum of the shock absorber defect is 0.09 < 1.11 < 2.13.

Table 11 Washing defect values and S/N ratio against trial numbers

Trial no.	Defects in experiment (%)				
	1	2	3	Average	S/N ratio
1	4.00	2.50	3.90	3.47	−10.96
2	1.33	2.50	1.30	1.71	−5.10
3	2.67	2.50	2.60	2.59	−8.27
4	1.33	2.50	1.30	1.71	−5.10
5	1.33	1.25	1.30	1.29	−2.24
6	1.33	1.25	1.30	1.29	−2.24
7	1.33	1.25	1.30	1.29	−2.24
8	1.33	1.25	2.60	1.73	−5.27
9	1.33	3.75	2.60	2.56	−8.77

Confirmation experiments

Three confirmation experiments were conducted at the optimum setting of the process parameters. The flowability was set at the second level (A2), coating thickness at the second level (B2), wheel speed at the second level (C2), and air pressure was kept at the second level (D2). The average of the respondent defects in each experiment is found to be 0.83%; the result was within the CI of the protected optimum of the casting defects. The confirming experiments results gave 0.83% < 2.13% (maximum of CI). Therefore, the selected parameters as well as their appropriate levels are significant enough to obtain the desired result.

Optimisation by genetic algorithm

Since the conventional methods of optimization do not work well, non-traditional optimization techniques such as genetic algorithm (GA), simulated annealing, scatter search algorithm, and ant colony algorithm have been widely applied to obtain better solution in the manufacturing industries. In this research work, genetic algorithm method has been applied on the process parameters which are earlier optimized by Taguchi method in order to obtain better solution. GA is a computerized search and optimization algorithm based on the mechanics of natural genetics and natural selection (Tewari et al. 2012). Genetic algorithm is mainly composed of three operators: reproduction, crossover, and mutation. In GA, the first step is to generate initial population of individuals randomly. The individuals in the genetic space are called chromosome. A collection of genes is called as chromosome, and the

Table 10 Process parameters with their ranges and values at three levels

Parameter designation	Process parameters	Range	Level 1	Level 2	Level 3
A	Total alkalinity (g/l)	40 to 80	40	70	80
B	Temperature (°C)	45 to 75	45	60	75
C	Rinsing water (pH)	5.5 to 8.5	5.5	6.5	8.5
D	Timing (min)	0 to 25	5	15	25

Table 12 Average values of washing process defects and S/N ratio at different levels

Factors	Level 1		Level 2		Level 3	
	Washing defects	S/N ratios	Washing defects	S/N ratios	Washing defects	S/N ratios
A	2.59	−8.11	1.43	−3.19	1.86	−5.43
B	2.16	−6.10	1.58	−4.20	2.15	−6.43
C	2.16	−6.16	2.00	−6.32	1.73	−4.25
D	2.44	−7.32	1.43	−3.19	2.00	−6.21

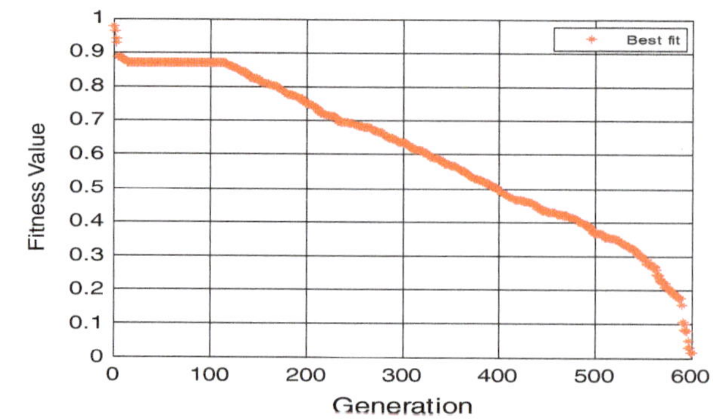

Figure 6 Evaluation of generations for washing process optimization.

chromosome potential is called its fitness function, which is evaluated by the objective function. In the selection process, low fitness chromosomes are eliminated using high-fitness chromosomes (Hafshejani et al. 2012). However, selection alone does not produce any new individuals into the population. Hence, cross over and mutation operations are involved. Crossover is the process for exchanging genes between two chromosomes using the one-point crossover, two-point crossover, or homologue crossover. In mutation, the genes may occasionally be altered, i.e., in binary code genes changing genes code from 0 to 1 or vice versa. The new population generated undergoes the further selection, crossover, and mutation till the termination criterion is not satisfied.

In this research work, through the application of MINITAB and based on Taguchi's orthogonal array, a mathematical model has been developed. The model has been used as objective function in the genetic algorithm. The objective function and constraints of the painting process are defined as follows:

$$Y = 3.72 – 0.095A + 0.0057B – 0.2123C + 0.0018D.$$

Minimize: $Y\ (A, B, C, D)$.
Constraints:

- Bounds on flowability (s): $15 \le A \le 25$.
- Bounds on coating thickness (μm): $35 \le B \le 45$.
- Bounds on pointage: $6 \le C \le 10$.
- Bounds on temperature (°C): $30 \le D \le 40$.

The problem of genetic algorithm has been solved using MATLAB. The parameters used in GA are population size - 100, cross over probability - 0.9, mutation probability - 0.1 and the number of generations - 600. The evaluation of generations and the best values obtained by GA are shown in Figures 4 and 5. Confirmation experiments have been conducted using the optimized parameters. It has been observed that the percentage of defects during the painting process is better than Taguchi method.

Results and discussion

The shock absorber experiments have been conducted, and subsequently, ANOVA is carried out using the results obtained from the experiments. Nevertheless, interpretation methods are used to obtain the percent contribution of each parameter and optimum levels of each parameter, and GA technique is applied on the parameters which have been optimized by Taguchi method:

1. The percent contribution of each parameter to the variation of painting defects and optimum parameter (under economic condition) is shown in Table 8.
2. The optimum levels of various parameters for minimum painting defects of shock absorber obtained by Taguchi method and genetic algorithm are shown in Table 9.
3. The predicted range of optimum painting defects is $0.09 < 1.11 < 2.13$.

Table 13 S/N ratios for washing process defects, including percent contribution

Source	Sum of squares	Degrees of freedom	Variance	*F* ratio	Expected SS'	Percent contribution (*P*)
A	36.25	2.00	18.13	112.84	35.93	33.20
B	8.54	2.00	4.27	26.58	8.22	7.59
C	7.85					
D	27.30	2.00	13.65	84.97	26.98	24.93
Error (pooled)	0.32	2.00	0.16	1.00	37.10	34.28
Total	80.26	8.00	36.21	225.39	108.23	100.00

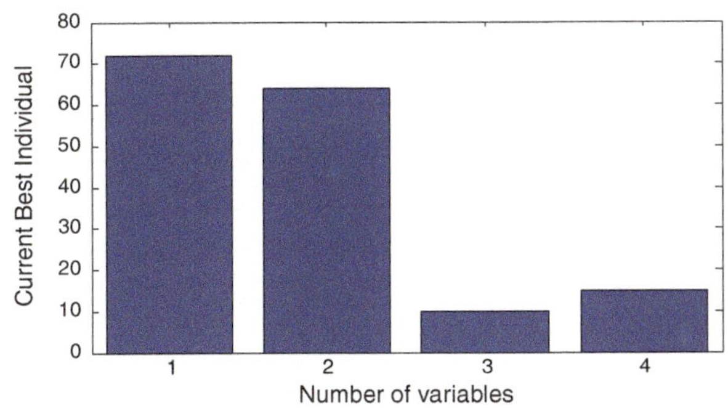

Figure 7 Best individual values obtained by GA for washing process optimisation.

In the same manner, calculations are carried out, and optimum values are found for washing process (Tables 10, 11, 12, and 13).

The mean for a selected trial condition for parameters at A2, B2, C3, and D2 is 0.90.

The 95% confidence interval of the predicted optimum of the shock absorber defect is $0.13 < 0.90 < 1.67$

Furthermore, to optimize the results of washing process, GA technique has been applied using the objective function and constraints as given below:

$$Y = 4.73 - 0.034A - 0.012B - 0.110C - 0.0248D.$$

Minimize: $Y\,(A, B, C, D)$.
Constraints:

- Bounds on total alkalinity (g/l): $65 \le A \le 75$.
- Bounds on temperature (°C): $55 \le B \le 65$.
- Bounds on pH value of rinsing water: $6 \le C \le 10$.
- Bounds on timing (s): $5 \le D \le 15$.

The evaluation of generations and the best values obtained by GA for washing process are shown in Figures 6 and 7. The optimized parameters obtained by Taguchi method and GA are shown in Table 14.

From the abovementioned analysis, it is observed that the percentage of defects drastically reduced after applying genetic algorithm technique in both painting and washing processes.

Table 14 Optimisation results obtained by Taguchi method and GA

	Total alkalinity (g/l)	Temperature (°C)	Rinsing water (pH)	Timing (s)	Percentage of defects
Taguchi method	70	60	7.5	10	1
Genetic algorithm	72	64	10	13	0.004

Conclusions

While the Taguchi method is an effective technique to achieve the optimum process parameters of shock absorber, the genetic algorithm technique, however, gives better solutions than Taguchi method to achieve the optimum process parameters of shock absorber. From the present study, the following results have been arrived.

1. Ishikawa diagram or cause and effect diagram is very effective to sort out all the possible causes affecting the quality of shock absorber.
2. Conventional FMEA has been applied to identify the most significant parameters (painting and washing processes) affecting the quality of the shock absorber.
3. It has been observed that the parameters pointage (painting) and timing (washing) significantly affect the quality of the shock absorber, which can be confirmed with the help of percent contribution.
4. The optimum range of the shock absorber defects has been predicted by the 95% confidence interval. The optimum range of painting process is $0.09 < 1.02 < 2.13$ and washing process is $0.13 < 0.90 < 1.67$.
5. The optimum levels of flowability, coating thickness, pointage, and temperature for painting process and total alkalinity, temperature, PH value of rinsing water, and timing for washing process were estimated.
6. The percentage of painting defects and washing defects obtained by Taguchi method are 0.83 and 1, respectively. After applying the genetic algorithm technique, the percentage of painting defects and washing defects had been drastically reduced to 0.001 and 0.004, respectively.

Abbreviations

α: Level of risk; μ: Mean; η eff: Effective number of replications; ANOVA: Analysis of variance; CI: Confidence interval; DOE: Design of experiments; $F(\alpha, 1, v_e)$: F ratio required at the level of risk; k_A: Number of levels for factor A; M: Overall mean percentage defects; n: Total number of experiments; N: Number of trials; P: Percent contribution; r: Number of tests in a trial; SS: Sum of square; T: Average values of defects at different levels; v: Degrees of freedom; v_A: Degrees of freedom for factor A; v_e: Degrees of freedom for the error; v_{LN}: Total degrees of freedom of the available orthogonal array; $v_{required}$: Total degrees of freedom required; V': Expected amount of variation; V_e: Error variance; V_{factor}: Variance of factor; y_i: Response value of observation in the ith test.

Competing interests

The authors declare that they have no competing interests.

Authors' contributions

AMJP carried out the literature study reviews, design, and experimental work. TSV worked on the computations. KPV helped to perform the experimental work. All authors read and approved the final manuscript.

Acknowledgments
The authors would like to thank the editor and anonymous reviewers for their valuable and constructive comments, which have led to a significant improvement in the manuscript.

Author details
[1]Department of Mechanical Engineering, Rajiv Gandhi College of Engineering and Technology, Kirumabakkam, Puducherry 607402, India. [2]Department of Mechanical Engineering, Pondicherry Engineering College, Pillaichavady, Puducherry 605014, India. [3]TENNECO Automotive India private Limited, Mettupalayam, Puducherry 605009, India.

References

Anawa EM, Olabi AG (2008) Using Taguchi method to optimize welding pool of dissimilar laser welded components. Optics & Laser Technology 40(2):379–388
Andrejkovic M, Hajduova L, Mixtaj E, Weiss R (2011) Improvement of processes in metallurgy company. Metalurgija 50(2):141–144
Arabian-Hoseynabadi H, Oraee H, Tavner PJ (2010) Failure modes and effects analysis (FMEA) for wind turbines. Elect Power and Energy Syst 32:817–824
Boldrin M, De Lorenzi A, Fiorentin A, Marcuzzi D, Peruzzo S, Pomaro N, Rigato W, Serianni G (2009) Potential failure mode and effects analysis for the ITER NB injector. Fusion Eng and Design 84:466–469
Ganapathy T, Murugesan K, Gakkhar RP (2009) Performance optimization of Jatropha bio diesel engine model using Taguchi approach. Appl Energy 86(11):2476–2486
Gologlu C, Sakarya N (2008) The effects of cutter path strategies on surface roughness of pocket milling of 1.2738 steel based on Taguchi method. J Mater Process Technol 206(1–3):7–15
Gopalsamy BM, Mondal B, Ghosh S (2009) Taguchi method and ANOVA: an approach for process parameters optimization of hard machining while machining hardened steel. J Sci Ind Res 68(8):686–695
Hafshejani KF, Valmohammadi C, Khakpoor A (2012) Retracted: using genetic algorithm approach to solve a multi-product EPQ model with defective items, rework, and constrained space. J Ind Eng Int 8(27):1–7
Hossain KA, Khan F, Hawboldt K (2010) Susdesign - an approach for a sustainable process system design and its application to a thermal power plant. Appl Thermal Energy 30(14–15):1896–1913
Ilie G, Ciocoiu CN (2010) Application of fishbone diagram to determine the risk of an event with multiple causes. Manag Res Pract 2(1):1–20
Kilickap E (2010) Optimization of cutting parameters on delamination based on Taguchi method during drilling of GFRP composite. Expert Syst Appl 37(8):6116–6122
Lakshminarayanan AK, Balasubramanian V (2008) Process parameters optimisation for friction stir welding of RDE- 40 aluminium alloy using Taguchi technique. Trans Non-ferrous Metals Soc China 18(3):548–554
Li TS, Chen SH, Chen HL (2009) Thermal-flow techniques for sub-35 nm contact-hole fabrication using Taguchi method in electron-beam lithography. Microelectron Eng 86:2170–2175
Liu Y-T, Changa W-C, Yamagata Y (2010) A study on optimal compensation cutting for an aspheric surface using the Taguchi method. CIRP J Manuf Sci Technol 3(1):40–48
Liu H-C, Liu L, Bian Q-H, Lin Q-L, Dong N, Peng Cheng X (2011) Failure mode and effects analysis using fuzzy evidential reasoning approach and grey theory. Expert Syst Appl 3(8):4403–4415
Mahapatra SS, Chaturvedi V (2009) Modelling and analysis of abrasive wear performance of composites using Taguchi approach. Int J Eng Sci Technol 1(1):123–135
Mahapatra SS, Patnaik A, Satapathy A (2008) Taguchi method applied to parametric appraisal of erosion behavior of GF-reinforced polyester composites. Wear 265(1–2):214–222
Mohamed H, Hisyam Lee M, Sarahintu M, Salleh S, Sanugi B (2008) Identifying factors affecting data delivery performance in mobile ad-hoc network routing protocol using a systematic approach. Matematika 24(1):43–51
Momeni M, Moayed MH, Davoodi A (2010) Tuning DOS measuring parameters based on double-loop EPR in H2SO4 containing KSCN by Taguchi method. Corros Sci 52(8):2653–2660
Moshat S, Datta S, Bandyopadhyay A, Pal P (2010) Optimization of CNC end milling process parameters using PCA-based Taguchi method. Int J Eng Sci Technol 2(1):92–102
Pishbina F, Simchi A, Ryana MP, Boccaccini AR (2010) A study of the electrophoretic deposition of Bioglass® suspensions using the Taguchi experimental design approach. J Eur Ceramic Soc 30(14):2963–2970
Prihandana GS, Mahardika M, Hamdi M, Wong YS, Kimiyuki M (2009) Effect of micro-powder suspension and ultrasonic vibration of dielectric fluid in micro-EDM Processes -Taguchi approach. Int J Mach Tools Manuf 49(12–13):1035–1041
Ross PJ (1988) Taguchi techniques for quality engineering. McGraw Hill, New York
Sen M, Shan HS (2005) Analysis of hole quality characteristics in the electro jet drilling process. Int J Mach Tools Manuf 45(15):1706–1716
Shiou FJ, Hsu CC (2008) Surface finishing of hardened and tempered stainless tool steel using sequential ball grinding, ball burnishing and ball polishing processes on a machining centre. J Mat Process Tech 205(1–3):249–258
Syrcos GP (2002) Die casting process optimization using Taguchi methods. J Mater Process Technol 135:68–74
Tewari PC, Khanduja R, Gupta M (2012) Performance enhancement for crystallization unit of a sugar plant using genetic algorithm technique. J Ind Eng Int 8(1):1–7
Tlale NS, Steyn JL, Govender S (2008) Technical cost modelling for a novel semi-solid metal (SSM) casting processes for automotive component manufacturing. In: International Association for Management of Technology. Nice, pp 1–20. 17–19 September 2008
Tsai FC, Yan BH, Kuan CY, Huang FY (2008) A Taguchi and experimental investigation into the optimal processing conditions for the abrasive jet polishing of SKD61 mold steel. Int J Mach Tools Manuf 48(7–8):932–945
Wang Y-M, Chin K-S, Poon GKK, Yang J-B (2009) Risk evaluation in failure mode and effects analysis using fuzzy weighted geometric mean. Expert Syst Appl 36(2):1195–1207
Wu H-W, Gu HW (2010) Analysis of operating parameters considering flow orientation for the performance of a proton exchange membrane fuel cell using the Taguchi method. J Power Sources 95(11):3621–3630
Yen Y-T, Fang T-H, Lin Y-C (2011) Optimization of screen-printing parameters of SN9000 ink for pinholes using Taguchi method in chip on film packaging. Robotics Comput Integrated Manuf 27(3):531–537
Zhang Z, Chu X (2011) Risk prioritization in failure mode and effects analysis under uncertainty. Expert Syst Appl 38(1):206–214

An electromagnetism-like metaheuristic for open-shop problems with no buffer

Bahman Naderi[1*], Esmaeil Najafi[2] and Mehdi Yazdani[3]

Abstract

This paper considers open-shop scheduling with no intermediate buffer to minimize total tardiness. This problem occurs in many production settings, in the plastic molding, chemical, and food processing industries. The paper mathematically formulates the problem by a mixed integer linear program. The problem can be optimally solved by the model. The paper also develops a novel metaheuristic based on an electromagnetism algorithm to solve the large-sized problems. The paper conducts two computational experiments. The first includes small-sized instances by which the mathematical model and general performance of the proposed metaheuristic are evaluated. The second evaluates the metaheuristic for its performance to solve some large-sized instances. The results show that the model and algorithm are effective to deal with the problem.

Keywords: Scheduling, Open shop with no buffer, Mixed integer linear programming, Electromagnetism algorithm

Introduction

Open-shop scheduling (OSS) can be defined by a set of n jobs that need to be processed by a set of m machines. Contrary to other shop scheduling, there are no predetermined processing routes for the jobs, which means that jobs could visit machines by any order. In other words, in open shops, there are two decisions to make: the determination of the processing routes of the jobs as well as job sequence on each machine (Pinedo 2008). The following assumptions are usually characterized to open shops. All the jobs are independent and available for their process at time 0. All machines are continuously available. Each machine can at most process one job at a time. Each job can be processed by at most one machine at a time. The process of a job on a machine cannot be interrupted. There are infinite buffers between all machines. There is no transportation time between machines. The objective function is the minimization of the total tardiness. The tardiness of each job is the amount of time that job is completed after its due date.

Bräsel et al. (2008) investigated OSS under minimization of mean flow time and proposed three types of heuristics based on matching heuristics, priority-dispatching rules, as well as insertion and appending algorithms combined with beam search. They also presented a new encoding scheme named rank matrices. Andresen et al. (2008) solved OSS with two metaheuristics, simulated annealing and genetic algorithms, with some operators. Like Bräsel et al. (2008), they used the rank matrix to encode a solution. Low and Yeh (2009) addressed OSS with some practical assumptions such as setup and removal times. To solve the problem, they presented a hybrid genetic algorithm.

The literature of scheduling is filled with different applied industrial assumptions to better represent the real nature of scheduling environments. One of the most prevailing and extremely favored assumptions by many researchers in real scheduling configurations is that there is no buffer between machines. In many industrial settings, due to characteristics of the jobs or the processing technology, the operations of a job must be performed without any interruption between machines. Applications of these scheduling problems can be found in hot-metal-rolling industries, where the heated metal has to undergo a set of operations at continuously high temperatures before it is cooled in order to prevent defects. Similarly, in the plastic molding and silverware production industries, a set of operations must be performed to immediately follow one another to prevent degradation. Other examples include chemical and pharmaceutical industries, food processing industries,

* Correspondence: Bahman_naderi62@yahoo.com
[1]Department of Industrial Engineering, Faculty of Engineering, University of Kharazmi, Karaj, Iran
Full list of author information is available at the end of the article

and advanced manufacturing environments. Hall and Sriskandarajah (1996) explained some other applications of no-buffer scheduling problems. Despite their practical applications, these problems are given far less attention than the other scheduling problems.

The two-machine open shop with no buffer (OSNB) was studied by Sidney and Sriskandarajah (1999). They introduced an approximation heuristic with a worst case performance ratio of 3:2 to solve the problem. Yao and Soewandi (2000) proposed four heuristics. One of them is a flowshop-based heuristic based on the Gilmore-Gomory algorithm. Two other heuristics are based on the 'matching' concept to minimize 'potential' idle time on either machine. The other one is a random search algorithm utilizing some features of other 'soflcomputing' metaheuristics. The experimental evaluation shows that the random search algorithm outperforms the others.

Yao and Lai (2002) proposed a genetic algorithm for the two-machine OSNB. The two-machine OSNB was also studied by Liaw et al. (2005). They developed a branch-and-bound with some dominance rules as well as a two-phase algorithm. OSNB with movable dedicated machines to minimize total occupation time of all the machines was considered by Lin et al. (2008). They introduced a mixed integer nonlinear program to model the problem. As far as we reviewed, almost none of the existing papers consider classical multi-machine OSNB. There is no attempt to linearly model the problem. There is only one metaheuristic (two-phase algorithm of Lin et al. (2008)) for a relevant problem.

The two-machine OSNB is strongly nondeterministic polynomial time-hard (NP-hard) (Sahni and Cho 1979). Therefore, the general multi-machine OSNB is NP-hard. The presentation of mathematical programs and heuristics is commonly used to solve scheduling problems (Stafford et al. 2005). We propose a mixed integer linear programming model to formulate the OSNB. We also carry out an experiment to analyze and compare the performance of the proposed model. Besides the model, the paper solves the large-sized instances by an effective metaheuristic based on an electromagnetism-like algorithm. Model's efficiency and metaheuristics' capability to solve the problem studied here are investigated on two computational evaluations including small- and large-sized instances.

The rest of the paper is organized as follows. The 'Problem formulation' section develops the mixed integer linear program. The 'Proposed electromagnetism algorithm' section introduces the proposed metaheuristic. The 'Computational evaluation' section describes the experimental design to evaluate the proposed method including the mathematical model and algorithm. Finally, the 'Conclusions and future studies' section gives some interesting conclusions and future studies.

Problem formulation

Even though mixed integer linear programming (MILP) models might not be an efficient solution method for all problem sizes, they are a natural way to attack scheduling problems (Stafford et al. 2005). This section presents a model to formulate OSNB problems. We analyze the model by the number of binary and continuous variables as well as the number of constraints required in formulating the problem. The following notations are used in the model:

Parameters:

- n The number of jobs.
- m The number of machines.
- $O_{j,i}$ The operation of job j on machine i.
- $p_{j,i}$ The processing time of $O_{j,i}$.
- d_j The due date of job j.
- M A large positive number.

Binary variables (BVs) used in this model show the relative sequence of different operations of a job as well as the relative order of the jobs on each machine. Due to the identity of these BVs, the model is called the sequence-based model. Note that for every machine we introduce, a dummy job 0 which precedes the first job on that machine is present and that for every job, a dummy machine 0 that precedes the first operation of that job is also present. It is necessary to indicate that the model requires the big M. The following notations are established:

Indices:

- ○ k Job index $\{0, 1, 2,\ldots, n\}$.
- ○ j Job index $\{1, 2,\ldots, n\}$.
- ○ l Machine index $\{0, 1, 2,\ldots, m\}$.
- ○ i Machine index $\{1, 2,\ldots, m\}$.

Variables:

- ○ $Y_{l,j,i}$ Binary variable that takes value 1 if $O_{j,i}$ is processed immediately after $O_{j,l}$, and 0 otherwise. $l \neq i$
- ○ $X_{k,j,i}$ Binary variable that takes value 1 if $O_{j,i}$ is processed immediately after $O_{k,i}$, and 0 otherwise. $k \neq j$
- ○ T_j Tardiness of job j.
- ○ $S_{j,i}$ The starting time of $O_{j,i}$.
- ○ B_j The starting time of job j.

The model formulates OSNB as follows:

$$\text{Minimize} \sum_{j=1}^{n} T_j,$$

which is subject to the following:

$$S_{j,i} - B_j \leq \sum_{l=1, l\neq i}^{m} p_{j,l} \qquad \forall_{j,i} \qquad (1)$$

$$S_{j,i} - B_j \geq p_{j,i} \qquad \forall_{j,i} \qquad (2)$$

$$S_{j,i} - S_{j,l} - p_{j,l} \geq M\left(Y_{l,j,i} - 1\right) \qquad \forall_{j,i,l\neq i} \qquad (3)$$

$$S_{j,i} - S_{k,i} - p_{k,i} \geq M\left(X_{k,j,i} - 1\right) \qquad \forall_{j,i,k \neq j} \qquad (4)$$

$$T_j - S_{j,i} \geq p_{j,i} - d_j \qquad \forall_{j,i} \qquad (5)$$

$$\sum_{k=0,k \neq j}^{n} X_{k,j,i} = 1 \qquad \forall_{j,i} \qquad (6)$$

$$\sum_{l=0,l \neq i}^{m} Y_{l,j,i} = 1 \qquad \forall_{j,i} \qquad (7)$$

$$\sum_{j=1,j \neq k}^{n} X_{k,j,i} \leq 1 \qquad \forall_{i,k \varepsilon \{1,2,..,n\}} \qquad (8)$$

$$\sum_{i=1,i \neq l}^{m} Y_{l,j,i} \leq 1 \qquad \forall_{j,l \varepsilon \{1,2,..,m\}} \qquad (9)$$

$$\sum_{j=1}^{n} X_{0,j,i} = 1 \qquad \forall_{i} \qquad (10)$$

$$\sum_{i=1}^{m} Y_{0,j,i} = 1 \qquad \forall_{j} \qquad (11)$$

$$X_{k,j,i}, Y_{l,j,i} \in \{0,1\} \qquad (12)$$

$$S_{j,i}, B_j, T_j \geq 0 \qquad (13)$$

Constraint sets (1) and (2) are to assure that no-buffer restrictions are held. Constraint set (3) ensures that $O_{j,i}$ cannot start before $O_{j,i}$ completes if $O_{j,i}$ precedes $O_{j,i}$. Constraint set (4) specifies that if $O_{j,i}$ is processed immediately after $O_{k,i}$, it cannot begin before $O_{k,i}$ completes. Constraint set (5) calculates tardiness of jobs. Constraint sets (6) and (7) state that every operation is scheduled once. Constraint sets (8) and (9) ensure that every operation must have at most one succeeding operation in the processing route of the jobs and in the job order of every machine. Constraint sets (10) and (11) enforce that dummy job 0 and machine 0 must have exactly one successor. Ultimately, Constraint sets (12) and (13) define the decision variables.

The model is analyzed with respect to the required numbers of binary variables, continuous variables, and constraints for a problem with n jobs and m machines. The model requires $nm(n + m)$ binary variables, i.e., $O(nm(n + m))$, $2n + nm$ continuous variables, i.e., $O(nm)$, and $nm\left(4 + \frac{3n}{2} + \frac{3m}{2}\right) + n + m$ constraints, i.e., $O(nm(n + m))$.

The proposed electromagnetism algorithm

Electromagnetism algorithm (or EA) is a solution method for optimization problems with bounded variables. It is a population-based metaheuristic firstly proposed by Bïrbïl and Fang (2003). EA is inspired from the attraction-repulsion mechanism of the electromagnetism theory. Each candidate solution is regarded as a particle whose charge is related to its objective function value. The direction and magnitude of attraction or repulsion over each candidate solution in the population are computed by this charge. The direction of this charge for candidate solution i is determined by vectorially adding the forces from each of the other solutions on candidate solution i. In this mechanism, a candidate solution with good objective function value attracts the other ones, whereas a candidate solution with worse objective function repels the other candidate solutions. The better (or worse) the objective function value is, the higher the size of attraction (or repletion). After the movement for each candidate solution in the current population is made, a new population is generated. The procedure is repeated until a stopping criterion is met.

Figure 1 shows the general outline of EA which includes four phases: *initialization* of algorithm, computation of *total force* exerted on each particle, *movement* along the direction of the force, and *local search.*

Encoding scheme and initialization

The encoding scheme is the procedure of making a solution recognizable for algorithms and plays an important role in the effectiveness of the algorithms. The decoding scheme is a complementary procedure that turns an encoded solution into a schedule. The proposed encoding scheme has two main parts: a permutation showing relative order of the jobs rather than the operations and a matrix whose rows determine the relative sequence of operations of each job. In this case, we just determine the relative order of the jobs and the processing route of each job. For each job permutation, we assign the random key of 0 to the first job and 1 to the last one. The jobs (i.e., n – 2 other jobs) in between are assigned real numbers of $\frac{1}{n-1}, \ldots, \frac{n-2}{n-1}$, respectively. For each job j, we assign the random key of 0 to the first operation and 1 to the last operation. The operations in between are assigned real numbers $\frac{1}{n_i-1}, \frac{2}{n_i-1}, \ldots, \frac{n_i-2}{n_i-1}$, respectively. For example, consider a problem with n = 5. One

Procedure ***Electromagnetism_algorithm***

Initialization
While ***stopping criterion is not met*** **do**
  Local search
  Computation of total forces
  Movement by forces
End while

Figure 1 The general outline of an EA.

```
Procedure Local_search

for k = 1 to popsize do
    iter = 0
    j = 1
    while j ≤ n do
        while iter < LSITER do
            The random key of job j in candidate solution k is randomly
regenerated
            if improvement is made then
                The new generated random key is accepted
                iter = LSITER
                j = n

            elseif
                iter = iter + 1

            endif
        endwhile
        j = j + 1
    endwhile
endfor
```

Figure 2 **The procedure of the local search.**

possible permutation is {4, 2, 5, 1, 3}. The real values assigned to jobs 4 and 3 are 0 and 1, respectively. For middle jobs, the values $\frac{1}{4}, \frac{2}{4}$, and $\frac{3}{4}$ is assigned to jobs 2, 5 and 1, respectively.

The decoding scheme turns an encoded solution into a schedule. We set the starting time of the first job in part 1 as 0. Then, to find the earliest possible starting time of the subsequent job, we first set the starting time to be 0. The first operation of that job is scheduled. If any overlap happens between this operation and previously scheduled operations on the same machine, the starting time is set to as the first possible starting time that avoids any overlap. The second operation in the processing route of the job is then scheduled. Again, if any overlap occurs between this operation and previously scheduled operations on that machine, we have to restart from the first operation in

```
Procedure Computation_of_total_forces

for k = 1 to popsize do
    calculate q(k)
    F_k = 0
endfor
for k = 1 to popsize do
    for l = 1 to popsize do
        if l = k & f(x_l) ≠ f(x_k) do
            if f(x_l) < f(x_k) then
                F_k = F_k + (x_l − x_k) (q_k · q_l) / ‖x_l − x_k‖²   (attraction)
            else
                F_k = F_k − (x_l − x_k) (q_k · q_l) / ‖x_l − x_k‖²   (repulsion)
            endif
        endif
    endfor
endfor
```

Figure 3 **The procedure of calculating *total forces*.**

```
Procedure Movement_by_total_forces

for k = 1 to popsize do
    if k ≠ best then
        β = random [0, 1]
        F^k = F_k / ||F_k||                          %Normalization of the total forces
        for j = 1 to n do
            if F_j^k > 0 then                        %F_j^k is j -th element of F^k
                x_kj = x_kj + β · F_j^k (1 + m_1 − x_kj)   %x_kj is random key of job j in solution
k
            else
                x_kj = x_kj + β · F_j^k (x_kj − 1)
            endif
        endfor
    endif
endfor
```

Figure 4 The procedure of *movement* by *total forces*.

order to hold no buffer constraints. The procedure repeats until all the operations of the job are scheduled.

Local search

To enhance the algorithm's performance, EA is hybridized with a local search. The procedure of this local search can be described as follows: for each candidate solution k, and then for a given job j, we generate a temporary random number between 0 and 1. The jobs are then sorted according to the random keys. Considering the previous example, the real values are $\{\frac{3}{4}, \frac{1}{4}, 1, 0, \frac{2}{4}\}$. Suppose the real value of job 2 is randomly regenerated and it becomes 47, therefore, the new solution is as follows: $\{\frac{3}{4}, \frac{4}{7}, 1, 0, \frac{2}{4}\}$.

If the new sequence results in a better objective function value, then the new permutation is accepted. Otherwise, it is left. This procedure iterates at most *LSITER* times. If any improvement is made in an iteration less than *LSITER*, then the search for the current candidate solution terminates. Figure 2 shows the procedure.

Table 1 The results obtained by the model for OSNB

$n \times m$	Instances	Time
3 × 3	3	0.07
3 × 4	3	1.26
4 × 3	3	2.71
4 × 4	3	218
5 × 3	1	753
5 × 4	0	-

Computation of total forces

To obtain the total force exerted from the other solutions on candidate solution k, we first need to calculate the charge of that candidate solution. The charge of each candidate solution k is denoted by q_k and calculated by the following formula:

$$q_k = \exp\left(-n \times m \frac{f(x_k) - f(x_{\text{best}})}{\sum_{l=1}^{\text{popsize}} (f(x_l) - f(x_{\text{best}}))}\right), \forall_k \quad k = 1, 2, \ldots, \text{popsize}$$

where x_k and x_{best} are candidate solution k and the best candidate solution. This formula ensures that solutions with better objective values are assigned higher charges. The *total force* (F_k) vector exerted on candidate solution k is also obtained by the sum of all the forces exerted from each of other solutions in the current population on candidate solution k. If solution l is better than solution k (i.e., if $f(x_l) < f(x_k)$), it attracts solution k by the

Table 2 The average RPD obtained by EA against TPSA in the Brucker et al. (1997) benchmark

Instance	EA	TPSA
j3	0.02	2.69
j4	0.82	2.17
j5	1.61	2.51
j6	1.79	3.22
j7	1.91	2.65
j8	2.43	4.73
Average	1.43	2.99

Table 3 The average RPD obtained by EA against TPSA in the Guéret and Prins (1998) benchmark

Instance	EA	TPSA
gp03	0.75	2.55
gp04	0.89	3.89
gp05	1.47	5.42
gp06	1.99	5.39
gp07	2.08	5.14
gp08	1.36	3.20
gp09	1.64	4.35
gp010	2.74	4.18
Average	1.62	4.26

force of $(x_l - x_k)\frac{q_k \cdot q_l}{\|x_l - x_k{}^2\|}$. Otherwise, (i.e., if $f(x_l) > f(x_k)$), it repulses solution k by the force of $(x_k - x_l)\frac{q_k \cdot q_l}{\|x_l - x_k{}^2\|}$. Figure 3 shows the procedure of calculation of *total force* vector exerted on candidate solutions.

Let us further illustrate the procedure by an example. Consider a problem with $n = 4$ and $m = 2$. We have popsize = 5 where $f(x_1) = 120$, $f(x_2) = 140$, $f(x_3) = 90$, $f(x_4) = 100$, and $f(x_5) = 110$. Therefore, we have $q_1 = 0.11$, $q_2 = 0.03$, $q_3 = 1$, $q_4 = 0.48$, and $q_5 = 0.23$. Suppose $x_1 = \{2.87, 2.05, 1.82, 2.52\}$ and $x_4 = \{1.22, 2.65, 1.43, 2.12\}$. The vector force exerted from x_4 on x_1 becomes as follows:

$$\left(\begin{bmatrix} 1.22 \\ 2.56 \\ 1.43 \\ 2.12 \end{bmatrix} - \begin{bmatrix} 2.87 \\ 2.05 \\ 1.82 \\ 2.52 \end{bmatrix}\right) \frac{0.11 \times 0.48}{1.84} = \begin{bmatrix} -0.06 \\ 0.01 \\ -0.01 \\ -0.02 \end{bmatrix}$$

Movement by total forces

To move according to the total forces obtained in the previous step, we first normalize the vector F_k. If $F_j^k > 0$ (F_j^k is the j th element of the normalized vector F_k), x_{kj} is increased (x_{kj} is the random key of job j in solution k). Otherwise, it is decreased. This move for each candidate solution is in direction of *total force* exerted on it by a random step length. This length is generated from a uniform distribution between [0, 1]. By selecting random length, we can guarantee that candidate solutions have a nonzero probability to move to the unvisited solution along this direction. Moreover by normalizing *total force* exerted on each candidate solution, we can avoid producing infeasible solutions. All the candidate solutions are moved with the exception of the current best solution. Figure 4 shows the outline of *movement* procedure.

Table 4 The average RPD obtained by EA against TPSA in the Taillard (1993) benchmark

Instance	EA	TPSA
Tail 4 × 4	0.42	3.16
Tail 5 × 5	2.20	3.58
Tail 7 × 7	1.86	2.49
Tail 10 × 10	2.10	4.58
Tail 20 × 20	2.03	6.10
Average	1.72	3.98

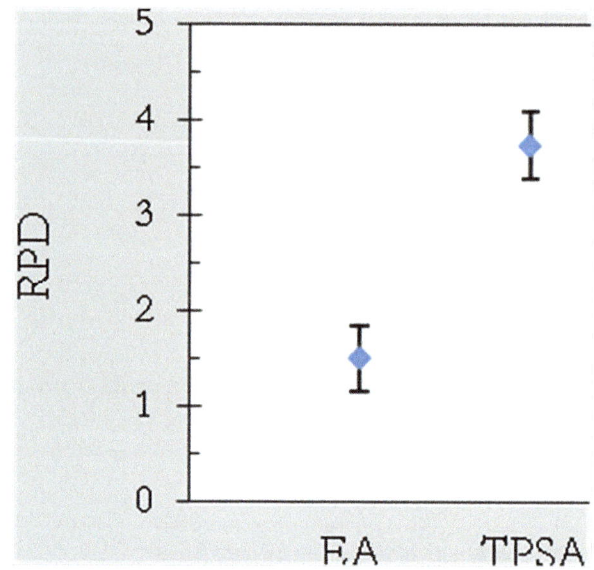

Figure 5 Mean plots and LSD intervals for the metaheuristics.

Computational evaluation

This section first evaluates the efficiency of the proposed MILP model on a computational experiment including small-sized instances. Afterwards, it investigates the general performance of the proposed metaheuristic against the optimal points obtained by the model. Moreover, it further evaluates the performance of the algorithms

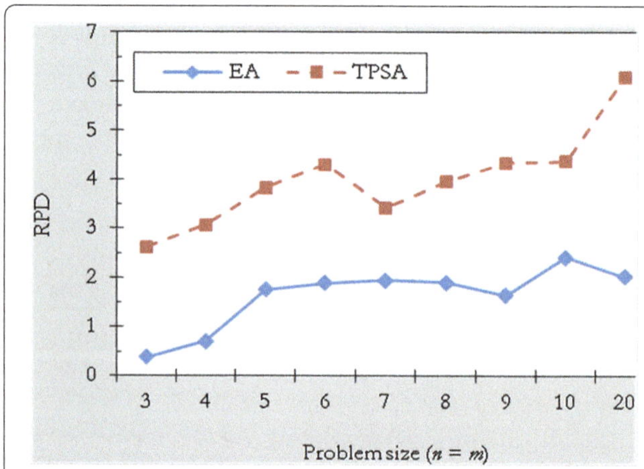

Figure 6 Mean plots for the metaheuristics versus the problem size.

against benchmarks in the literature of basic open shops. Since existing optimal solutions of the benchmarks are for basic open shops, they cannot be used for the evaluation. In this case, we use a performance measure named relative percentage deviation (RPD) obtained by the following formula:

$$\mathrm{RPD} = \frac{\mathrm{Alg}_{\mathrm{sol}} - \mathrm{Min}_{\mathrm{sol}}}{\mathrm{Min}_{\mathrm{sol}}} \times 100,$$

where $\mathrm{Min}_{\mathrm{sol}}$ and $\mathrm{Alg}_{\mathrm{sol}}$ are the lowest total tardiness for a given instance obtained by any of the algorithms and the solution obtained by a given algorithm. We implement the MILP models in CPLEX 10 and the other algorithms in Borland C++ and run on a PC with 2.0-GHz Intel Core 2 Duo and 2 GB of RAM memory. The stopping criterion used when testing all instances with the metaheuristics is set to a computational time limit fixed to $n \times m \times 0.5$ s. This stopping criterion permits for more time as the number of jobs or machines increases. Our proposed electromagnetism algorithm includes only one parameter of popsize. To set this parameter, we consider the sizes of 5, 10, and 20. We randomly generate 30 instances in different sizes and solve them by the three EA obtained by different levels of popsize. The popsize of 10 obtained the best results.

Evaluation of the MILP model

This section first evaluates the efficiency of the MILP model to solve NW-OSS problems. We generate a set of different instances as follows. We have 10 problem sizes ranging from $(n \times m) = (3 \times 3)$ up to (5×4). The processing times are randomly distributed over (1, 99). For each problem size, we generate three instances. The MILP model is allowed a maximum of 900 s (15 min) of computational time. Table 1 shows the results obtained by the model including the number of instances solved to optimality by the model in each problem size and the average computational time required to solve the instances. The model is capable of solving instances up to (5×3).

Evaluation of the metaheuristic

This section evaluates the algorithm against the optimal solution obtained by the model in the previous subsection. EA optimally solves 12 instances out of 13 instances. After having investigated the general performance of the proposed electromagnetism algorithm, we further evaluate the proposed algorithm against the adaptation of a relevant algorithm, the two-phase solution algorithm (TPSA) of Lin et al. (2008). To do so, we use standard benchmarks in the literature, such as those of Taillard (1993), Brucker et al. (1997), and Guéret and Prins (1998).

We use RPD measure to compare the algorithms. Tables 2, 3, and 4 show the results of the experiments for benchmarks of Brucker et al. (1997), Guéret and Prins (1998), and Taillard (1993), respectively. EA outperforms TPSA in all the three benchmarks. It obtains the RPD of 1.43%, 1.62%, and 1.72% in Brucker et al. (1997), Guéret and Prins (1998), and Taillard (1993) benchmarks, respectively. To further statistically analyze the results, we carry out an 'analysis of variance' or ANOVA. The results demonstrate that there are significant differences between the two algorithms with the p value very close to 0. Figure 5 shows the mean plots and least significant difference or LSD intervals at 95% confidence level for the different algorithms. As could be seen, EA statistically outperforms TPSA. It is also interesting to plot the performance of the algorithms versus the problem size. Figure 6 shows the means obtained by the algorithms in the different problem sizes. As shown, EA significantly performs well in all the problem sizes.

Conclusions and future studies

This paper studied the problem of open-shop scheduling, with no intermediate buffer to minimize total tardiness. To optimally solve the problem, one mathematical model in the form of mixed integer linear program is developed. Next, we proposed a metaheuristic in form of an electromagnetism algorithm to solve the large-sized problems in an acceptable computational time. We carried out two computational experiments to evaluate the performances of model and metaheuristic. In the first one, we had some small-sized instances by which we assessed the mathematical model and evaluated the general performance of the proposed metaheuristic. In the second experiment, the proposed algorithm was further evaluated against an algorithm in the literature. All the results supported that the model and metaheuristic were effective to tackle open-shop problems with no intermediate buffer between machines. As an interesting future research, one can study the multi-objective case of the problem under consideration. Another impressive research is to present a branch-and-bound or any other exact method for the problem.

Competing interests

The authors declare that they have no competing interests.

Authors' contributions

BN, EN, and MY formulated the problem of scheduling no-wait open shops with no intermediate buffer. To solve the problem, we proposed a metaheuristic in the form of electromagnetism-like algorithm. All authors read and approved the final manuscript.

Author details

[1]Department of Industrial Engineering, Faculty of Engineering, University of Kharazmi, Karaj, Iran. [2]Department of Industrial Engineering, Science &

Research Branch, Islamic Azad University, Tehran, Iran. [3]Department of Industrial Engineering, Qazvin Branch, Islamic Azad University, Qazvin, Iran.

References

Andresen M, Bräsel H, Morig M, Tusch J, Werner F, Willenius P (2008) Simulated annealing and genetic algorithms for minimizing mean flow time in an open shop. Math Comput Model 48(7–8):1279–1293

Bïrbïl I, Fang SC (2003) An electromagnetism-like mechanism for global optimization. J Glob Optim 25:263–282

Bräsel H, Herms A, Morig M, Tautenhahn T, Tusch T, Werner F (2008) Heuristic constructive algorithms for open shop scheduling to minimize mean flow time. Eur J Oper Res 189(3):856–870

Brucker P, Hurink J, Jurisch B, Wöstmann BA (1997) A branch and bound algorithm for the open-shop problem. Discret Appl Math 76:43–59

Guéret C, Prins C (1998) Classical and new heuristics for the open shop problem: a computational evaluation. Eur J Oper Res 107:306–314

Hall NC, Sriskandarajah CA (1996) Survey of machine scheduling problems with blocking and no-wait in process. Oper Res 44:510–525

Liaw CF, Cheng CY, Chen M (2005) Scheduling two-machine no-wait open shops to minimize makespan. Comput Oper Res 32:901–917

Lin HT, Lee HT, Pan WJ (2008) Heuristics for scheduling in a no-wait open shop with movable dedicated machines. Int J Prod Econ 111(2):368–377

Low C, Yeh Y (2009) Genetic algorithm-based heuristics for an open shop scheduling problem with setup, processing, and removal times separated. Robot Comput Integr Manuf 25(2):314–322

Pinedo ML (2008) Scheduling: theory, algorithms, and systems, 3rd edn. Springer Science + Business Media, New York

Sahni S, Cho Y (1979) Complexity of scheduling shops with no-wait in process. Math Oper Res 4:448–457

Sidney JB, Sriskandarajah C (1999) A heuristic for the two-machine no-wait open shop scheduling problem. Nav Res Logist 46:129–145

Stafford EF Jr, Tseng FT, Gupta JND (2005) Comparative evaluation of MILP flowshop models. J Oper Res Soc 56:88–101

Taillard E (1993) Benchmarks for basic scheduling problems. Eur J Oper Res 64:278–285

Yao MJ, Lai CW (2002) A genetic algorithm for the two machine open shop scheduling problem with blocking. In: The second Japanese-Sino optimization meeting (JSOM 2002), Rihga Royal Hotel Kyoto, Kyoto, 25–27 Sept 2002

Yao MJ, Soewandi H (2000) Simple heuristics for the two machine open shop problem with blocking. J Chin Inst Ind Eng 17(5):537–547

A multiple objective approach for joint ordering and pricing planning problem with stochastic lead times

Zeinab Hosseini[1], Reza Ghasemy Yaghin[2] and Maryam Esmaeili[1*]

Abstract

The integration of marketing and demand with logistics and inventories (supply side of companies) may cause multiple improvements; it can revolutionize the management of the revenue of rental companies, hotels, and airlines. In this paper, we develop a multi-objective pricing-inventory model for a retailer. Maximizing the retailer's profit and the service level are the objectives, and shortage is allowed. We present the model under stochastic lead time with uniform and exponential distributions. Since pricing is important and influences demand, the demand is considered as a general function of price. The multiple-objective optimization model is solved using the weighting method as well as the L-P metric method. Concerning the properties of a nonlinear model, a genetic algorithm is taken into account to find the optimal solutions for the selling price, lot size, and reorder point. Finally, numerical examples with sensitivity analysis regarding key parameters are provided.

Keywords: Multi-objective nonlinear optimization; Pricing; Stochastic lead time; L-P metric method; Genetic algorithm

Introduction

Planning and inventory control is one of the major issues in industrial engineering; it is also one of the inevitable activities in each organization. Therefore, there exist several studies in this research area. Here, we investigate two general groups of papers. The first group considers inventory models with pricing. The integration of inventory and pricing policies leads to the joint optimization of the whole system and maximization of the total profit. Selling price consumer selection can gravitate to the customers. Lee (2011) considered two pricing policies based on two service levels. The author concluded that increasing the price leads some customers towards low service levels with lower prices. Whitin (1955) was the first researcher to formulate a newsboy model with price effect. In that model, the probability distribution of demand depended on the unit selling price where price was a decision variable. Abad (2003, 2008), Dye (2007), and Dye and Hsieh (2010) assumed deteriorating items and allowable shortage. They presented models that considered the price as a decision variable. Abad (2003, 2008) and Dye (2007) considered demand as a general function of price. Mukhopadhyay et al. (2004) and Esmaeili (2009) studied infinite planning horizon where shortage was not allowed. Mukhopadhyay considered demand to be a nonlinear function of price, while in the model of Esmaeili, demand was a general function. The annual profit of the manufacturer is maximized to determine the selling price, marketing expenditure, and lot size. Sana (2011) presented a stochastic inventory in which demand was considered to be dependent on the random selling price. Dye and Hsieh (2013) studied an advanced sales system with deteriorating items where prices were dependent on demand. They showed that advanced sales price is lower than the spot sales price. Sadjadi et al. (2012) used geometric programming (GP) to obtain optimal lot sizing, pricing, and marketing decisions such that the profit is maximized. However, using a single objective function is the major shortcoming of the models.

Most inventory models aggregate several cost concepts and service requirements into a single objective and use

* Correspondence: esmaeili_m@alzahra.ac.ir
[1]Department of Industrial Engineering, Alzahra University, Tehran, Iran
Full list of author information is available at the end of the article

Table 1 Comparison between the most relevant models in literature

Research papers	Objective function	Objective function	Pricing	Demand	Lead time	Solution method
Sheikh Sajadieh and Akbari Jokar (2009)	Single	Cost minimization	Not considered	Constant	Stochastic (uniform)	Exact
Sheikh Sajadieh et al. (2009)	Single	Cost minimization	Not considered	Constant	Stochastic (exponential)	Exact
Esmaeili (2009)	Single	Profit maximization	Considered	Price dependent	Constant	Exact
Tsou (2008)	Multiple	Cost minimization, frequency of stock out occasions, and number of items stocked out	Not considered	Stochastic	Constant	TOPSIS and MOPSO algorithm
Dye and Hsieh (2013)	Single	Profit maximization	Considered	Price dependent	Constant	Exact
Proposed model	Multiple	Profit maximization and service level	Considered	Price dependent	Stochastic (uniform and exponential)	L-P metric, Weighting and GA algorithm

traditional methods to solve them. Nonetheless, one of the significant characteristics of modern business is the multiple criteria context of retail industries.

In the second group, the decision maker seeks to maximize or minimize two or more objectives simultaneously. This group of models has been applied in several fields, but few of these multi-objective problems have dealt with inventory control optimization.

Padmanabhan and Vart (1990) solved a multi-objective inventory model with deteriorating items and stock-dependent demand by a nonlinear goal programming method. Agrell (1995) proposed a multi-objective inventory model with three objective functions. These objectives included minimization of average total relevant annual cost, annual average frequency of stock out occasions, and annual average number of times stock out. Considering lot size and safety factor as decision variables, it is assumed that planning horizon is infinite, demand has a normal distribution, and shortage is allowed. Later Tsou (2008, 2009) and Moslemi and Zandieh (2011) considered Agrell's model. However, nondominated solutions of a reorder point and order size have been obtained using multi-objective particle swarm optimization algorithm (MOPSO). Moreover, TOPSIS was used to rank the nondominated solutions using the preference of decision makers (Tsou 2008). Tsou (2009) involved MOPSO and multi-objective electromagnetism-like optimization (MOEMO) algorithms to obtain nondominated solutions of lot size and safety stock. Moslemi and Zandieh (2011) created some strategies based on the MOPSO algorithm in continuous review stochastic inventory control system.

Many researchers have considered deterioration in a multi-item multi-objective inventory model under a

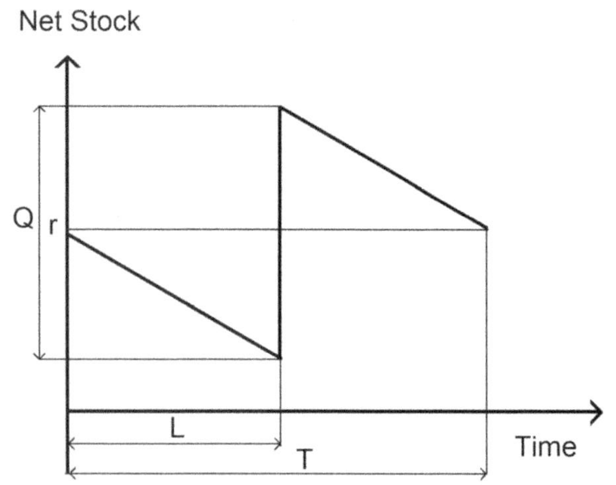

Figure 1 Case1 $r > bD$.

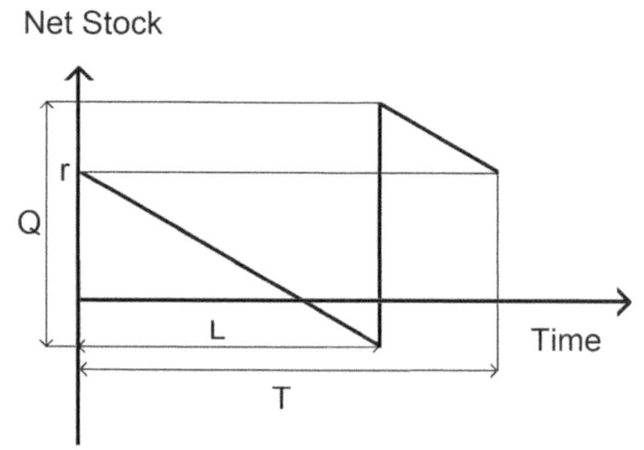

Figure 2 Case2 $r \leq bD$.

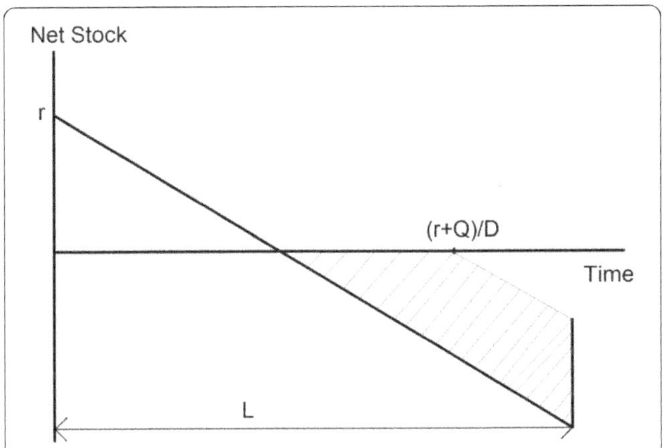

Figure 3 $L > T$.

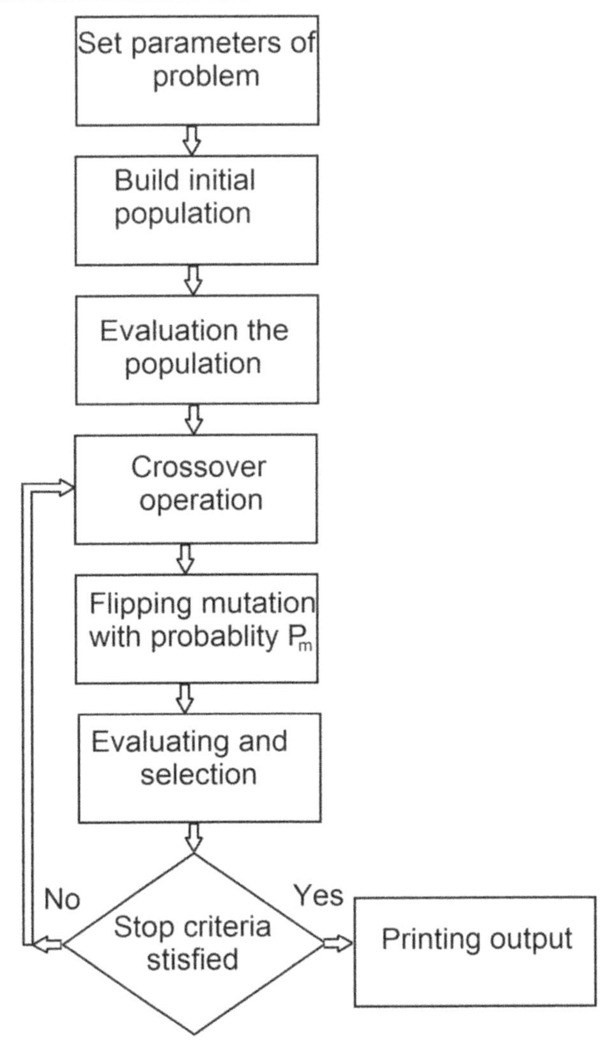

Figure 4 Proposed genetic algorithm.

Table 2 Optimal results of model with uniform lead time

Method	Weights of objectives	P*	Q*	r*	Z*	SL*
Weighting		89	107	32	64,444	0.4055
L-P metric, $p = 1$	W1 = 0.2	88	197	79	63,476	0.9987
L-P metric, $p = 2$	W2 = 0.8	88	181	79	63,508	0.9987
Weighting		90	107	31	65,107	0.3938
L-P metric, $p = 1$	W1 = 0.4	90	198	78	64,808	0.9909
L-P metric, $p = 2$	W2 = 0.6	90	179	78	64,845	0.9909
Weighting		90	108	29	65,108	0.3684
L-P metric, $p = 1$	W1 = 0.6	90	196	78	64,812	0.9909
L-P metric, $p = 2$	W2 = 0.4	90	176	78	64,850	0.9909

fuzzy environment. For instance, Roy and Maiti (1998) maximized profit and minimized wastage cost; while the demand was dependent on the inventory level, planning horizon was finite and shortage was allowed. To obtain optimal solution, they used fuzzy nonlinear programming (FNLP) and fuzzy additive goal programming (FAGP) methods considering budget and space constraints. However, Mandal et al. (2005) included storage space, number of orders, and production cost in their model and achieved optimal solution by applying GP. In addition, Maity and Maiti (2008) and Islam (2008) presented a multi-item multi-objective inventory model under fuzzy inflation and discounting. Maity and Maiti (2008) assumed that time horizon is finite, shortage is allowed, and demand is dependent on advertisement. They have used utility function method (UFM) and generalized reduced gradient (GRG) methods to obtain optimal solution. Islam (2008) considered an infinite planning horizon under the limitation of space capacity and total shortage cost constraints with demand uniformly distributed and dependent on the marketing

Table 3 Optimal results of model with exponential lead time

Method	Weights of objectives	P*	Q*	r*	Z*	SL*
Weighting		89	107	32	64,444	0.4055
L-P metric, $p = 1$	W1 = 0.2	88	197	79	63,476	0.9987
L-P metric, $p = 2$	W2 = 0.8	88	181	79	63,508	0.9987
Weighting		90	107	31	65,107	0.3938
L-P metric, $p = 1$	W1 = 0.4	90	198	78	64,808	0.9909
L-P metric, $p = 2$	W2 = 0.6	90	179	78	64,845	0.9909
Weighting		90	108	29	65,108	0.3684
L-P metric, $p = 1$	W1 = 0.6	90	196	78	64,812	0.9909
L-P metric, $p = 2$	W2 = 0.4	90	176	78	64,850	0.9909

Table 4 Sensitivity analysis of the uniform model with respect to *C* - weighting method

$c =$	6	9	12	15
P^*	82	88	94	100
Q^*	110	108	108	107
r^*	30	30	29	28
Z^*	63,038	64,602	66,095	67,515
SL^*	0.3738	0.3792	0.3720	0.3646

Table 5 Sensitivity analysis of the uniform model with respect to *h* - weighting method

$h =$	2	4	6	8
P^*	84	88	92	95
Q^*	160	120	101	92
r^*	40	31	27	23
Z^*	61,251	63,827	66,377	68,244
SL^*	0.5008	0.3919	0.3447	0.2958

Table 6 Sensitivity analysis of the uniform model with respect to *π* - weighting method

$\pi =$	15	25	35	45
P^*	60	80	100	120
Q^*	121	112	106	100
r^*	13	26	31	35
Z^*	43,543	58,314	71,505	83,105
SL^*	0.1539	0.3224	0.4036	0.4797

Table 7 Sensitivity analysis of the uniform model with to respect *α* - weighting method

$\alpha =$	0.5	1.5	2.5	3.5
P^*	90	90	90	90
Q^*	118	111	105	98
r^*	36	32	27	23
Z^*	75,859	68,691	61,524	54,395
SL^*	0.3927	0.3854	0.3629	0.3498

Table 8 Sensitivity analysis of the uniform model with respect to *C* - L-P metric method (*p* = 2)

$c =$	6	9	12	15
P^*	82	87	94	100
Q^*	178	178	172	167
r^*	80	79	77	75
Z^*	62,774	63,670	65,846	67,280
SL^*	0.9968	0.9963	0.9878	0.9766

Table 9 Sensitivity analysis of the uniform model with respect to *h* - L-P metric method (*p* = 2)

$h =$	2	4	6	8
P^*	83	88	92	95
Q^*	178	176	172	169
r^*	80	79	77	76
Z^*	60,507	63,645	66,050	67,757
SL^*	0.9992	0.9987	0.9829	0.9774

Table 10 Sensitivity analysis of the uniform model with respect to *π* - L-P metric method (*p* = 2)

$\pi =$	15	25	35	45
P^*	60	80	100	120
Q^*	192	188	169	161
r^*	84	80	76	72
Z^*	43,197	58,020	71,271	82,902
SL^*	0.9943	0.9921	0.9896	0.9868

Table 11 Sensitivity analysis of the uniform model with respect to α - L-P metric method (*p* = 2)

$\alpha =$	0.5	1.5	2.5	3.5
P^*	90	90	90	90
Q^*	203	183	163	143
r^*	91	82	73	64
Z^*	75,549	68,422	61,295	54,167
SL^*	0.9926	0.9875	0.9812	0.9732

Table 12 Sensitivity analysis of the uniform model with respect to *C* - L-P metric method (*p* = 1)

$c =$	6	9	12	15
P^*	82	87	94	100
Q^*	201	198	193	188
r^*	80	79	77	75
Z^*	62,730	63,632	65,806	67,241
SL^*	0.9968	0.9963	0.9878	0.9766

Table 14 Sensitivity analysis of the uniform model with respect to *π* - L-P metric method (*p* = 1)

$\pi =$	15	25	35	45
P^*	60	80	100	120
Q^*	256	201	192	178
r^*	84	80	76	71
Z^*	43,065	57,955	71,228	82,876
SL^*	0.9943	0.9921	0.9869	0.9731

cost. To minimize inventory, marketing, and production costs, the optimal solution was obtained using the GP method. A significant shortcoming of all these models is that they only regard a deterministic lead time. However, the lead time in the real world is usually a random variable. Moreover, the lead time has an effective role in determining the optimal policy of inventory models. Recently, Hosseini et al. (2012) presented a multi-objective model with uniformly distributed lead time to optimize retailing activities.

Price and service level are important factors in attracting customers and increasing their satisfaction (Liang et al. 2008). Therefore, in this paper, a multi-objective inventory model is presented which includes the retailer's profit and service level. With infinite planning horizon and allowable shortage, we assume that lead time has uniform and exponential distribution while demand is a general function of price. Selling price, lot size, and reorder point are obtained by maximizing both the retailer's profit and service level. Genetic algorithm (GA) is used since the model is complex and nonlinear. In the end, a numerical example is provided along with sensitivity analysis on key parameters including shortage, purchasing and holding costs, and demand elasticity. A comparison between the proposed model and the most relevant models in the literature has been provided in Table 1.

This paper is organized as follows: In the 'Model formulation and assumption' section, notation, assumptions, and mathematical model are provided. Solution algorithm is presented in the 'Solution procedure' section. The 'Numerical example and sensitivity analysis' section includes the numerical example and sensitivity analysis on the key parameters of the model. Finally, the suggestions and results obtained from this study are presented in the 'Conclusions' section.

Table 13 Sensitivity analysis of the uniform model with respect to *h* - L-P metric method (*p* = 1)

$h =$	2	4	6	8
P^*	83	88	92	95
Q^*	201	198	193	192
r^*	80	79	77	76
Z^*	60,497	63,614	66,000	67,680
SL^*	0.9992	0.9987	0.9829	0.9774

Model formulation and assumptions

This section introduces the notations, assumptions, decision variables, and input parameters of our model.

Notation

The following is the notation:

- P selling price (decision variable)
- Q lot size (decision variable)
- r reorder point (decision variable)
- C purchase cost (US$ per unit)
- A ordering cost (US$ per order)
- h holding cost (US$ per unit)
- π shortage cost (US$ per unit)
- L lead time
- T duration of inventory cycle
- $D(P)$ demand rate; for simplicity, we let $D \equiv D(P)$
- Sr sales revenue
- TC expected total cost
- Z retailer's profit
- SL service level

Assumptions

The proposed models are based on the following assumptions:

1- Planning horizon is infinite.
2- Shortage is allowed and completely back-ordered.
3- Similar to the models proposed by Abad (2003, 2008), Dye (2007), and Esmaeili (2009), demand is represented by a general function of price.
4- The lead time is stochastic and follows uniform and exponential distribution.
5- Inventory is continuously reviewed.

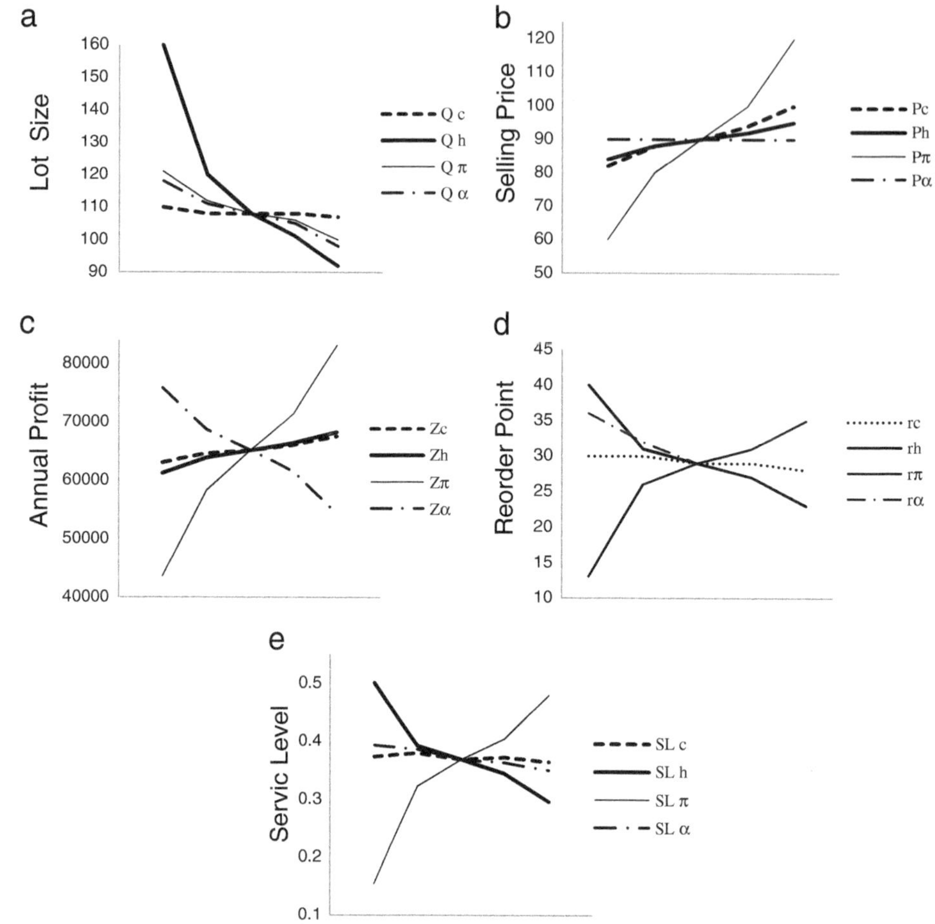

Figure 5 Effect of key parameters on (a) Q, (b) P, (c) Z, (d) r, and (e) SL (uniform model, weighting method).

6- The customers are myopic and thus make a purchase immediately if the price is below their willingness to pay without considering future prices.

Mathematical model

In this section, the multi-objective inventory model, including the retailer's profit and service level with uniform and exponential lead time, is presented.

Modeling with uniform distribution

Consider a retailer who is going to maximize the profit and attract the customers to increase their satisfaction. Therefore, the model would be a multi-objective inventory model with two objectives. By maximizing the service level and the retailer's profit, the optimal selling price, lot size, and reorder point (*P*, *Q*, and *r*) are to be obtained. The costs include purchasing, ordering, holding, and shortage. The lead time is stochastic and follows a uniform distribution with parameters *a* and *b* ($L \sim U$ [a, b]). Therefore, its probability density is as follows:

$$f_l(L) = \frac{1}{b-a} \qquad a \leq 1 \leq b. \tag{1}$$

Since the lead time is assumed to be a random variable, two cases can occur during each cycle time (Sheikh Sajadieh and Akbari Jokar 2009; Taleizadeh et al. 2010). In the first case, the reorder point is greater than the maximum demand during lead time. Therefore, the retailer does not face any shortage (Figure 1). In the second case, the ordering point is smaller than or equal to the maximum demand during lead time; it is probable to face shortage (Figure 2).

The annual profit function of the model is expressed as the retailer's profit = sales revenue − purchase cost − ordering cost − holding cost − shortage cost, which respectively are as follows:

$$S_r = PD \tag{2}$$

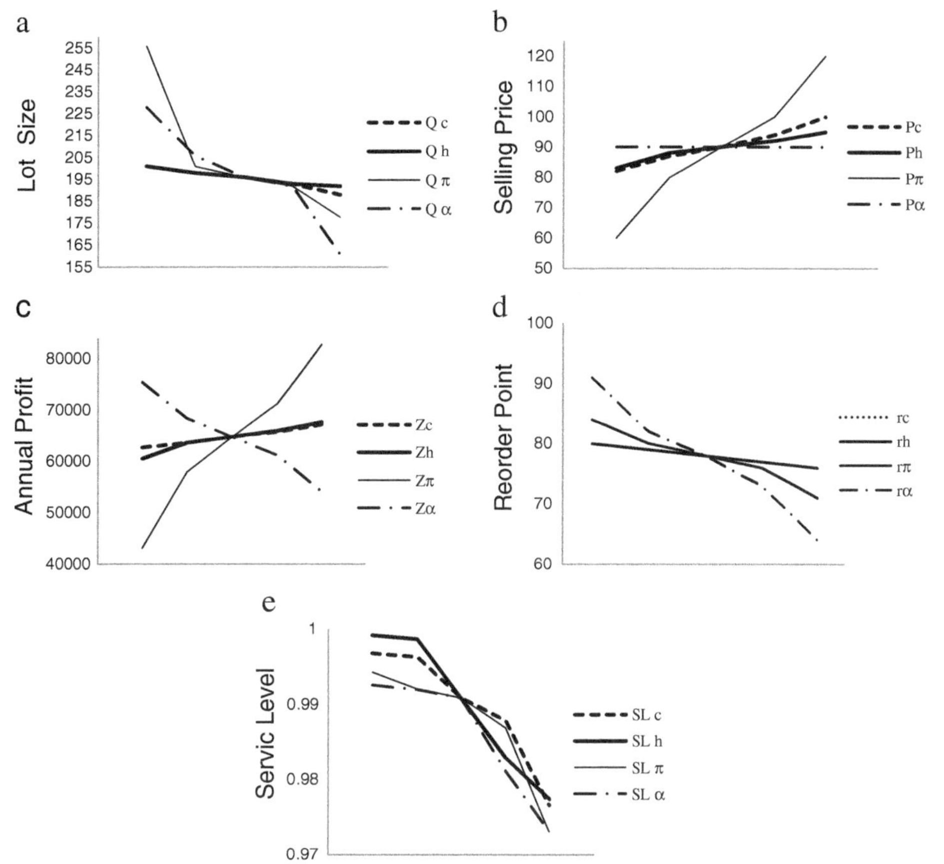

Figure 6 Effect of key parameters on (a) Q, (b) P, (c) Z, (d) r, and (e) SL (uniform model, L-P metric method *p* = 1).

$$\begin{aligned} TC(P,Q,r) &= CD + \frac{DA}{Q} + \frac{Dh}{Q}\int_a^{r/D}\left(rl - \frac{Dl^2}{2}\right)f_l(L)dl \\ &+ \frac{rh}{Q}\int_a^b (Q-D)f_l(L)dl + \frac{h}{2Q}\int_a^b (Q-Dl)^2 f_l(L)dl \\ &+ \frac{r^2h}{2Q}\int_{r/D}^b f_l(L)dl + \frac{\pi}{2Q}\int_{r/D}^b (Dl-r)^2 f_l(L)dl \end{aligned} \tag{3}$$

By substituting Equation 1 in Equation 3, we have

$$\begin{aligned} Z(P,Q,r) = PD - CD - \frac{DA}{Q} - hr \\ - \frac{h\left(2r^2 - 3rD^2a^2 + D^3a^3\right)}{6DQ(b-a)} + \frac{hrD(a+b)}{2Q} \\ - \frac{h(Q-Db-Da)}{2} - \frac{hD^2\left(a^2+ab+b^2\right)}{6Q} \\ - \frac{hr^2(Db-r)}{2DQ(b-a)} - \frac{\pi(Db-r)^3}{6DQ(b-a)} \end{aligned} \tag{4}$$

Table 15 Sensitivity analysis of the uniform model with respect to *a* - L-P metric method (*p* = 1)

$a =$	0.5	1.5	2.5	3.5
P^*	90	90	90	90
Q^*	228	206	192	161
r^*	91	82	73	64
Z^*	75,499	68,378	61,240	54,136
SL^*	0.9926	0.9875	0.9812	0.9732

Table 16 Sensitivity analysis of the exponential model with respect to *C* - the weighting method

$c =$	6	9	12	15
P^*	82	88	94	100
Q^*	127	127	127	127
r^*	32	30	28	27
Z^*	62,903	64,470	65,965	67,388
SL^*	0.5495	0.5316	0.5129	0.5049

Table 17 Sensitivity analysis of the exponential model with respect to *h* - the weighting method

$h =$	2	4	6	8
P^*	84	88	92	95
Q^*	187	143	122	110
r^*	49	33	24	18
Z^*	61,172	63,711	66,235	68,084
SL^*	0.7068	0.5658	0.4581	0.3705

Table 18 Sensitivity analysis of the exponential model with respect to *π* - weighting method

$\pi =$	15	25	35	45
P^*	60	80	100	120
Q^*	129	127	127	125
r^*	13	24	33	38
Z^*	4,345	5,819	7,136	8,294
SL^*	0.2649	0.4485	0.5765	0.6471

Table 19 Sensitivity analysis of the exponential model with respect to *α* - weighting method

$\alpha =$	0.5	1.5	2.5	3.5
P^*	90	90	90	90
Q^*	144	136	126	118
r^*	35	30	26	21
Z^*	75,701	68,551	61,403	54,255
SL^*	0.5339	0.5144	0.5028	0.4720

Table 20 Sensitivity analysis of the exponential model with respect to *C* - L-P metric method (*p* = 2)

$c =$	6	9	12	15
P^*	82	88	94	100
Q^*	421	411	398	383
r^*	168	164	159	153
Z^*	61,792	63,394	64,936	66,415
SL^*	0.9848	0.9842	0.9831	0.9814

Table 21 Sensitivity analysis of the exponential model with respect to *h* - L-P metric method (*p* = 2)

$h =$	2	4	6	8
P^*	84	88	92	95
Q^*	512	418	388	384
r^*	199	167	155	153
Z^*	60,697	62,875	64,998	66,345
SL^*	0.9931	0.9853	0.9809	0.9804

Table 22 Sensitivity analysis of the exponential model with respect to *π* - L-P metric method (*p* = 2)

$\pi =$	15	25	35	45
P^*	60	80	100	120
Q^*	446	432	398	341
r^*	178	172	159	136
Z^*	42,156	57,104	70,349	82,190
SL^*	0.9852	0.9860	0.9841	0.9759

Table 23 Sensitivity analysis of the exponential model with respect to *α* - L-P metric method (*p* = 2)

$\alpha =$	0.5	1.5	2.5	3.5
P^*	90	90	90	90
Q^*	512	441	388	384
r^*	201	176	154	127
Z^*	74,296	67,374	60,399	53,473
SL^*	0.9875	0.9856	0.9841	0.9790

Table 24 Sensitivity analysis of the exponential model with respect to *C* - L-P metric method (*p* = 1)

$c =$	6	9	12	15
P^*	82	88	94	100
Q^*	512	432	421	416
r^*	215	195	192	187
Z^*	61,340	62,528	64,717	66,168
SL^*	0.9935	0.9928	0.9927	0.9923

Table 25 Sensitivity analysis of the exponential model with respect to *h* - L-P metric method (*p* = 1)

h =	2	4	6	8
P^*	84	88	92	95
Q^*	478	443	389	385
r^*	215	199	175	173
Z^*	60,696	62,701	64,786	66,183
SL^*	0.9954	0.9935	0.9885	0.9883

The second objective is the retailer's service level. It is the probability of not facing shortage during the lead time, which presented is as follows:

$$\mathrm{SL}(P,r) = p(lD \le r) = \int_0^D f_1(L)dl = \frac{r}{D(b-a)}. \tag{5}$$

Modeling with exponential distribution

In this section, we obtain the optimal selling price, lot size, and reorder point (*P*, *Q*, and *r*) while the lead time has an exponential distribution with parameter λ ($L \sim \exp(\lambda)$) and the following probability density function:

$$f_L(l) = \lambda e^{-\lambda l}\ 0 \le 1 < \infty \tag{6}$$

Therefore, in addition to the two cases shown in Figures 1 and 2, there is a probability that the delivery is received after the cycle time which is depicted in Figure 3 (Sheikh Sajadieh et al. 2009).

The expected total cost is given by

$$\begin{aligned}\mathrm{TC}(P,Q,r) = CD + \frac{DA}{Q} + h\int_0^{r/D}\left(\frac{Q}{2}+r-DL\right)f_L(l)dl \\ + \int_{\frac{r}{D}}^{\frac{(r+Q)}{D}}\left[\frac{\pi(DL-r)^2 + h(Q+r-DL)^2}{2Q}\right]f_L(l)dl \\ + \int_{\frac{(r+Q)}{D}}^{\infty}\pi\left(DL-r-\frac{Q}{2}\right)f_L(l)dl.\end{aligned} \tag{7}$$

Table 26 Sensitivity analysis of the exponential model with respect to *π* - L-P metric method (*p* = 1)

π =	15	25	35	45
P^*	60	80	100	120
Q^*	512	441	416	392
r^*	209	198	187	176
Z^*	41,843	56,861	70,168	81,873
SL^*	0.9929	0.9926	0.9923	0.9920

Table 27 Sensitivity analysis of the exponential model with respect to *α* - L-P metric method (*p* = 1)

α =	0.5	1.5	2.5	3.5
P^*	90	90	90	90
Q^*	512	454	389	347
r^*	230	204	175	156
Z^*	74,152	67,204	60,288	53,267
SL^*	0.9934	0.9926	0.9909	0.9913

Substituting Equation 6 in Equation 7, we have

$$\begin{aligned}\mathrm{TC}(P,Q,r) = CD + \frac{DA}{Q} + h\left(r + \frac{Q}{2} - \frac{D}{\lambda}\right) \\ + \frac{D^2(\pi+h)}{\lambda^2 Q}\left(e^{-\frac{r\lambda}{D}} - e^{-\frac{(r+Q)\lambda}{D}}\right).\end{aligned} \tag{8}$$

Thus, the annual profit is calculated as follows:

$$\begin{aligned}Z(P,Q,r) = D(P-C) - \frac{DA}{Q} - h\left(r + \frac{Q}{2} - \frac{D}{\pi}\right) \\ - \frac{D^2(\pi+h)}{\lambda^2 Q}\left(e^{-\frac{r\lambda}{D}} - e^{-\frac{(r+Q)\lambda}{D}}\right).\end{aligned} \tag{9}$$

In order to calculate the service level, we can write

$$\begin{aligned}\mathrm{SL}(P,r) = P(lD \le r) = \int_0^{\frac{r}{D}} f_l(L)dl \\ = 1 - \exp\left(-\frac{\lambda r}{D}\right).\end{aligned} \tag{10}$$

Solution procedure

There are different methods to change a multi-objective optimization to a single-objective optimization. In this paper, we apply the weighting method that is useful and simple in concept and implementation. Also, the L-P metric method which is one of the famous methods of MCDM is applied in this paper. In this technique, the objective functions have the most proximity to their ideal values. These methods are explained respectively.

Weighting method

Considering a weight for each objective, the new objective function is obtained from the sum of objective functions with the corresponding weights. Therefore, the model for uniform and exponential lead time would respectively be as follows:

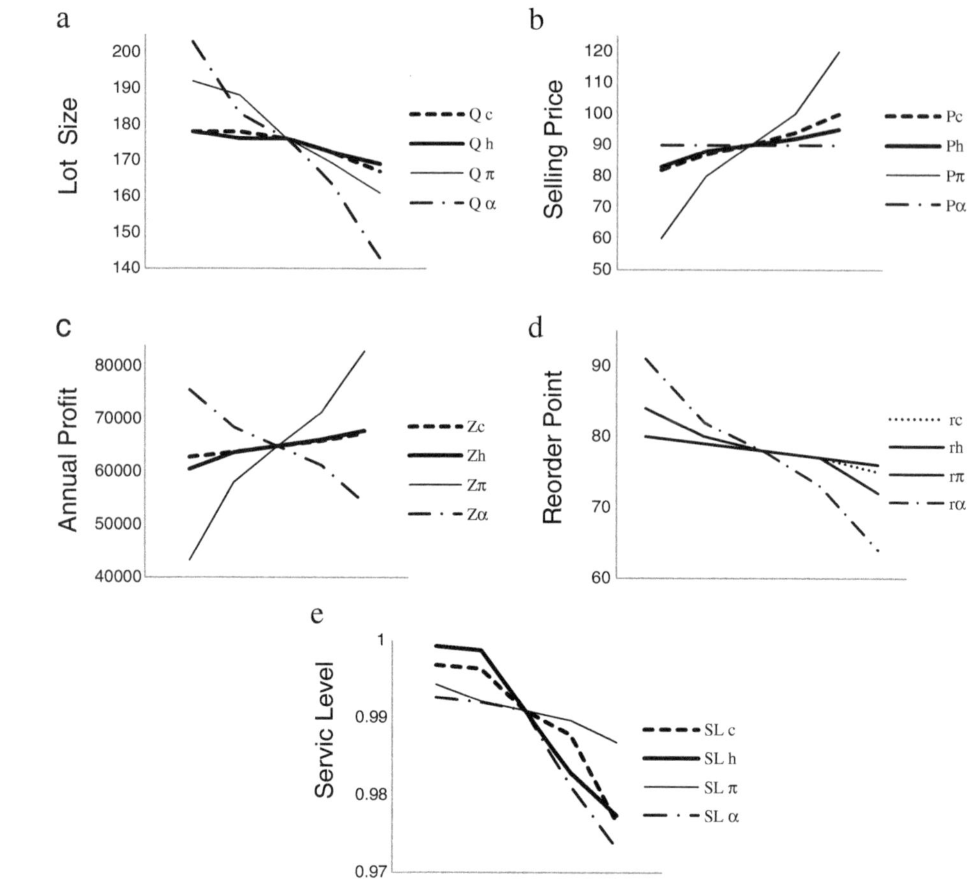

Figure 7 Effect of key parameters on (a) Q, (b) P, (c) Z, (d) r, and (e) SL (uniform model, L-P metric method $p = 2$).

$$\text{Max}: W_1.\left[D(P-C)-\frac{DA}{Q}-\frac{h}{2}(Q+2r-D(a+b)) + \frac{h\left(D^3a^3+3r^2Db-3rD^2a^2-r^3\right)}{6QD(b-a)} + \frac{3hrD(a+b)-hD^2\left(a^2+b^2+ab\right)}{6Q} - \frac{\pi(Db-r)^3}{6QD(b-a)}\right] + W_2.\left[\frac{r}{D(b-a)}\right]$$

$$S.t\ P, Q, r \geq 0 \text{ and } W_1 + W_2 = 1. \tag{11}$$

$$\text{Max}: W_1.\left[D(P-C)-\frac{DA}{Q}-h\left(r+\frac{Q}{2}-\frac{D}{\pi}\right) - \frac{D^2(\pi+h)}{\lambda^2 Q}\left(e^{-\frac{r\lambda}{D}}-e^{-\frac{(r+Q)\lambda}{D}}\right)\right] + W_2.\left[1-\exp\left(-\frac{\lambda r}{D}\right)\right]$$

$$S.t\ P, Q, r \geq 0 \text{ and } W_1 + W_2 = 1. \tag{12}$$

L-P metric method

In the L-P metric method, the distance of any present solution from the ideal solution is minimized (Banke et al. 2008):

$$L-P = \left\{\sum_{j=1}^{k} Y_j\left(f_j\left(x^{*j}\right)-f_j(x)\right)^p\right\}^{1/p} \tag{13}$$

where x^{*j} shows the ideal solution for optimizing the jth objective, x is the assumed solution, and Y_j indicates the significance degree for the jth objective. $1 \leq p \leq \infty$ is the parameter that specifies the L-P family. The p value indicates the emphasis level on the existing deviations. Therefore, the larger the p, the more emphasis will be there on the largest deviations. The values $p = 1$, $p = 2$, and $p = \infty$ are commonly used. In the present approach, we consider $p = 1$ and $p = 2$. When $p = 1$, the deviation is simply summed over all attributes, and when $p = 2$, the metric measures the shortest geometric distance between two points, which is a straight line. Other values of p are not as easily interpreted, but they may be reasonable choices in a

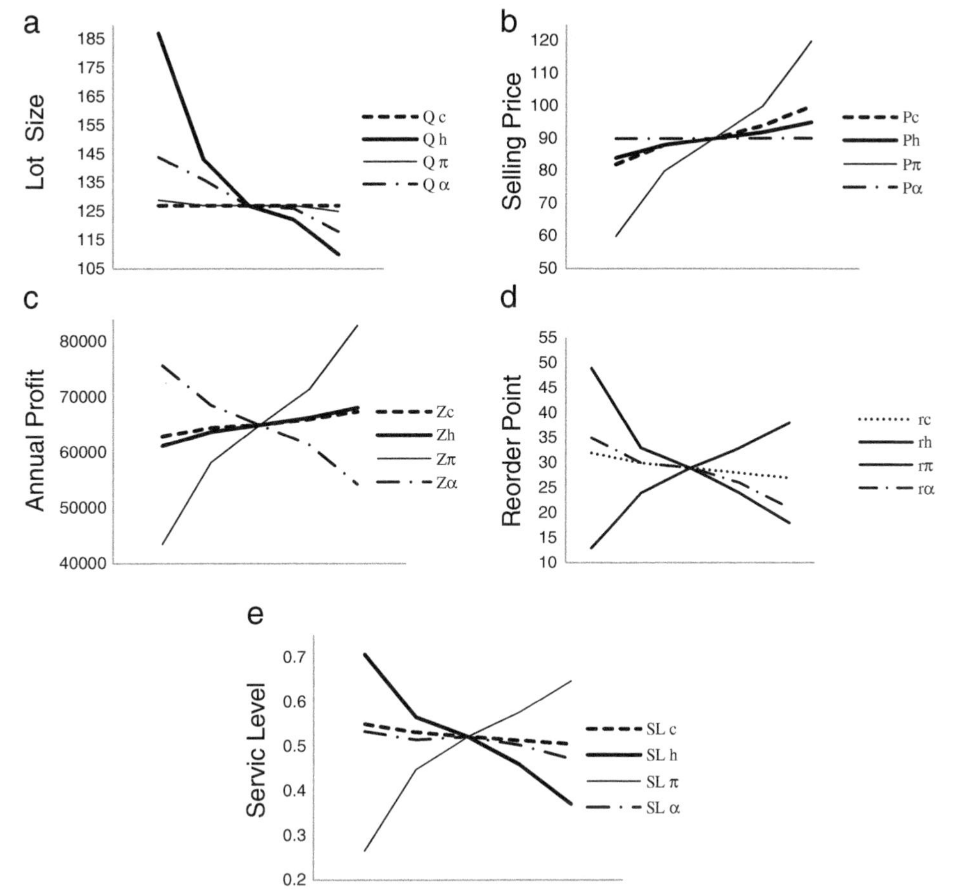

Figure 8 Effect of key parameters on (a) Q, (b) P, (c) Z, (d) r, and (e) SL (exponential model, weighting method).

given application. The p value depends on decision maker measures. The L-P method is affected by the objective measurement scale. Thus, the following formula is used:

$$L-P=\left\{\sum_{j=1}^{k}\gamma_j\left[\frac{f_j(x^{*j})-f_j(x)}{f_j(x^{*j})}\right]^p\right\}^{1/p}. \tag{14}$$

Therefore, the model for uniform lead time is given by

$$\text{Min } L\text{-}P=\left[w_1\left[\frac{Z^*-\left[\begin{array}{l}D(P-C)-\frac{DA}{Q}-\frac{h}{2}(Q+2r-D(a+b))\\+\frac{h(D^3a^3+3r^2Db-3rD^2a^2-r^3)}{6QD(b-a)}\\+\frac{3hrD(a+b)-hD^2(a^2+b^2+ab)}{6Q}\\-\frac{\pi(Db-r)^3}{6QD(b-a)}\end{array}\right]}{Z^*}\right]^p+W_2\left(\frac{\mathrm{SL}^*-\frac{r}{D(b-a)}}{\mathrm{SL}^*}\right)^p\right]^{1/p}$$

If lead time is exponential, we have the following objective function:

$$\text{Min } L\text{-}P=\left[w_1\left[\frac{Z^*-\left[\begin{array}{l}(P-C)-\frac{DA}{Q}-h\left(r+\frac{Q}{2}-\frac{D}{\pi}\right)\\-\frac{D^2(\pi+h)}{\lambda^2Q}\left(e^{-\frac{r\lambda}{D}}-e^{-\frac{(r+Q)\lambda}{D}}\right)\end{array}\right]}{Z^*}\right]^p+W_2\left(\frac{\mathrm{SL}^*-\left(1-\exp\left(-\frac{\lambda r}{D}\right)\right)}{\mathrm{SL}^*}\right)^p\right]^{1/p}$$

Genetic algorithm

Most researchers have used genetic algorithm to solve optimization problems (Taleizadeh et al. 2010; Maiti et al. 2009; Pasandideh et al. 2011). Considering the complexity of the nonlinear model, genetic algorithm is applied to find the optimal solution. GA was first

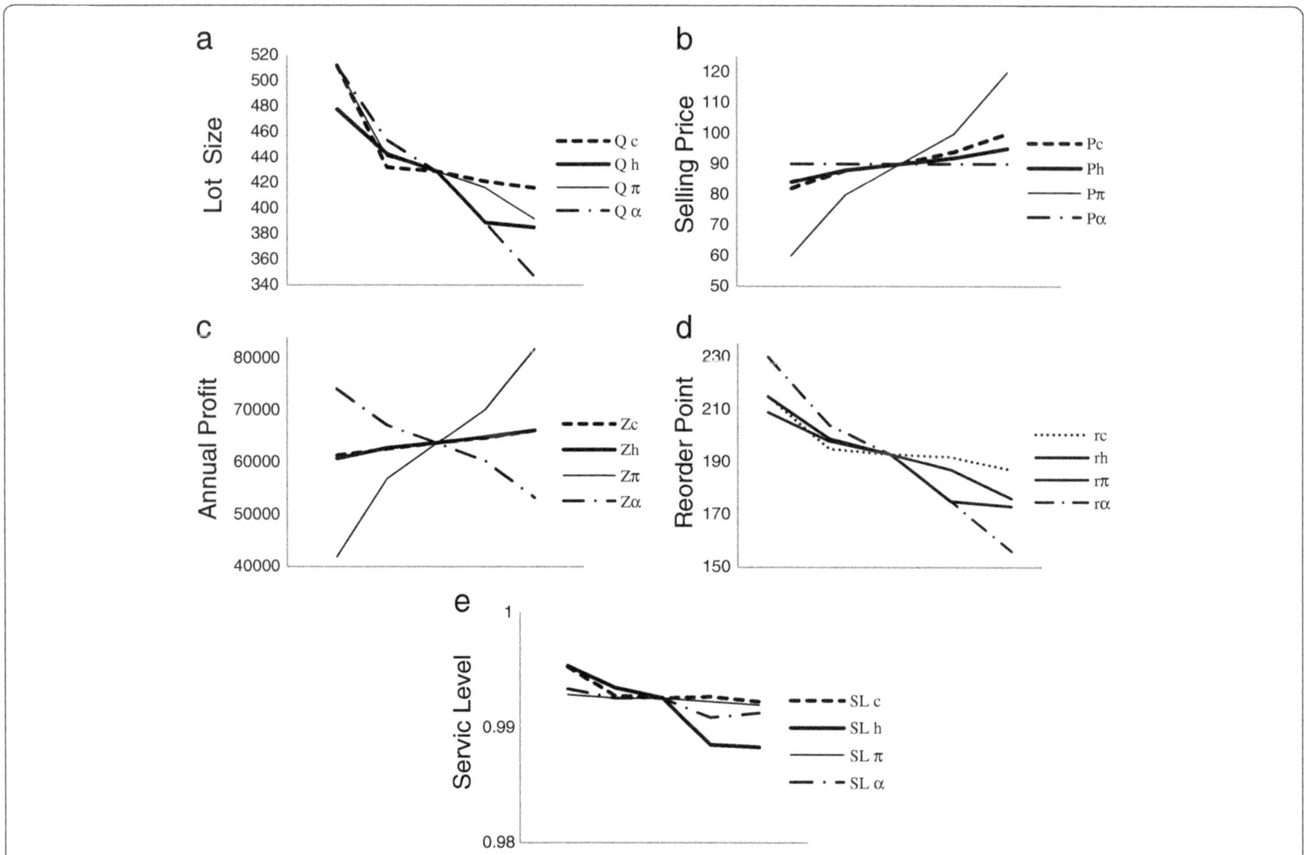

Figure 9 Effect of key parameters on (a) Q, (b) P, (c) Z, (d) r, and (e) SL (exponential model, L-P metric method $p = 1$).

presented by Holland in 1975 and was developed based on the principles of genetics and evolution (Haupt and Haupt 2004). Genetic algorithm begins to work with an initial population of solutions (chromosomes). New solutions are developed by crossover and mutation operators. To form the new population, the best solutions will be selected from the existing population using a fitness function. The solutions improve from one generation to another so that the desirable solution is obtained gradually.

Chromosome
Chromosome is a series of bits in which the coded forms of all suitable or unsuitable are placed. A suitable design of chromosome structure is an important part of genetic algorithm. In our algorithm, a string is designed with a length of k in which the first, second, and third one-thirds indicate reorder point, selling price, and lot size, respectively.

Population
A group of chromosomes is called a population. The initial population is generated completely randomly, and the number of chromosomes in each population becomes the population size (N). The value of N is important and must be specified based on the type of problem and its coding. In the present paper, the population size is set to 400 ($N = 400$).

Crossover
Crossover operator is applied on two parents, and a new child is generated as a result. We consider a two-point crossover in the manner that two crossover points are randomly chosen from the string, and then the two parent chromosomes are interchanged between these points to produce two new children.

Mutation
The second operator in the genetic algorithm is the mutation operator which prevents the algorithm to fall in the local optimum. In our algorithm, we use random mutation.

Stopping criterion
Stopping criterion is the final stage in the genetic algorithm. There are several indexes in this regard. Here, the maximum generation reproduction rule is used, i.e., once the generation counter reaches a certain number, the algorithm will stop. This rule has been used by several researchers. Figure 4 depicts the proposed genetic algorithm.

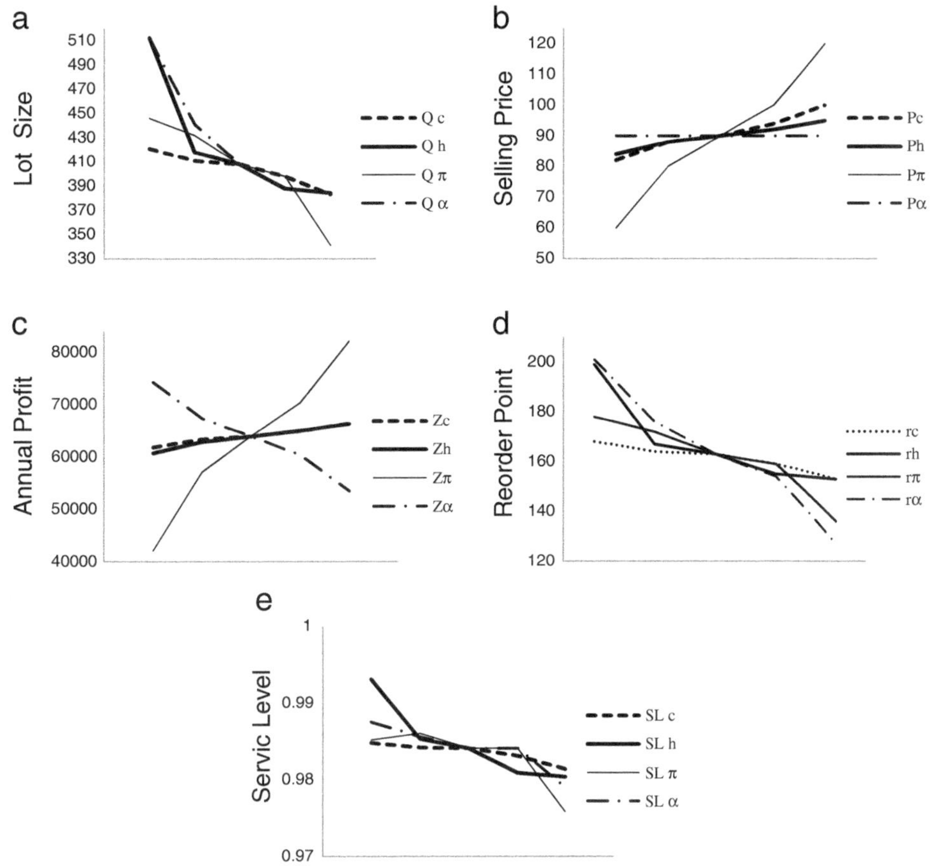

Figure 10 Effect of key parameters on (a) Q, (b) P, (c) Z, (d) r, and (e) SL (exponential model, L-P metric method $p = 2$).

Numerical example and sensitivity analysis

In this section, we illustrate the quality of our model by presenting some examples. We will also perform sensitivity analysis for the key parameters (π, C, h, and α) of the model. Assume that the retailer faces a linear demand function of $D = 1{,}000 - \alpha p$, where α is the demand elasticity coefficient which is equal to 2. The cost of each ordering is US\$25, while the holding cost for each unit of items is US\$5/year and the shortage cost of each unit of items is US\$30/year; $a = 0$ and $b = 35$ (days) if the demand is distributed uniformly, and $\lambda = 17.5$ if it is distributed exponentially. Therefore, the average lead time is the same in both cases.

We consider $p = 1$ and $p = 2$ for the L-P metric method and different objective weights for the weighting method. The obtained optimal solution is shown in Tables 2 and 3.

Sensitivity analysis

To select a suitable strategy for the retailer, we consider the effect of parameters π, C, h, and α on decision variables and objective functions. The optimal solutions are shown in Tables 3, 4, 5, 6, 7, 8, 9, 10, 11, 12, 13, and 14 while the results of sensitivity analysis are shown in Figures 4, 5, and 6. In addition, the weights of the first and second objective functions are considered 0.6 and 0.4, respectively.

As it can be seen in the figures and tables, by increasing the cost parameters, the selling price (p^*) and profit (Z^*) will increase. Moreover, when demand elasticity increases, the lot size (Q^*) will decrease. This happens because the higher the sensitivity of items to the selling price, the more will be the tangible demand decrease against high price. Moreover, reorder point has a direct relation with service level.

As Tables 4, 5, 6, 7, 8, 9, 10, 11, 12, 13, 14, 15, 16, 17, 18, 19, 20, 21, 22, 23, 24, 25, 26, and 27 and Figures 6, 7, 8, 9, and 10 show, the service level in the L-P metric method is larger while the annual profit of the weighting method is larger. In addition, in the L-P metric method, the profit is larger for larger p, but this does not hold for service level.

Conclusions

In this paper, a multi-objective inventory model is presented which includes the maximization of the retailer's profit along with the maximization of customer service level. The demand is assumed as a general function of price. In addition, planning horizon is considered infinite. The stochastic lead time is assumed to be uniformly and exponentially distributed, and the shortage is allowed. The proposed model is a complex (a multi-objective nonlinear model). Therefore, the optimal solution of the selling price, lot size, and reorder point is obtained using the genetic algorithm. The weighting and L-P metric methods are used to change a multi-objective function to a single-objective function. Numerical examples and sensitivity analysis on the key parameters (π, C, h, and α) of the model are presented. The results show that the retailer's profit in uniform distribution is larger, while lot size and reorder point of the exponential model is larger. In addition, increasing the cost parameters will increase the selling price (p^*) and the profit (Z^*). Moreover, when the demand elasticity (α) increases, the lot size (Q^*) will decrease. In addition, the reorder point has a direct relation with the service level. In addition, using the L-P metric method, a higher service level is obtained, but the annual profit of the weighting method is larger.

In the future, it may be interesting to examine a scenario in which the system deals with stochastic consumer demand as well as stochastic lead time in order to define the system more accurately. Considering a multi-period system and planning of the prices of the product as a dynamic pricing may be a scope for future research. The proposed GA variants that employ various crossover and mutation operations could be another area of future interest. New solution methodology based on tabu search or heuristic methods can be developed to obtain new optimal solutions for the multi-objective problem. In this case, conducting more numerical tests to justify the developed algorithm would be necessary. Additionally, uncertainty of costs and demand parameters can be taken into account in the model, and new solution methodologies including uncertainty can be developed via fuzzy models.

Authors' contributions

Zeinab Hosseini drafted the manuscript. She formulated the stochastic joint pricing and inventory problem and designed the solution procedure. Reza Ghasemy Yaghin defined the research area, carried out the finalization of joint pricing and inventory model in multiple objective environment. He supposed multiple objective optimization solution procedures in order to solve the aforementioned model. Maryam Esmaeili supposed and analyzed the mathematical properties of the developed model. Numerical studies had been provided by her. She also supervised the research.

Acknowledgements

The authors would like to thank anonymous referees for their valuable comments and helpful suggestions on improving the quality of the paper.

Author details

[1]Department of Industrial Engineering, Alzahra University, Tehran, Iran.
[2]Department of Industrial Engineering, Amirkabir University of Technology, Tehran, Iran.

References

Abad PL (2003) Optimal pricing and lot-sizing under conditions of perishability, finite production and partial backordering and lost sale. Eur J Oper Res 144:677–685

Abad PL (2008) Optimal price and order size under partial backordering incorporating shortage, backorder and lost sale costs. Int J Prod Econ 114:179–186

Agrell PJ (1995) A multicriteria framework for inventory control. International Journal of Production Economics 41:59–70

Branke J, Deb K, Miettinen K, Slowinski R (editors) (2008) Multi objective optimization. Interactive and evolutionary approaches. Springer, Berlin, Germany

Dye C-Y (2007) Joint pricing and ordering policy for a deteriorating inventory with partial backlogging. Omega 35:184–189

Dye C-Y, Hsieh T-P (2010) A particle swarm optimization for solving joint pricing and lot-sizing problem with fluctuating demand and unit purchasing cost. Comput Math Appl 60:1895–1907

Dye C-Y, Hsieh T-P (2013) Joint pricing and ordering policy for an advance booking system with partial order cancellations. Appl Math Model 37(6):3645–3659

Esmaeili M (2009) Optimal selling price, marketing expenditure and lot size under general demand function. Int J Adv Manuf Tech 45:191–198

Hosseini Z, Esmaeili M, Ghasemy Yaghin R (2012) A multiple objective approach for joint inventory and pricing planning model with stochastic lead times. Paper presented at the 8th International Industrial Engineering Conference. Amirkabir University of Technology, Tehran. 15–16 February 2012

Haupt RL, Haupt SE (2004) Practical genetic algorithms. John Wiley & Sons, Hoboken, New Jersey

Islam S (2008) Multi-objective marketing planning inventory model: a geometric programming approach. Appl Math Comput 205:238–246

Lee J (2011) Inventory control by different service levels. Appl Math Model 35:497–505

Liang S-K, Chu P, Yang K-L (2008) Improved periodic review inventory model involving lead time with crashing components and service level. Int J Syst Sci 39(4):421–426

Mukhopadhyay S, Mukherjee RN, Chaudhuri KS (2004) Joint pricing and ordering policy for a deteriorating inventory. Comput Ind Eng 47:339–349

Moslemi H, Zandieh M (2011) Comparisons of some improving strategies on MOPSO for multi-objective (r, Q) inventory system. Expert Syst Appl 38:12051–12057

Mandal NK, Roy TK, Maiti M (2005) Multi-objective fuzzy inventory model with three constraints: a geometric programming approach. Fuzzy Set Syst 150:87–106

Maity K, Maiti M (2008) A numerical approach to a multi-objective optimal inventory control problem for deteriorating multi-items under fuzzy inflation and discounting. Comput Math Appl 55:1794–1807

Maiti AK, Maiti MK, Maiti M (2009) Inventory model with stochastic lead-time and price dependent demand incorporating advance payment. Appl Math Model 33:2433–2443

Padmanabhan G, Vrat P(1990) Analysis of multi-item inventory systems under resource constraints: A non-linear goal programming approach. Engineering Cost and Production Economics 20:121–127

Pasandideh SHR, Akhavan Niaki ST, Tokhmehchi N (2011) A parameter-tuned genetic algorithm to optimize two-echelon continuous review inventory systems. Expert Syst Appl 38:11708–11714

Roy TK, Maiti M (1998) Multi objective inventory models of deteriorating items with some constraints in a fuzzy environment. Comput Oper Res 25(12):1085–1095

Sana SS (2011) The stochastic EOQ model with random sales price. Appl Math Comput 218:239–248

Sadjadi SJ, Yazdian SA, Shahanaghi K (2012) Optimal pricing, lot-sizing and marketing planning in a capacitated and imperfect production system. Comput Ind Eng 62:349–358

Sheikh Sajadieh M, Akbari Jokar MR (2009) An integrated vendor-buyer cooperative model under stochastic supply lead-time. Int J Adv Manuf Tech 41:1043–1050

Sheikh Sajadieh M, Akbari Jokar MR, Modarres M (2009) Developing a coordinated vendor–buyer model in two-stage supply chains with stochastic lead-times. Comput Oper Res 36:2484–2489

Taleizadeh A, Akhavan Niaki ST, Aryanezhad M-B, Fallah Tafti A (2010) A genetic algorithm to optimize multiproduct multi constraint inventory control systems with stochastic replenishment intervals and discount. Int J Adv Manuf Tech 51:311–323

Tsou C-S (2008) Multi-objective inventory planning using MOPSO and TOPSIS. Expert Syst Appl 35:136–142

Tsou C-S (2009) Evolutionary Pareto optimizers for continuous review stochastic inventory systems. Eur J Oper Res 195:364–371

Whitin T-M (1955) Inventory control and price theory. Manage Sci 2:61–68

20

Mathematical solution of multilevel fractional programming problem with fuzzy goal programming approach

Kailash Lachhwani[1*] and Mahaveer Prasad Poonia[2]

Abstract

In this paper, we show a procedure for solving multilevel fractional programming problems in a large hierarchical decentralized organization using fuzzy goal programming approach. In the proposed method, the tolerance membership functions for the fuzzily described numerator and denominator part of the objective functions of all levels as well as the control vectors of the higher level decision makers are respectively defined by determining individual optimal solutions of each of the level decision makers. A possible relaxation of the higher level decision is considered for avoiding decision deadlock due to the conflicting nature of objective functions. Then, fuzzy goal programming approach is used for achieving the highest degree of each of the membership goal by minimizing negative deviational variables. We also provide sensitivity analysis with variation of tolerance values on decision vectors to show how the solution is sensitive to the change of tolerance values with the help of a numerical example.

Keywords: Multilevel fractional programming, Fuzzy goal programming, Membership function, Tolerance values

Background

Hierarchical optimization or multilevel programming problems (MLPPs) have the following common characteristics: interactive decision making units exist within predominantly hierarchical structures; the execution of decision is sequential from higher level to lower level; each decision-making unit independently controls a set of decision variables and is interested in maximizing its own objective but is affected by the reaction of lower level decision makers (DMs). Due to their dissatisfaction with the decision of the higher level DMs, decision deadlock arises frequently in the decision-making situation. Multilevel fractional programming problems (MLFPPs) involve objective functions in fractional form, i.e., $f(X) = \frac{N(X)}{D(X)}$ at each level with the assumption that the denominator of objectives remains positive at each level in the feasible region. Some important existing solution approaches such as the extreme point search, the procedure based on the Karush-Kuhn Tucker condition, and the decent method (Anandilingam 1988; Anandilingam and Apprey 1991; Biswas and Pal 2005; Bellmann 1957; Charnes and Cooper 1962; Craven and Mond 1975; Lai 1996) are effective only for solving simple types of multilevel programming problems. Initially, fuzzy approach was used to handle multiobjective optimization problems (Chakraborty and Gupta 2002; Jimenez and Bilbas 2009). Lai and Hwang (1993) at first developed an effective fuzzy approach using the concept of tolerance membership functions for solving MLPPs in 1996. Shih et al. (1996) extended Lai's concept using a non-compensatory maximum-minimum aggregation operator for solving MLPPs. Shih and Lee (2000) further extended Lai's concept by introducing the compensatory fuzzy operator for solving MLPPs. Sinha (2003a,b) studied alternative MLP techniques based on fuzzy mathematical programming (FMP). The basic concept of these fuzzy approaches is the same, and evaluation of the problem again and again by redefining the elicited membership values is essentially needed in the solution search process to obtain a satisfactory solution. So, computational load is also inherently involved in the fuzzy approaches developed so far. In the FMP techniques of Sinha (2003a,b), the last (lower) level is the most important, and the decision of the lowest level remains either unchanged or closest to individual best decisions, which leads

* Correspondence: kailashclachhwani@yahoo.com
[1]Department of Mathematics, Government Engineering College, Bikaner 334004, India
Full list of author information is available at the end of the article

to the paradox that the decision power of the lowest level DM dominates the higher level DM. To overcome such difficulties, the fuzzy goal programming (FGP) approach to multidecision-making problems was introduced by Mohamed (1997) which is extended by Pramanik and Roy (2007) to solve MLPPs. Baky (2009) used fuzzy goal programming to solve decentralized bilevel multiobjective programming problems. Chang (2009) suggested goal programming approach for fuzzy multiobjective fractional programming problems. Recently, Pal and Gupta (2009) studied a genetic algorithm to fuzzy goal programming formulation of fractional multiobjective decision-making problems.

In real-world decision-making situations, decision makers sometimes may be faced with the decision to optimize inventory/sales, actual cost/standard cost, output/employee, etc. with respect to some constraints. Such type of problems in a large hierarchical organization from higher level to lower level and their sequential decisions on complex and conflicting objectives formulate the MLFPPs. Practical optimization situations involving multilevel with fractional objectives have been rare, but such problems can be encountered in the most complex design, pattern recognition, control theory, and resource allocation situations (Mohamed 1997). Motivated by the concept of interactive fuzzy goal programming and fractional programming, an effort has been made to examine the possibility of unifying the level-wise (hierarchical) and stage-wise operations with the assumption of a positive denominator of objective functions at each level. The aim of this paper is to present a procedure to solve multilevel fractional programming problems. Our proposed methodology involves the fuzzy goal levels of the numerator and denominator part of each objective as well as decision vectors controlled by the higher level DMs, which are determined by individual optimal solutions. Then, the fuzzy goals are characterized by the associated membership functions which are transformed into fuzzy flexible membership goals by means of introducing negative and positive deviational variables and assigning a higher membership value (unity) as aspiration level to each of them. Since overdeviation from any fuzzy goal implies the full achievement of the membership values, we assign only negative deviational variables to the achievement function and minimize negative deviational variables to get a compromise optimal solution. To illustrate our proposed method, we solve a numerical example and compare the results with the change in tolerance limits.

The paper is organized as follows: In the 'Formulation of MLPP' section, we discuss formulation of MLFPP and the related terminology. In the 'Fuzzy programming formulation of MLFPP' section, we characterize the linear membership functions for the numerator and denominator of objective functions at each level as well as decision vectors controlled by the higher level DMs. In the next section, we discuss the proposed FGP approach to tackle MLFPPs and formulate different mathematical models related to it. In the 'Selection of compromise solution' section, selection criteria of compromise optimal solution are described. To illustrate the proposed methodology, a numerical example is considered and sensitivity analysis is performed with the change in tolerance limits in the 'Numerical example' section. Concluding remarks are given in the last sections.

Results and discussion

Formulation of MLFPP

We consider a T-level fractional programming problem of maximization-type objectives at each level. Mathematically, we can state it as follows:

$$\max_{X_1} Z_1(\bar{X}) = \frac{\bar{C}_{11}\ \bar{X}_1 + \bar{C}_{12}\ \bar{X}_2 + \ldots + \bar{C}_{1T}\ \bar{X}_T + \alpha_1}{\bar{D}_{11}\ \bar{X}_1 + \bar{D}_{12}\ \bar{X}_2 + \ldots + \bar{D}_{1T}\ \bar{X}_T + \beta_1}$$

$$\max_{X_2} Z_2(\bar{X}) = \frac{\bar{C}_{21}\ \bar{X}_1 + \bar{C}_{22}\ \bar{X}_2 + \ldots + \bar{C}_{2T}\ \bar{X}_T + \alpha_2}{\bar{D}_{21}\ \bar{X}_1 + \bar{D}_{22}\ \bar{X}_2 + \ldots + \bar{D}_{2T}\ \bar{X}_T + \beta_2}$$

$$\max_{X_T} Z_T(\bar{X}) = \frac{\bar{C}_{T1}\ \bar{X}_1 + \bar{C}_{T2}\ \bar{X}_2 + \ldots + \bar{C}_{TT}\ \bar{X}_T + \alpha_T}{\bar{D}_{T1}\ \bar{X}_1 + \bar{D}_{T2}\ \bar{X}_2 + \ldots + \bar{D}_{TT}\ \bar{X}_T + \beta_T}$$

subject to

$$\bar{A}_{i1}\ \bar{X}_1 + \bar{A}_{i2}\ \bar{X}_2 + \ldots\ldots + \bar{A}_{iT}\ \bar{X}_T\ (\le, =, \ge) b_i \quad \forall i = 1, 2, \ldots, m$$

and (1)

$$\bar{X}_1 \ge 0,\ \bar{X}_2 \ge 0,\ \ldots,\ \bar{X}_T \ge 0$$

$\bar{X}_1 = \{X_1^1, X_1^2, \ldots\ldots, X_1^{N_1}\}'$ decision variables are under the control of the first level DM; $\bar{X}_T = \{X_T^1, X_T^2, \ldots\ldots, X_T^{N_T}\}'$ decision variables are under the control of the t-level DM.

Where $'$ denotes transposition, $\bar{A}_{ij}$ $i = 1, 2, \ldots, m$, and $j = 1, 2, \ldots, T$ are m row vectors, each with dimension $(1 \times N_j)$. $\bar{A}_{it}\ \bar{X}_t$, $t = 1, 2, \ldots, T$ is a column vector of dimension $(M \times 1)$. $\bar{C}_{11}, \bar{C}_{21}, \ldots, \bar{C}_{T1}$ are row vectors of dimension $(1 \times N_1)$. Similarly, $\bar{C}_{1T}, \bar{C}_{2T}, \ldots, \bar{C}_{TT}$ and $\bar{D}_{1T}, \bar{D}_{2T}, \ldots, \bar{D}_{TT}$ are row vectors of dimension $(1 \times N_T)$. We take $\bar{X} = \bar{X}_1 \cup \bar{X}_2 \cup \ldots \cup \bar{X}_T$ and $N = N_1 + N_2 + \ldots\ldots + N_T$. Here, one DM is located on each level. Decision vector $\bar{X}_t$, $t = 1, 2, \ldots, T$ is the control of the t-th level DM having N_t number of decision variables. Here, it is assumed that the denominator of objective functions is positive at each level for all the values of decision variables in the constraint region.

Fuzzy programming formulation of MLFPP

To formulate the fuzzy programming model of MLFPP, the objective numerator $f_{iN}(\bar{X}) + \alpha_i, \forall t = 1, 2, \ldots, T$ and objective denominator $f_{iD}(\bar{X}) + \alpha_i,\ \ \forall t = 1, 2, \ldots, T$ at each level and the decision vector $\bar{X}_t$, $(t = 1, 2, .., T-1)$ would be transformed into fuzzy goals by means of assigning an aspiration level to each of them. Then, they are to be characterized by the associated membership functions by defining tolerance limits for the achievement of the aspired levels of the corresponding fuzzy goals.

Characterization of membership function of MLFPP

In the decision-making context, each DM is interested in maximizing his or her own objective function; the optimal solution of each DM when calculated in isolation would be considered as the best solution, and the associated objective values can be considered as the aspiration level of the corresponding fuzzy goal. Let $\bar{X}_t^B$ be the best solution of the t-th level DM. It is quite natural that objective values which are equal to or larger than $Z_t^B = Z_t\left(\bar{X}_t^B\right) = \underset{\bar{X}\in X}{Max}\, Z_t(\bar{X}), t = 1, 2, \ldots, T$ should be absolutely satisfactory to the t-th level DM. If the individual best solutions $\bar{X}_t^B, t = 1, 2, \ldots, T$ are the same, then a satisfactory optimal solution of the system is reached. However, this rarely happens due to the conflicting nature of the objectives. To obtain a satisfactory solution, the higher DM should give some tolerance (relaxation), and the relaxation of the decision of the higher level DM depends on the needs, desires, and practical situations in the decision-making situation. Then, the fuzzy goals take the form

$$Z_t(\bar{X}) \gtrsim Z_t\left(\bar{X}_t^B\right), t = 1, 2, \ldots, T$$

and $\bar{X}_t \cong \bar{X}_t^B, t = 1, 2, \ldots, (T-1)$.

To build membership functions, fuzzy goals and tolerance should be determined first. However, they could hardly be determined without meaningful supporting data. Using the individual best solution, we find the values of all the numerator objective functions and denominator objective functions at each best solution and construct a payoff matrix as follows:

$$\begin{bmatrix} X_1^B & N_1 & N_2 & \cdot & \cdot & N_T \\ \bar{X}_1^B & N_1(\bar{X}_1^B) & N_2(\bar{X}_1^B) & \cdot & \cdot & N_T(\bar{X}_1^B) \\ \bar{X}_2^B & N_1(\bar{X}_2^B) & N_2(\bar{X}_2^B) & \cdot & \cdot & N_T(\bar{X}_2^B) \\ \cdot & \cdot & \cdot & \cdot & \cdot & \cdot \\ \cdot & \cdot & \cdot & \cdot & \cdot & \cdot \\ \bar{X}_T^B & N_1(\bar{X}_T^B) & N_2(\bar{X}_T^B) & \cdot & \cdot & N_T(\bar{X}_T^B) \end{bmatrix}$$

and (2)

$$\begin{bmatrix} X_1^B & D_1 & D_2 & \cdot & \cdot & D_T \\ \bar{X}_1^B & D_1(\bar{X}_1^B) & D_2(\bar{X}_1^B) & \cdot & \cdot & D_T(\bar{X}_1^B) \\ \bar{X}_2^B & D_1(\bar{X}_2^B) & D_2(\bar{X}_2^B) & \cdot & \cdot & D_T(\bar{X}_2^B) \\ \cdot & \cdot & \cdot & \cdot & \cdot & \cdot \\ \cdot & \cdot & \cdot & \cdot & \cdot & \cdot \\ \bar{X}_T^B & D_1(\bar{X}_T^B) & D_2(\bar{X}_T^B) & \cdot & \cdot & D_T(\bar{X}_T^B) \end{bmatrix}$$

Here, $\bar{X}_t (t = 1, 2, \ldots, T)$ are assumed to be the main decision vectors. The maximum value of each column $N_t(\bar{X}_t^B)$ and $D_t(\bar{X}_t^B)$ give upper tolerance limit or aspired level of achievement for the t-th numerator objective function and denominator objective function, respectively, where $N_t^B = N_t\left(\bar{X}_t^B\right) = \underset{\bar{X}\in X}{\text{Max}}\, N_t(\bar{X}), \quad t = 1, 2, \ldots, T$. The minimum value of each column gives the lower tolerance limit or lowest acceptable level of achievement for the t-th numerator objective function and denominator objective function, respectively, where $N_t^L = \underset{\bar{X}\in X}{\text{Min}} \{N_t(\bar{X}_1^B), N_t(\bar{X}_2^B), \ldots, N_t(\bar{X}_T^B)\},. t = 1, 2, \ldots, T$. Then, the linear membership functions for the defined fuzzy goals are as follows (see also Figures 1 and 2):

$$\mu_{Z_t}(N_t(\bar{X})) = \begin{cases} 1 & if\ N_t(\bar{X}) \geq N_t^B \\ \dfrac{N_t(\bar{X}) - N_t^L}{N_t^B - N_t^L} & if\ N_t^L \leq N_t(\bar{X}) \leq N_t^B \\ 0 & if\ N_t(\bar{X}) \leq N_t^L \\ & \forall t = 1, 2, \ldots, T \end{cases} \tag{3}$$

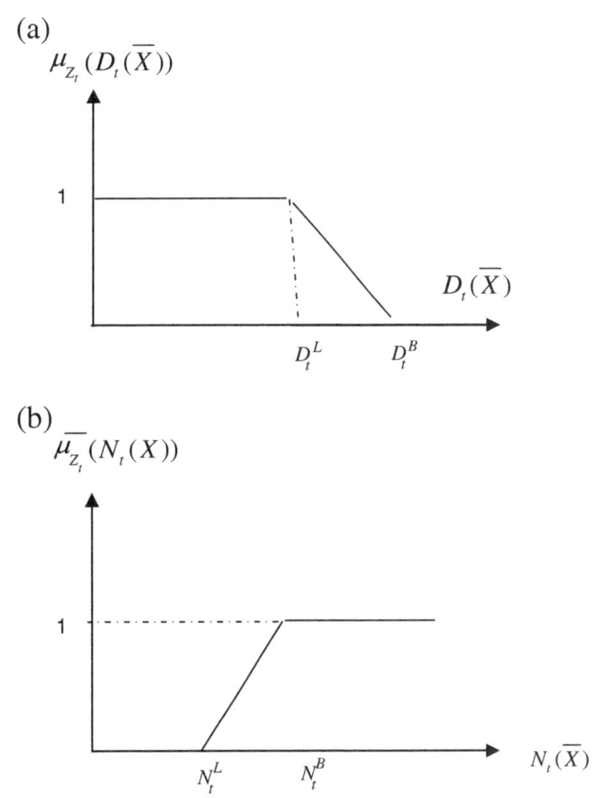

Figure 1 (a): Membership function for $\mu_{Z_t}(D_i(\bar{X}))$ (b): Membership function for $\mu_{Z_t}(N_i(\bar{X}))$.

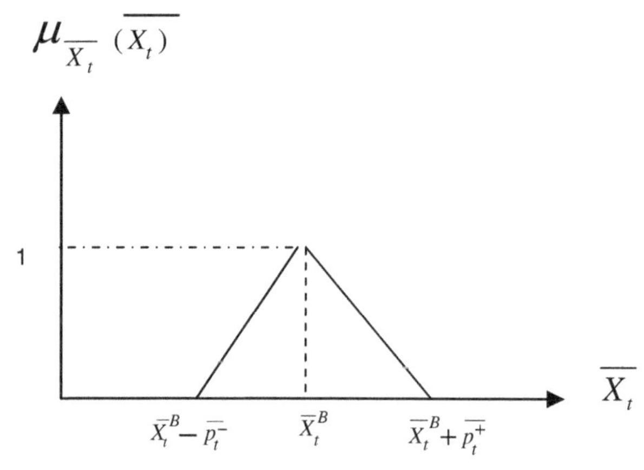

Figure 2 Membership functions of decision vector $\overline{X}_t$ (t = 1, 2, .., T−1).

$$\mu_{Z_t}(D_t(\bar{X})) = \begin{cases} 0 & \text{if } D_t(\bar{X}) \geq D_t^B \\ \dfrac{D_t^B - D_t(\bar{X})}{D_t^B - D_t^L} & \text{if } D_t^L \leq D_t(\bar{X}) \leq D_t^B \\ 1 & \text{if } D_t(\bar{X}) \leq D_t^L \\ & \forall t = 1,2,\ldots,T \end{cases} \tag{4}$$

Here, linear membership functions are more suitable than nonlinear functions as less computational difficulties arise in models due to it. Let $\overline{p_t^-}, \overline{p_t^+}$ $(t = 1,2,\ldots,T-1)$ be the negative and positive tolerance values on decision vectors $\overline{X}_t$ considered by the t-th level DM. This is a triangular fuzzy number. Then, the linear membership functions for decision vectors $\overline{X}_t$ can be formulated as follows:

$$\mu_{X_t}(\overline{X}_t) = \begin{cases} \dfrac{\overline{X}_t - \left(\overline{X}_t^B - \overline{p_t^-}\right)}{\overline{p_t^-}} & \text{if } \left(\overline{X}_t^B - \overline{p_t^-}\right) \leq \overline{X}_t \leq \overline{X}_t^B \\ \dfrac{\left(\overline{X}_t^B + \overline{p_t^+}\right) - \overline{X}_t}{\overline{p_t^+}} & \text{if } \overline{X}_t^B \leq \overline{X}_t \leq \left(\overline{X}_t^B + \overline{p_t^+}\right) \\ 0 & \text{otherwise} \\ & \forall t = 1,2,\ldots,T \end{cases} \tag{5}$$

Here, $\overline{p_t^-}$ and $\overline{p_t^+}$ are the negative and positive tolerance vectors; $\overline{p_t^-}$ and $\overline{p_t^+}$ are not necessarily same. Generally, $\overline{X}_t$ lies between $\overline{X}_t^B - \overline{p_t^-}$ and $\overline{X}_t^B + \overline{p_t^+}$. DMs may prefer to shift the range of $\overline{X}_t^B$ which may be the left of $\overline{X}_t^B$ or the right of $\overline{X}_t^B$, only depending on the needs and desires of the higher level DMs in the decision-making situation. Then, the membership function becomes one-sided. For example, if $\overline{X} = \overline{0}$, then $\overline{X}$ should lie on the right of $\overline{0}$. Then, the DM should assign $\overline{p_t^-} \leq \overline{0}$, $\overline{p_t^+} \geq \overline{0}$ and $|\overline{p_t^-}| \leq |\overline{p_t^+}|$. If the DM wants the shift towards the left of $\overline{X}_t^B$, then $\overline{p_t^-}$ should be assigned a positive value while $\overline{p_t^+}$ should be assigned a negative value, i.e., $\overline{p_t^-} \geq \overline{0}$, $\overline{p_t^+} \leq \overline{0}$, and $|\overline{p_t^-}| \geq |\overline{p_t^+}|$. Similarly, if the shift is required to the right of $\overline{X}_t^B$, then the DM should assign $\overline{p_t^-} \leq \overline{0}, \overline{p_t^+} \geq 0$, and $|\overline{p_t^-}| \leq |\overline{p_t^+}|$. We may treat the tolerance as variables with the restrictions that $\overline{p_t^-} \leq \overline{X}_t^B$ (so that the value of the variables remain non-negative).

FGP solution approach

FGP is an extension of conventional goal programming (GP) introduced by Charnes and Cooper (1962). GP has been extensively studied and widely circulated in literature (Arora and Gupta 2009; Pramanik and Roy 2007). In this paper, GP approach to fuzzy multiobjective decision-making problems introduced by Mohamed (1997) is extended to solve MLFPP problems. In a decision-making situation, the aim of each DM is to achieve the highest membership value (unity) of the associated fuzzy goal in order to obtain the absolute satisfactory solution. However, in real practice, achievement of all membership values to the highest degree (unity) is not possible due to conflicting objectives. Therefore, the decision policy for minimizing the regrets of the DMs for all the levels should be taken into consideration. Then, each DM should try to maximize his or her membership function by making them as close as possible to unity by minimizing its negative deviational variables. Therefore, in effect, we are simultaneously optimizing all the objective functions. So, for the defined membership functions in Equations 3, 4, and 5, the flexible membership goals having the aspired level unity can be represented as follows:

$$\mu_{Z_t}(N_t(\bar{X})) + D_{t1}^- - D_{t1}^+ = 1;\ t = 1,2,\ldots,T \tag{6}$$

$$\mu_{Z_t}(D_t(\bar{X})) + D_{t2}^- - D_{t2}^+ = 1;\ t = 1,2,\ldots,T \tag{7}$$

$$\mu_{\overline{X}_t}(\overline{X}_t) + \overline{D}_{t3}^- - \overline{D}_{t3}^+ = \overline{I};\ t = 1,2,\ldots,(T-1) \tag{8}$$

Here, D_{t1}^-, D_{t2}^- are negative deviational variables, and D_{t1}^+, D_{t2}^+ are positive deviational variables; $\overline{D}_{t3}^+, \overline{D}_{t3}^-$ represent the vector of negative deviational and positive deviational variables. It is to be noted that any overdeviation from a fuzzy goal implies the full achievement value. Then, Equations 6, 7, and 8 can be written as follows:

$$\mu_{Z_t}(N_t(\bar{X})) + D_{t1}^- \geq 1\ ;\ t = 1,2,\ldots,T \tag{9}$$

$$\mu_{Z_t}(D_t(\bar{X})) + D_{t2}^- \geq 1\ ;\ t = 1,2,\ldots,T \tag{10}$$

$$\mu_{\overline{X}_t}(\overline{X}_t) + \overline{D}_{t3}^- \geq \overline{I}\ ;\ t = 1,2,\ldots,(T-1) \tag{11}$$

FGP formulation can be presented as follows:

Model I: Find $\bar{X}$ so as to minimize λ subject to

$$
\begin{aligned}
&\mu_{Z_t}(N_t(\bar{X}))+D_{t1}^{-}\geq 1;\; t=1,2,\ldots,T \\
&\mu_{Z_t}(D_t(\bar{X}))+D_{t2}^{-}\geq 1;\; t=1,2,\ldots,T \\
&\mu_{\bar{X}_t}(\bar{X}_t)+\bar{D}_{t3}^{-}\geq \bar{I};\; t=1,2,\ldots,(T-1) \\
&\lambda\geq D_{t1}^{-}; t=1,2,\ldots,T \\
&\lambda\geq D_{t2}^{-};\; t=1,2,\ldots,T \\
&\lambda\geq \bar{D}_{t3}^{-}\bar{I};\; t=1,2,\ldots,(T-1) \\
&D_{t1}^{-}\geq 0; t=1,2,\ldots,T \\
&D_{t2}^{-}\geq 0; t=1,2,\ldots,T \\
&\bar{D}_{t3}^{-}\geq \bar{0}; t=1,2,\ldots,(T-1) \\
&\bar{A}_{i1}\;\bar{X}_1+\bar{A}_{i2}\;\bar{X}_2+\ldots\ldots\ldots\ldots+\bar{A}_{iT}\;\bar{X}_T\,(\leq,=,\geq)b_i \\
&\qquad\forall i=1,2,\;\ldots,m
\end{aligned}
\tag{12}
$$

and $\bar{X}_1\geq 0,\bar{X}_2\geq 0,\ldots\ldots\ldots,\bar{X}_T\geq 0$

The above problem can be rewritten as follows:

Minimize λ subject to

$$
\begin{aligned}
&\frac{N_t(\bar{X})-N_t^L}{N_t^B-N_t^L}+D_{t1}^{-}\geq 1;\; t=1,2,\ldots,T \\
&\frac{D_t^B-D_t(\bar{X})}{D_t^B-D_t^L}+D_{t2}^{-}\geq 1;\; t=1,2,\ldots,T \\
&\bar{X}_t-\left(\bar{X}_t^B-\bar{p}_t^{-}\right)\bar{p}_t^{-}+\bar{D}_{t31}^{-}\geq \bar{I};\; t=1,2,\ldots,(T-1) \\
&\left(\bar{X}_t^B+\bar{p}_t^{+}\right)-\bar{X}_t\bar{p}_t^{+}+\bar{D}_{t32}^{-}\geq \bar{I};\; t=1,2,\ldots,(T-1) \\
&\lambda\geq D_{t1}^{-};\; t=1,2,\ldots,T \\
&\lambda\geq D_{t2}^{-};\; t=1,2,\ldots,T \\
&\lambda\geq \bar{D}_{t31}^{-}\bar{I};\; t=1,2,\ldots,(T-1) \\
&\lambda\geq \bar{D}_{t32}^{-}\bar{I};\; t=1,2,\ldots,(T-1) \\
&D_{t1}^{-}\geq 0;\; t=1,2,\ldots,T \\
&D_{t2}^{-}\geq 0;\; t=1,2,\ldots,T \\
&D_{t31}^{-}\geq \bar{0};\; t=1,2,\ldots,\bar{(}T-1) \\
&D_{t32}^{-}\geq \bar{0};\; t=1,2,\ldots,\bar{(}T-1) \\
&\bar{A}_{i1}\;\bar{X}_1+\bar{A}_{i2}\;\bar{X}_2+\ldots\ldots\ldots\ldots+\bar{A}_{iT}\;\bar{X}_T\,(\leq,=,\geq)b_i \\
&\qquad\forall i=1,2,\;\ldots,m
\end{aligned}
\tag{13}
$$

and

$\bar{X}_1\geq 0,\bar{X}_2\geq 0,\ldots\ldots\ldots,\bar{X}_T\geq 0$

Here, D_{t1}^{-},D_{t2}^{-} are negative deviational variables. $\bar{D}_{t31}^{-},\bar{D}_{t32}^{-}$ represent the vector of underdeviational variables. $\bar{I}$ is the column vector having all components equal to 1, and its dimension depends on $\bar{X}$.

Model IIa: Find $\bar{X}$ so as to minimize

$$\lambda=\sum_{t=1}^{T}W_{t1}^{-}D_{t1}^{-}+\sum_{t=1}^{T-1}W_{t2}^{-}D_{t2}^{-}+\sum_{t=1}^{T-1}\bar{W}_{t3}^{-}\bar{D}_{t3}^{-}$$

Model IIb: Find $\bar{X}$ so as to minimize

$$\lambda=\sum_{t=1}^{T}D_{t1}^{-}+\sum_{t=1}^{T-1}D_{t2}^{-}+\sum_{t=1}^{T-1}\bar{D}_{t3}^{-}$$

subject to

$$
\begin{aligned}
&\mu_{Z_t}(N_t(\bar{X}))+D_{t1}^{-}\geq 1;\; t=1,2,\ldots,T \\
&\mu_{Z_t}(D_t(\bar{X}))+D_{t2}^{-}\geq 1;\; t=1,2,\ldots,T \\
&\mu_{\bar{X}_t}(\bar{X}_t)+\bar{D}_{t3}^{-}\geq \bar{I};\; t=1,2,\ldots,(T-1) \\
&D_{t1}^{-}\geq 0;\; t=1,2,\ldots,T \\
&D_{t2}^{-}\geq 0;\; t=1,2,\ldots,T \\
&\bar{D}_{t3}^{-}\geq \bar{0};\; t=1,2,\ldots,(T-1) \\
&\bar{A}_{i1}\;\bar{X}_1+\bar{A}_{i2}\;\bar{X}_2+\ldots\ldots\ldots\ldots+\bar{A}_{iT}\;\bar{X}_T\,(\leq,=,\geq)b_i \\
&\qquad\forall i=1,2,\;\ldots,m
\end{aligned}
\tag{14}
$$

and

$\bar{X}_1\geq 0,\bar{X}_2\geq 0,\ldots\ldots\ldots,\bar{X}_T\geq 0$

Here, D_{t1}^{-},D_{t2}^{-} are negative deviational variables. $\bar{D}_{t3}^{-}$ represents the vector of underdeviational variables. The numerical weights are taken as $W_{t1}^{-}={}^{1}\big/_{(N_t^B-N_t^L)}$, $W_{t2}^{-}={}^{-}1/_{(D_t^B-D_t^L)}$, and $\bar{W}_{t3}^{-}=\left[{}^{1}\big/_{(\bar{p}_t^{-})},{}^{1}\big/_{(\bar{p}_t^{+})}\right]$. $\bar{I}$ is the column vector having all components equal to 1, and its dimension depends on $\bar{X}$. The above problem in models IIa and IIb can be rewritten as follows:

Minimize

$$\lambda=\sum_{t=1}^{T}W_{t1}^{-}D_{t1}^{-}+\sum_{t=1}^{T-1}W_{t2}^{-}D_{t2}^{-}+\sum_{t=1}^{T-1}\bar{W}_{t31}^{-}\bar{D}_{t31}^{-}+\sum_{t=1}^{T-1}\bar{W}_{t32}^{-}\bar{D}_{t32}^{-}$$

Minimize

$$\lambda=\sum_{t=1}^{T}D_{t1}^{-}+\sum_{t=1}^{T-1}D_{t2}^{-}+\sum_{t=1}^{T-1}\bar{D}_{t31}^{-}+\sum_{t=1}^{T-1}\bar{D}_{t32}^{-}$$

subject to

$$
\begin{aligned}
&\frac{N_t(\bar{X})-N_t^L}{N_t^B-N_t^L}+D_{t1}^{-}\geq 1\;;\; t=1,2,\ldots,T \\
&\frac{D_t^B-D_t(\bar{X})}{D_t^B-D_t^L}+D_{t2}^{-}\geq 1\;;\; t=1,2,\ldots,T \\
&\bar{X}_t-\left(\bar{X}_t^B-\bar{p}_t^{-}\right)\bar{p}_t^{-}+\bar{D}_{t31}^{-}\geq \bar{I}\;;\, \mathrm{t}=1,2,\ldots,(T-1) \\
&\left(\bar{X}_t^B+\bar{p}_t^{+}\right)-\bar{X}_t\bar{p}_t^{+}+\bar{D}_{t32}^{-}\geq \bar{I}\;;\; t=1,2,\ldots,(T-1) \\
&D_{t1}^{-}\geq 0\;;\; t=1,2,\ldots,T \\
&D_{t2}^{-}\geq 0\;;\; t=1,2,\ldots,T \\
&\bar{D}_{t3}^{-}\geq \bar{0}\;;\; t=1,2,\ldots,(T-1) \\
&\bar{A}_{i1}\;\bar{X}_1+\bar{A}_{i2}\;\bar{X}_2+\ldots\ldots\ldots\ldots+\bar{A}_{iT}\;\bar{X}_T\,(\leq,=,\geq)b_i \\
&\qquad\forall i=1,2,\;\ldots,m
\end{aligned}
$$

and

$\bar{X}_1\geq 0,\bar{X}_2\geq 0,\ldots\ldots\ldots,\bar{X}_T\geq 0$

(15)

Here, D_{t1}^{-}, D_{t2}^{-} are negative deviational variables. $\bar{D}_{t31}^{-}$ and $\bar{D}_{t32}^{-}$ represent the vector of underdeviational variables. $W_{t1}^{-} = {}^{1}\big/_{(N_t^B - N_t^L)}$, $W_{t2}^{-} = {}^{-1}\big/_{(D_t^B - D_t^L)}$ and $\bar{W}_{t31}^{-} = {}^{1}\big/_{(\bar{p}_t^{-})}$; $\bar{W}_{t31}^{-} = {}^{1}\big/_{(\bar{p}_t^{+})}$. $\bar{I}$ is the column vector having all components equal to 1, and its dimension depends on $\bar{X}$. By solving Equation 15, if the DMs are satisfied with this solution, then a satisfying solution is reached. Otherwise, higher level DMs should provide new tolerance limits for the control variable until a satisfying solution is reached. In general, considering a set of positive relaxation offered by the higher level DMs, the solution of Equation 15 becomes satisfying for all the level DMs.

Numerical example

Let us consider the following tri-level linear fractional programming problem as

$$\underset{x_1,x_2}{\text{Max}}\ Z_1 = \frac{7x_1 + 3x_2 - 4x_3 + 2x_4}{x_1 + x_2 + x_3 + 1}$$

$$\underset{x_3}{\text{Max}}\ Z_2 = \frac{x_2 + 3x_3 + 4x_4}{x_1 + x_2 + x_3 + 2}$$

$$\underset{x_4}{\text{Max}}\ Z_3 = \frac{2x_1 + x_2 + x_3 + x_4}{x_1 + x_2 + x_3 + 3}$$

subject to

$$x_1 + x_2 + x_3 + x_4 \le 5$$

$$x_1 + x_2 - x_3 - x_4 \le 2$$

$$x_1 + x_2 + x_3 \ge 1$$

$$x_1 - x_2 + x_3 + 2x_4 \le 4$$

$$x_1 + 2x_3 + 2x_4 \le 3$$

$$x_4 \le 2$$

and $x_1 \ge 0, x_2 \ge 0, x_3 \ge 0, x_4 \ge 0.$

We find the best optimal solution

$N_1^L = -6$ at $(0, 0, 1.5, 0)$,

$N_1^B = 17$ at $(2.3333, 0, 0,0.3333)$,

$N_2^B = 9.5$ at $(0,3.5,\ 0,1.5)$,

$N_2^L = 0$ at $(1,\ 0,\ 0,0)$,

$N_3^L = 1$ at $(0,\ 1,\ 0,0)$, and

$N_3^B = 5$ at $(2.3333, 0,0.3333,0)$.

Similarly, $D_1^B = 6$ at $(0, 3.5,1.5,0)$,

$D_1^L = 2$ at $(1, 0, 0,0)$,

$D_2^B = 7$ at $(0, 3.5,\ 1.5, 0)$,

$D_2^L = 3$ at $(1, 0, 0,0)$,

$D_3^B = 8$ at $(0,3.5,1.5,0)$,

and $D_3^L = 4$ at $(1, 0, 0,0)$.

Let the first level DM decide that $x_1 = 2.3333$ with −2 (negative) and +2 (positive) tolerance limits and $x_2 = 0$ with −6.43 (negative) and +6.43 (positive) tolerance limits. Let the second level DM decide that $x_3 = 0$ with −1 (negative) and +1 (positive) tolerance limits.

Then, following the procedure, FGP model I gives the problem as follows:

Minimize λ subject to

$$\frac{N_t(\bar{X}) - N_t^L}{N_t^B - N_t^L} + D_{t1}^{-} \ge 1;\ t = 1, 2, \ldots, T$$

$$\Rightarrow 7x_1 + 3x_2 - 4x_3 + 2x_4 + 23D_{11}^{-} \ge 17$$

$$\Rightarrow x_2 + 3x_3 + 4x_4 + 9.5D_{21}^{-} \ge 9.5$$

$$\Rightarrow 2x_1 + x_2 + x_3 + x_4 + 4D_{31}^{-} \ge 5$$

$$\frac{D_t^B - D_t(\bar{X})}{D_t^B - D_t^L} + D_{t2}^{-} \ge 1\ ;\ t = 1, 2, \ldots, T$$

$$\Rightarrow x_1 + x_2 + x_3 - 4D_{12}^{-} \ge 1$$

$$\Rightarrow x_1 + x_2 + x_3 - 4D_{22}^{-} \le 1$$

$$\Rightarrow x_1 + x_2 + x_3 - 4D_{32}^{-} \le 1$$

$$\bar{X}_t - \left(\bar{X}_t^B - \bar{p}_t^{-}\right)\bar{p}_t^{-} + \bar{D}_{t31}^{-} \ge \bar{I};$$

$$\left(\bar{X}_t^B + \bar{p}_t^{+}\right) - \bar{X}_t\bar{p}_t^{+} + \bar{D}_{t32}^{-} \ge \bar{I};\ t = 1, 2, \ldots, (T-1)$$

$$\Rightarrow -0.5x_1 + D_{1311}^{-} \ge -1.1666$$

$$\Rightarrow -0.5x_1 + D_{1321}^{-} \ge -1.1666$$

$$\Rightarrow -0.1555x_2 + D_{1312}^{-} \ge 0$$

$$\Rightarrow -0.1555x_2 + D_{1322}^{-} \ge 0$$

$$\Rightarrow -x_3 + D_{2311}^{-} \ge 0$$

$$\Rightarrow -x_3 + D_{2321}^{-} \ge 0$$

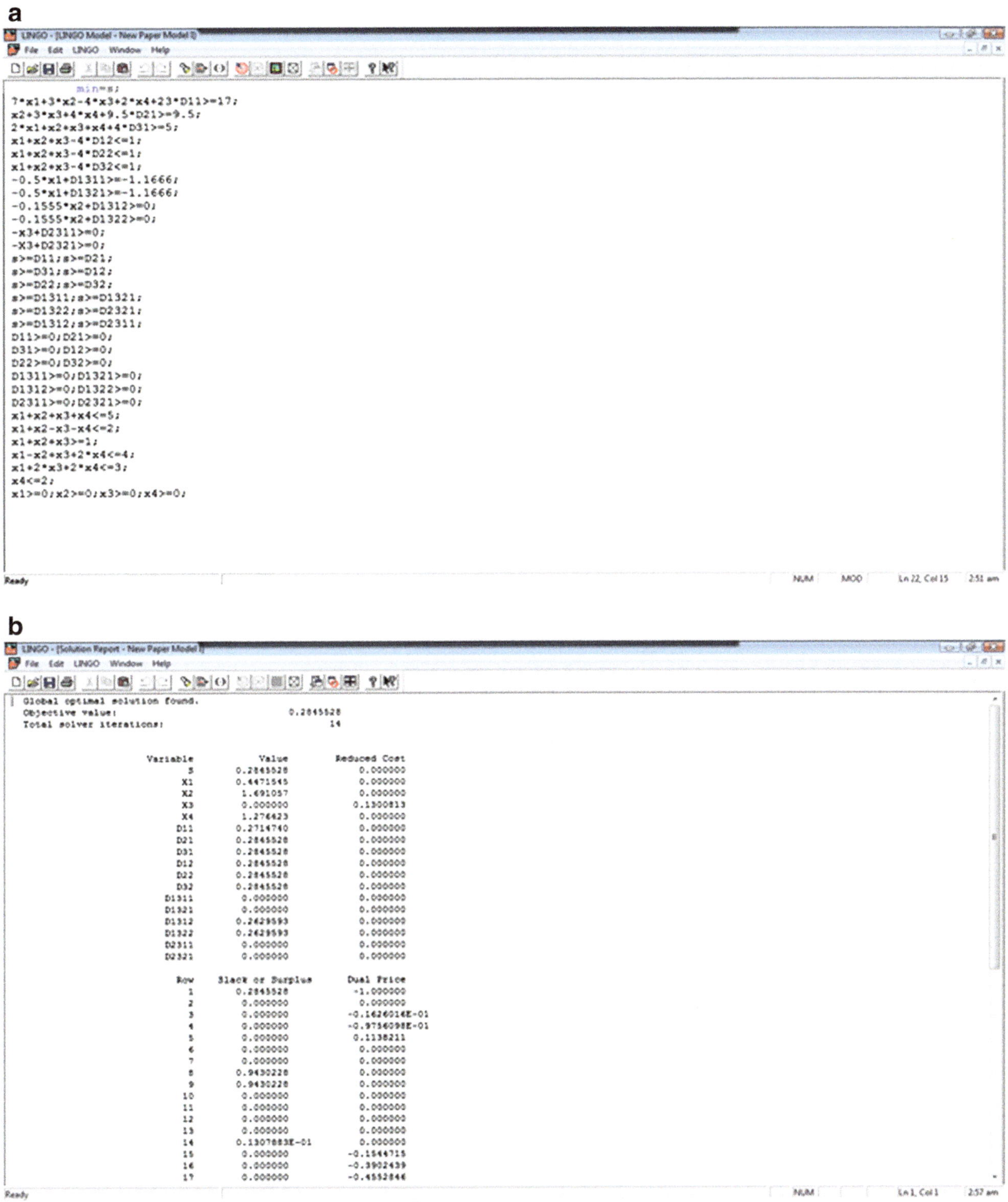

Figure 3 Description (a) and solution (b) of example 1 with model I in LINDO 10.0 (trial version).

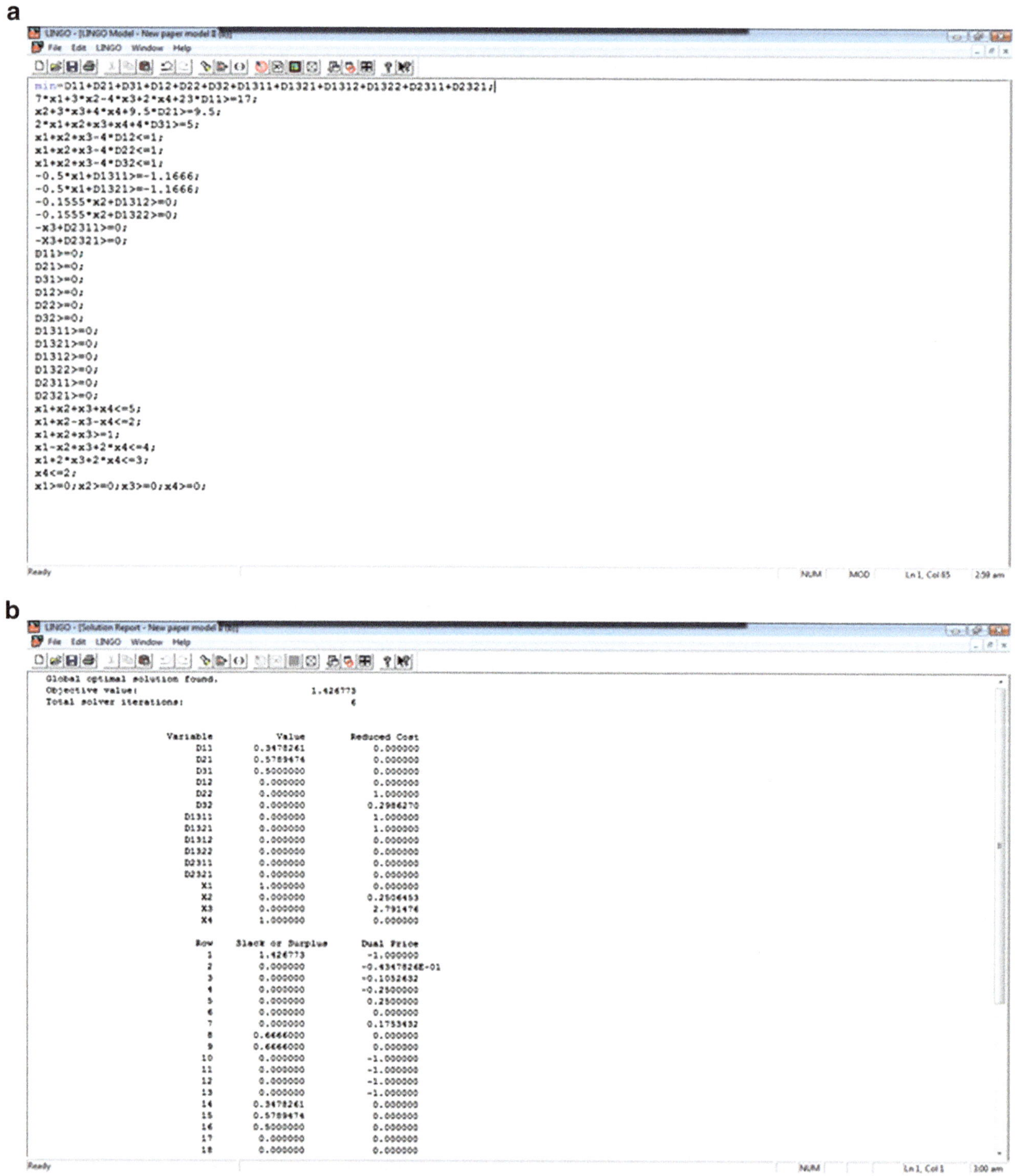

Figure 4 Description (a) and solution (b) of example 1 with model IIb in LINDO 10.0 (trial version).

$\lambda \geq D_{t1}^{-};\ t = 1, 2, \ldots, T$

$\Rightarrow \lambda \geq D_{11}^{-},\ \ \lambda \geq D_{21}^{-},\ \lambda \geq D_{31}^{-}$

$\lambda \geq D_{t2}^{-};\ t = 1, 2, \ldots, T$

$\Rightarrow \lambda \geq D_{12}^{-},\ \ \lambda \geq D_{22}^{-},\ \lambda \geq D_{32}^{-}$

$\lambda \geq \bar{D}_{t31}^{-}\bar{I}, \lambda \geq \bar{D}_{t32}^{-}\bar{I};\ t = 1, 2, \ldots, (T-1)$

$\Rightarrow \lambda \geq D_{1311}^{-},\ \ \lambda \geq D_{1321}^{-},\ \lambda \geq D_{1312}^{-},$
$\lambda \geq D_{1322}^{-},\ \lambda \geq D_{2311}^{-},\ \lambda \geq D_{2321}^{-}$

$D_{t1}^{-} \geq 0;\ t = 1, 2, \ldots, T$

$\Rightarrow D_{11}^{-} \geq 0, D_{21}^{-} \geq 0, D_{31}^{-} \geq 0$

$D_{t2}^{-} \geq 0;\ t = 1, 2, \ldots, T$

$\Rightarrow D_{12}^{-} \geq 0, D_{22}^{-} \geq 0, D_{32}^{-} \geq 0$

$\bar{D}_{t31}^{-} \geq \bar{0}, \bar{D}_{t32}^{-} \geq \bar{0};\ t = 1, 2, \ldots, (T-1)$

$\Rightarrow D_{1311}^{-} \geq 0, D_{1321}^{-} \geq 0, D_{1312}^{-} \geq 0, D_{1322}^{-} \geq 0,$
$D_{2311}^{-} \geq 0, D_{2321}^{-} \geq 0$

$A_{i1}\ \bar{X}_1 + A_{i2}\ \bar{X}_2 + \ldots\ldots\ldots\ldots + A_{iT}\ \bar{X}_T\ (\leq, =, \geq) b_i \quad \forall i = 1, 2, \ldots, m$

$\Rightarrow x_1 + x_2 + x_3 + x_4 \leq 5$
$x_1 + x_2 - x_3 - x_4 \leq 2$

$x_1 + x_2 + x_3 \geq 1$

$x_1 - x_2 + x_3 + 2x_4 \leq 4$

$x_1 + 2x_3 + 2x_4 \leq 3$

$x_4 \leq 2$

and $\bar{X}_1 \geq 0, \bar{X}_2 \geq 0, \ldots\ldots\ldots, \bar{X}_T \geq 0$

$\Rightarrow x_1 \geq 0, x_2 \geq 0, x_3 \geq 0, x_4 \geq 0.$

Solving the above programming problem using nonlinear packages (as shown in Figure 3a,b), we obtain solution $x_1 = 0.4471,\ \ x_2 = 1.69105, x_3 = 0.0000,$ and $x_4 = 1.2764,\ \ \lambda = 0.2845$; and $Z_1 = 3.42738,\ \ Z_2 = 1.642437$, and $Z_3 = 0.7515643$.

Model IIb: Find $\bar{X}$ so as to minimize

$$\lambda = \sum_{t=1}^{T} D_{t1}^{-} + \sum_{t=1}^{T-1} D_{t2}^{-} + \sum_{t=1}^{T-1} \bar{D}_{t3}^{-}$$

$$\min \lambda = \left(D_{11}^{-} + D_{21}^{-} + D_{31}^{-}\right) + \left(D_{12}^{-} + D_{22}^{-} + D_{32}^{-}\right) + \left(D_{1311}^{-} + D_{2311}^{-} + D_{1321}^{-} + D_{1312}^{-} + D_{2321}^{-} + D_{1322}^{-}\right)$$

subject to

$\mu_{Z_t}(N_t(\bar{X})) + D_{t1}^{-} \geq 1; t = 1, 2, \ldots, T$

$\Rightarrow 7x_1 + 3x_2 - 4x_3 + 2x_4 + 23D_{11}^{-} \geq 17$
$\Rightarrow x_2 + 3x_3 + 4x_4 + 9.5D_{21}^{-} \geq 9.5$
$\Rightarrow 2x_1 + x_2 + x_3 + x_4 + 4D_{31}^{-} \geq 5$

$\mu_{Z_t}(D_t(\bar{X})) + D_{t2}^{-} \geq 1;\ t = 1, 2, \ldots, T$

$\Rightarrow x_1 + x_2 + x_3 - 4D_{12}^{-} \geq 1$

$\Rightarrow x_1 + x_2 + x_3 - 4D_{22}^{-} \leq 1$

$\Rightarrow x_1 + x_2 + x_3 - 4D_{32}^{-} \leq 1$

$\mu_{\bar{X}_t}(\bar{X}_t) + \bar{D}_{t3}^{-} \geq \bar{I};\ t = 1, 2, \ldots, (T-1)$

$\bar{X}_t - \left(\bar{X}_t^B - \bar{p}_t^{-}\right)\bar{p}_t^{-} + \bar{D}_{t31}^{-} \geq \bar{I};$

$\left(\bar{X}_t^B + \bar{p}_t^{+}\right) - \bar{X}_t \bar{p}_t^{+} + \bar{D}_{t32}^{-} \geq \bar{I};\ t = 1, 2, \ldots, (T-1)$

$\Rightarrow -0.5x_1 + D_{1311}^{-} \geq -1.1666$
$\Rightarrow -0.5x_1 + D_{1321}^{-} \geq -1.1666$
$\Rightarrow -0.1555x_2 + D_{1312}^{-} \geq 0$
$\Rightarrow -0.1555x_2 + D_{1322}^{-} \geq 0$
$\Rightarrow -x_3 + D_{2311}^{-} \geq 0$
$\Rightarrow -x_3 + D_{2321}^{-} \geq 0$

$D_{t1}^{-} \geq 0;\ t = 1, 2, \ldots, T$

$\Rightarrow D_{11}^{-} \geq 0, D_{21}^{-} \geq 0, D_{31}^{-} \geq 0$

$D_{t2}^{-} \geq 0;\ t = 1, 2, \ldots, T$

$\Rightarrow D_{12}^{-} \geq 0, D_{22}^{-} \geq 0, D_{32}^{-} \geq 0, \bar{D}_{t3}^{-} \geq \bar{0};\ t = 1, 2, \ldots, (T-1)$

$\Rightarrow D_{1311}^{-} \geq 0, D_{1321}^{-} \geq 0, D_{1312}^{-} \geq 0,$
$D_{1322}^{-} \geq 0, D_{2311}^{-} \geq 0, D_{2321}^{-} \geq 0$

$A_{i1}\ \bar{X}_1 + A_{i2}\ \bar{X}_2 + \ldots\ldots\ldots\ldots + A_{iT}\ \bar{X}_T\ (\leq, =, \geq) b_i \quad \forall i = 1, 2, \ldots, m$

$\Rightarrow x_1 + x_2 + x_3 + x_4 \leq 5$

$x_1 + x_2 - x_3 - x_4 \leq 2$

$x_1 + x_2 + x_3 \geq 1$

$x_1 - x_2 + x_3 + 2x_4 \leq 4$

$x_1 + 2x_3 + 2x_4 \leq 3$

$x_4 \leq 2$ and $\bar{X}_1 \geq 0, \bar{X}_2 \geq 0, \ldots\ldots\ldots, \bar{X}_T \geq 0$

$\Rightarrow x_1 \geq 0, x_2 \geq 0, x_3 \geq 0, x_4 \geq 0.$

Solving the above FGP model problem using nonlinear packages (as shown in Figure 4a,b), we obtain solution $x_1 = 1.0000$, $x_2 = 0.0000$, $x_3 = 0.0000$, and $x_4 = 1.0000$,

Table 1 Comparison of euclidean distance for the solution of example 1

Method	(x_1, x_2, x_3, x_4)	(Z_1, Z_2, Z_3)	$(\mu_{z_1}(N_1(\bar{X}), \mu_{z_1}(D_1(\bar{X}), \mu_{z_2}(N_2(\bar{X}), \mu_{z_2}(D_2(\bar{X}), \mu_{z_3}(N_3(\bar{X}), \mu_{z_3}(D_3(\bar{X}))$	Euclidean distance
Proposed method I	(0.4471, 1.69105, 0.0000, 1.2764)	(3.42738, 1.642437, 0.7515643)	(0.7285, 0.7155, 0.7154, 0.7155, 0.7154, 0.7154)	0.6918
Proposed method IIb	(1.0000, 0.0000, 0.0000,1.0000)	(4.5, 1.3333, 0.75)	(0.6521, 1, 0.42105, 1, 0.5000, 1)	0.706163

Based on proposed FGP methods I and IIb.

Table 2 Solution set (for example 1) based on the tolerance on the decision variables for model

(p_1^-, p_1^+)	(p_1^-, p_1^+)	(p_2^-, p_2^+)	$(x_1, x_2, x_3, x_4, \lambda)$	(Z_1, Z_2, Z_3)
(−2, 2)	(−6.43, 6.43)	(−1, 1)	(0.4471, 1.69105, 0.0000, 1.2764, 0.2845)	(3.42738, 1.642437, 0.7515643)
(−2, 2)	(−5.5, 5.5)	(−1, 1)	(0.46383, 1.623581, 0.0000, 1.268083, 0.295167)	(3.450697, 1.638180, 0.7507402)
(−2, 2)	(−5, 5)	(−1, 1)	(0.48427, 1.54088, 0.0000, 1.25786, 0.3081761)	(3.480241, 1.632814, 0.7496851)
(−2, 2)	(−4.5, 4.5)	(−1, 1)	(0.50655, 1.450743, 0.0000, 1.24672, 0.322355)	(3.513862, 1.626774, 0.7485059)
(−2, 2)	(−4, 4)	(−1, 1)	(0.5310345, 1.351724, 0.0000, 1.234481, 0.337931)	(3.552630, 1.619892, 0.7471748)

$\lambda = 1.426773$; and $Z_1 = 3.42738$, $Z_2 = 1.642437$, and $Z_3 = 0.7515643$ (see Table 1 for a comparison of Euclidean distance and Table 2 for the solution set of example 1).

Conclusion

An effort has been made to solve the multilevel fractional programming problem based on the fuzzy set theory and goal programming approach. The main advantage of the proposed approach is that computational complexity is reduced by defining separate membership functions for numerator and denominator functions of objectives at each level, but tolerance values (positive and negative) are chosen so that the satisfying solution is forced towards the optimal solutions. However, the main difficulties of these methods are the following: (1) Preference information required from the DM cannot easily be given, particularly when the information thus required is large and complex and/or the objective functions in fractional form are measured on different scales, and (2) if DMs are not satisfied with solution, then higher level DMs should provide new tolerance limits for the control variables until a satisfying solution is reached.

Besides this, the proposed methodology can be easily reduced to the solution procedure as given by Sinha (2003b) and Pramanik and Roy (2007) for the multilevel linear programming case by avoiding the goal membership functions for denominators of objectives and corresponding goal variables.

Methods

Selection of compromise solution

The concept of utopia point (the ideal point) and the use of distance function for group decision analysis was first studied by Yu (1973). Biswas and Pal (2005) used the Euclidean distance function to select the appropriate priority structure in the application of fuzzy goal programming technique to land use planning in the agricultural system. Here, different FGP models provide different optimal solutions. The Euclidean distance function is used only to identify which FGP model (model I, model IIa, and model IIb) offers a better optimal solution. In the FGP formulation, since the aspired level of each of the membership function goals is unity, the point consisting of the highest membership value of each of the goals would represent the ideal point. The Euclidean distance function can be defined in this case as follows:

$$D^2 = \left[\sum_{t=1}^{T}\left\{\left[1-\mu_{Z_t}(N_t(\bar{X}_t))\right]^2 + \left[1-\mu_{Z_t}(D_t(\bar{X}_t))\right]^2\right\}\right]^{1/2},$$

where $\mu_{Z_t}(N_t(\bar{X}_t)$ and $\mu_{Z_t}(D_t(\bar{X}_t)$ represent the achieved membership value of the t-th numerator and denominator objective goals, respectively. The solution for which D^2 is minimum would be the most satisfying solution.

Competing interests

The authors declare that they have no competing interests.

Authors' contributions

KL carried out discussion related to formulation of MLFPP, related terminology, Characterization of linear membership functions for the numerator and denominator functions at each level as well as decision vector. KL proposed FGP approach to tackle MLFPPs and formulate different mathematical models related to it and also illustrate proposed methodology with a numerical example and show sensitivity analysis performed on it. MPP described the use of selection criteria of compromise solution for MLFPPs. All authors read and approved the final manuscript.

Acknowledgements

The authors are indebted to the anonymous referee for the encouraging comments and valuable suggestions, which have improved the presentation of the paper.

Author details
[1]Department of Mathematics, Government Engineering College, Bikaner 334004, India. [2]Department of Mechanical Engineering, Government Engineering College, Bikaner 334004, India.

References

Anandilingam G (1988) A mathematical programming model of decentralized multi-level system. J Oper Res Soc 39(11):1021–1033
Anandilingam G, Apprey V (1991) Multi-level programming and conflicting resolution. European Journal of Operational Research 51:233–247
Arora SR, Gupta R (2009) Interactive fuzzy goal programmings approach for bilevel programming problem. European Journal of Operational Research 194:368–376
Baky IA (2009) Fuzzy goal programming algorithm for solving decentralized bilevel multiobjective programming problems. Fuzzy Sets and Systems 160:2701–2710
Bellmann RE (1957) Dynamic programming. Princeton University Press, Princeton
Biswas A, Pal BB (2005) Application of fuzzy goal programming technique to land use planning in agriculture systems. Omega 33:391–398
Chakraborty M, Gupta S (2002) Fuzzy mathematical programming for multi objective linear fractional programming problem. Fuzzy Sets and Systems 125:335–342
Chang CT (2009) A goal programming approach for fuzzy multiobjective fractional programming problems. International Journal of System Sciences 40:867–874
Charnes A, Cooper WW (1962) Programming with linear fractional functional. Nov Res Logistics Quart 9:181–186
Craven BD, Mond B (1975) On fractional programming and equivalence. Nov Res Logistics Quart 22:405–410
Jimenez M, Bilbas A (2009) Pareto optimal solutions in fuzzy multi-objective linear programming. Fuzzy Sets and Systems 160:2714–2721
Lai YJ (1996) Hierarchical optimization: a satisfactory solution. Fuzzy Sets and Systems 77:321–335
Lai YJ, Hwang CL (1993) Fuzzy mathematical programming methods and applications. Springer, Berlin
Mohamed RH (1997) The relationship between goal programming and fuzzy programming. Fuzzy Sets and Systems 89:215–222
Pal BB, Gupta S (2009) A genetic algorithm to fuzzy goal programming formulation of fractional multiobjective decision making problem. In: Proceeding of 1st International Conference on Advanced Computing (ICAC). Chennai, 13-15 December 2009
Pramanik S, Roy TK (2007) Fuzzy goal programming approach to multilevel programming problems. European Journal of Operational Research 176:1151–1166
Shih HS, Lee ES (2000) Compensatory fuzzy multiple decision making. Fuzzy Sets and Systems 14:71–87
Shih HS, Lai YJ, Lee ES (1996) Fuzzy approach for multi-level programming problems. Computer and Operation Research 23:73–91
Sinha S (2003a) Fuzzy mathematical programming applied to multi-level programming problems. Computers & Operations Research 30:1259–1268
Sinha S (2003b) Fuzzy programming approach to multi-level programming problems. Fuzzy Sets and Systems 136:189–202
Yu PL (1973) A class of solutions for group decision problems. Management Sciences 19:936–946

Analysis of unreliable bulk queue with state dependent arrivals

Charan Jeet Singh[1*], Madhu Jain[2] and Binay Kumar[3]

Abstract

In this paper, we investigate a single-server Poisson input queueing model, wherein arrivals of units are in bulk. The arrival rate of the units is state dependent, and service time is arbitrary distributed. It is also assumed that the system is subject to breakdown, and the failed server immediately joins the repair facility which takes constant duration to repair the server. By using supplementary variable technique, we obtain the probability generating function of the number of units in the system which is further used to establish some performance indices such as the mean number of units in the system, mean waiting time, etc. Special cases are also discussed. In order to obtain approximate values of system state probabilities, the principle of maximum entropy is employed. Numerical results are also presented to validate the analytical formulae.

Keywords: Bulk queue; Arbitrary service time; Breakdown; Supplementary variable; Queue size

Introduction

In many congestion problems of the queueing systems, it is a common phenomenon that the server is always available in the service station without any interruption and the service system never fails. But in some real life examples, these assumptions are not realistic. In many practical activities, we face the situation where service stations may fail and can be repaired by the repairman available in the system. For example, in manufacturing systems, the machine may break down due to some physical problems. Similarly, many other examples are available in the area of computer communication networks and flexible manufacturing systems where the performance of such systems may be affected by service station breakdown.

There is extensive literature on the $M/G/1$ queue which has been studied in various forms by numerous researchers; some of them are mentioned by Levy and Yechiali (1976), Keilson and Servi (1986), Madan (1999), and many others. The unreliable service systems with repair facility of the server or some other type of service interruptions can be experienced in manufacturing and production process. Among some earlier papers on service interruptions, we refer the papers by Avi-Itzhak and Naor (1963) and Mitrany and Avi-Itzhak (1968) for some fundamental works. Madan (1989), Sengupta (1990), Tang (1997), and Takine and Sengupta (1998) have studied some queueing systems with interruptions, wherein one of the underlying assumptions is that the server immediately joins repair facility as soon as it fails. Li et al. (1997) obtained the reliability analysis of $M/G/1$ queueing system with server breakdowns and vacations. The concept of essential service and optional service in $M/G/1$ queue with server breakdowns and repairs was discussed by Wang (2004). Ke and Pearn (2004) described the management policy for single-server model of queueing system wherein the server is unreliable and the arrival of the units depends on server's current status. The cost analysis is also considered to obtain the optimal results of service quality. The homogeneous finite source queue with server subject to breakdowns and repairs was explored by Almasi et al. (2005). Ke (2007) discussed the operating characteristics of single server non-Markovian queueing model with server breakdowns. It is assumed that the arrival of units is in batches and server may breakdowns.

$M^x/G/1$ queue with random set-up time with a Bernoulli vacation schedule under a restricted admissibility policy to obtain queue size distribution was generalized by Choudhury (2008). Falin (2010) investigated a retrial queueing system, wherein the customer may leave the system on server breakdowns and may join the retrial group

* Correspondence: cjsmath@gmail.com
[1]Department of Mathematics, Guru Nanak Dev University, Amritsar 143005, India
Full list of author information is available at the end of the article

for getting service. In his investigation, he studied the steady-state behavior of the generally distributed system. Choudhury and Tadj (2011) considered an unreliable bulk queue vacation model with two phase service, wherein it is assumed that server may breakdowns during any phase of service. It is also considered that server can start the service when he finds at least N customers in the system and wait for getting the service.

Recently, Choudhury and Ke (2012) analysed an unreliable $M^x/G/1$ queueing system under Bernoulli schedule vacation policy, wherein it is assumed that customer may join the system as retrial customer if his service is not completed on first time service due to any reason and it is also assumed that server may suffer from unpredictable breakdown and delay repair.

The main objective of the maximum entropy principle is to provide a method of estimating an unknown probability distribution which has been widely applied in the field of queueing theory and computer performance analysis. The complex queueing systems by using maximum entropy principle were analysed by several researchers such as Ferdinand (1970), Shore (1982), El-Affendi and Kouvatsos (1983), Kouvatsos (1986), Wu and Chan (1989), Arizono et al. (1991), and the references cited therein. Wang et al. (2007) utilized the MEP technique to analyse a queueing system with multiple vacations and server breakdowns to obtained the approximate values of performance indices. Single server non-Markovian model with removable and unreliable server under (p, N) -policy was discussed by Wang and Huang (2009). In this investigation, they have obtained the expressions for approximate values of queue length distribution for the number of customers in the system and the expected waiting time. By using improved maximum entropy principle, randomized queueing situation with N - policy was considered by Wang et al. (2011) and carried out comparative analysis between the exact and approximate result obtained and showed that it can provide enough accurate result for practical purpose.

Madan (2003) studied the steady-state behavior of $M/G/1$ queueing system, wherein the units arrive at the system one by one according to Poisson process with uniform arrival rate and the service system is subject to random breakdowns. In many congestion situations, the arrival of units may be in bulk. Due to sudden breakdowns of the server, the arrivals of units may also be affected. For example, if a public telephone system breaks down, some of the customers in the queue depart from the telephone booth without getting the service. Such type of models has also applications in call centers and other computer centers where different types of facilities for arrived customers are available. In fact, the server may be subject to lengthy and unpredictable breakdowns while serving a customer because there may be sudden major fault in the system and the arrival rate of the jobs that arrive during the breakdown period may be different from the arrival rate in the operating state.

The facts discussed above have motivated us to extend the model presented by Madan (2003) to single-server model, wherein the arrival of units is in bulk and follows Poisson process with varying arrival rates depending upon server status which may be in idle state, operating state, and repair state. The layout of the investigation is as follows: The mathematical description of the model is presented in the 'Model description' section. The set of steady state equations governing the model is constructed in 'Governing equations' section. In the 'Analysis' section, we analyse the model to obtain various performance measures. Special cases are deduced in the 'Special cases' section by taking appropriate parameter values. In the 'Maximum entropy results' section, maximum entropy principle is discussed to find the queue size distribution. Next, the 'Numerical illustration' section is devoted to numerical results and sensitivity analysis. Finally, the conclusion is drawn in the 'Conclusions' section.

Model description

We consider an $M^x/G/1$ queueing system wherein the customers arrive according to a Poisson process with state-dependent rates. The service system is subject to random breakdowns in working or idle state. As soon as the server fails, it immediately joins the repair facility, which performs repair work in constant duration. The service times of the units follow arbitrary distribution. The life time of the server is exponentially distributed. The notations used in the formulation of the model are as follows:

- λ_1, λ_2, λ_3: Mean arrival rate of the units when the server is in idle state, busy state, and under repair server, respectively.
- $B(v)$, $b(v)$: Distribution and density functions of the service time, respectively.
- $\mu(x)dx$: The conditional probability of completion of service during the interval $(x, x + dx)$ with elapsed time x of the unit which is in service.
- $\alpha\ dt$: The probability at which the system may fail during time interval $(t, t + dt)$.
- d: The duration of the repair time of the failed system.
- ρ: Traffic intensity
- N: Number of units in the queue, excluding the unit which one is in service (≥ 0) with elapsed service time x.
- $W_n(t, x)$: The probability of n units in the queue excluding one being served with elapsed time x when server is busy at time t.
- $R_n(t)$: The probability of n units in the queue at time t when the broken down server is under repair.
- $Q(t)$: The probability that there are no units in the system and the server is in idle state at time t.

- K_r: The probability of r arrivals during a repair time period.
- C_i: Probability mass function of batch size X.
- $C_r^{(k)}$: The probability that there are r units during k batches arrives.
- $P_q(z)$: The probability generating function of the queue length, whether the server is in operating state or in breakdown state.
- $P(z)$, $C(z)$: The probability generating functions of the number of units in the system and for bulk size, respectively.
- X, $E(X)$, $E(X^{(2)})$: The random variable of batch size of the arriving units, the first and second factorial moments of batch size of the arrived units, respectively.
- $E(B)$, $E(B^2)$: First and second moments, respectively, of the service time distribution.

Denote

$$W_n(x) = \lim_{t\to\infty} W_n(t,x); \quad W_n = \lim_{t\to\infty} W_n(t);$$
$$R_n = \lim_{t\to\infty} R_n(t); \qquad Q = \lim_{t\to\infty} Q(t).$$

Also,

$$\mu(x) = \frac{b(x)}{1\text{–}B(x)}; \tag{1}$$

$$b(v) = \mu(v)\exp.\left[-\int_0^v \mu(x)dx\right]; \tag{2}$$

$$W_n(t) = \int_0^\infty W_n(x,t)dx; \tag{3}$$

$$K_r = \sum_{k=0}^{r} \frac{e^{-\lambda_3 d}(\lambda_3 d)^k}{k!} C_r{}^{(k)}, \quad r = 0,1,2... \tag{4}$$

The steady-state equations

In this section, we formulate the steady state equations of the system by using the probability reasoning as follows:

$$\frac{d}{dx}W_n(x) + (\lambda_2 + \alpha + \mu(x))W_n(x) = \lambda_2 \sum_{k=1}^{n} C_k W_{n\text{-}k}(x); \; n \geq 1 \tag{5a}$$

$$\frac{d}{dx}W_0(x) + (\lambda_2 + \alpha + \mu(x))W_0(x) = 0; \tag{5b}$$

$$(\lambda_1 + \alpha)Q = \int_0^\infty W_0(x)\mu(x)dx + R_0K_0; \tag{5c}$$

$$R_n = \alpha W_{n\text{-}1}; n \geq 1; \tag{5d}$$

$$R_0 = \alpha Q. \tag{5e}$$

For the solution of above set of equations, the following boundary conditions are

$$W_n(0) = \int_0^\infty W_{n+1}(x)\mu(x)dx + R_0K_{n+1} + R_1K_n$$
$$+ \ldots + R_{n+1}K_0 + \lambda_1 C_{n+1}Q; \; n \geq 0. \tag{6}$$

We define the following probability generating functions

$$W(x,z) = \sum_{n=0}^\infty W_n(x)z^n; \; W(z) = \sum_{n=0}^\infty W_n z^n; R(z) = \sum_{n=0}^\infty R_n z^n; \tag{7a}$$

$$C(z) = \sum_{k=1}^\infty C_k z^k. \tag{7b}$$

Analysis

In this section, we obtain the probability generating functions of the number of units in the system. By using Equations (5a) and (5b), we have

$$\frac{d}{dx}W(x,z) + (p_\alpha + \mu(x))W(x,z) = 0. \tag{8}$$

Similarly, Equations (5d) and (5e) give

$$R(z) = \alpha(Q + zW(z)). \tag{9}$$

Also, from Equation (6), we have

$$zW(0,z) = \int_0^\infty W(x,z)\mu(x)dx + R(z)e^{-r_\alpha}\text{–}K_0R_0$$
$$+ \lambda_1 C(z)Q\text{–}\int_0^\infty W_0(x)\mu(x)dx, \tag{10}$$

where $\sum_{n=0}^\infty K_n z^n = e^{-r_\alpha}$.

By using Equation (5c), Equation (10) can be written as

$$zW(0,z) = \int_0^\infty W(x,z)\mu(x)dx + R(z)e^{-r_\alpha}\text{–}q_\alpha Q. \tag{11}$$

On integration between the limits 0 and x, Equation (8) gives

$$W(x,z) = W(0,z)\exp.(-p_\alpha x)\exp.\left\{-\int_0^x \mu(t)dt\right\}, \tag{12}$$

where $W(0, z)$ can be obtained from Equation (11).

On multiplying Equation (12) by $\mu(x)$ and integrating with respect to x and by taking limit 0 to ∞ and using Equation (2), we get

$$\int_0^\infty W(x,z)\mu(x)dx = W(0,z)\int_0^\infty \exp.(-p_\alpha x)b(x)dx = W(0,z)\bar{b}(p_\alpha), \quad (13)$$

where $\bar{b}(p_\alpha) = \int_0^\infty \exp(-p_\alpha x)b(x)dx$ is the Laplace transform of the service time density function $b(x)$.

Here,

$$p_\alpha = (\lambda_2 - \lambda_2 C(z) + \alpha),\quad q_\alpha = (\lambda_1 - \lambda_1 C(z) + \alpha),$$
$$r_\alpha = \lambda_3 d(1 - C(z)).$$

Theorem 1: The probability generating function of the queue length, whether the server is in operating state or in broken down state is

$$P_q(z) = \frac{[[1-\bar{b}(p_\alpha)][\alpha e^{-r_\alpha} - q_\alpha] + \alpha\{z(p_\alpha - q_\alpha) + (zq_\alpha - p_\alpha)\bar{b}(p_\alpha)\}]Q}{p_\alpha\{z-\bar{b}(p_\alpha)\} - \{1-\bar{b}(p_\alpha)\}\alpha z e^{-r_\alpha}}$$

where Q is given by Equation (32).

Proof: By using the above mentioned steady-state equations and boundary conditions, we obtain the probability generating function of the queue length.

[For proof, see Appendix 1].

Theorem 2: The probability generating function of the number of units in the system is

$$P(z) = (1+\alpha z)\left[\frac{z(p_\alpha - q_\alpha) + (zq_\alpha - p_\alpha)\bar{b}(p_\alpha)}{p_\alpha\{z-\bar{b}(p_\alpha)\} - \{1-\bar{b}(p_\alpha)\}\alpha z e^{-r_\alpha}}\right]Q.$$

Proof: [For proof, see Appendix 2].

The expected number of units in the queue can be obtained as

$$L_q = \lim_{z\to 1}\frac{d}{dz}P_q(z) = \frac{D'(1)N''(1) - N'(1)D''(1)}{2[D'(1)]^2}, \quad (14)$$

where

$$N'(1) = \begin{bmatrix}\{\alpha\lambda_3 dE(X) + \lambda_1 E(X) - \alpha(\lambda_2 - \lambda_1)E(X)\} \\ +\{\alpha E(X)(\lambda_2 - \lambda_1) + \alpha^2 - \alpha\lambda_3 dE(X) - \lambda_1 E(X)\}\bar{b}(\alpha)\end{bmatrix}Q;$$

$$N''(1) = \Bigg[\begin{Bmatrix}\alpha\lambda_3{}^2 d^2(E(X))^2 + \alpha\lambda_3 dE(X^{(2)}) + \lambda_1 E(X^{(2)}) \\ -\alpha(\lambda_2 - \lambda_1)(2E(X) + E(X^{(2)})\end{Bmatrix} + 2\lambda_2 E(X)\{\alpha\lambda_3 dE(X) + \lambda_1 E(X) - \alpha^2 + \alpha E(X)(\lambda_1 - \lambda_2)\}\bar{b}'(\alpha) - \begin{Bmatrix}\alpha E(X^{(2)})(\lambda_1 - \lambda_2) + 2\alpha\lambda_1 E(X) + \alpha(\lambda_3 dE(X))^2 \\ +\alpha\lambda_3 dE(X^{(2)}) + \lambda_1 E(X^{(2)})\end{Bmatrix}\bar{b}(\alpha)\Bigg]Q$$

$$D'(1) = [\bar{b}(\alpha)(\alpha + \lambda_2 E(X) + \lambda_3\alpha dE(X)) - (\lambda_2 + \lambda_3\alpha d)E(X)];$$

$$D''(1) = \Big[-\{2\lambda_2 E(X) + \lambda_2 E(X^{(2)}) + 2\alpha\lambda_3 dE(X) + \alpha(\lambda_3 dE(X))^2 + \alpha\lambda_3 dE(X^{(2)})\} - 2\{\alpha + \lambda_2 E(X) + \alpha\lambda_3 dE(X)\}\lambda_2 E(X)\bar{b}'(\alpha) + \{\lambda_2 E(X^{(2)}) + 2\alpha\lambda_3 dE(X) + \alpha\lambda_3{}^2 d^2(E(X))^2 + \alpha\lambda_3 dE(X^{(2)})\}\bar{b}(\alpha)\Big]$$

with

$$Q = \frac{\{\alpha\bar{b}(\alpha) - (1-\bar{b}(\alpha))(\lambda_2 + \lambda_3 d\alpha)E(X)\}}{(1+\alpha)\{(1-\bar{b}(\alpha))(\lambda_1 - \lambda_2)E(X) + \alpha\bar{b}(\alpha)\}};$$

$$\rho = 1 - Q.$$

- The average waiting time in the queue can be obtained by using Little's formula as

$$W_q = \frac{L_q}{\lambda_{eff}E(X)};$$

$$\lambda_{eff} = \{\lambda_1 Q + \lambda_2 W(1) + \lambda_3 R(1)\}. \quad (15a)$$

- The units in the system can be obtained as

$$L_s = L_q + \rho. \quad (15b)$$

Special cases

In this section, we establish key performance measures of some models as special cases of our model.

Unreliable $M^X/M/1$ queue with state-dependent arrival rates

In this case, the queue length is obtained by using

$$L_q = \lim_{z\to 1}\frac{d}{dz}P_q(z) = \frac{D'(1)N''(1) - N'(1)D''(1)}{2[D'(1)]^2}, \quad (16)$$

where

$$N'(1) = \begin{bmatrix}\{\alpha\lambda_3 dE(X) + \lambda_1 E(X) - \alpha(\lambda_2 - \lambda_1)E(X)\} \\ +\{\alpha E(X)(\lambda_2 - \lambda_1) + \alpha^2 - \alpha\lambda_3 dE(X) - \lambda_1 E(X)\}\left(\frac{\mu}{\alpha+\mu}\right)\end{bmatrix}Q;$$

Table 1 $L_q(L_s)$ **by varying** μ **and** λ **for** $\alpha = .21$, $d = 1.5$, $E(X) = 3$, $\lambda_1 = 1.2\lambda$, $\lambda_2 = \lambda$, $\lambda_3 = .8\lambda$

μ/λ	6.00		6.25		6.50		6.75		7.00	
	$k = 1$	$k = 4$	$k = 1$	$k = 4$	$k = 1$	$k = 4$	$k = 1$	$k = 4$	$k = 1$	$k = 4$
1.1	9.7(10.5)	9.6(10.4)	8.6(9.3)	8.5(9.2)	7.7(8.4)	7.6(8.3)	6.9(7.6)	6.8(7.5)	6.3(7.0)	6.2(6.9)
1.2	13.4(14.2)	13.4(14.2)	11.5(12.3)	11.4(12.2)	10.0(10.8)	9.9(10.7)	8.9(9.7)	8.8(9.6)	8.0(8.7)	7.9(8.6)
1.3	19.5(20.4)	19.9(20.7)	16(16.8)	16.0(16.9)	13.5(14.3)	13.5(14.3)	11.7(12.5)	11.6(12.4)	10.3(11.1)	10.2(11)
1.4	32.0(32.9)	33.6(34.6)	23.9(24.8)	24.5(25.4)	19.1(20)	19.3(20.1)	15.9(16.8)	15.9(16.8)	13.6(14.5)	13.5(14.4)
1.5	70(70.9)	83.5(84.4)	41.6(42.5)	44.8(45.7)	29.6(30.5)	30.7(31.6)	23.0(23.9)	23.3(24.2)	18.8(19.6)	18.8(19.7)

$$
\begin{aligned}
N''(1) = \Big[& \{\alpha\lambda_3{}^2 d^2 (E(X))^2 + \alpha\lambda_3 dE(X^{(2)}) + \lambda_1 E(X^{(2)}) \\
& -\alpha(\lambda_2 - \lambda_1)(2E(X) + E(X^{(2)})\} \\
& +2\lambda_2 E(X)\{\alpha\lambda_3 dE(X) + \lambda_1 E(X) - \alpha^2 \\
& +\alpha E(X)(\lambda_1 - \lambda_2)\}\left(\frac{-\mu}{(\alpha+\mu)^2}\right) - \{\alpha E(X^{(2)}) \\
& \times(\lambda_1 - \lambda_2) + 2\alpha\lambda_1 E(X) + \alpha(\lambda_3 dE(X))^2 \\
& +\alpha\lambda_3 dE(X^{(2)}) + \lambda_1 E(X^{(2)})\}\left(\frac{\mu}{\alpha+\mu}\right)\Big] Q;
\end{aligned}
$$

$$
D'(1) = \left[(\alpha + \lambda_2 E(X) + \lambda_3 \alpha dE(X))\left(\frac{\mu}{\alpha+\mu}\right) - (\lambda_2 + \lambda_3 \alpha d)E(X)\right];
$$

$$
\begin{aligned}
D''(1) = \Big[& -\{2\lambda_2 E(X) + \lambda_2 E(X^{(2)}) + 2\alpha\lambda_3 dE(X) \\
& +\alpha(\lambda_3 dE(X))^2 + \alpha\lambda_3 dE(X^{(2)})\} - 2\{\alpha + \lambda_2 E(X) \\
& +\alpha\lambda_3 dE(X)\}\lambda_2 E(X)\left(\frac{-\mu}{(\alpha+\mu)^2}\right) + \{\lambda_2 E(X^{(2)}) \\
& +2\alpha\lambda_3 dE(X) + \alpha\lambda_3{}^2 d^2 (E(X))^2 \\
& +\alpha\lambda_3 dE(X^{(2)})\}\left(\frac{\mu}{\alpha+\mu}\right)\Big];
\end{aligned}
$$

$$
Q = \frac{\{\mu - (\lambda_2 + \lambda_3 d\alpha)E(X)\}}{(1+\alpha)\{(\lambda_1 - \lambda_2)E(X) + \mu\}}.
$$

Reliable $M^X/G/1$ ***queue with state-dependent arrival rates***

For this model, the queue length becomes

$$
L_q = \frac{\lambda_1\{\lambda_2 (E(X))^2 E(B^2) + E(X^{(2)})E(B)\}}{2(1 - \lambda_2 E(X)E(B))(1 + (\lambda_1 - \lambda_2)E(X)E(B))}. \tag{17}
$$

Reliable $M^X/G/1$ ***queue with homogeneous arrival rates***

In this case, the queue length is given by

$$
L_q = \frac{\{\lambda^2 (E(X))^2 E(B^2) + \lambda E(X^{(2)})E(B)\}}{2(1 - \lambda E(X)E(B))}. \tag{18}
$$

Reliable M/G/1 queue with state-dependent arrival rates

The queue length can be obtained by setting

$$
L_q = \frac{\lambda_1 \lambda_2 E(B^2)}{2(1 - \lambda_2 E(B))(1 + (\lambda_1 - \lambda_2)E(B))}. \tag{19}
$$

Unreliable M/G/1 queue with homogeneous arrival rates

In this case, queue length can be determined as

$$
L_q = \lim_{z \to 1} \frac{d}{dz} P_q(z) = \frac{D'(1)N''(1) - N'(1)D''(1)}{2[D'(1)]^2}, \tag{20}
$$

where

$$
\begin{aligned}
N'(1) &= \left[\{\alpha\lambda d + \lambda\} + \{\alpha^2 - \alpha\lambda_3 d - \lambda_1\}\bar{b}(\alpha)\right] Q \\
N''(1) &= \left[\alpha\lambda^2 d^2 + 2\lambda\{\alpha\lambda d + \lambda - \alpha^2\}\bar{b}'(\alpha) - \{2\alpha\lambda + \alpha\lambda^2\}\bar{b}(\alpha)\right] Q \\
D'(1) &= \left[\bar{b}(\alpha)(\alpha + \lambda + \lambda\alpha d) - (\lambda + \lambda\alpha d)\right] \\
D''(1) &= \Big[-\{2\lambda + 2\alpha\lambda d + \alpha(\lambda d)^2\} - 2\{\alpha + \lambda + \alpha\lambda d\}\lambda\bar{b}'(\alpha) \\
&\quad + \{2\alpha\lambda d + \alpha\lambda_3{}^2 d^2\}\bar{b}(\alpha)\Big]
\end{aligned}
$$

$$
Q = \frac{\{\alpha\bar{b}(\alpha) - (1 - \bar{b}(\alpha))(\lambda + \lambda d\alpha)\}}{\alpha(1+\alpha)\bar{b}(\alpha)}
$$

In this case, our model reduces to $M/G/1$ queue with uniform arrival rates with time-homogeneous breakdowns and deterministic repair times [see Madan (2003)].

Table 2 $W_q(W_s)$ **by varying** μ **and** λ **for** $\alpha = .21$, $d = 1.5$, $E(X) = 3$, $\lambda_1 = 1.2\lambda$, $\lambda_2 = \lambda$, $\lambda_3 = .8\lambda$

μ/λ	6.00		6.25		6.5		6.75		7.0	
	$k = 1$	$k = 4$	$k = 1$	$k = 4$	$k = 1$	$k = 4$	$k = 1$	$k = 4$	$k = 1$	$k = 4$
1.1	2.9(3.7)	2.9(3.7)	2.6(3.3)	2.5(3.3)	2.3(3.0)	2.2(3.0)	2.0(2.8)	2.0(2.7)	1.9(2.6)	1.8(2.5)
1.2	3.7(4.5)	3.7(4.5)	3.2(4.0)	3.1(3.9)	2.8(3.5)	2.7(3.5)	2.4(3.2)	2.4(3.2)	2.2(2.9)	2.2(2.9)
1.3	5.1(5.9)	5.1(6.0)	4.1(4.9)	4.1(5.0)	3.5(4.3)	3.4(4.3)	3(3.8)	3.0(3.8)	2.6(3.4)	2.6(3.4)
1.4	7.7(8.7)	8.2(9.1)	5.8(6.6)	5.9(6.8)	4.6(5.4)	4.6(5.5)	3.8(4.6)	3.8(4.6)	3.2(4.1)	3.2(4.0)
1.5	16(16.9)	19.1(20.0)	9.4(10.4)	10.2(11.1)	6.7(7.6)	6.9(7.8)	5.2(6.0)	5.2(6.1)	4.2(5.1)	4.2(5.1)

Reliable M/G/1 queue with homogeneous arrival rates
The queue length is

$$L_q = \frac{\lambda^2 E(B^2)}{2(1-\lambda E(B))}. \tag{21}$$

Equation (21) is a well-known Pollaczec-Khinchine formula for $M/G/1$ queue [see Kashyap and Chaudhary (1988) and Medhi (1982)].

Maximum entropy results

In this section, we describe the maximum entropy principle to obtain the estimate of the probability distributions of the queue size and determine the steady-state probabilities for complex queuing systems where the exact probability distribution cannot be found. We formulate the maximum entropy model as follows:

Consider entropy function Z of the bulk arrival queue with unreliable server of the form

$$Z = -Q\log Q - \sum_{n=0}^{\infty} W_n \log W_n - \sum_{n=0}^{\infty} R_n \log R_n. \tag{22}$$

The maximum entropy solutions for the model are obtained by maximizing Equation (22) subject to the following constraints:

$$Q + \sum_{n=0}^{\infty} W_n + \sum_{n=0}^{\infty} R_n = 1; \qquad Q = \eta_0 = 1-\rho; \tag{23a}$$

$$\sum_{n=0}^{\infty} W_n = \eta_1 = W(1); \tag{23b}$$

$$\sum_{n=0}^{\infty} R_n = \eta_2 = R(1); \tag{23c}$$

$$\sum_{n=1}^{\infty} nW_n + \sum_{n=1}^{\infty} nR_n = L_q. \tag{23d}$$

Theorem 3: The steady-state probabilities of the state-dependent bulk arrival queue with unreliable server for different states are given as

$$W_n = \eta_1 \left(\frac{\eta_1 + \eta_2}{\eta_1 + \eta_2 + L_q}\right)\left(\frac{L_q}{\eta_1 + \eta_2 + L_q}\right)^n; \tag{24a}$$

$$R_n = \eta_2 \left(\frac{\eta_1 + \eta_2}{\eta_1 + \eta_2 + L_q}\right)\left(\frac{L_q}{\eta_1 + \eta_2 + L_q}\right)^n. \tag{24b}$$

Proof: [For proof, see Appendix 3].

Numerical illustration

In this section, we present the numerical illustration to evaluate the queue size distribution for unreliable bulk queueing model by using analytical results derived in previous section. It is assumed that the batch size of the units follows a geometrical distribution with $E(X) = \frac{q}{p}$, $E(X^2) = \frac{q(1+q)}{p^2}$; $q = 1 - p$. The service time is generally distributed with deterministic repair time $\left(\frac{1}{d} > \alpha\right)$. To be more specific, we assume that the service time is Erlangian (E_k) queuing model with $E(B) = \frac{1}{\mu}$, $E(B^2) = \frac{k+1}{k\mu^2}$, $\bar{b}(\alpha) = \left(\frac{k\mu}{\alpha+k\mu}\right)^k$, $\bar{b}'(\alpha) = \frac{-k(k\mu)^k}{(\alpha+k\mu)^{k+1}}$. For computation purpose, we set default parameters as α =. 21, d = 1.5, $\lambda_1 = 1.2\lambda$, $\lambda_2 = \lambda$, $\lambda_3 = .8\lambda$. In Tables 1 and 2, we observe that queue length (L_q) and waiting time (W_q) decrease with the increase in service time but increase with the increase in the arrival rate of the units. It is also noticed that the queue length (L_q) and waiting time (W_q) increase significantly with the increase in the number of service phases (k). The approximate values of steady-state probabilities W_n and R_n by using maximum entropy principle are summarized in Table 3. The value of Q is obtained as .4,468 by using the formula given by Equation (32).

Figure 1 exhibits the queue length (L_q) by varying homogenous arrival rates ($\lambda_1 = \lambda_2 = \lambda_3 = \lambda$) and heterogeneous arrival rates ($\lambda_1 = 1.2\lambda$, $\lambda_2 = \lambda$, $\lambda_3 = .8\lambda$) with mean batch size 1(2) for default parameters α =.21, d = 1.5, μ = 6, k = 4. This figure reveals that the queue length (L_q) increases with the increase in arrival rate (λ) as well as mean batch size $E(X)$. Figure 2 exhibits the queue length (L_q) by varying service time (μ) and mean batch size $E(X)$ for default parameters α = .21, d = 1.5, λ_1 = 1.5, λ_2 = 1.25, λ_3 = 1, k = 4. It is observed that the queue length (L_q) decreases with the increase in service time (μ) but it increases with the increase in mean batch size $E(X)$. Figure 3

Table 3 Steady-state probabilities W_n and R_n(α =.21, d = 1.5, Q = .4,468, $E(X)$ = 2, λ_1 = 1.2, λ_2 = 1, λ_3 =. 8, k = 4, μ = 6)

n	0	1	2	3	4	5	6	7	8	9
W_n	0.0781	0.0620	0.0493	0.0391	0.0311	0.0247	0.0196	0.0156	0.0124	0.0098
R_n	0.0357	0.0284	0.0225	0.0179	0.0142	0.0113	0.0090	0.0071	0.0057	0.0045
n	10	11	12	13	14	15	16	17	18	19
W_n	0.0078	0.0062	0.0049	0.0039	0.0031	0.0025	0.0020	0.0016	0.0012	0.0010
R_n	0.0036	0.0028	0.0023	0.0018	0.0014	0.0011	0.0009	0.0007	0.0006	0.0005
n	20	21	22	23	24	25	26	27	28	29
W_n	0.0008	0.0006	0.0005	0.0004	0.0003	0.0002	0.0002	0.0002	0.0001	0.0001
R_n	0.0004	0.0003	0.0002	0.0002	0.0001	0.0001	0.0001	0.0001	0.0001	0.0000

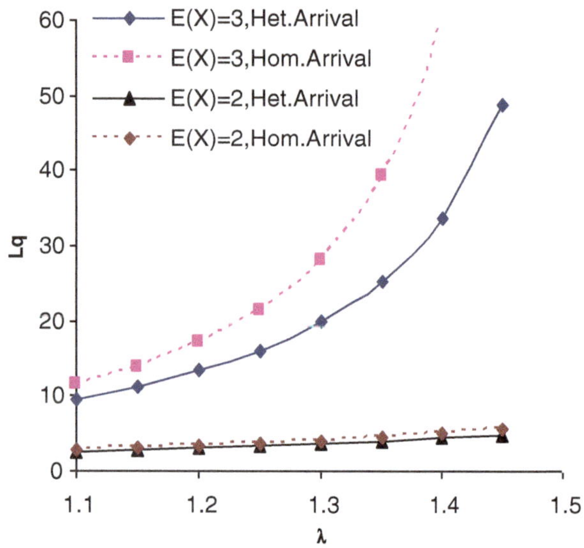

Figure 1 L_q **on** λ **vs.** $E(X)$ **for** $M/E4/1$ **model.**

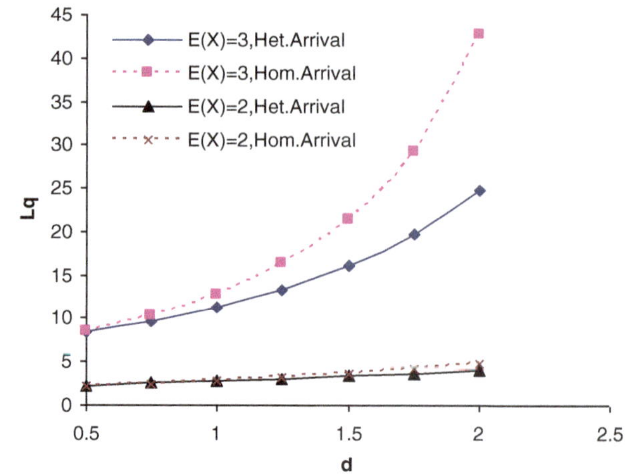

Figure 3 L_q **on** d **vs.** $E(X)$ **for** $M/E4/1$ **model.**

exhibits the queue length (L_q) by varying repair time (d) and mean batch size $E(X)$ for default parameters fixed as $\alpha = .21$, $\lambda_1 = 1.5$, $\lambda_2 = 1.25$ $\lambda_3 = 1$, $k = 4$, $\mu = 6$. It is noticed that the queue length (L_q) increases with the increase in repair time (d) and batch size mean $E(X)$.

Finally, we conclude that

- When arrival rate (service rate) increases, the queue length and waiting time increase (decrease); this situation matches with our expectation in various real life congestion problems.
- The queue length increases with the increase in repair time as well as in batch size of the arriving units.
- As many situations can be observed in production/manufacturing systems where the server is unreliable and arrival of units are in bulk, the queue length increases with the increase in the number of service phases.

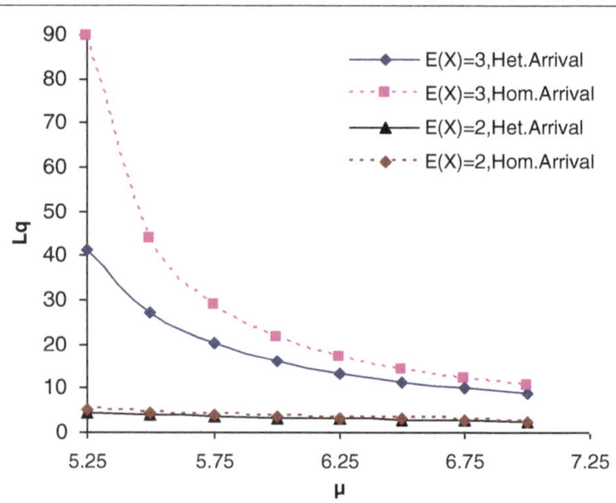

Figure 2 L_q **on** μ **vs.** $E(X)$ **for** $M/E4/1$ **model.**

Conclusions

In this investigation, we have studied the single server queueing model with vacation, wherein arrivals of units are in bulk. By incorporating the state-dependent arrival rate of units and server breakdowns, our model deals with more versatile and realistic scenario of congestion problems. The suggested method determines the queue length and other performance indices in explicit form. For the real life situations, where the arrival of units depends on the server status, such performance indices established may be very helpful in the designing and development of many systems in the field of computer system, manufacturing system, telecommunication networks, etc. The model studied can be further extended for retrial queue and N - policy queue.

Appendices

Appendix 1 Proof of Theorem 1

From Equations (11) and (13), we have

$$W(0,z) = \frac{R(z)e^{-r_\alpha} - q_\alpha Q}{z - \bar{b}(p_\alpha)}. \tag{25}$$

On integrating Equation (12) with respect to x, we have

$$W(z) = W(0,z)\left[\frac{1-\bar{b}(p_\alpha)}{p_\alpha}\right]. \tag{26}$$

By using Equation (25), Equation (26) becomes

$$W(z) = \left[\frac{1-\bar{b}(p_\alpha)}{p_\alpha}\right]\left[\frac{R(z)e^{-r_\alpha} - q_\alpha Q}{z - \bar{b}(p_\alpha)}\right]. \tag{27}$$

On using Equation (9), Equation (27) becomes

$$W(z) = \left[\frac{1-\bar{b}(p_\alpha)}{p_\alpha}\right]\left[\frac{\alpha(Q+zW(z))e^{-r_\alpha}-q_\alpha Q}{z-\bar{b}(p_\alpha)}\right]. \quad (28)$$

On simplification, Equation (28) gives

$$W(z) = \frac{[1-\bar{b}(p_\alpha)][\alpha e^{-r_\alpha}-q_\alpha]Q}{p_\alpha[z-\bar{b}(p_\alpha)]-[1-\bar{b}(p_\alpha)]\alpha z e^{-r_\alpha}}. \quad (29)$$

To obtain the value of Q, we use the normalizing condition

$$W(1)+R(1)+Q=1. \quad (30)$$

From Equation (29), we have

$$W(1) = \lim_{z\to 1} W(z) = \frac{(1-\bar{b}(\alpha))(\lambda_3\alpha d+\lambda_1)E(X)Q}{\alpha\bar{b}(\alpha)-(1-\bar{b}(\alpha))(\lambda_2+\lambda_3 d\alpha)E(X)},$$

where $\bar{b}(\alpha) = E(e^{-\alpha x}) = \int_0^\infty e^{-\alpha x} b(x)dx$. From Equation (9), we have

$$R(1) = \frac{\alpha Q\{\alpha\bar{b}(\alpha)+(1-\bar{b}(\alpha))(\lambda_1-\lambda_2)E(X)\}}{[\alpha\bar{b}(\alpha)-(1-\bar{b}(\alpha))(\lambda_2+\lambda_3 d\alpha)E(X)]}. \quad (31)$$

From Equation (30), we get

$$Q = \frac{\{\alpha\bar{b}(\alpha)-(1-\bar{b}(\alpha))(\lambda_2+\lambda_3 d\alpha)E(X)\}}{(1+\alpha)\{(1-\bar{b}(\alpha))(\lambda_1-\lambda_2)E(X)+\alpha\bar{b}(\alpha)\}}. \quad (32)$$

On putting the value of Q in Equation (29), we have

$$W(z) = \left[\frac{\{1-\bar{b}(p_\alpha)\}\{\alpha e^{-r_\alpha}-q_\alpha\}}{p_\alpha\{z-\bar{b}(p_\alpha)\}-\{1-\bar{b}(p_\alpha)\}\alpha z e^{-r_\alpha}}\right]\cdot\left[\frac{\{\alpha\bar{b}(\alpha)-(1-\bar{b}(\alpha))(\lambda_2+\lambda_3 d\alpha)E(X)\}}{(1+\alpha)\{(1-\bar{b}(\alpha))(\lambda_1-\lambda_2)E(X)+\alpha\bar{b}(\alpha)\}}\right]. \quad (33)$$

On using Equations (32) and (33) in Equation (9), we obtain

$$R(z) = \left[\frac{\alpha\{z(p_\alpha-q_\alpha)+(zq_\alpha-p_\alpha)\bar{b}(p_\alpha)\}}{p_\alpha\{z-\bar{b}(p_\alpha)\}-\{1-\bar{b}(p_\alpha)\}\alpha z e^{-r_\alpha}}\right] \times \left[\frac{\{\alpha\bar{b}(\alpha)-(1-\bar{b}(\alpha))(\lambda_2+\lambda_3 d\alpha)E(X)\}}{(1+\alpha)\{(1-\bar{b}(\alpha))(\lambda_1-\lambda_2)E(X)+\alpha\bar{b}(\alpha)\}}\right]. \quad (34)$$

On adding Equations (33) and (34), we have

$$P_q(z) = W(z)+R(z);$$

$$\frac{[[1-\bar{b}(p_\alpha)][\alpha e^{-r_\alpha}-q_\alpha]+\alpha\{z(p_\alpha-q_\alpha)+(zq_\alpha-p_\alpha)\bar{b}(p_\alpha)\}]Q}{p_\alpha\{z-\bar{b}(p_\alpha)\}-\{1-\bar{b}(p_\alpha)\}\alpha z e^{-r_\alpha}}, \quad (35)$$

where Q is given by Equation (32).

Appendix 2 Proof of Theorem 2

By using Equations (32) and (35), we get

$$P(z) = Q+zP_q(z); \quad (36)$$

$$(1+\alpha z)\left[\frac{z(p_\alpha-q_\alpha)+(zq_\alpha-p_\alpha)\bar{b}(p_\alpha)}{p_\alpha\{z-\bar{b}(p_\alpha)\}-\{1-\bar{b}(p_\alpha)\}\alpha z e^{-r_\alpha}}\right]Q, \quad (37)$$

where Q is given by Equation (32).

Appendix 3 Proof of Theorem 3

By using Lagrange's multipliers method, the entropy function (22) and set of constraints (23a) to (23d) give

$$Z = -Q\log Q-\sum_{n=0}^{\infty}W_n\log W_n-\sum_{n=0}^{\infty}R_n\log R_n-\alpha_1 \times\left(Q+\sum_{n=0}^{\infty}W_n+\sum_{n=0}^{\infty}R_n-1\right)-\alpha_2\left(\sum_{n=0}^{\infty}W_n-\eta_1\right) -\alpha_3\left(\sum_{n=0}^{\infty}R_n-\eta_2\right)-\alpha_4\left(\sum_{n=1}^{\infty}nW_n+\sum_{n=1}^{\infty}nR_n-L_q\right), \quad (38)$$

where α_1, α_2, α_3, and α_4 are the Lagrangian multipliers corresponding to constraints (23a) to (23d), respectively.

By taking the partial derivatives of Z with respect to Q, W_n, and R_n and setting the results equal to zero, we have

$$\frac{\partial Z}{\partial Q} = -1-\log Q-\alpha_1 = 0; \quad (39)$$

$$\frac{\partial Z}{\partial W_n} = -(1+\log W_n)-\alpha_1-\alpha_2-\alpha_4 n = 0; \quad n\geq 0; \quad (40)$$

$$\frac{\partial Z}{\partial R_n} = -(1+\log R_n)-\alpha_1-\alpha_3-\alpha_4 n = 0; \quad n\geq 0. \quad (41)$$

On simplifying, we have

$$Q = e^{-(1+\alpha_1)}; \quad W_n = e^{-(1+\alpha_1)}e^{-\alpha_2}e^{-\alpha_4 n}; \quad R_n = e^{-(1+\alpha_1)}e^{-\alpha_3}e^{-\alpha_4 n}. \quad (42)$$

Let $\xi_1 = e^{-(1+\alpha_1)}, \xi_2 = e^{-\alpha_2}, \xi_3 = e^{-\alpha_3}, \xi_4 = e^{-\alpha_4}$. From Equation (42), we have

Equations (23b) and (23c) give

$$Q = \xi_1,\ W_n = \xi_1\xi_2\xi_4{}^n, R_n = \xi_1\xi_3\xi_4{}^n. \quad (43)$$

$$\xi_1\xi_2 = \eta_1(1-\xi_4);$$

$$\xi_1\xi_3 = \eta_2(1-\xi_4).$$

From Equation (23d), we have

$$\xi_4 = \frac{L_q}{\eta_1+\eta_2+L_q}. \quad (44)$$

Also,

$$\xi_2 = \frac{\eta_1}{Q}\left(\frac{\eta_1+\eta_2}{\eta_1+\eta_2+L_q}\right); \xi_3 = \frac{\eta_2}{Q}\left(\frac{\eta_1+\eta_2}{\eta_1+\eta_2+L_q}\right) \quad (45)$$

Using Equations (44) and (45) in Equation (43), we get the required results.

Competing interests
The authors declare that they have no competing interests.

Authors' contributions
CJS has worked on the analysis of stochastic model of queueing system by using the approach based on supplementary variable and generating function method. The queue size distribution and various performance measures via. queue theoretic approach have been obtained by MJ. BK has obtained numerical results and carried out sensitivity analysis by taking an illustration. All authors read and approved the final manuscript.

Acknowledgements
The authors are thankful to the learned reviewers and Editor-in-Chief for their valuable comments and suggestions for the improvement of the paper.

Author details
[1]Department of Mathematics, Guru Nanak Dev University, Amritsar 143005, India. [2]Department of Mathematics, Indian Institute of Technology, Roorkee 247667, India. [3]Department of Mathematics, M.L.U. DAV College, Phagwara 144402, India.

References
Almasi B, Roszik J, Sztrik J (2005) Homogeneous finite-source queues with server subject to breakdowns and repairs. Mathematics and Computer Modeling 42(5–6):673–682
Arizono I, Cui Y, Ohta H (1991) An analysis of *M/M/s* queueing systems based on the maximum entropy principle. J Oper Res Soc 42:69–73
Avi-Itzhak B, Naor P (1963) Some queueing problems with the service station subject to breakdowns. Oper Res 11:303–320
Choudhury G (2008) A note on the $M^X/G/1$ queue with a random set-up time under a restricted admissibility policy with a Bernoulli vacation schedule. Statistical Methodology 5:21–29
Choudhury G, Ke JC (2012) A batch arrival retrial queue with general retrial times under Bernoulli vacation schedule for unreliable server and delaying repair. Appl Math Model 36(1):255–269
Choudhury G, Tadj L (2011) The optimal control of an $M^X/G/1$ unreliable server queue with two phases of service and Bernoulli vacation schedule. Math Comput Model 54(1–2):673–688
El-Affendi MA, Kouvatsos DD (1983) A maximum entropy analysis of the *M/G/1* and *G/M/1* queueing systems at equilibrium. Acta Information 19:339–355
Falin G (2010) An *M/G/1* retrial queue with an unreliable server and general repair times. Perform Eval 67(7):569–582
Ferdinand AE (1970) A statistical mechanical approach to systems analysis. IBM J Res Dev 14:539–547
Kashyap BRK, Chaudhary ML (1988) An Introduction to Queueing Theory. A & A publications, Kingston, Canada
Ke JC (2007) Batch arrival queues under vacation policies with server breakdown and startup/closedown times. Appl Math Model 31:1282–1292
Ke JC, Pearn WL (2004) Optimal management policy for heterogeneous arrival queueing system with server breakdowns and vacations. Quality Technology and Quantitative Management 1:149–162
Keilson J, Servi ND (1986) Oscillating random walk models for *GI/G/1* vacation systems with Bernoulli schedules. J Appl Probab 23:790–802
Kouvatsos DD (1986) Maximum entropy and the *G/G/1/N* queue. Acta Information 23:545–565
Levy Y, Yechiali U (1976) An *M/M/s* queue with server vacations. Infor 14(2):153–163
Li W, Shi D, Chao X (1997) Reliability analysis of *M/G/1* queueing system with server breakdowns and vacation. J Appl Probab 34:546–555
Madan KC (1989) A single channel queue with bulk service subject to interruptions. Microelectron Reliab 29(5):813–818
Madan KC (1999) An *M/G/1* queue with optional deterministic server vacations. Metron LVII:83–95
Madan KC (2003) An *M/G/1* queue with time homogeneous breakdown and deterministic repair times. Soochow Journal of Mathematics 29:103–110
Medhi J (1982) Stochastic Processes. Wiley Eastern, India
Mitrany IL, Avi-Itzhak B (1968) A many server queue with service interruptions. Oper Res 16:628–638
Sengupta B (1990) A queue with service interruptions in an alternating random environment. Oper Res 38:308–318
Shore JE (1982) Information theoretic approximations for M/G/1 and G/G/1 queueing systems. Acta Information 17:43–61
Takine T, Sengupta B (1998) A single server queue with service interruptions. Queueing Systems 26:285–300
Tang Y (1997) A single server *M/G/1* queueing system subject to breakdowns - some reliability and queueing problems. Microelectron Reliab 37:315–321
Wang J (2004) An *M/G/1* queue with second optional service and server breakdowns. Computers and Mathematics with Applications 47:1713–1723
Wang KH, Chan MC, Ke JC (2007) Maximum entropy analysis of the $M^{[X]}/M/1$ queueing system with multiple vacations and server breakdowns. Computers & Industrial Engineering 52:192–202
Wang KH, Huang KB (2009) A maximum entropy approach for the (p, N) -policy M/G/1 queue with removable and unreliable server. Applied Mathematical Modelling 33:2024–2034
Wang KH, Yang DY, Pearn WL (2011) Comparative analysis of a randomized *N*-policy queue: an improved maximum entropy method. Expert Systems with Applications 38(8):9461–9471
Wu JS, Chan WC (1989) Maximum entropy analysis of multiple-server queueing systems. J Oper Res Soc 40:815–825

22

Constrained consumable resource allocation in alternative stochastic networks via multi-objective decision making

Seyed Saeid Hashemin[1*] and Seyed Mohammad Taghi Fatemi Ghomi[2]

Abstract

Many real projects complete through the realization of one and only one path of various possible network paths. Here, these networks are called alternative stochastic networks (ASNs). It is supposed that the nodes of considered network are probabilistic with exclusive-or receiver and exclusive-or emitter. First, an analytical approach is proposed to simplify the structure of the network. This approach transforms the network into a simpler equivalent one. This paper discusses the constrained consumable resource allocation problem in an ASN. Many recent researchers apply heuristic and simulation methods to solve the constrained resource allocation in these problems. In this paper, we propose an analytical approach based on multi-objective modeling. The objective functions of this model are the cumulative distribution function of the completion time of ASN paths. These functions must be maximized within the desired network completion time. Lexicographic method is used to solve the proposed multi-objective model. The proposed method is illustrated by an example.

Keywords: Stochastic network, Resource allocation, Multi-objective, Decision making, Lexicographic method, Gaussian quadrature formula, Conditional simulation

Introduction

In many real world projects, the occurrence of activities and their durations are stochastic. This is why these projects are formulated as a stochastic network (Pritsker and Happ 1966). On the other hand, the completion of projects on time has a significant effect on its cost, revenue, and usefulness. Therefore, the main objective of project managers is to avoid any delay. To achieve this goal, consuming extra resources can shorten the duration of each individual activity.

To the best of our knowledge, many recent researchers apply heuristic and simulation methods to solve the constrained resource allocation in ASNs. Constrained resource allocation in ASNs is dependent on the estimation of completion time of networks. Analytical methods for this subject have been introduced in (Pritsker and Happ 1966; Pritsker and Whitehouse 1966, 1969; Pritsker 1966; Whitehouse 1973). Furthermore, simulation methods have also been introduced in Whitehouse (1973). Efficient Monte Carlo simulation methods to estimate ASN characteristics such as project time, and project cost have been proposed in Kurihara and Nishiuchi (2002). An algorithm to fulfill the equivalent simplifying transformations of the structure of ASN has been described in (Shibanov 2003). A two-level decision-making model to control stochastic projects has been proposed in Golenko-Ginzburg (1993). Golenko-Ginzburg et al. (1996) have developed a hierarchical three-level decision-making model. These levels are upper level (company level), medium level (project level), and subnetwork level. The main goal has been to develop a unified three-level decision-making model and to indicate planning and control action and optimization problems for all levels. When the constrained resources are nonconsumable, Golenko-Ginzburg and Gonik (1997), using a zero–one integer programming, have maximized the total contribution of accepted activities to the expected project duration. The contribution of each activity is the product of the average duration of the activity and its probability of being on the critical path. A new heuristic control algorithm for

* Correspondence: s_s_hashemin@yahoo.com
[1]Department of Industrial Engineering, Ardabil Branch, Islamic Azad University, Ardabil, 56157-31567, Iran
Full list of author information is available at the end of the article

stochastic network projects has been presented in Golenko-Ginzburg and Gonik (1998a). The developed control algorithm is essentially more efficient than the step-by-step control procedures. This algorithm has reduced computational time and has provided better solutions than the ones which would be attained using online sequential statistical analysis. Golenko-Ginzburg and Gonik (1998b) have developed a look over heuristic algorithm for allocation of resource-constrained in program evaluation and review technique type networks. Each activity is of random duration, depending on the resource amounts assigned to that activity. The aim has been to minimize the expected project duration. An optimization procedure to maximize the probability confidence for project due dates under budget constraints or to minimize the project budget under due dates chance constraints has been developed in Golenko-Ginzburg et al. (2000). The study of Golenko-Ginzburg et al. (2003) has presented a resource-constrained scheduling simulation model for alternative stochastic network projects when several renewable activity-related resources, such as machines and manpower, are imbedded in the model. Each type of resources is limited. The activity duration is a random variable with given density function. The aim of the problem is to minimize the expected project duration.

Up to now, because of the complexity of computations, only simulation and heuristic methods have been used to allocate the resources to ASN. Among our investigations, we have not found an analytical approach to the problem.

However, this paper proposes an analytical stochastic model based on multi-objective decision-making (MODM) model. This model has some advantages. First, this model, unlike previous researches, has no limitations related to the type of random variables of activity durations. Second, in the studied problem, the number of feasible allocations can be very great, especially in large scale networks. Evaluation of all allocations requires tedious computations. The proposed method prevents us from evaluating all of feasible solutions because of solving the problem in several stages (using MODM model). Furthermore, for solving the proposed model, simulation is combined with analytical method (conditional Monte Carlo simulation method). Also, one of the most accurate numerical methods (generalization of Gaussian quadrature formula) is applied for solving the model.

The paper has the following structure. The 'Problem description' section describes the problem. The 'Analytical approach' section introduces MODM model. Solving the MODM model is described in 'Solving the MODM model' section. 'Example' section gives a numerical example to demonstrate how the proposed method works. 'Conclusion' section is devoted to conclusions and recommendations for future studies.

The problem description

Suppose that a project is formulated as an ASN and has the following characteristics:

1. The network has a single source node and it can have one or more sink nodes.
2. The network contains only exclusive-or probabilistic nodes (nodes with exclusive-or receiver and exclusive-or emitter).
3. The network does not contain any loop.
4. Activity implementation requires only one kind of consumable (non-renewable) resource.
5. The amount of available resource is limited and deterministic.
6. The resource allocation for activities is performed discretely. In other words, the amount of resource allocated for each activity is limited to some specific levels.
7. The duration of network activities is arbitrary continuous random variable or they can have constant values.
8. Probability density function of activity durations is dependent on the amount of resource allocated to the activity and varies as this amount changes. By increasing the allocated resource to each activity, completion time will be shorter.
9. The due date of the project is constant and known value. We want the project completion time be smaller than or equal with the due date.
10. The objective is to allocate the total constrained resource among the activities such that the cumulative distribution function (CDF) of the project completion time is being maximized for the due date. Since by maximizing the CDF of the project completion time, we also maximize the probability of project completion on time.

Analytical approach

In this section we develop an analytical approach to allocate the resource among the activities of the projects. This analytical approach uses a multi-objective decision-making model. The following symbols introduce the necessary notations before MODM model is explained.

Notations

N:	Number of activities (arcs)
M:	Number of sink nodes
n_i:	Number of paths which start from source node and terminate in i-th sink node
$F_{ij}(t)$:	CDF of j-th path which terminates in i-th sink node
P_{ij}:	Occurrence probability of j-th path which terminates in i-th sink node

$F_i(t)$:	CDF of occurrence time of i-th sink node, given that this node has occurred
p_i:	Occurrence probability of i-th sink node when $t \to +\infty$
$\overline{P}_k$:	Accomplishment probability of k-th activity, given that start node of this activity has occurred
$P_i(t)$:	Occurrence probability of i-th sink node in t
S_{ij}:	Activity set of j-th path which terminates in i-th sink node
t_k:	Duration time random variable of k-th activity
$S_{[\gamma]}$:	The set of activities of a path with preference γ
$F_{[\gamma]}(t)$:	CDF of a path with preference γ
Z:	Number of network paths
s_{l_K}:	The amount of resource allocated to k-th activity
T:	Due date of project network
RS:	The available amount of limited resource
v_k:	Number of discrete values which indicates the amount of resource allocated to k-th activity
$s^*_{l_K}$:	The optimal amount of resource allocated to k-th activity
$F^*_{[\gamma]}(T)$:	The optimal value of CDF of a path with preference γ in T
Q:	The set of activities which the optimal amount of resource allocated to them is determined
$f_{t_r}(s_{l_r}, t_r)$:	Probability density function of r-th activity completion time, when the allocated resource is s_{l_r}
$t_{K,S_{l_k}}$:	Duration time random variable of k-th activity when allocated resource is s_{l_k}
$F(t)$:	CDF of network completion time
$T_{[\gamma]}$:	Completion time of a path with preference γ
$F_{t_r}(s_{l_r}, t_r)$:	CDF of r-th activity when allocated resource is s_{l_r}
$t^{(q)}_{K,S_{l_k}}$:	Duration time of k-th activity in q-th simulation run when the allocated resource is s_{l_k}
$\overline{Q}$:	Number of simulation runs

A multi-objective decision-making model

Based on conditional probability, we can transform the above-mentioned ASN to the graphical evaluation and review technique networks with $\sum_{i=1}^{M}\sum_{j=1}^{n_i} 1$ parallel paths. This transformation has been described in Hashemin and Fatemi Ghomi (2005). Based on the study, we can write:

$$F_i(t) = \frac{\sum_{j=1}^{n_i} P_{ij}F_{ij}(t)}{\sum_{j=1}^{n_i} P_{ij}} \qquad i = 1, 2, \ldots, M \tag{1}$$

Where $P_{ij} = \prod_{k \in S_{ij}} \overline{P}_k$

For shorter times (when t is remarkably smaller than $+\infty$), we have

$$P_i(t) = \sum_{j=1}^{n_i} P_{ij}F_{ij}(t) \qquad i = 1, 2, \ldots, M \tag{2}$$

and $\lim_{t \to +\infty} P_i(t) = P_i$ is evident.

After the above transformation, we order the paths and their CDF on the basis of priority. The most probable path is the path with the highest priority.

This ordering is performed by a simple algorithm described in Appendix B. Ordered CDF of ASN paths are shown by $F_{[1]}(t), \ldots, F_{[Z]}(T)$.

The proposed MODM model with Z objective function is as follows:

$$\text{Max} \quad \{F_{[1]}(T), F_{[2]}(T), \ldots, F_{[Z]}(T)\}$$

$$\text{s.t.} \quad \sum_{K \in S_{ij}} s_{l_k} \leq RS \quad i = 1, \ldots, M, j = 1, \ldots n_i$$

$$s_{l_K} = s_1, \ldots, s_{v_K} \qquad k = 1, \ldots, N$$

The decision variables of this model are s_{l_K}, $k = 1, \ldots, N$. One of the discrete values $s_1, \ldots s_{v_K}$ can be the value of these variables. Each objective function is the sum of some continuous random variables. Hence, the proposed model is a stochastic multi-objective model with discrete decision variables. Lexicographic method (Hwang and Masud 1979) is used to solve the model because the CDF of completion time of the path with the highest occurrence probability has the highest effectiveness in network completion time CDF. Furthermore, execution of lexicographic method is simple in practice because as each activity is realized, it becomes evident that some paths are unlikely to occur. Then, some paths will be eliminated and the network can be smaller. Consequently, the remaining resources will be allocated to this reduced network. This trend continues until the problem is solved. This process is introduced in the succeeding section.

Solving the MODM model

Let's suppose that, before project implementation, the allocated resource to each activity should be predetermined. To determine each one of the objective functions of the model, it is required to determine the CDF of the sum of some random variables. Fatemi Ghomi and Hashemin (1999) have generalized the numerical integration with Gaussian quadrature formula to determine the CDF of completion time of stochastic networks.

The conditional Monte Carlo simulation to determine CDF of completion time of stochastic networks is presented in Burt and Garman (1971). Here, these two methods are converted and introduced in such a way that they would be compatible with requisitions

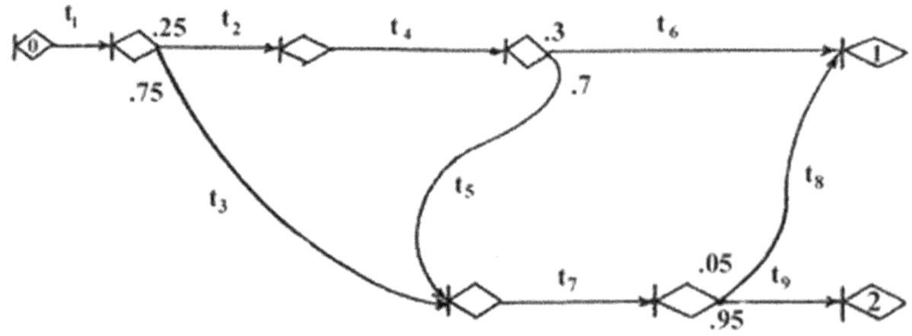

Figure 1 Network of example.

and aims of the paper (see Appendix A). To solve the MODM model using lexicographic method, the following algorithm is devised.

Algorithm 1

Step1. Obtain the optimal solution of the following problem for $\gamma = 1$.

$$Max\ F_{[\gamma]}(T)$$

s.t.

$$\sum_{k \in S_{[\gamma]}} s_{l_K} \leq RS$$

$$\sum_{k \in S_{[\gamma]} \cap S_{[W]}} s_{l_K} \leq RS - \sum_{k \in S_{[w]} - S_{[\gamma]}} \min s_{l_K} \quad W = \gamma + 1, \ldots, Z$$

$$\forall k \in S_{[\gamma]} \quad s_{l_K} = s_1, \ldots, s_{v_K}$$

Suppose the optimal values of s_{l_K} for $k \in S_{[\gamma]}$ are denoted by $s^*_{l_K}$. These values are obtained with computing $F_{[\gamma]}(T)$ for all feasible values s_{l_K}, $k \in S_{[\gamma]}$ and determining optimal value of $F_{[\gamma]}(T)$, namely, $F^*_{[\gamma]}(T)$. If for each $k = 1, \ldots, N$, $s^*_{l_K}$ has been determined, stop. Otherwise, set $Q = S_{[\gamma]}$ and go to step 2.

Step 2. Set $\gamma \leftarrow \gamma + 1$ and obtain the optimal solution of the following problem.

$$Max\ \ F_{[\gamma]}(T)$$

$$s.t. \quad s_{l_K} = s^*_{l_K} \ \ \forall k \in Q$$

$$\sum_{k \in S_{[\gamma]} - Q} s_{l_K} \leq RS - \sum_{k \in S_{[\gamma]} \cap Q} s^*_{l_K}$$

$$\sum_{k \in S_{[\gamma]} \cap S_{[W]} - Q} s_{l_K} \leq RS - \sum_{k \in S_{[W]} - S_{[\gamma]} - Q} \min s_{l_K} - \sum_{k \in S_{[W]} \cap Q} s^*_{l_K}$$

$$W = \gamma + 1, \ldots, Z$$

$$\forall k \in S_{[\gamma]} \quad s_{l_K} = s_1, \ldots, s_{v_K}$$

If for each $k = 1, \ldots, N$, $s^*_{l_K}$, has been determined, stop. Otherwise, set $Q \leftarrow Q \cup S_{[\gamma]}$ and repeat step 2.

Note that in step 2, if the amount of resource allocated to one activity is unknown, then this amount would be the greatest feasible number that satisfies the inequality $s_{l_K} \leq RS - \sum_{k \in S_{[\gamma]} \cap Q} s^*_{l_K}$ (with $k \in S_{[\gamma]} - Q$). In other words, the optimal value can be found without the solution procedure being performed.

In all steps of the above algorithm, to find the optimal solution of problem, the objective function of problem is computed for all feasible values s_{l_K} using one of the introduced methods in Appendix A. Feasible values of s_{l_K} which maximize the objective function would be the optimal solution of the problem.

In some practical situations, it may not be necessary to determine the amount of resource allocated to each activity before the project is started. In other words, the constrained resource allocation and project implementation can be done simultaneously. In such situations, the gathered information resulting from the previous activities can be helpful in the resource allocation to the succeeding activities. The following algorithm is devised for such situations.

Algorithm 2

Step1. Since the network under study has a single source node and all network nodes are exclusive-or and probabilistic type, only one activity can be implemented in the beginning of project. Determine the amount of resource allocated to this activity using step 1 of algorithm described in 'Analytical approach' section. Start the implementation of this activity. When the end node

Table 1 Information of activity durations for the example

l_1	s_{l_1}	$f_1(s_{l_1}, t_1)$		l_2	s_{l_2}	$f_2(s_{l_2}, t_2)$
1	4	$4t_1^3$	$0 < t_1 < 1$	1	1	$\frac{1}{2}e^{-\frac{1}{2}t_2}$
2	5	$4(1-t_1)^3$	$0 < t_1 < 1$	2	2	e^{-t_2}
l_3	s_{l_3}	$f_3(s_{l_3}, t_3)$		l_6	s_{l_6}	$f_6(s_{l_6}, t_6)$
1	5	$\frac{1}{2}e^{-\frac{1}{2}t_3}$	$t_3 > 0$	1	5	1
2	6	e^{-t_3}	$t_3 > 0$	2	6	$\frac{1}{2}$
l_8	s_{l_8}	$f_8(s_{l_8}, t_8)$		l_9	s_{l_9}	$f_9(s_{l_9}, t_9)$
1	1	$3t_8^2$	$0 < t_8 < 1$	1	1	$\frac{1}{3}$
2	2	$3(1-t_8)^2$	$0 < t_8 < 1$	2	2	$\frac{2}{3}$
l_4	s_{l_4}	$t_{4,s_{l_4}}$	l_5	s_{l_5}	$t_{5,s_{l_5}}$	l_7
1	1	3	1	1	2	1
2	2	2	2	2	1	2

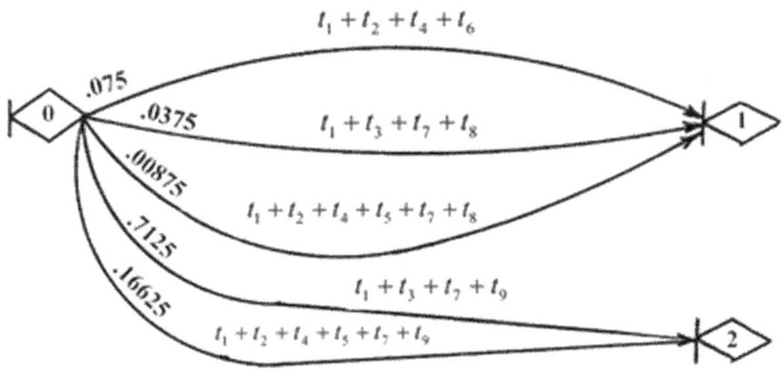

Figure 2 Transformed network of example.

of the activity occurs and if this node is one of the sink nodes of the network, stop. Otherwise, go to step 2.

Step 2. Set T ← T—implemented activity duration time and RS ← RS—the amount of consumed resource for implemented activity.

Among all activities emanating from the last realized node, determine the activity which must be realized and omit the other activities, and hence, the paths of network which have no possibility of occurrence. If $T > 0$, return to step 1. Otherwise, if ($T \leq 0$), conclude that the project has not been completed in T.

Example

Consider the network in Figure 1. The amount of limited resource is 14 units (RS = 14) and the due date of project network is 9 units of time ($T = 9$).

First, we suppose that before project implementation, the allocated resource to each activity should be predetermined.

Results and discussion

The duration times of activities 4, 5, and 7 are not random variables. They are constant values but dependent on the amount of resource allocated to the corresponding activities.

The duration times of remaining activities are continuous random variables and their probability density functions are dependent on the amount of resource allocated to them. Table 1 contains the information of the activities duration times.

Paths and their occurrence probabilities are

$$t_1+t_3+t_7+t_9$$
$$P_{21}=\bar{P}_1\bar{P}_3\bar{P}_7\bar{P}_9=(1)(.75)(1)(.95)=.7125$$
$$t_1+t_2+t_4+t_5+t_7+t_9$$
$$P_{22}=\bar{P}_1\bar{P}_2\bar{P}_4\bar{P}_5\bar{P}_7\bar{P}_9=(1)(.25)(1)(.7)(1)(.95)=.16625$$
$$t_1+t_2+t_4+t_6$$
$$P_{11}=\bar{P}_1\bar{P}_2\bar{P}_4\bar{P}_6=(1)(.25)(1)(.3)=.075$$
$$t_1+t_3+t_7+t_8$$
$$P_{12}=\bar{P}_1\bar{P}_3\bar{P}_7\bar{P}_8=(1)(.75)(1)(.05)=.0375$$
$$t_1+t_2+t_4+t_5+t_7+t_8$$
$$P_{13}=\bar{P}_1\bar{P}_2\bar{P}_4\bar{P}_5\bar{P}_7\bar{P}_8=(1)(.25)(1)(.7)(1)(.05)=.00875.$$

According to the preferences introduced in Appendix B, we have

$$S_{[1]}=S_{21}=\{1,3,7,9\}$$
$$S_{[2]}=S_{22}=\{1,2,4,5,7,9\}$$
$$S_{[3]}=S_{11}=\{1,2,4,6\}$$
$$S_{[4]}=S_{12}=\{1,3,7,8\}$$
$$S_{[5]}=S_{13}=\{1,2,4,5,7,8\}.$$

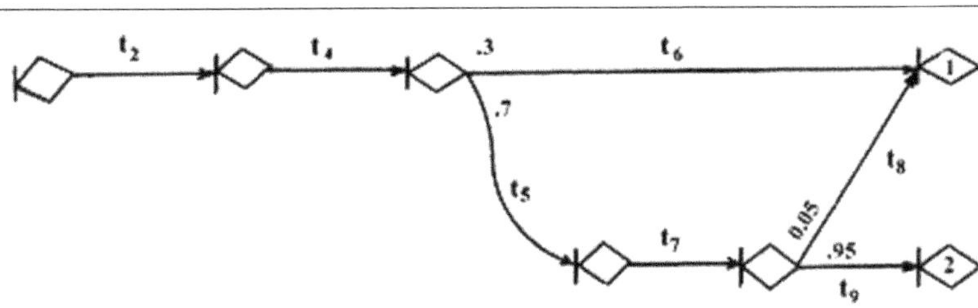

Figure 3 Remaining network in the first iteration of step 2 of algorithm 2 (first subnetwork).

So, the transformed network can be illustrated as Figure 2.

The MODM model of problem would be as follows:

$$\begin{array}{lll} Max & \{F_{[1]}(9), F_{[2]}(9), F_{[3]}(9), F_{[4]}(9), F_{[5]}(9)\} & \\ s.t. & \sum_{k \in S_{[1]}} s_{l_K} \leq 14 & S_{l_1} = 4,5 \\ & & S_{l_2} = 1,2 \\ & \sum_{k \in S_{[2]}} s_{l_K} \leq = 14 & S_{l_3} = 5,6 \\ & & S_{l_4} = 1,2 \\ & \sum_{k \in S_{[3]}} S_{l_K} \leq = 14 & \\ & & S_{l_5} = 1,2 \\ & \sum_{k \in S_{[4]}} S_{l_K} \leq 14 & \\ & & S_{l_6} = 5,6 \\ & & S_{l_7} = 3,4 \\ & \sum_{k \in S_{[5]}} S_{l_K} \leq 14 & S_{l_8} = 1,2 \\ & & S_{l_9} = 1,2 \end{array}$$

In step 1, the problem should be solved in the following way:

$$\begin{array}{lll} & Max \quad F_{[1]}(9) & \\ s.t. & s_{l_1} + s_{l_3} + s_{l_7} + s_{l_9} \leq 14 & s_{l_1} = 4,5 \\ & s_{l_1} + s_{l_7} + s_{l_9} \leq 11 & s_{l_3} = 5,6 \\ & s_{l_1} \leq 7 & s_{l_7} = 3,4 \\ & s_{l_1} + s_{l_3} + s_{l_7} \leq 13 & s_{l_9} = 1,2 \\ & s_{l_1} + s_{l_7} \leq 10 & \end{array}$$

The optimal solution of problem, using the generalized Gaussian quadrature formula introduced in Appendix A, is found as below:

$s_{l_1}^*$	$s_{l_3}^*$	$s_{l_7}^*$	$s_{l_9}^*$	$F_{[1]}^*(9)$
4	6	3	1	.9035535

In step 2, $\gamma = 2$ and $Q = S_{[1]} = \{1, 3, 7, 9\}$, and the following problem should be solved.

$$\begin{array}{lll} & Max \quad F_{[2]}(9) & \\ s.t. & s_{l_1} = 4, s_{l_3} = 6, & s_{l_7} = 3, s_{l_9} = 1 \\ & s_{l_2} + s_{l_4} + s_{l_5} \leq 6 & s_{l_2} = 1,2 \\ & s_{l_2} + s_{l_4} \leq 5 & s_{l_4} = 1,2 \\ & & s_{l_5} = 1,2 \end{array}$$

The optimal solution, using generalized Gaussian quadrature formula, is as follows:

$s_{l_2}^*$	$s_{l_4}^*$	$s_{l_5}^*$	$F_{[2]}^*(9)$
2	2	2	.1687556

On path $S_{[3]}$, only the amount of allocable resource to activity 6 has not been determined. Its optimal value will be $s_{l_6}^* = 6$ and $F_{[3]}^*(9) = .9823442$. On path $S_{[4]}$, only the amount of allocable resource to activity 8 has not been determined. Its optimal value will be $s_{l_8}^* = 1$ and $F_{[4]}^*(9) = .9882507$.

Now, the amount of allocated resource to all activities $S_{[5]}$ has been determined and $F_{[5]}^*(9) = .7584068$.

Utilizing the formula 1 in the 'A multi-objective decision-making model' section, $F_1(9)$ and $F_2(9)$ would be:

$$F_1(9) = .96801$$
$$F_2(9) = .76454$$

Utilizing the formula 2 in 'A multi-objective decision-making model' section, $P_1(9)$ and $P_2(9)$ would be:

$$P_1(9) = .11737$$
$$P_2(9) = .67184$$

Finally, the network completion time distribution function for $T = 9$ can be computed as follows:

$$F(9) = P_1F_1(9) + P_2F_2(9) = P_1(9) + P_2(9) = .78921$$

Now, suppose that, it may not be necessary to determine the amount of resource allocated to each activity before the project is started.

The algorithm 2 is described in one of the cases, which can happen for the example in the 'Example' section. In step 1, the amount of resource allocated to activity 1, as explained before, is 4 units.

Assume that activity 1 is implemented using 4 units of limited resource ($s_{l_1}^* = 4$) and this activity is completed

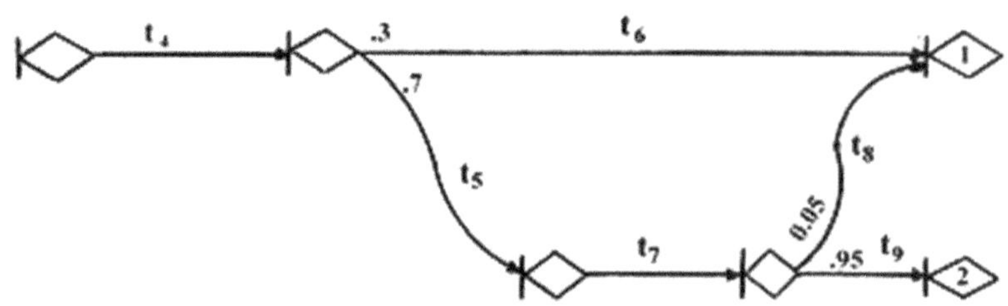

Figure 4 Remaining network in the second iteration of step 2 of algorithm 2 (second subnetwork).

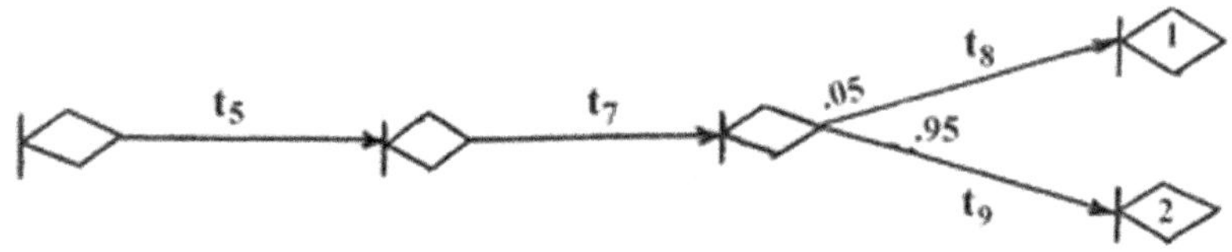

Figure 5 **Remaining network in the third iteration of step 2 of algorithm 2 (third subnetwork).**

in 0.7 units of time and the end node of activity is realized. Since this latter node is not the sink node of network, we go to step 2.

In step 2, we set $T = 9–0.7 = 8.3$ and $RS = 14–4 = 10$. Assume that, among the activities emanating from the end node of activity 1, activity 2 has been realized. So, the activity 3 will never occur and the remaining network is as Figure 3.

Above results are as follows.

Step 1	Step 2
$s^*_{l_1} = 4, t_1 = 0.7$	$T = 8.3$, RS = 10

With $T = 8.3$ and $RS = 10$, steps 1 and 2 of algorithm are repeated for subnetwork of Figure 3. Results are shown below.

Step 1	Step 2
$s^*_{l_2} = 1, t_2 = 1.3$	$T = 7$, RS = 9

By omitting the implemented activity, the remaining network would be as Figure 4.

With $T = 7$ and $RS = 9$, steps 1 and 2 of algorithm are applied for the second subnetwork, results are shown as follows.

Step 1	Step 2
$s^*_{l_4} = 2, t_4 = 2$	$T = 5$, RS = 7

Assume that, among the activities emanating from the end node of activity 4, activity 5 is realized. So, the activity 6 is not realized and the remaining network is as seen in Figure 5.

With $T = 5$ and $RS = 7$, the steps 1 and 2 of algorithm are repeated. For the third subnetwork, results are shown below.

Step 1	Step 2
$s^*_{l_5} = 2, t_5 = 1$	$T = 4$, RS = 5

By omitting the implemented activity, the remaining network would be as Figure 6.

With $T = 4$ and $RS = 5$, the steps 1 and 2 of algorithm are repeated. For the fourth subnetwork, results are shown in the following way.

One unit of limited resource remains for one of the realized activity which will be emanated from the end node of activity 7. If activity 8 is realized, the allocated resource will be $s^*_{l_8} = 1$. Otherwise, the allocated resource will be $s^*_{l_9} = 1$. In this approach, the allocation of limited resource is performed simultaneously with the project implementation, based on the last available information. In the given example, the allocation of resource was explained for the special cases that the path {1,2,4,5,7,8} or {1,2,4,5,7,9} was realized. For the other cases of path occurrences, the similar limited resource allocation can be performed.

Step 1	Step 2
$s^*_{l_7} = 4, t_7 = 2$	$T = 2$ RS = 1

Conclusions

The following conclusions can be made:

1. The proposed model, unlike previous researches, has no limitations related to the type of random variables of activity durations.
2. In the problem under study, the number of feasible allocations can be very great, especially in large scale networks. Evaluation of all of these allocations requires tedious computations. The proposed method of this paper prevents us from evaluating all

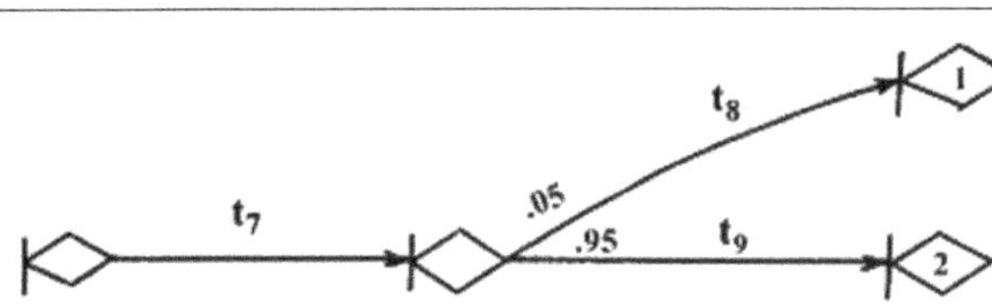

Figure 6 **Remaining network in the fourth iteration of step 2 of algorithm 2 (fourth subnetwork).**

of feasible solutions because of solving the problem in several stages (using MODM model).

3. In solving the example of paper, generalized Gaussian quadrature formula and conditional Monte Carlo simulation have been used, respectively. Both methods have desirable accuracy. But in comparison, generalized Gaussian quadrature formula is more accurate.

Recommendations

1. This paper utilizes the path occurrence probability as a criterion to determine path preference. If, in some networks, the sink nodes have some preferences to each other, this matter can be considered as a new criterion to determine path preference. This new criterion can be combined with the current criterion of this paper. This combination of criteria is recommended for future studies.
2. Other methods concerning the optimization of multi-objective problems can be studied to solve the current problem of this paper.
3. In this paper, proposed methods were designed to be applicable for alternative stochastic networks with exclusive-or, probabilistic nodes. Constrained resource allocation in ASN with other types of nodes is recommended as an area for future studies. In these networks, using a critical chain is suggested.
4. The solution procedure of this paper can be extended for the case where several types of limited resources are concerned.
5. Optimal resource allocation can be a very interesting subject to study, when some resources are of renewable types and some others are of non-renewable types.
6. In some networks, the allocation of limited resource to activities might be performed continuously. In this case, development of a method for optimal allocation of resource to the activities would be valuable.
7. It is evident that each combination of recommendations 1–6 can be utilized for future studies.

Appendix A

To compute the CDF completion time of path $S_{[y]} = S_{ij}$ we can write

$$T_{[y]} = \sum_{k\in S_{[y]}} t_{k,s_{l_K}}$$

$$P(T_{[y]} \le T) = P\left(\sum_{k\in S_{[y]}} t_{k,s_{l_K}} \le T\right)$$

If $r \in S_{[y]}$, then $P(T_{[y]} \le T \,|\, t_{k,s_{l_k}}, k \in S_{[y]}, k \neq r)$

$$= P\left(t_{r,s_{l_r}} \le T - \sum_{\substack{k \in S_{[y]} \\ k \neq r}} t_{k,s_{l_K}}\right)$$

So, we can write $F_{[y]}\left(T \,\middle|\, t_{k,s_{l_K}}, k \in S_{[y]}, k \neq r\right)$

$$= F_{t_r}\left(s_{l_r}, T - \sum_{\substack{k \in S_{[y]} \\ k \neq r}} t_{k,s_{l_K}}\right)$$

Based on the above equality, two algorithms, A and B, are developed to apply conditional Monte Carlo simulation and generalized Gaussian quadrature formula, respectively.

Algorithm A

Compute $F_{[y]}(T)$ by using the following steps for all sets of feasible values $\{s_{l_K} | k \in S_{[y]}\}$. The set $\{s_{l_K} | k \in S_{[y]}\}$ is feasible, if it satisfies the constraints of mathematical model of step1 or step 2 of algorithm 1. That set which provides the greatest value for $F_{[y]}(T)$, indicates the optimal allocation of limited resource to activities of path $S_{[y]}$.

Step 1. Set q=1.

Step 2. Set $L_{[y]} = 0$.

Step 3. For each $k \in S_{[y]}$, $k \neq r$, generate random variables $t^{(q)}_{k,s_{l_k}}$.

Step 4. $L_{[y]} \leftarrow L_{[y]} + F_{t_r}\left(s_{l_r}, T - \sum_{\substack{k \in S_{[y]} \\ k \neq r}} t^{(q)}_{k,s_{l_K}}\right)$

Step 5. Set $q \leftarrow q+1$. If $q \le \bar{Q}$, go to step 3. Otherwise, go to step 6.

Step 6. $F_{[y]}(T) = \frac{L_{[y]}}{\bar{Q}}$

Algorithm B

Compute $F_{[y]}(T)$ by using the following equality for all sets of feasible values $\{s_{l_K} | k \in S_{[y]}\}$. Here, the feasible set is defined the same as defined in algorithm A. That set which provides the greatest value for $F_{[y]}(T)$, indicates the optimal allocation of limited resource to the activities of path $S_{[y]}$.

$$F_{[y]}(T) = \underset{\sum_{k\in S_{[y]}} t_k \le T}{\iint\cdots\int} F_{t_r}\left(s_{l_r}, T - \sum_{\substack{k \in S_{[y]} \\ k \neq r}} t_k\right) \prod_{\substack{k \in S_{[y]} \\ k \neq r}} f_{t_k}(s_{l_K}, t_k)\, dt_k$$

Except for special cases, the above analytical computation is not so much easy job. So, now application of Gaussian quandrature formula generalized for stochastic networks is being proposed by Fatemi Ghomi and Hashemin (1999).

Appendix B

The sets $S_{[1]}$ to $S_{[Z]}$ and functions $F_{[1]}(t)$ to $F_{[Z]}(t)$ are determined using the following steps.

Step 1. Set $\gamma=1$ and $A = \phi$

Step 2. If $P_{\alpha\beta} = \max\{\{P_{ij}|i=1,\ldots,M,j=1,\ldots,n_i\}-A\}$, then $S_{[\gamma]} = S_{\alpha\beta}$ and $F_{[\gamma]}(t) = F_{\alpha\beta}(t)$.

Step 3. If $\gamma = Z$, stop. Otherwise set $\gamma \leftarrow \gamma + 1$ and $A \leftarrow A\cup\{P_{\alpha\beta}\}$ and return to step 2.

The aim of implementing the above steps is ordering the paths based on their criticality indices in descending manner.

Competing interests

The authors declare that they have no competing interests.

Author details

[1]Department of Industrial Engineering, Ardabil Branch, Islamic Azad University, Ardabil, 56157-31567, Iran. [2]Department of Industrial Engineering, Amirkabir University of Technology, Tehran, 15988-46611, Iran.

References

Burt JM, Garman MB (1971) Conditional Monte Carlo: a simulation technique for stochastic network analysis. Manag Sci 18:207–217

Fatemi Ghomi SMT, Hashemin SS (1999) A new analytical algorithm and generation of Gaussian quadrature formula for stochastic network. Eur J Oper Res 114:610–625

Golenko-Ginzburg D (1993) A two-level decision making model for controlling stochastic projects. Int J Prod Econ 32(1):117–127

Golenko-Ginzburg D, Gonik A, Kesler S (1996) Hierarchical decision making model for planning and controlling stochastic projects. Int J Prod Econ 46–47:39–54

Golenko-Ginzburg D, Gonik A (1997) Stochastic network project scheduling with non-consumable limited resources. Int J Prod Econ 48(1):29–37

Golenko-Ginzburg D, Gonik A (1998a) High performance heuristic algorithm for controlling stochastic network projects. Int J Prod Econ 54(3):235–245

Golenko-Ginzburg D, Gonik A (1998b) A heuristic network project scheduling with random activity duration depending on the resource allocation. Int J Prod Econ 55:149–162

Golenko-Ginzburg D, Gonik A, Sitniakorsky S (2000) Resource supportability model for stochastic network projects under a chance constraint. Commun Dependability Qual Manage 3:89–102

Golenko-Ginzburg D, Gonik A, Laslo Z (2003) Resource constrained scheduling simulation model for alternative stochastic network projects. Math Comput Simul 63(2):105–117

Hashemin SS, Fatemi Ghomi SMT (2005) A hybrid method to find cumulative distribution function of completion time of GERT networks. J Ind Eng Int 1(1):1–9

Hwang CL, Masud ASM (1979) Multiple objective decision making methods and applications. Springer, New York

Kurihara K, Nishiuchi N (2002) Efficient Monte Carlo simulation method of GERT-type network for project management. Comput Ind Eng 42:521–531

Pritsker AAB () GERT: graphical evaluation and review technique. Rand Memorandum RM-4973-NASA. Rand Corporation, California

Pritsker AAB, Happ WW (1966) GERT: graphical evaluation and review technique, part I, fundamentals. J Ind Eng 17(5):267–274

Pritsker AAB, Whitehouse GE (1966) GERT: graphical evaluation and review technique, part II, probabilistic and industrial engineering applications. J Ind Eng 17(6):293–300

Shibanov AP (2003) Finding the distribution density of the time taken to fulfill the GERT network on the basis of equivalent simplifying transformation. Autom Remote Control 64:279–287

Whitehouse GA (1973) System analysis and design using network techniques. Prentice-Hall, Inc., New Jersey

Whitehouse GE, Pritsker AAB (1969) GERT-generating functions, conditional distributions, counters, renewal times and correlations. AIIE Transaction 1(1):45–50

23

Multiple vacation policy for $M^X/H_k/1$ queue with un-reliable server

Madhu Jain[1], Richa Sharma[2*] and Gokul Chandra Sharma[3]

Abstract

This paper studies the operating characteristics of an $M^X/H_k/1$ queueing system under multiple vacation policy. It is assumed that the server goes for vacation as soon as the system becomes empty. When he returns from a vacation and there is one or more customers waiting in the queue, he serves these customers until the system becomes empty again, otherwise goes for another vacation. The breakdown and repair times of the server are assumed to follow a negative exponential distribution. By using a generating function, we derive various performance indices. The approximate formulas for the probability distribution of the waiting time of the customers in the system by using the maximum entropy principle (MEP) are obtained. This approach is accurate enough for practical purposes and is a useful method for solving complex queueing systems. The sensitivity analysis is carried out by taking a numerical illustration.

Keywords: Batch arrival; k-type hyper-exponential distribution; State-dependent rates; Maximum entropy principle; Long-run probabilities; Un-reliable server; Queue length

Background

Server vacation models are useful for queueing systems in which the server wants to utilize his idle time for different purposes. The vacation mechanism considered in this paper is termed as 'multiple vacation policy'. That is, the server upon returning from a vacation leaves immediately for another one if the system is empty at that moment. Applications of the server with multiple vacation models can be found in manufacturing systems, designing of computer and communication systems, etc. Queueing systems with multiple server vacations have attracted the attention of numerous researchers. Baba (1986) studied batch-arrival $M^X/G/1$ queueing systems with multiple vacations. A discrete-time Geo/G/1 queue with multiple vacations was studied by Tian and Zhang (2002). An $M^X/G(a,b)/1$ queue with multiple vacations including closedown time has been studied by Arumuganathan and Jeykumar (2004). Kumar and Madheswari (2005) analyzed a Markovian queue with two heterogeneous servers and multiple vacations. By using the matrix geometric method, they derived the stationary queue length distribution and mean system size. Wu and Takagi (2006) investigated an M/G/1 queue with multiple vacations and exhaustive service discipline such that the server works with different rates rather than completely stopping the service during vacation. Ke (2007) studied an $M^X/G/1$ queueing system under a variant vacation policy where the server takes at most *j* vacations. He derived the system size distribution as well as waiting time distribution in the queue. Ke and Chang (2009) considered an $M^X/(G_1,G_2)/1$ retrial queue with general retrial times, where the server provides two phases of heterogeneous service to all customers under Bernoulli vacation schedules. They constructed the mathematical model and derived the steady-state distribution of the server state and the number of customers in the system/orbit. Ke et al. (2009) studied the vacation policy for a finite buffer M/M/c queueing system with an un-reliable server. Threshold *N*-policy for an $M^X/H_2/1$ queueing system with an un-reliable server and vacations was studied by Sharma (2010). Moreover, Singh et al. (2012) investigated an M/G/1 queueing model with vacation and used the generating function method for obtaining various performance measures. Very recently, an un-reliable bulk queue with state-dependent arrival rates was examined by Singh et al. (2013).

Queueing models with an un-reliable server under multiple vacation policy are more realistic representation

* Correspondence: aligarh.richa@gmail.com
[2]Department of Mathematics, JK Lakshmipat University, Jaipur 302026, India
Full list of author information is available at the end of the article

of the systems. The service of the components may be interrupted when the operator encounters unpredicted breakdowns, and it is to be immediately recovered with a random time. When the repair is completed, the server immediately returns for service. Wang et al. (1999) extended Wang's model to the N-policy for an $M/H_2/1$ queueing system and focused on single-arrival Erlangian service time queueing model with an un-reliable server. Wang et al. (2004) considered an $M/H_k/1$ queueing system with a removable and un-reliable server under N-policy and presented the optimal operating policy. Wang (2004) considered an M/G/1 queue with an un-reliable server and second optional service. Using the supplementary variable method, he obtained transient and steady-state solutions for both queueing and reliability measures of interest. Ke (2005) studied a modified T vacation policy for an M/G/1 queueing system where an un-reliable server may take at most J vacations repeatedly until at least one customer appears in the queue upon returning from vacation, and the server needs a startup time before starting each of his service periods. Li et al. (2007) proposed a single-server vacation queue with two policies, working vacation and service interruption. Choudhury and Deka (2008) studied an M/G/1 retrial queue with an additional second phase of optional service subject to breakdowns occurring randomly at any instant while serving the customers. Further, Wang and Xu (2009) obtained the solution of an M/G/1 queue with second optional service and server breakdown using the method of functional analysis. The work on an M/G/1 queue with second optional service and server breakdown has been done by Choudhury and Tadj (2009). They derived the Laplace-Stieltjes transform of busy period distribution and waiting time distribution. Further, an un-reliable server queue with multi-optional services and multi-optional vacations was analyzed by Jain et al. (2013).

Approximate results of many complex queueing models have been developed by several authors by applying the technique of maximum entropy principle. Jain and Singh (2000) used the principle of maximum entropy to analyze the optimal flow control of a G/G/c finite capacity queue. Jain and Dhakad (2003) provided the steady-state queue size distribution of an $M^X/G/1$ queue using the maximum entropy approach in which the constraints are expressed in terms of mean arrival rates, mean service rates, and mean number of customers in the system. Further, Ke and Lin (2006) employed the principle of maximum entropy to derive the approximate formulas for the steady-state probability distributions of the queue length. Ke and Lin (2008) suggested the maximum entropy principle to examine the $M^X/G/1$ queueing system in different frameworks. Omey and Gulck (2008) did maximum entropy analysis of an $M^X/M/1$ queueing model with multiple vacations and server breakdowns. Wang and Huang (2009) analyzed a single removable and un-reliable server M/G/1 queue under the (p,N)-policy. They did the maximum entropy analysis to obtain the approximate formulas for the probability distributions of the number of customers and the expected waiting time in the system. Maximum entropy approach has been applied for an un-reliable server vacation queueing model by Jain et al. (2012).

The main objective of our study is to develop an $M^X/H_k/1$ queueing model with an un-reliable server under multiple vacation policy. In this paper, an $M^X/H_k/1$ queue has been analyzed including more features, namely (1) bulk arrival, (2) server breakdown, and (3) multiple vacation. The various approximate results for waiting time distribution have been analyzed using the maximum entropy principle which was not considered in the previous study. Now, we cite a real-life situation of a given model wherein all the features are encountered simultaneously. To highlight the application, we cite an example of production of *heat transfer equipment*. In the production of these equipment, the raw material of these equipment arrives in group (batch arrival). The production of these equipment has been done by the machine in phases (called k-type hyper-exponential distribution). During the production, the production of the equipment may be interrupted due to some machinery faults called server breakdown. After interruption in the service, the machine is immediately sent for repairing. After repairing, the machine renews and works as a new one. As the raw material of these equipment is finished, the operator may take multiple vacations till the raw material arrives again. More realistic assumptions incorporated in our model provide a new dimension in the area of queueing systems.

Our main objective of this paper is to develop an $M^X/H_k/1$ queueing model with an un-reliable server under multiple vacation policy. Further, we intend to determine the approximate results for the steady-state probability distributions of the queue length using the maximum entropy approach. The rest of the paper is organized as follows. In the 'Model description' section, we describe the model and construct the steady-state equations governing the model. Afterwards, in the 'Probability generating function' section, we obtain the queue size distribution by using the probability generating function technique. In the 'Performance measures' section, we derive various performance indices. The principle of maximum entropy is described in the 'Maximum entropy principle' section to establish the approximate results for the expected system size and expected waiting time. In the 'Numerical illustration and sensitivity analysis' section, numerical illustrations and sensitivity analysis are presented to validate the analytical results. Finally, conclusion has been drawn in the 'Conclusion' section.

Model description

Consider a single un-reliable, removable server queue with state-dependent rates. We assume that the states of the system are described by the triplet (i,j,n), where $i = 0, 1,..., k$; $j = V, B, D$; and $n = 0, 1, 2,...$. Here $i = 0$ denotes that the customer is not in service, and $i = 1, 2,..., k$ denotes that the customer is in the ith phase service; $j = V, B, D$ represents that the server is on vacation, busy, and under repair after failure, respectively; and n denotes the number of customers present in the system. The service time is assumed to follow the k-type hyper-exponential distribution. It is assumed that μ_i $(i = 1, 2,..., k)$ is the service rate of ith phase service. Let the probability that the next customer to enter in the service is of type i be q_i $(i = 1, 2,..., k)$ and $\sum_{i=1}^{k} q_i = 1$. Other assumptions made to construct the mathematical model are as follows:

- The customers arrive in batches according to the Poisson process with state-dependent arrival rate λ_j given by

$$\lambda_j = \begin{cases} \lambda_0, & \text{if the server is on vacation} \\ \lambda_1, & \text{if the server is turned on and is in operation} \\ \lambda_2, & \text{if the server is turned on and is under repair} \end{cases}$$

  Let A be the random variable denoting the batch size, and then the batch size distribution is given by

$$c_j = \Pr[A = j], \ j = 1, 2, ..., d$$

  Furthermore, the generating function for the batch size distribution is $A(z) = \sum_{j=1}^{\infty} c_j z^j$. It follows that E $(A) = A'(1)$ and $E[A(A-1)] = A''(1)$.
- When the breakdown occurs, the server is unable to render service to the customers, but after completing repair provided by a repairman, it works as efficiently as before the failure. The life time and repair time of the server are negative exponentially distributed with mean $1/\alpha$ and $1/\beta$, respectively.
- The customers are served according to the first come, first served (FCFS) discipline.

Let us denote the steady-state probabilities depicting the system status as follows:

- $P_{0,V}(n)$: Probability that there are n $(n = 0, 1,...)$ customers in the system and the server is on vacation.
- $P_{i,B}(n)$: Probability that there are n $(n = 1, 2,...)$ customers in the system and the customer in service is in phase i $(i = 1, 2,..., k)$, when the server is turned on and is in busy state.
- $P_{i,D}(n)$: Probability that there are n $(n = 1, 2,...)$ customers in the system and the customer in service is in phase i $(i = 1, 2,..., k)$, when the server is turned on and is in a breakdown state.

The steady-state equations governing the model are given as follows:

$$(\lambda_0 + \nu)\, P_{0,V}(n) = \lambda_0 \sum_{k=1}^{n} P_{0,V}(n-k) c_k; \quad n \geq 1 \tag{1}$$

$$\lambda_0 P_{0,V}(0) = \sum_{j=1}^{k} \mu_j P_{0,V}(1) \tag{2}$$

$$(\lambda_1 + \alpha + \mu_i) P_{i,B}(1) = q_i \sum_{j=1}^{k} \mu_j P_{j,B}(2) + \beta P_{i,D}(1); \quad 1 \leq i \leq k \tag{3}$$

$$(\lambda_1 + \alpha + \mu_i) P_{i,B}(n) = q_i \sum_{j=1}^{k} \mu_j P_{j,B}(n+1) + \beta P_{i,D}(n) + \lambda_1 \sum_{k=1}^{n-1} P_{i,B}(n-k) c_k; \ n \geq 2, \ 1 \leq i \leq k \tag{4}$$

$$(\lambda_2 + \beta) P_{i,D}(1) = \alpha P_{i,B}(1); \quad 1 \leq i \leq k \tag{5}$$

$$(\lambda_2 + \beta) P_{i,D}(n) = \alpha P_{i,B}(n) + \lambda_2 \sum_{k=1}^{n-1} P_{i,D}(n-k) c_k; \quad n \geq 2, \ 1 \leq i \leq k \tag{6}$$

Probability generating function

In this section, we present the probability generating function (PGF) technique to obtain the analytical solution of Equations 1 to 6. Let us define the following partial generating functions:

$$G_{0,V}(z) = \sum_{n=0}^{\infty} z^n P_{0,V}(n), \qquad |z| \leq 1 \tag{7}$$

$$G_{i,B}(z) = \sum_{n=1}^{\infty} z^n P_{i,B}(n), \quad 1 \leq i \leq k, \quad |z| \leq 1 \tag{8}$$

$$G_{i,D}(z) = \sum_{n=1}^{\infty} z^n P_{i,D}(n), \quad 1 \leq i \leq k, \quad |z| \leq 1 \tag{9}$$

Lemma 1. *The expressions for the partial generating functions are obtained as follows*:

$$G_{0,V}(z) = \frac{1}{1-\rho_v A(z)} P_{0,V}(0) \tag{10}$$

$$G_{i,B}(z) = \frac{N_i(z)}{D(z)} P_{0,V}(0), \qquad i = 1, 2, ..., k \tag{11}$$

$$G_{i,D}(z) = \frac{\alpha}{(\lambda_2 + \beta - \lambda_2 A(z))} G_{i,B}(z), \ 1 \le i \le k \tag{12}$$

where

$$\rho_v = \frac{\lambda_0}{\lambda_0 + v}; D(z) = \prod_{i=1}^{k} \theta_i(z) + \sum_{i=1}^{k} \left[q_i \mu_i \prod_{j \ne i}^{k} \theta_j(z) \right]$$

$$N_i(z) = \prod_{j \ne i}^{k} \theta_j(z) q_i \lambda_0 z;$$

$$\theta_i(z) = \left[\lambda_1 z A(z) - \left(\lambda_1 + \alpha + \mu_i + \frac{\alpha\beta}{(\lambda_2 C(z) - \lambda_2 - \beta)} \right) z \right], \qquad i = 1, 2, ..., k$$

Proof. For proof, see 'Proof of Lemma 1' in the Appendix.

Lemma 2. *The probability $P_{0,V}(0)$ is*

$$P_{0,V}(0) = \left[\frac{1}{1-\rho_v} + \sum_{i=1}^{k} \frac{\rho_i q_i \left(1 + \frac{\alpha}{\beta}\right) \left[i\beta - (\lambda_1 \beta E[A] + \alpha \lambda_2 E[A]) \sum_{i=1}^{k-1} \frac{1}{\mu_i} \right]}{\left[(\lambda_1 \beta E[A] + \alpha \lambda_2 E[A]) \sum_{i=1}^{k} \frac{q_i}{\mu_i} - \beta \right]} \right]^{-1} \tag{13}$$

Also, the stability condition is given by

$$\frac{\alpha}{\beta} < \frac{(1 - \lambda_1 E[A]) \sum_{i=1}^{k} \frac{q_i}{\mu_i}}{\lambda_2 E[A] \sum_{i=1}^{k} \frac{q_i}{\mu_i}} \tag{14}$$

Proof. For proof, see 'Proof of Lemma 2' in the Appendix.

Theorem 1. *The probability generating function of the number of customers in the system is given by*

$$G(z) = \left[\frac{1}{1-\rho_v C(z)} + \sum_{i=1}^{k} \left[1 + \frac{\alpha}{(\lambda_2 + \beta - \lambda_2 C(z))} \right] \frac{N_i(z)}{D(z)} \right] P_{0,V}(z) \tag{15}$$

Proof. For proof, see 'Proof of Theorem 1' in the Appendix.

Performance measures

In order to predict the system characteristics under variant circumstances, it is worthwhile to explore key aspects by establishing analytical formulae. In this section, some performance measures in terms of steady-state probabilities are obtained. The long-run probabilities of the server being on vacation, busy, and breakdown are denoted by P_V, P_B, and P_D, respectively. Thus, we obtain

$$P_V = \sum_{n=1}^{\infty} z^n P_{0,V}(n) = G_{0,V}(1)^{0,V0,V} = \frac{1}{1-\rho_v} P_{0,V}(0) \tag{16}$$

$$P_B = \sum_{n=1}^{\infty} \sum_{i=1}^{k} z^n P_{i,B}(n) = \sum_{i=1}^{k} G_{i,B}(1)$$

$$= \sum_{i=1}^{k} \frac{\rho_i q_i \left(1 + \frac{\alpha}{\beta}\right) \left[i\beta - (\lambda_1 \beta E[A] + \alpha \lambda_2 E[A]) \sum_{i=1}^{k-1} \frac{1}{\mu_i} \right]}{\left[(\lambda_1 \beta E[A] + \alpha \lambda_2 E[A]) \sum_{i=1}^{k} \frac{q_i}{\mu_i} - \beta \right]} P_{0,V}(0) \tag{17}$$

$$P_D = \sum_{n=1}^{\infty} \sum_{i=1}^{k} z^n P_{i,D}(n) = \sum_{i=1}^{k} G_{i,D}(1)$$

$$= \sum_{i=1}^{k} \frac{\alpha \rho_i q_i \left(1 + \frac{\alpha}{\beta}\right) \left[i\beta - (\lambda_1 \beta E[A] + \alpha \lambda_2 E[A]) \sum_{i=1}^{k-1} \frac{1}{\mu_i} \right]}{\beta \left[(\lambda_1 \beta E[A] + \alpha \lambda_2 E[A]) \sum_{i=1}^{k} \frac{q_i}{\mu_i} - \beta \right]} P_{0,V}(0) \tag{18}$$

Theorem 2. *The expected number of customers in the system (L_N) is given by*

$$L_N = G'(1) = \left[\frac{\rho_v E[A]}{(1-\rho_v)^2} + \sum_{i=1}^{k} \left[\frac{(\alpha + \beta)(N_i''(1) D'(1) - N_i'(1) D''(1)}{2\beta D'(1)^2} + \frac{N_i'(1) \lambda_2 E[A]}{D'(1) \beta^2} \right] \right] P_{0,V}(0) \tag{19}$$

where

$$N_i'(1) = \prod_{j \ne i}^{k} (1)^{i+1} \lambda_0 \mu_j q_i \left(1 - \sum_{i=1}^{k} a_i \right), \qquad i = 1, 2, ..., k$$

$$N_i''(1) = (-1)^i q_i \lambda_0 \left[b \sum_{i=1}^{k-1} \prod_{\substack{j=1 \\ j \neq i}}^{k-1} \mu_j + 2 \prod_{i=1}^{k-1} \mu_i \left\{ \sum_{i=1}^{k-1} a_i - \sum_{i=1}^{k-1} a_i \sum_{\substack{j=1 \\ j \neq i}}^{k-1} a_j \right\} \right], \quad i = 1, 2, \ldots, k$$

$$D'(1) = \prod_{i=1}^{k} \mu_i \sum_{i=1}^{k} a_i q_i, \qquad i = 1, 2, \ldots, k$$

$$D''(1) = b \sum_{i=1}^{k} q_i \prod_{j \neq i}^{k} \mu_j - 2 \prod_{i=1}^{k} \mu_i a_i \sum_{i=1}^{k} \frac{1-q_i}{a_i}, \qquad i = 1, 2, \ldots, k$$

$$a_i = \frac{\lambda_1 E[A] + \delta}{\mu_i} - 1, \quad i = 1, 2, \ldots, k, \qquad \delta = \frac{\alpha \lambda_2 E[A]}{\beta}$$

$$b = 2\lambda_1 E[A] + 2\delta + E[A(A-1)] \left(\lambda_1 + \frac{\alpha}{\beta} \lambda_2 \right) - \frac{2}{\alpha} \delta^2$$

Proof. For proof, see 'Proof of Theorem 2' in the Appendix.

Maximum entropy principle

Exact probabilities of the system states of an $M^X/H_k/1$ queueing system with multiple vacations and an un-reliable server have not been found earlier to the best information of the authors. In order to evaluate approximate results for the steady-state probabilities, we employ the maximum entropy approach. It is well established that the principle of maximum entropy can be used for estimating probabilistic information measures which is further used to obtain the queue size distribution of various complex queueing systems in different frameworks. In order to obtain the steady-state probabilities $P_{0,V}(n)$, $P_{i,B}(n)$, and $P_{i,D}(n)$ by using the principle of maximum entropy, we formulate the maximum entropy model as follows.

The maximum entropy model

Following El-Affendi and Kouvatsos (1983), the entropy function y can be mathematically formulated as

$$y = -\sum_{n=0}^{\infty} P_{0,V}(n) \log P_{0,V}(n) - \sum_{n=0}^{\infty} \sum_{i=1}^{k} P_{i,B}(n) \log P_{i,B}(n) - \sum_{n=0}^{\infty} \sum_{i=1}^{k} P_{i,D}(n) \log P_{i,D}(n) \tag{20}$$

subject to the following constraints:

1. Normalizing condition

$$\sum_{n=0}^{\infty} P_{0,V}(n) + \sum_{n=1}^{\infty} \sum_{i=1}^{k} P_{i,B}(n) + \sum_{n=1}^{\infty} \sum_{i=1}^{k} P_{i,D}(n) = 1 \tag{21}$$

2. The probability that the server being busy

$$\sum_{n=1}^{\infty} \sum_{i=1}^{k} P_{i,B}(n) = \sum_{i=1}^{k} G_{i,B}(1) = \sum_{i=1}^{k} A_i \tag{22}$$

3. The probability that the server is in a breakdown state

$$\sum_{n=1}^{\infty} \sum_{i=1}^{k} P_{i,D}(n) = \sum_{i=1}^{k} G_{i,D}(1) = \sum_{i=1}^{k} E_i \tag{23}$$

4. The expected number of customers in the system

$$\sum_{n=0}^{\infty} n P_{0,V}(n) + \sum_{n=1}^{\infty} \sum_{i=1}^{k} n P_{i,B}(n) + \sum_{n=1}^{\infty} \sum_{i=1}^{k} n P_{i,D}(n) = L_N \tag{24}$$

where $A_i = G_{i,B}(1)$, $E_i = G_{i,D}(1)$, $(1 \le i \le k)$, and L_N are given by Equations 17, 18, and 19, respectively.

After introducing Lagrange's multipliers corresponding to constraints (21) to (24), we construct Lagrange's function as

$$\begin{aligned} y = &-\sum_{n=0}^{\infty} P_{0,V}(n) \log P_{0,V}(n) - \sum_{n=0}^{\infty} \sum_{i=1}^{k} P_{i,B}(n) \log P_{i,B}(n) - \sum_{n=0}^{\infty} \sum_{i=1}^{k} P_{i,D}(n) \log P_{i,D}(n) \\ &- \theta_1 \left[\sum_{n=0}^{\infty} P_{0,V}(n) + \sum_{n=1}^{\infty} \sum_{i=1}^{k} P_{i,B}(n) + \sum_{n=1}^{\infty} \sum_{i=1}^{k} P_{i,D}(n) - 1 \right] \\ &- \sum_{i=1}^{k} \eta_i \left[\sum_{n=1}^{\infty} P_{i,B}(n) - A_i \right] - \sum_{i=1}^{k} \xi_i \left[\sum_{n=1}^{\infty} P_{i,D}(n) - E_i \right] \\ &- \xi_{k+1} \left[\sum_{n=0}^{\infty} n P_{0,V}(n) + \sum_{n=1}^{\infty} \sum_{i=1}^{k} n P_{i,B}(n) + \sum_{n=1}^{\infty} \sum_{i=1}^{k} n P_{i,D}(n) - L_N \right] \end{aligned} \tag{25}$$

where θ_1, η_i $(1 \le i \le k)$ *and* ξ_i $(1 \le i \le k+1)$ are Lagrange's multipliers corresponding to constraints (21) to (24), respectively.

The maximum entropy analysis

The maximum entropy results are obtained by taking the partial derivatives of y w.r.t. $P_{0,V}(n)$, $P_{i,B}(n)$, and $P_{i,D}(n)$ and equating to zero. Thus, we get

$$\frac{\partial y}{\partial P_{0,V}(n)} = -\log P_{0,V}(n) - 1 - \theta_1 - \xi_{k+1} n = 0 \tag{26}$$

$$\frac{\partial y}{\partial P_{i,B}(n)} = -\log P_{i,B}(n) - 1 - \theta_1 - \eta_i - \xi_{k+1} n = 0, \quad 1 \le i \le k \tag{27}$$

$$\frac{\partial y}{\partial P_{i,D}(n)} = -\log P_{i,D}(n) - 1 - \theta_1 - \xi_i - \xi_{k+1} n = 0, \quad 1 \le i \le k \tag{28}$$

From Equations 26 to 28, we get

$$P_{0,V}(n) = e^{-(1+\theta_1)} e^{-\xi_{k+1} n}, \quad n = 0, 1, \ldots \tag{29}$$

$$P_{i,B}(n) = e^{-(1+\theta_1+\eta_i)} e^{-\xi_{k+1} n}, \quad n = 1, 2, \ldots, \quad 1 \le i \le k \tag{30}$$

$$P_{i,D}(n) = e^{-(1+\theta_1+\xi_i)} e^{-\xi_{k+1} n}, \quad n = 1, 2, \ldots, \quad 1 \le i \le k \tag{31}$$

Denote $\phi_1 = e^{-(1+\theta_1)}, \psi_i = e^{-\eta_i}, 1 \le i \le k$ *and* $\delta_i = e^{-\xi_i}$, $1 \le i \le k+1$.

Then, Equations 29 to 31 can be written as

$$P_{0,V}(n) = \phi_1 \delta_{k+1}^n, \quad n = 0, 1, \ldots \tag{32}$$

$$P_{i,B}(n) = \phi_1 \psi_i \delta_{k+1}^n, \quad n = 1, 2, \ldots, \quad 1 \le i \le k \tag{33}$$

$$P_{i,D}(n) = \phi_1 \delta_i \delta_{k+1}^n, \quad n = 1, 2, \ldots, \quad 1 \le i \le k \tag{34}$$

On substituting the values of $P_{0,V}(n)$, $P_{i,B}(n)$, and $P_{i,D}(n)$ from Equations 32 to 34 into Equations 21 to 23, we obtain

$$\sum_{n=0}^{\infty} \phi_1 \delta_{k+1}^n = \frac{\phi_1}{1-\delta_{k+1}} = 1 - \sum_{i=1}^{k} (A_i + E_i) \tag{35}$$

$$\sum_{n=1}^{\infty} \phi_1 \psi_i \delta_{k+1}^n = \frac{\phi_1 \psi_i \delta_{k+1}}{1-\delta_{k+1}} = A_i, \quad 1 \le i \le k \tag{36}$$

$$\sum_{n=1}^{\infty} \phi_1 \delta_i \delta_{k+1}^n = \frac{\phi_1 \delta_i \delta_{k+1}}{1-\delta_{k+1}} = E_i, \quad 1 \le i \le k \tag{37}$$

It follows from Equations 35 to 37 that

$$\phi_1 = (1-\rho)(1-\delta_{k+1}) \tag{38}$$

$$\psi_i = \frac{A_i}{(1-\rho)\delta_{k+1}}, \quad 1 \le i \le k \tag{39}$$

$$\delta_i = \frac{E_i}{(1-\rho)\delta_{k+1}}, \quad 1 \le i \le k \tag{40}$$

where $\rho = \sum_{i=1}^{k} (A_i + E_i)$.

On substituting the values of ϕ_1, ψ_i *and* $\delta_i (1 \le i \le k)$ from Equations 38 to 40 into Equation 24 and after doing some algebraic manipulations, we obtain

$$\delta_{k+1} = \frac{L_N - \rho}{1 + L_N - \rho} \tag{41}$$

On substituting the values of ϕ_1, ψ_i, and δ_i from Equations 38 to 40 into Equations 32 to 34 and using Equation 40, we finally get

$$P_{0,V}(n) = \left(\frac{1-\rho}{1+L_{N_1}-\rho}\right)\left(\frac{L_N-\rho}{1+L_N-\rho}\right)^n, \quad n = 0, 1, \ldots \tag{42}$$

$$P_{i,B}(n) = \left(\frac{A_i}{1+L_{N_1}-\rho}\right)\left(\frac{L_N-\rho}{1+L_N-\rho}\right)^{n-1}, \quad n = 1, 2, \ldots, \quad 1 \le i \le k \tag{43}$$

$$P_{i,D}(n) = \left(\frac{E_i}{1+L_{N_1}-\rho}\right)\left(\frac{L_N-\rho}{1+L_N-\rho}\right)^{n-1}, \quad n = 1, 2, \ldots, \quad 1 \le i \le k \tag{44}$$

The expected waiting time in the system

Let W_S *and* $\hat{W}_S$ denote the exact and the expected waiting time in the system, respectively. Then,

$$W_S = \frac{L_N}{\lambda_{eff}} \tag{45}$$

where $\lambda_{eff} = [\lambda_0 P_V + \lambda_1 P_B + \lambda_2 P_D] E[A]$.

Following the work of Wang et al. (2007), the approximate expected waiting time in the system is given by the approximate expected waiting time in

$$\hat{W}_S = \sum_{i=1}^{k}\sum_{n=1}^{\infty}\left[\frac{nq_i}{\mu_i}+\frac{1}{\nu}+\frac{q_i}{2\mu_i}\left(\frac{E[A^2]}{E[A]}-1\right)\right]P_{0,V}(n)$$
$$+\sum_{i=1}^{k}\sum_{n=0}^{\infty}\left[\frac{nq_i}{\mu_i}+\frac{q_i}{2\mu_i}\left(\frac{E[A^2]}{E[A]}-1\right)\right]P_{i,B}(n)$$
$$+\sum_{i=1}^{k}\sum_{n=0}^{\infty}\left[\frac{nq_i}{\mu_i}+\frac{1}{\beta}+\frac{q_i}{2\mu_i}\left(\frac{E[A^2]}{E[A]}-1\right)\right]P_{i,D}(n) \quad (46)$$

Substituting the values of $P_{0,V}(n), P_{i,B}(n)$ *and* $P_{i,D}(n)$ from Equations 42 to 44 into Equation 46, the approximate expected waiting time in the system is given by

$$\hat{W}_S = \frac{1-\sum_{i=1}^{k}A_i-\sum_{i=1}^{k}E_i}{\nu}$$
$$+\sum_{i=1}^{k}\frac{q_i}{\mu_i}\left[L_N+\frac{1}{2}\left(\frac{E[A^2]}{E[A]}-1\right)\right]+\frac{\sum_{i=1}^{k}E_i}{\beta} \quad (47)$$

Numerical illustration and sensitivity analysis

In this section, we present a numerical simulation by taking the illustration of production of heat transfer equipment (HTE) discussed in the 'Background' section. For developing the code of computational program, we have used the 'MATLAB' software. For computation purposes, we assume that the raw materials arrive in batches of fixed batch size $k = 3$. The arrival rates are chosen as $\lambda_1 = 0.5$, $\lambda_2 = 0.9$, and $\lambda_3 = 0.7$. The service times of the machine when producing these equipment are $\mu_1 = 1$, $\mu_2 = 2$, and $\mu_3 = 3$. The processing of the equipment may be interrupted with rate $\alpha = 0.8$ and again becomes available for processing with rate $\beta = 2$. Further, the server may go for multiple vacations with rate $\nu = 0.09$. The expected number of these equipment in the system is obtained by using Equations 1 to 19 as $L_N = 36$.

Now, we present the numerical results to demonstrate the effects of different parameters on various performance indices. The accuracy of numerical results is examined by comparing the exact waiting time (W_S) obtained in the 'Maximum entropy principle' section using the probability generating function approach with the approximate waiting time ($\hat{W}_S$) obtained by the maximum entropy principle (MEP) of the $M^X/H_K/1$ queueing system under multiple vacation policy. Relative percentage error is tabulated for this purpose. The variations of different parameters on the average queue length are shown in Figures 1 and 2 graphically.

Table 1 summarizes the numerical results for long-run fraction of time of the server being in different states by varying the parameters λ, μ, α, β, and ν for two different cases of q_i (case 1: $(q_1, q_2, q_3) = (0.6, 0.3, 0.1)$ and case 2: $(q_1, q_2, q_3) = (0.7, 0.2, 0.1)$). For the sake of convenience, we choose the default parameters $\lambda_1 = 0.9$, $\lambda_2 = 0.8$, $\lambda_3 = 0.7$, $\mu = 1$, $\mu_1 = 1\,\mu$, $\mu_2 = 2\,\mu$, $\mu_3 = 3\,\mu$, $\alpha = 0.2$, $\beta = 1$, and $\nu = 0.9$. It is noticed that P_V shows a decreasing trend with respect to the increasing values of λ, α, and ν, but an increasing trend has been found with other parameters for both cases. Similarly, P_B and P_D increase as we increase the values of λ, α, and ν and decrease with increasing values of μ and β for both cases. Table 2 presents the comparison between W_S and $\hat{W}_S$ for case 1: $(q_1, q_2, q_3) = (0.5, 0.1, 0.4)$ and case 2: $(q_1, q_2, q_3) = (0.4, 0.3, 0.3)$. We fix default parameters for numerical results summarized in Table 2 as $\lambda_1 = 0.8$, $\lambda_2 = 0.7$, $\lambda_3 = 0.6$, $\mu = 1$, $\mu_1 = 1\,\mu$, $\mu_2 = 2\,\mu$, $\mu_3 = 3\,\mu$, $\alpha = 0.2$, $\beta = 3$, and $\nu = 0.01$. For both cases, W_S increases as we increase the values of λ_1 and α but decreases with increasing values of λ_0, μ, β, and ν. As we increase the values of λ_1, μ, β, and ν, it is seen that $\hat{W}_S$ decreases but increases with λ_0 and α for both cases. It can be observed easily from Table 2 that the relative percentage error varies from 0 % to 3 % which is reasonably less.

Figure 1a,b,c depicts the effect of different parameters on average queue length for various sets of heterogeneous arrival rate ($\lambda_0 = 1.8\lambda$, $\lambda_1 = 0.5\lambda$, $\lambda_2 = 0.6\lambda$) shown by discrete lines and homogeneous arrival rate ($\lambda_0 = \lambda_1 = \lambda_2 = \lambda$) shown by continuous lines. We observe that the average queue length is higher for the heterogeneous arrival rate in comparison to the homogeneous arrival rate on increasing the breakdown rate of the server. The queue length shows a gradual decreasing trend on increasing the repair rate and the vacation rate. Further, Figure 2a,b,c visualizes the effect of batch size on the average queue length. It is observed that the average queue length reveals an increasing trend with increasing values of α and β while shows a decreasing trend with increasing values of ν. The tractability of numerical results shows that our model can be easily implemented for the quantitative assessment of the performance of many real-time congestion systems.

Methods

In this paper, an $M^X/H_k/1$ queue under multiple vacations and an un-reliable server with varying arrival rates is studied. The probability generating function technique is used to determine various performance measures in explicit form. Then, MEP is further employed to compare the approximate results with exact results. For validating the analytical results, the sensitivity analysis is also carried out.

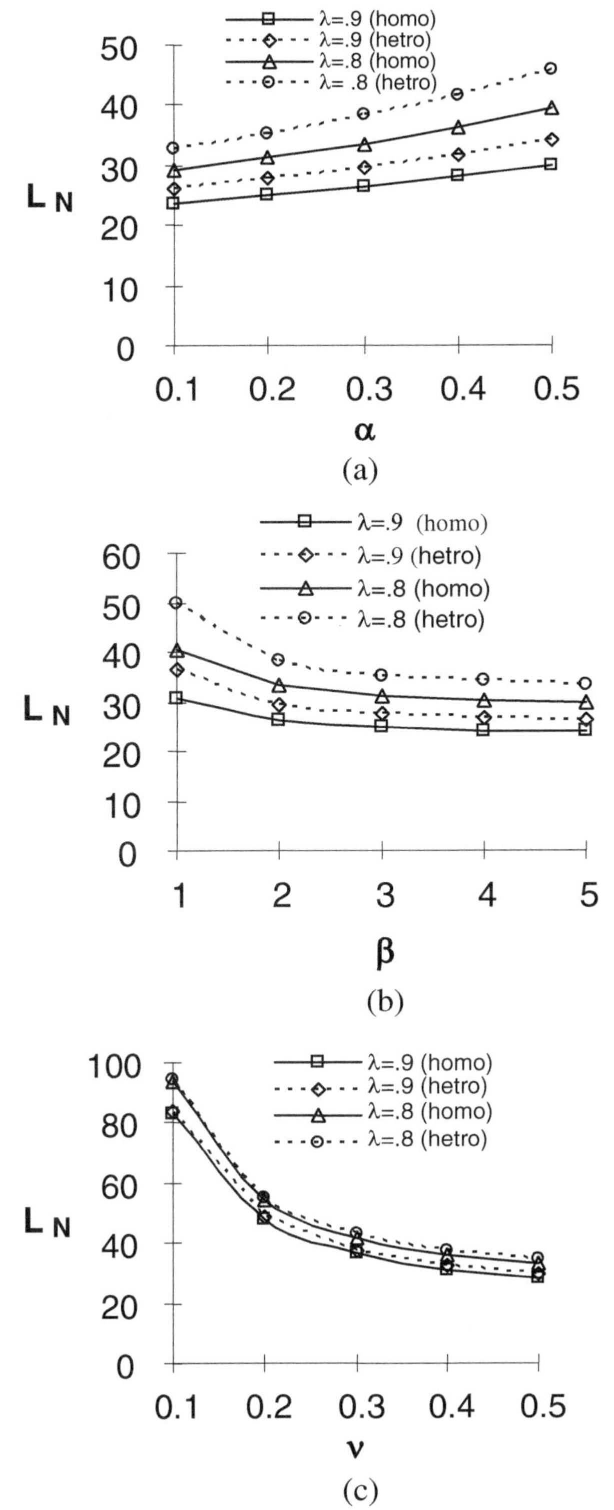

Figure 1 Average queue length vs. (a) α, (b) β, and (c) v for homogeneous and heterogeneous arrival rates.

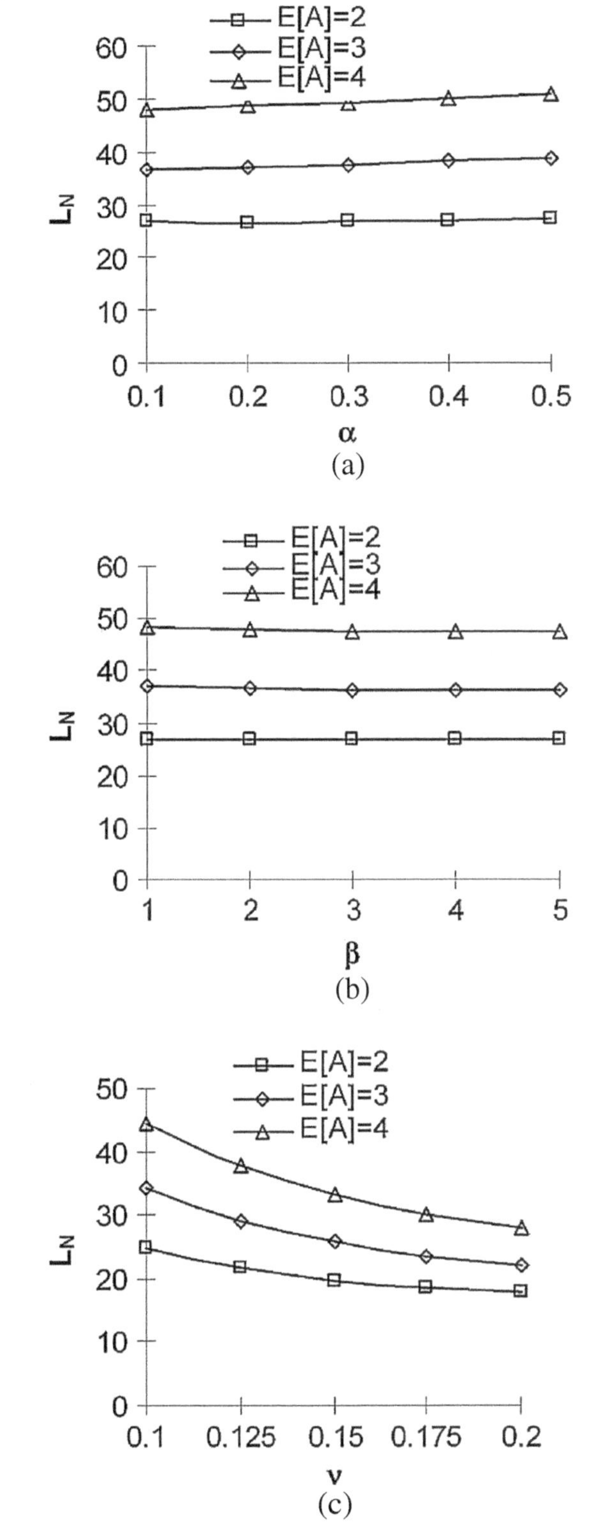

Figure 2 Average queue length vs. (a) α, (b) β, and (c) v for different batch sizes.

Table 1 Effect of various parameters on long-run fraction of time of the server being in different states

		Case 1: $(q_1, q_2, q_3) = (0.6, 0.3, 0.1)$			Case 2: $(q_1, q_2, q_3) = (0.7, 0.2, 0.1)$		
		P_V	P_B	P_D	P_V	P_B	P_D
λ	0.700	0.997	0.002	0.001	0.909	0.090	0.018
	0.725	0.948	0.051	0.010	0.852	0.147	0.029
	0.750	0.896	0.103	0.020	0.790	0.209	0.041
	0.775	0.840	0.159	0.031	0.725	0.274	0.054
	0.800	0.781	0.218	0.043	0.657	0.342	0.068
μ	1.000	0.226	0.773	0.154	0.187	0.812	0.162
	1.050	0.412	0.587	0.117	0.342	0.657	0.131
	1.100	0.586	0.413	0.082	0.489	0.510	0.102
	1.150	0.742	0.257	0.051	0.623	0.376	0.075
	1.200	0.878	0.121	0.024	0.742	0.257	0.051
α	0.100	0.947	0.052	0.005	0.848	0.151	0.015
	0.200	0.810	0.189	0.037	0.688	0.311	0.062
	0.300	0.650	0.349	0.104	0.509	0.490	0.147
	0.400	0.479	0.520	0.208	0.323	0.676	0.270
	0.500	0.308	0.691	0.345	0.144	0.855	0.427
β	1.000	0.810	0.189	0.037	0.688	0.311	0.062
	1.250	0.868	0.131	0.021	0.755	0.244	0.039
	1.500	0.904	0.095	0.012	0.798	0.201	0.026
	1.750	0.929	0.070	0.008	0.827	0.172	0.019
	2.000	0.947	0.052	0.005	0.848	0.151	0.015
v	0.100	0.955	0.044	0.008	0.916	0.083	0.016
	0.200	0.921	0.078	0.015	0.858	0.141	0.028
	0.300	0.895	0.104	0.020	0.815	0.184	0.036
	0.400	0.874	0.125	0.025	0.782	0.217	0.043
	0.500	0.856	0.143	0.028	0.755	0.244	0.048

Table 2 Comparison between the exact and the approximate results of waiting times

		Case 1: $(q_1, q_2, q_3) = (0.5, 0.1, 0.4)$			Case 2: $(q_1, q_2, q_3) = (0.4, 0.3, 0.3)$		
		W_S	$\hat{W}_S$	% error	W_S	$\hat{W}_S$	% error
λ_0	0.600	103.237	101.411	1.769	103.058	101.424	1.586
	0.700	102.776	101.480	1.261	102.622	101.490	1.103
	0.800	102.430	101.548	0.861	102.295	101.556	0.723
	0.900	102.160	101.615	0.533	102.041	101.621	0.411
	1.000	101.945	101.683	0.256	101.837	101.686	0.148
λ_1	0.600	101.947	101.616	0.324	101.821	101.622	0.196
	0.700	102.054	101.616	0.428	101.931	101.621	0.303
	0.800	102.160	101.615	0.533	102.041	101.621	0.411
	0.900	102.267	101.615	0.637	102.151	101.620	0.519
	1.000	102.374	101.615	0.742	102.261	101.620	0.626
μ	1.000	102.160	101.615	0.533	102.041	101.621	0.411
	2.000	101.006	100.803	0.201	100.920	100.805	0.114
	3.000	100.659	100.534	0.124	100.596	100.536	0.059
	4.000	100.491	100.400	0.090	100.441	100.401	0.039
	5.000	100.391	100.320	0.071	100.350	100.321	0.028
α	0.100	102.068	101.552	0.505	101.951	101.556	0.388
	0.200	102.160	101.615	0.533	102.041	101.621	0.411
	0.300	102.254	101.681	0.561	102.132	101.688	0.434
	0.400	102.350	101.748	0.588	102.224	101.757	0.457
	0.500	102.447	101.817	0.615	102.318	101.828	0.479
β	1.000	102.543	101.887	0.640	102.412	101.900	0.500
	2.000	102.254	101.680	0.560	102.131	101.688	0.434
	3.000	102.160	101.615	0.533	102.041	101.621	0.411
	4.000	102.114	101.584	0.519	101.996	101.588	0.399
	5.000	102.087	101.565	0.511	101.969	101.569	0.392
v	0.010	102.160	101.615	0.533	102.041	101.621	0.411
	0.012	82.475	81.499	1.183	82.319	81.508	0.984
	0.015	69.456	68.088	1.969	69.264	68.100	1.679
	0.017	60.246	58.509	2.882	60.018	58.524	2.489
	0.020	53.416	51.326	3.913	53.153	51.341	3.407

Results and discussion

The role of queueing analysis based on MEP lies in the fact that it helps basically system designers and managers to take decisions based on the performance indices determined using the probability distribution of the system size in terms of available information by MEP approach. The numerical illustration presented demonstrates that maximum entropy analysis is a simple approach to deal with complex scenarios for real-life congestion situations and can be easily applied to complex queueing scenarios for which performance measures are not easily obtained by using a classical approach. Based on the numerical experiment and sensitivity analysis carried out, we overall conclude that the comparative analysis of approximate results with exact results has demonstrated that the results obtained by MEP are reasonably good. The average queue length is higher for heterogeneous arrival rate in comparison to homogeneous arrival rate. The trends are more perceptible for larger batch size, which is quite obvious as the congestion increases significantly if the batch size of arriving customers is large.

Conclusion

In this paper, an $M^X/H_k/1$ queue with multiple vacations and an un-reliable server has been studied in order to facilitate various performance indices in explicit form by using an analytical approach based on the generating function method and maximum entropy principle. The incorporation of some more realistic features such as multiple vacations, un-reliable server, and batch arrival makes our model closer to real-life congestion situations. This model depicts many real-time embedded systems, namely production system, computer system, data communication system, etc. The numerical results and sensitivity analysis obtained provide an insight into how the system can be made more efficient by controlling the sensitive parameters. The queueing model studied can be further extended by taking the concept of k-phase optional repair. The concept of working vacation, *N*-policy, and multi-repair can also be included which is the topic of our future research work.

Appendix

Proof of Lemma 1

Multiplying Equation 1 by appropriate powers of z and then summing over n, we get

$$(\lambda_0 + \nu - \lambda_0 \mathrm{A}(z))G_{0,V}(z) = (\lambda_0 + \nu)P_{0,V}(0) \tag{48}$$

Substituting $\rho_\nu = \frac{\lambda_0}{\lambda_0+\nu}$ in Equation 48, we have

$$G_{0,V}(z) = \frac{1}{1-\rho_\nu \mathrm{A}(z)} P_{0,V}(0) \tag{49}$$

Multiplying Equation 2 by $q_i z$, Equation 3 by z^2, and Equation 4 by z^{n+1} and then adding these equations term by term for all possible values of n, finally, we obtain

$$\begin{aligned}&[\lambda_1 z\mathrm{A}(z) - (\lambda_1 + \alpha + \mu_i)z]G_{i,B}(z)\\ &\quad + q_i \sum_{j=1}^{k} \mu_j G_{j,B}(z) + \beta z G_{i,D}(z)\\ &\quad = q_i z \lambda_0 P_{0,V}(0)^{0,V0,V}, 1 \le i \le k\end{aligned} \tag{50}$$

After multiplying Equation 5 by z and Equation 6 by z^n and then adding these equations term by term for all possible values of n, thus, we obtain

$$\begin{aligned}(\lambda_2 + \beta)G_{i,D}(z) &= \alpha G_{i,B}(z)\\ &\quad + \lambda_2 \mathrm{A}(z) G_{i,D}(z), \quad 1 \le i \le k\end{aligned} \tag{51}$$

Using Equation 51, we have

$$G_{i,D}(z) = \frac{\alpha}{(\lambda_2 + \beta - \lambda_2 \mathrm{A}(z))} G_{i,B}(z), \quad 1 \le i \le k \tag{52}$$

Now put $i = 1$ in Equation 50 and using Equation 52 into Equation 50, we have

$$\begin{aligned}&\left[\lambda_1 z A(z) - \left(\lambda_1 + \alpha + \mu_1 + \frac{\alpha\beta}{(\lambda_2 A(z) - \lambda_2 - \beta)}\right)z + q_1\mu_1\right] G_{1,B}(z)\\ &\quad + q_1 \sum_{j=2}^{k} \mu_j G_{j,B}(z) = q_1 z \lambda_0 P_{0,V}(0)\end{aligned} \tag{53}$$

Again put $i = 2$ in Equation 50 and using Equation 52 into Equation 50, we obtain

$$\begin{aligned}&q_2\mu_1 G_{1,B}(z) + \left[\lambda_1 z A(z) - \left(\lambda_1 + \alpha + \mu_2 + \frac{\alpha\beta}{(\lambda_2 A(z) - \lambda_2 - \beta)}\right)z + q_2\mu_2\right] G_{2,B}(z)\\ &\quad + q_2 \sum_{j=3}^{k} \mu_j G_{j,B}(z) = q_2 z \lambda_0 P_{0,V}(0)\end{aligned} \tag{54}$$

Similarly, repeating this process for $i = k$, we get

$$\begin{aligned}&q_k \sum_{j=1}^{k-1} \mu_j G_{j,B}(z)\\ &\quad + \left[\lambda_1 z A(z) - \left(\lambda_1 + \alpha + \mu_k + \frac{\alpha\beta}{(\lambda_2 A(z) - \lambda_2 - \beta)}\right)z + q_k\mu_k\right] G_{k,B}(z)\\ &\quad = q_k z \lambda_0 P_{0,V}(0)\end{aligned} \tag{55}$$

We use Cramer's rule to solve Equations 53 to 55. Now we get

$$G_{i,B}(z) = \frac{N_i(z)}{D(z)} P_{0,V}(0), \qquad i = 1, 2, \ldots, k \tag{56}$$

Proof of Lemma 2

Using Lemma 1, we obtain

$$G_{0,V}(1) = \lim_{z\to1} G_{0,V}(z) = \frac{1}{1-\rho_\nu} P_{0,V}(0) \tag{57}$$

$$\begin{aligned}G_{i,B}(1) &= \lim_{z\to1} G_{i,B}(z)\\ &= \frac{\rho_i q_i \left[i\beta - (\lambda_1 \beta E[A] + \alpha\lambda_2 E[A]) \sum_{i=1}^{k-1} \frac{1}{\mu_i}\right]}{\left[(\lambda_1 \beta E[A] + \alpha\lambda_2 E[A]) \sum_{i=1}^{k} \frac{q_i}{\mu_i} - \beta\right]} P_{0,V}(0), i = 1, 2, \ldots, k\end{aligned} \tag{58}$$

$$\begin{aligned}G_{i,D}(1) &= \lim_{z\to1} G_{i,D}(z) = \frac{\alpha}{\beta} G_{i,B}(1), i\\ &= 1, 2, \ldots, k\end{aligned} \tag{59}$$

where $\rho_i = \frac{\lambda_0}{\mu_i}$.

The L-Hospital rule has been applied to compute the above results.

To determine $P_{0,V}(0)$, we use the normalizing condition given by

$$\mathrm{G}(1) = G_{0,V}(1) + \sum_{i=1}^{k} (G_{i,B}(1) + G_{i,D}(1)) \tag{60}$$

On substituting the values of $G_{0,V}(1)$, $G_{i,B}(1)$, and $G_{i,D}(1)$ from Equations 57 to 59 into Equation 60, we obtain the value of $P_{0,V}(0)$.

For finding the result of the stable condition, we use the condition

$$0 < P_{0,V}(0) < 1. \quad (61)$$

Using Equation 13 into Equation 61, we have

$$0 < (\lambda_1\beta + \alpha\lambda_2)E[A]\sum_{i=1}^{k}\frac{q_i}{\mu_i}\beta < 1. \quad (62)$$

After some algebraic manipulations, Equation 62 provides the result given in Equation 14.

Proof of Theorem 1

In order to prove Equation 15, we have

$$\begin{aligned} G(z) &= \sum_{n=0}^{\infty} z^n P_{0,V}(n) + \sum_{n=1}^{\infty}\sum_{i=1}^{k} z^n P_{i,B}(n) \\ &+ \sum_{n=1}^{\infty}\sum_{i=1}^{k} z^n P_{i,D}(n) \\ &= G_{0,V}(z) + \sum_{i=1}^{k}\left[G_{i,B}(z) + G_{i,D}(z)\right] \end{aligned} \quad (63)$$

On substituting the values of $G_{0,V}(z)$, $G_{i,B}(z)$, and $G_{i,D}(z)$ from Lemma 1 into Equation 63, we get Equation 15.

Proof of Theorem 2

The average system size is computed using

$$L_N = \lim_{z\to 1} G'(z).$$

The L-Hospital rule is applied twice to compute the results given in Equation 19.

Competing interests

The authors declare that they have no competing interests.

Authors' contributions

MJ has worked on MX/Hk/1 queueing system under multiple vacation policy. Various performance indices including queue size using generating function and the approximate formulas for waiting time of the customers using the maximum entropy principle have been obtained by RS. GC has performed numerical results by taking an illustration.

Acknowledgements

The authors are thankful to the learned referee and editor for their valuable comments and suggestions for the improvement of the paper.

Author details

[1]Department of Mathematics, IIT, Roorkee 247667, India. [2]Department of Mathematics, JK Lakshmipat University, Jaipur 302026, India. [3]Department of Mathematics, I.B.S., Khandari, Agra 282002, India.

References

Arumuganathan R, Jeykumar S (2004) Analysis of bulk queue with multiple vacations and closedown times. Int J Inform Manage Sci 15(1):45–60
Baba Y (1986) On the $M^X/G/1$ queue with vacation time. Oper Res Lett 5:93–98
Choudhury G, Deka K (2008) An M/G/1 retrial queueing system with two phases of service subject to server breakdown and repair. Perf Eval 65(10):714–724
Choudhury G, Tadj L (2009) An M/G/1 queue with two phases of service subject to the server breakdown and delayed repair. Appl Math Comp 33(6):2699–2709
El-Affendi MA, Kouvatsos DD (1983) A maximum entropy analysis of the M/G/1 and G/M/1 queueing systems at equilibrium. Acto Info 19:339–355
Jain M, Dhakad MJ (2003) Queueing distribution $M^X/G/1$ model using principle of maximum entropy. J Dec Math Sci 8(1–3):37–44
Jain M, Sharma GC, Sharma R (2012) Maximum entropy approach for un-reliable server vacation queueing model with optional bulk service. JKAU: Eng Sci 23(2):89–113
Jain M, Sharma GC, Sharma R (2013) Unreliable server M/G/1 queue with multi-optional services and multi-optional vacations. Int J Math Oper Res 5(2):145–169
Jain M, Singh M (2000) Diffusion process for optimal flow control of G/G/c finite capacity queue. J MACT 33:80–88
Ke JC, Chang FM (2009) $M^X/(G_1, G_2)/1$ retrial queue under Bernoulli vacation schedules with general repeated attempts and starting failures. Appl Math Model 32(4):443–458
Ke JC (2005) Modified T vacation policy for an M/G/1 queueing system with unreliable server and startup. Appl Math Model 41:1267–1277
Ke JC (2007) Operating characteristic analysis on the $M^X/G/1$ system with variant vacation policy and balking. Appl Math Model 31(7):1321–1337
Ke JC, Lin CH (2006) Maximum entropy solutions for batch-arrival with an un-reliable server and delaying vacations. Appl Math Comp 183(2):1328–1340
Ke JC, Lin CH (2008) Maximum entropy approach for batch-arrival queue under N policy with an un-reliable server and single vacation. Comp Appl Math 221(1):1–15
Ke JC, Lin CH, Yang JY, Zhang ZG (2009) Optimal (d, c) vacation policy for a finite buffer M/M/c queue with unreliable servers and repairs. Appl Math Model 33(10):3949–3962
Kumar BK, Madheswari SP (2005) An M/M/2 queueing system with heterogeneous servers and multiple vacations. Math Comp Model 41(13):1415–1429
Li JH, Tian NS, Ma ZY (2007) Performance analysis of GI/M/1 queue with working vacations. Oper Res Lett 35(5):595–600
Omey E, Gulck SV (2008) Maximum entropy analysis of the $M^X/M/1$ queueing system with multiple vacation and server breakdowns. Comp Indus Eng 54(4):1078–1086
Sharma R (2010) Threshold N-policy for $M^X/H_2/1$ queueing system with un-reliable server and vacations. J Int Acd Phy Sci 14:1–11
Singh CJ, Jain M, Kumar B (2012) Analysis of M/G/1 queueing model with state dependent arrival and vacation. J Indus Eng Int 8(2):1–8
Singh CJ, Jain M, Kumar B (2013) Analysis of unreliable bulk queue with state dependent arrival. J Indus Eng Int 9(21):1–9
Tian N, Zhang ZG (2002) The discrete-time GI/Geo/1 queue with multiple vacations. Queue Sys 40(3):283–294
Wang J (2004) An M/G/1 queue with second optional service and server breakdowns. Comp Math Appl 47(10–11):1713–1723
Wang KH, Chan MC, Ke JC (2007) Maximum entropy analysis of $M^{[X]}/M/1$ queueing system with multiple vacations and server breakdowns. Comp Indus Eng 52:192–202
Wang KH, Chang KW, Sivazlian BD (1999) Optimal control of a removable and non-reliable in an infinite and a finite $M/H_2/1$ queueing system. Appl Math Model 23:651–666
Wang KH, Huang KB (2009) A maximum entropy approach for the p, N-policy M/G/1 queue with removable and unreliable server. Appl Math Model 33(4):2024–2034
Wang KH, Kao HT, Chen G (2004) Optimal management of a removable and non-reliable server in an infinite and a finite $M/H_k/1$ queueing system. Qual Tech Quant Mang 1(2):325–339
Wang WL, Xu GQ (2009) The well-posedness of an M/G/1 queue with second optional service and server breakdown. Comp Math Appl 57(5):729–739
Wu DA, Takagi H (2006) An M/G/1 queues with multiple working vacations and exhaustive service discipline. Perf Eval 63(7):654–681

Distribution network design under demand uncertainty using genetic algorithm and Monte Carlo simulation approach: a case study in pharmaceutical industry

Arman Izadi and Ali mohammad Kimiagari*

Abstract

Distribution network design as a strategic decision has long-term effect on tactical and operational supply chain management. In this research, the location-allocation problem is studied under demand uncertainty. The purposes of this study were to specify the optimal number and location of distribution centers and to determine the allocation of customer demands to distribution centers. The main feature of this research is solving the model with unknown demand function which is suitable with the real-world problems. To consider the uncertainty, a set of possible scenarios for customer demands is created based on the Monte Carlo simulation. The coefficient of variation of costs is mentioned as a measure of risk and the most stable structure for firm's distribution network is defined based on the concept of robust optimization. The best structure is identified using genetic algorithms and 14% reduction in total supply chain costs is the outcome. Moreover, it imposes the least cost variation created by fluctuation in customer demands (such as epidemic diseases outbreak in some areas of the country) to the logistical system. It is noteworthy that this research is done in one of the largest pharmaceutical distribution firms in Iran.

Keywords: Distribution network design; Facility location; Genetic algorithms; Monte carlo simulation

Introduction

A supply chain is a set of facilities, suppliers, customers, products, and methods of purchasing, monitoring inventory, and distribution. In a supply chain, the flow of goods between suppliers and customers goes across several stages and each stage may include some facilities (Sabri and Beamon 2000). Supply chain design is the most fundamental decision that will impact the entire supply chain, and it has been obtaining much importance recently due to increasing competitiveness in markets (Simchi-Levi and Kaminsky 2004; Thomas and Griffin 1996). Firms are forced to keep high customer service levels while they are obliged to reduce costs and maintain profit margins simultaneously.

Distribution networks are considered as the main profitability key because they directly affect both the cost of supply chains and satisfaction of the customers (Chopra 2003). The high cost of buying assets and building facilities has made locating or relocating the facilities a long-term investment, and the deployed facilities are expected to carry out their activities for a long period of time. Therefore, it seems essential to design a supply chain network which performs well with some parameters variation across time (Santosa 2003). General structure of the supply chain is shown in Figure 1.

Supply chain design includes strategic decisions such as determining the number and location and capacity of facilities in each echelon of the supply chain. The location of facilities and the allocation of demands to them have been a substantial research area. The typical location-allocation problem involves selecting p facilities among n potential locations and allocating demand to the open facilities to minimize the total cost (Hakimi 1965). In addition, distances, times, or costs among customers and facilities can be considered (Drezner and Hamacher 2004; ReVelle and Eiselt 2005). This is p-median problem which

* Correspondence: kimiagar@aut.ac.ir
Department of industrial engineering, Amirkabir university of technology, Hafez street, NO 424, 15875-4413, Tehran, Iran

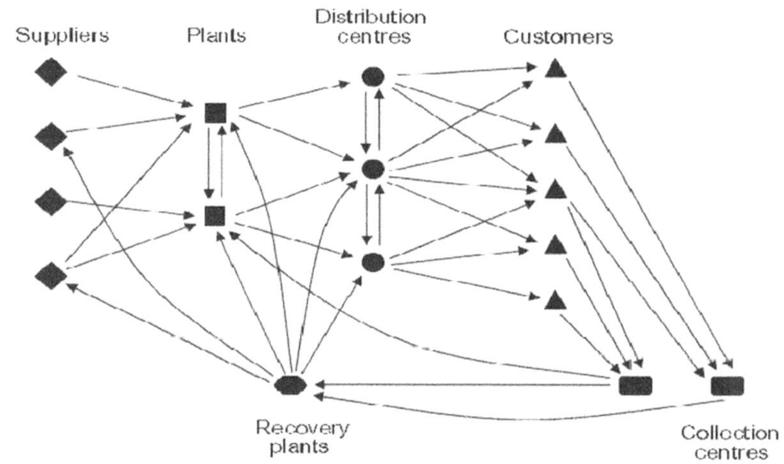

Figure 1 Typical supply chain structure (Melo et al. 2009).

has attracted much attention in the literature (Daskin 1995; Mladenovic et al. 2007).

Different techniques have been employed to solve problems relating to facility location. The *p*-median problem is typically formulated as an integer program (IP). Enumeration and heuristics methods were the primary techniques which are proposed to solve this problem. Also, several methods have been used to solve LP relaxations of the IP formulation problems including branch-and-bound, dual ascent, subgradient optimization, and Lagrangian relaxation.

Metaheuristics and approximation algorithms are the prominent techniques in the literature over the last few years. The most common metaheuristics are genetic algorithms, simulated annealing, variable neighborhood search, tabu search, and neural networks. Moreover, some approximation algorithms have been created which attain different bounds on the favorable solution quality.

In fact, the configuration stability of a supply chain network is highly desirable because of great investments normally dedicated with this type of decisions. In this paper, the goal is to obtain a distribution network design that can cope with the future demand variations and achieve a 'robust' network design. In this case study, actual data is used from a company which is one of the largest pharmaceutical distribution firms in Iran. Because drug and medicine are considered as strategic commodity in all around the world, minimizing the total cost of pharmaceutical supply chain is not the entire objective and achieving a robust supply chain design is more important than having the least cost network.

In this paper, the goal is to determine the optimal number and location of distribution centers. Moreover, the allocation of customer demands to distribution centers under demand uncertainty is defined in order to minimize the total annual cost including warehouse fixed costs, inventory costs, and transportation costs while maintaining the required service level.

This paper is organized as follows: In the 'Problem statement' section, the mathematical model and background on methodology is stated. The proposed genetic algorithm is given in the third section, and the approach to deal with uncertainty is discussed in the fourth section. Numerical results are presented in the 'Results and analysis' section, and finally, the paper is concluded in the 'Conclusion' section.

Problem statement

The problem considered in this paper is designing the distribution network for pharmaceutical distribution companies in Iran. Existing in the pharmaceutical industry entails accepting a series of requirements and commitments related to the Department of Health. For example, manufacturers and importers in this industry do not have the permission to distribute their own products and they are obliged to distribute their products through distribution companies. Actually, the pharmaceutical distribution firms are the connecting chain between suppliers and retailers.

Based on the scope of activity, pharmaceutical distribution companies are divided into provincial and national firms. Provincial distribution companies are licensed to operate in particular region bounds whereas national distribution companies are forced by the Ministry of Health to distribute drugs in all provinces of the country. This company is one of the national distribution firms in Iran which should cover the demands of all cities in country with its own products.

According to the latest published statistics, the market share of this company is 15% of the medicines and medical equipment market in Iran. The suppliers of this company include 32 manufactures and 46 importers of medicine and medical equipment which are the main suppliers of the country. These infrastructures make this company as the second pharmaceutical distribution company in Iran.

The customers of this company include pharmacies, hospitals, and medical centers in most of the cities in Iran whose numbers roughly equate 10,000. It is obvious that this optimization problem needs a very large amount of data. Therefore, data aggregation is the first essential step that should be done. The efficient approach which is used for this purpose is classifying the pharmacies, hospitals, and medical centers in each city in one cluster.

Then, all customers located in each group are replaced by a single demand point in the center of cluster. Even if suitable technology exists to solve the problem with large initial data, data aggregation would be useful. Aggregation causes reduce the data variability and forecasting with aggregated data is significantly more precise compared with non-aggregated data (Simchi-Levi and Kaminsky 2004).

This company approximately owns 1,900 different drugs and medical equipment in its product portfolio, and the composition of them has been determined based on market demands. Due to the relatively large number and small size of the products, these are integrated to one product. Moreover, what is significant matter for the company is how much space each customer demands occupies the content of vehicles in the transportation process. So, according to the length, width, and height of product packages, the total volume of demand products is calculated, and it is used as the measure of customer demands. So, this problem is considered as a single product and multi-level supply chain design problem (Sousa et al. 2011).

The current logistical structure of the company includes a warehouse in Tehran and 18 distribution centers in different provinces. The company wants to redesign its distribution network to satisfy all customer demands and tolerate the minimum total costs caused by demand fluctuations.

Location-allocation model

The design of the distribution network for this company is modeled in the form of a mixed integer linear programming problem in both location and allocation phases. In the location phase, the objective is to determine the optimal number of distribution centers and define which customers are supplied with each distribution center. In the allocation phase, the objective is to specify the amount of inventory which is transferred from the warehouse to distribution centers and from distribution centers to the customers (Amiri 2006; Gebennini et al. 2009; Jayaraman and Ross 2008; Jayaraman and Pirkul 2001). The assumptions considered in this model are

- The delivery time of transferring goods from warehouses to distribution centers and from distribution centers to customers is considered constant.
- The cost of transportation is calculated based on the distance between two points and the shipping cost of a product unit in each path.
- The total demand of each customer must be provided by a distribution center.
- The number of customers and their demands are known.
- The numbers and capacities of potential distribution centers and warehouses are known.

Set I represents the set of customer zones, J is the set of potential distribution centers, K is the set of potential warehouse locations, and L is the set of products. The location model intends to minimize three types of costs. The first type of costs is fixed costs associated with open warehouses (V_K) and distribution centers (F_J).

Second, external transportation cost to transfer products from the warehouses to distribution centers (C_{jkl}). The third type is internal transportation costs to transfer products from distribution centers to customer zones (C_{ijl}). The customer demands for products (a_{il}), the capacity of distribution centers (b_j), and warehouses (d_k) are all given as input parameter to the model.

The decision variables track the set of customer zones that are dedicated to distribution centers for products (X_{ijl}), the set of open distribution centers that are dedicated to open warehouses (Y_{jkl}), and the set of distribution centers (Z_j) and warehouses (P_K) which are open or closed.

M_1: Location model

$$\text{MIN } Z_1 = \sum_j F_j Z_j + \sum_k V_k P_K + \sum_i \sum_j \sum_l C_{ijl} X_{ijl} + \sum_j \sum_k \sum_l C_{jkl} Y_{jkl} \tag{1}$$

Subject to:

$$\sum_j X_{ijl} = 1 \quad \forall\, i \text{ and } l \tag{2}$$

$$\sum_k Y_{jkl} = Z_j \quad \forall\, j \text{ and } l \tag{3}$$

$$\sum_i \sum_l a_{il}\, X_{ijl} \le b_j \quad \forall\, j \tag{4}$$

$$\sum_j \sum_l b_j Y_{jkl} \le d_k \quad \forall\, k \tag{5}$$

$$X_{ijl} \le Z_j \quad \forall\, i, j, \text{and } l \tag{6}$$

$$Y_{jkl} \le P_k \quad \forall\, j, k, \text{and } l \tag{7}$$

$$X_{ijl} \in \{0, 1\} \quad \forall\, i, j, \text{and } l \tag{8}$$

$$Y_{jkl} \in \{0, 1\} \quad \forall\, j, k, \text{and } l \tag{9}$$

$$Z_j \in \{0, 1\} \quad \forall\, j \tag{10}$$

$$P_k \in \{0, 1\} \quad \forall\, k \tag{11}$$

In allocation model, (S_{ijl}) shows the cost to supply one unit of product L from distribution center J which would be used by customer I, (S_{jkl}) is transportation cost for one unit of product L from warehouse K to distribution center J, (TC_{kl}) is the transportation cost to deliver one unit of product L to warehouse K and a_{il} is the demand from customer I for product L. The decision variables are represented by B_{kl} which is the quantity of product L sent to warehouse k, Q_{jkl} is the quantity of product L sent from warehouse k to distribution center J, and V_{ijl} is the quantity of product L sent from distribution center J to customer I (Jang et al. 2002; Shu et al. 2010; Tragantalerngsak et al. 2000).

$$\text{MIN } Z_2 = \sum_k \sum_l TC_{kl}B_{kl} + \sum_i \sum_j \sum_l S_{ijl}V_{ijl} + \sum_j \sum_k \sum_l S_{jkl}Q_{jkl} \quad (12)$$

M_2: Allocation model
Subject to:

$$B_{kl} = \sum_j Q_{jkl} \quad \forall\, k \text{ and } l \quad (13)$$

$$B_{kl} = \sum_j Q_{jkl} \quad \forall\, k \text{ and } l \quad (14)$$

$$\sum_j V_{ijl} = a_{il} \quad \forall\, i \text{ and } l \quad (15)$$

$$B_{kl} \geq 0 \quad \forall\, k \text{ and } l \quad (16)$$

$$V_{ijl} \geq 0 \quad \forall\, i, j, \text{ and } l \quad (17)$$

$$Q_{jkl} \geq 0 \quad \forall\, j, k, \text{ and } l \quad (18)$$

Backgrounds on methodology

Over the past 40 years, a great number of combinatorial problems have emerged from computer science and engineering (such as the location-allocation problem studied here). It has been proved that these combinatorial problems belong to the class of NP-complete problems (Garey and Johnson 1979) and with increase in number of variables, the time required to find the optimal solution are enhanced exponentially (Garey and Johnson 1979). Thus, solving large scale of these problems to achieve optimality in a reasonable amount of computational time is not possible.

To solve these problems, heuristic algorithms were developed which provide acceptable solutions in proper computational time. One of the major defects of heuristic algorithms is the high risk of being trapped in local optimal solutions. To improve these algorithms, new approaches were presented in 1970s. They were meta-heuristic algorithms which explore the solution area to find optimality with diversification and intensification search (Talbi 2009).

These algorithms are very efficient and successful in solving complex optimization problems which classical methods are unable to solve. As a meta-heuristic algorithm, genetic algorithms have been used widely as a tool in simulation and exploring the solution area in various contexts. The high applicability and simplicity usage are the main reasons for the widespread use of this algorithm.

Proposed genetic algorithm

Genetic algorithms (GA) are based on Darwin's theory, and two main ideas are considered in this theory. First, next generations are created from the evolution of the previous generations. Second, members who are stronger in the struggle for survival have greater living chance. So, GA is an algorithm which utilizes initial population, and each member of the population is considered as a solution (Arostegui et al. 2006; Melo et al. 2012).

Representation

The first step to solve a problem by genetic algorithm is defining the structure of representation and the way to convert a chromosome to a solution. Representation is one of the most important subjects that influence the performance of genetic algorithms. The first application of GAs in distribution problems was carried out by Michalewicz et al. (1991). One known way to represent network problems is tree-based representation. Basically, there are three ways for representing a tree: edge-based encoding, vertex-based encoding, and edge-and-vertex-based encoding (Gen and Cheng 2000). In this paper, matrix-based representation which belongs to edge-based encoding category is used to represent the solutions (Altiparmak et al. 2009; Drezner and Wesolowsky 2003; Syarif et al. 2002).

For this purpose, the ($n \times m$) binary matrix is used which n indicates the number of distribution centers and m indicates the number of demand points. There are two constraints in the representation of solutions which associate with assumptions. First, the summation of the columns of the representation matrix should certainly be one which means that the demand of each city should be covered only by a distribution center. Second, the summation of each row of this matrix must not be zero which means each distribution center should cover the demand of at least one customer. It is noticeable that the number of initial population size be considered 250 with regards to the problem largeness.

Fitness function

Fitness function indicates the quality of the solutions with a numerical value and guides the Searching process to better solutions. The fitness function in this problem consists of three costs. The first cost is external transportation which is defined as the cost of transportation from warehouses to distribution centers. The second is internal transportation cost which is defined as the cost of transportation from distribution centers to customers and the third cost is the inventory holding costs in distribution centers (Talbi 2009).

- External transportation cost
  The location of each distribution center (DC) is identified based on logistical structure and the latitude and longitude of each demand point. After the calculation of distances between each DC to the warehouse, the required vehicles for transportation of products are determined with regard to the demand of the cities. Then, the external transportation cost is calculated based on the unit cost of transportation in each path.
- Internal transportation cost
  According to the specific location of each DC, the distance between DCs and cities are calculated and with regard to the demand of each city, the required vehicles and the internal transportation cost are determined.
- Inventory holding costs
  The number of distribution centers to cover all demand points is determined by logistical structure of each solution. Based on the financial department documents related to the costs of DCs, a strong linear relationship between the amount of inventory in storage facilities and the available labors in DCs and annual cost of DCs are found. Thus, the required labor and annual cost of each DC is estimated, and accordingly, the inventory holding cost is calculated.

Genetic operators

In each generation, the chromosomes are evaluated by measuring their related fitness functions. The chromosomes of next generation which are produced through two ways mentioned below are called offspring.

1. Merging two chromosomes using crossover operator
2. Changing chromosomes using mutation operator

Crossover is an operator used to explore new solutions in search space, and it is responsible to exchange parts of chromosomes among selected parents. In this study, two-point crossover operator is employed which these two points is randomly selected among the corresponding parts of parents. It is noteworthy that crossover probability has been considered 0.8 in this operator.

Mutation is another operator that is responsible for modifying genes within a chromosome at low rates and creates random variations in some chromosomes. The main role of mutation is identifying some areas of the search space whose possibility of being explored is low. Also, this operator reduces the probability of trapping populations in local optimal and prevents the premature convergence (Jaramillo et al. 2002). In the proposed algorithm, uniform mutation is employed and mutation probability is considered 0.1 in this operator.

Selection strategy

The most important question that should be answered to determine the selection strategy is which parents should be selected for creating the next generation to move toward a better fitness function. In this paper, roulette wheel method is used to determine the parents. In this method, the selection probability of each chromosome is proportional to the degree they deserve. The reason for using this method is to maintain diversification in the population and provide even low chance of being selected for parents which have low fitness function (Talbi 2009).

Control parameters

This parameter impacts on the stopping criteria of algorithms and can be decreased when the convergence in solution occur. Also, this parameter can be increased when significant improvement was not observed after a certain number of iterations in order to enhance the diversification in the population. It is noticeable that the optimal solution is not known in advance in this problem and the stopping criteria are set to achieve a certain number of iterations (Talbi 2009). In this paper, the number of generations is set to 1,000 and the algorithm stops if it has not any improvement over 400 generations.

Regarding to the proposed mathematical model and being in the NP-hard category, there is no rational computational time for solving medium and large scale of this problem. To validate the proposed algorithm, some small scale problems are solved by exact solutions and the answers are compared with the answers of the proposed genetic algorithm. It is noticeable that 'Gams 21.7' is the software which is used to solve the mathematical model with exact approach.

The results of solving small scale problems with exact algorithms demonstrate that they are identical with the answers resulting from the proposed genetic algorithm. Exact algorithms find the optimal solution with exploring all the search space of the problems. This equality indicates that the proposed genetic algorithm can achieve optimality in problems with small size and this demonstrates the validation of the algorithm. Therefore, it can be concluded that the proposed algorithm is appropriate for solving large and medium problems.

The approach to deal with uncertainty

With studying the data of customer demands in the past 40 months, it becomes clear that the demands' behavior do not follow any specific probability distribution function. This uncertainty occurs due to the strong bullwhip effect which exists in most of the industries, but it is very impressive in pharmaceutical industry in Iran. Figure 2 illustrates the behavior of the demands quantity in an instance city in the last 40 months.

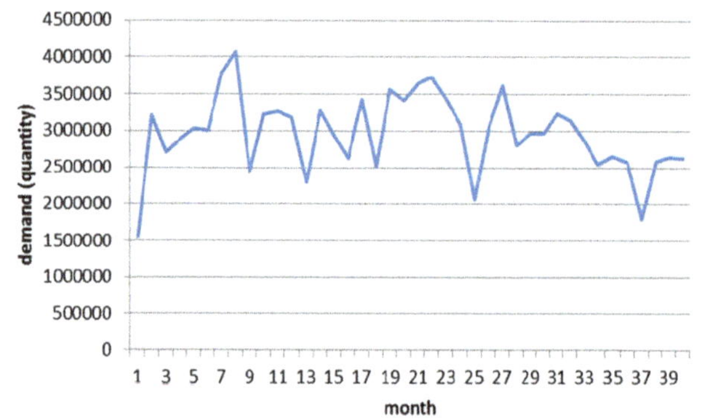

Figure 2 Behavior of pharmaceutical product demands in the 40 past months in Tehran.

The main reason for the demand fluctuations and the bullwhip effect exacerbation is false orders in periods of drug shortages. Such orders are so common because the customers think that the future supply of drugs and medicines will encounter by reduction. Also, the fluctuation in the price of medicines is a cause to intensify this phenomenon (Simchi-Levi and Kaminsky 2004).

An important step in the process of decision making is risk analysis. If there is a lot of uncertainty in a parameter or there is not enough availability or accessibility to the previous information, the future cannot be predicted accurately. Monte Carlo simulation provides the possibility to distinguish all the consequences of decisions and evaluate the related risk of each factor and helps to make better decisions under uncertain conditions. This technique is used in a wide variety of fields such as insurance, financial management, project management, manufacturing, research development, and transportation (Deleris and Erhun 2005; Flores et al. 2006).

Monte Carlo simulation conducts risk analysis based on the outputs with replacing the values of uncertain variables. In this technique, a random set of values between the upper and lower bound for each variable is obtained from its probability distribution function. Monte Carlo simulation can contain even several thousands of calculations with regard to the number of uncertain variables and their variation ranges. Thus, a set of patterns are generated in each scenario which represent one of the possible conditions. The output of the Monte Carlo simulation illustrates the frequency of objective function values which is associated with occurrence of different scenarios (Kalos and Whitlock 2008; Klibi et al. 2010). It is obvious that the higher numbers of scenarios, the more likely conditions are examined in the available space.

Results and analysis

As was mentioned, to consider the uncertainty in customer demands, some scenarios are created based on Monte Carlo simulation. For each scenario, the optimal number and location of distribution centers and the allocation of customers demand to distribution centers is determined. Figure 3 shows the location of 209 demand points which are the integration of pharmacies, hospitals, and health centers in the intended cities with their geographical specifications (longitude and latitude).

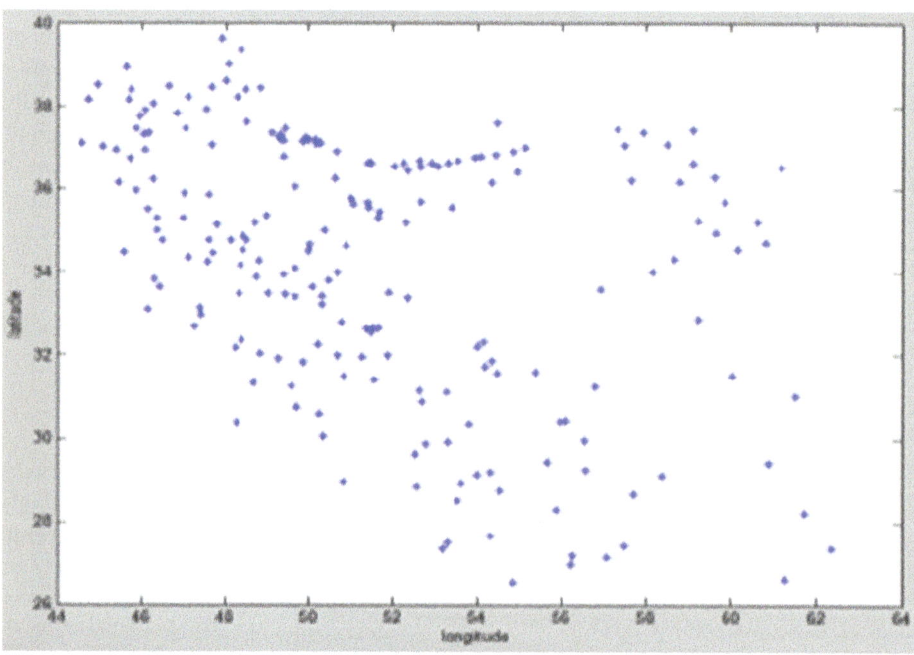

Figure 3 Location of the demand points.

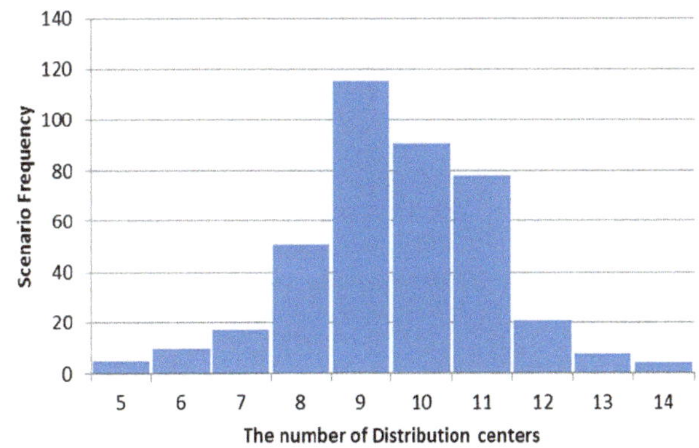

Figure 4 Frequency histogram of the demand scenarios.

Figure 4 demonstrates the histogram of the optimal number of distribution centers based on the 400 scenarios of customer demands. In each scenario, the best number of distribution centers is calculated using the proposed genetic algorithm. This figure illustrates that the logistical structures with 9, 10, and 11 distribution centers gain the most frequencies among the structures. In the next step, the optimal logistical structures for candidate solutions are determined. This time, a great number of scenarios (10,000 scenarios) for customer demands are created, and the average and standard deviation of costs for these three structures are compared and shown in Table 1.

The results show that logistical structure with nine distribution centers has the least coefficient of variation among others. In this paper, the coefficient of variation is mentioned as an index of risk, so it can be concluded that this structure is more stable against uncertainty in customer demands. Figure 5 illustrates the logistical structure with nine distribution centers which is selected as the best distribution network for this company. It is remarkable that the total cost in current distribution network is 9.97×10^8 monthly and the proposed distribution network leads to 14% reduction in total cost compared with the current distribution network.

Table 1 Comparisons of average and standard deviation of the logistical structure costs

Candidate solutions	Average cost	STD cost	CV
Logistical structure with 9 DCs	8.74×10^8	0.36×10^8	0.041
Logistical structure with 10 DCs	8.52×10^8	0.41×10^8	0.048
Logistical structure with 11 DCs	8.82×10^8	0.44×10^8	0.049

To prove the significance of the difference between the costs of the structures, ANOVA test is used in SPSS 20.0 software. This test is done to compare the mean cost of these three groups and to study the influence of independent variables on dependent variable.

In this problem, the independent variables include the number of distribution centers in each candidate logistical structure, and the dependent variable is the logistical costs associated with the structures. The result of this test is shown in Table 2. As it can be shown, Sig is smaller than 0.05 which means H_0 is rejected and there are significant differences between the average costs of these structures.

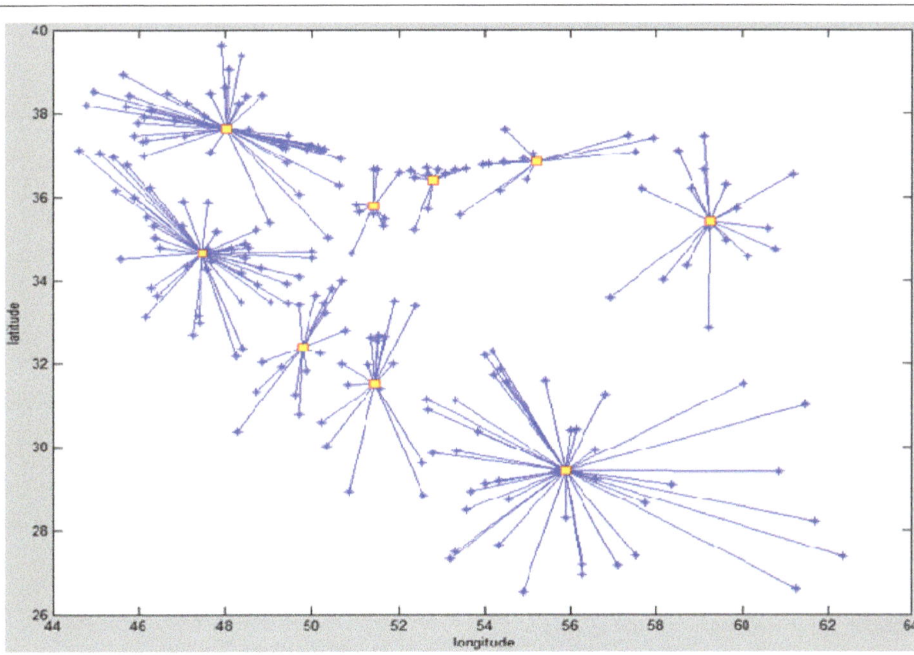

Figure 5 Logistical structure with nine distribution centers.

Table 2 Result of ANOVA test

ANOVA test	Sum of squares	DF	Mean square	*F*	Sig
Between groups	1.43×10^{19}	2	7.144×10^{18}	922.647	.000
Within groups	2.32×10^{20}	29,997	7.743×10^{15}		
Total	2.47×10^{20}	29,997			

Conclusion

In this study, a model of location-allocation problem in a three echelon supply chain with the aim of locating an unknown number of distribution centers to distribute pharmaceutical products from a specific warehouse to customers (pharmacies, hospitals, and medical centers in specific cities) is provided. The objective function of this model was minimizing the external and internal transportation costs and holding costs of inventory in distribution centers.

On the one hand, increasing the number of distribution centers is a cause to enhance the service levels and reducing the transportation costs from the distribution centers to the customers. On the other hand, decreasing the number of distribution centers leads to inventory integration and take advantage of risk pooling and reduce the inventory holding cost. Thus, the questions which are raised in this paper are how the optimal number of distribution centers should be determined, where distribution centers should be established and which customers must be assigned to each distribution center.

We are witnessing the customer demand uncertainty in pharmaceutical industry, and it cannot be considered in any specific probability distribution function. This uncertainty is due to the intense bullwhip effect created by retailers and other factors. To consider the uncertainty in this study, a set of possible scenarios is created based on the Monte Carlo simulation method, and for each scenario, the distribution network has been optimized using genetic algorithm.

According to the frequency of optimal distribution center numbers specified from scenarios, three structures with 9, 10, and 11 distribution centers which have the highest frequency are introduced as candidate solutions. The mean and standard deviation of the costs of structures are compared, and the significant difference between them was proved. The results show that logistical structure with nine distribution centers leads to 14% reduction in total distribution network cost compared to current network. Moreover, it has fewer coefficient of variation than two others. In this paper, the coefficient of variation is mentioned as an index of risk, and with considering the risk of solutions, it can be concluded that this structure is more robust than other structures and in case of happening turbulence in customer demands such as epidemic diseases outbreak in some areas of the country, fewer costs will be imposed to this company distribution network.

Competing interests

The authors declare that they have no competing interests.

Authors' contributions

AI, AmK The main contribution of this paper is the simulation based optimization approach which is mentioned in solving the distribution network design problem. It is noticeable that the uncertainty in pharmaceutical product demands is high duo to available intense bullwhip effect in Iran. Thus, in this paper, customer demand uncertainty is considered through creating scenarios with Monte Carlo simulation method and Genetic algorithm is employed to find the optimal solution. In addition, the concept of robust optimization is regarded to obtain the most stable logistical structure for one of the greatest pharmaceutical distribution companies in Iran. It is remarkable that the related holding company (TPICO) has decided to implement this research in other Subordinate distribution companies. Both authors read and approved the final manuscript.

Acknowledgements

I would like to express my deep gratitude to Dr. Ali Mohammad Kimiagari, the research supervisor, for his patient guidance and enthusiastic encouragement of this research work. I would also like to extend my thanks to the technicians of the laboratory of the Industrial engineering department in Amirkabir university of technology for their help in offering me the resources in running the program. Finally, I wish to thank my parents for their support and encouragement throughout my study.

References

Altiparmak F, Gen M, Lin L, Karaoglan I (2009) A steady-state genetic algorithm for multi-product supply chain network design. Comput Ind Eng 56:521–537

Amiri A (2006) Designing a distribution network in a supply chain system: formulation and efficient solution procedure. Eur J Oper Res 171:567–576

Arostegui MA, Jr, Kadipasaoglub SN, Khumawala BM (2006) An empirical comparison of tabu search, simulated annealing, and genetic algorithms for facilities location problem. Int J Prod Econ 103:742–754

Chopra S (2003) Designing the distribution network in a supply chain. Transp Res E 39:123–140

Daskin MS (1995) Network and Discrete Location: Models, Algorithms, and Applications. John Wiley & Sons, New York

Deleris LA, Erhun F (2005) Risk management in supply networks using Monte-Carlo simulation. In: Simulation Conference, Proceedings of the Winter, Orlando, FL, 4–7 Dec. 2005

Drezner Z, Hamacher HW (2004) Facility Location: Applications and Theory. Springer, New York

Drezner Z, Wesolowsky GO (2003) Network design: selection and design of links and facility location. Transp Res A 37:241–256

Flores C, Makiyama F, Nassar S, Freitas P, Jacinto C (2006) Risk Analysis using Monte Carlo Simulation and Bayesian Networks. In: Simulation Conference, WSC 06. Proceedings of the Winter, Monterey, 3–6 Dec. 2006

Garey M, Johnson D (1979) Computers and intractability: A guide to the theory of NP-completeness. W. H Freeman & Co, San Francisco

Gebennini E, Gamberini R, Manzini R (2009) An integrated production–distribution model for the dynamic location and allocation problem with safety stock optimization. Int J Prod Econ 122:286–304

Gen M, Cheng R (2000) Genetic algorithms and engineering optimization. John Wiley & Sons, New York

Hakimi SL (1965) Optimum distribution of switching centers in a communication network and some related graph theoretic problems. Oper Res 13:462–475

Jang YJ, Jang SY, Chang BM, Park J (2002) A combined model of network design and production/distribution planning for a supply network. Comput Ind Eng 43:263–281

Jaramillo JH, Bhadury J, Batta R (2002) On the use of genetic algorithms to solve location problems. Comput Oper Res 29:761–779

Jayaraman V, Pirkul H (2001) Planning and coordination of production and distribution facilities for multiple commodities. Eur J Oper Res 133:394–408

Jayaraman V, Ross A (2008) An evaluation of new heuristics for the location of cross-docks distribution centers in supply chain network design. Comput Ind Eng 55:64–79
Kalos MH, Whitlock PA (2008) Monte Carlo methods. John Wiley & Sons, New York
Klibi W, Martel A, Guitouni A (2010) The design of robust value-creating supply chain networks: a critical review. Eur J Oper Res 203(2):283–293
Melo MT, Nickel S, Saldanha-Da-Gama F (2009) Facility location and supply chain management - a review. Eur J Oper Res 196:401–412
Melo MT, Nickel S, Saldanha-da-Gama F (2012) A Tabu search heuristic for redesigning a multi-echelon supply chain network over a planning horizon. Int J Prod Econ 136:218–230
Michalewicz Z, Vignaux GA, Hobbs M (1991) A non-standard genetic algorithm for the nonlinear transportation problem. ORSA J Comput 3:307–316
Mladenovic N, Brimberg J, Hansen P, Moreno-perez JA (2007) The p-median problem: a survey of meta-heuristic approaches. Eur J Oper Res 179(3):927–939
ReVelle CS, Eiselt HA (2005) Location analysis: a synthesis and survey. Eur J Oper Res 165:1–19
Sabri EH, Beamon BM (2000) A multi-objective approach to simultaneous strategic and operational planning in supply chain. Omega 28(5):581–598
Santosa T (2003) A comprehensive model and efficient solution algorithm for the design of global supply chains under uncertainty. Dissertation, Georgia Institute of Technology, pp 22–44
Shu J, Ma Q, Li S (2010) Integrated location and two-echelon inventory network design under uncertainty. Ann Oper Res 181:233–247
Simchi-Levi D, Kaminsky P (2004) Managing the supply chain: The definitive guide for the business professional. Irwin McGraw-Hill, Boston
Sousa RT, Liu S, Papageorgiou LG, Shaha N (2011) Global supply chain planning for pharmaceuticals. Chem Eng Res Des 89:2396–2409
Syarif A, Yun YS, Gen M (2002) Study on multi-stage logistic chain network: a spanning tree-based genetic algorithm approach. Comput Ind Eng 43:299–314
Talbi EG (2009) METAHEURISTICS, from design to Implementation. John Wiley & Sons, Hoboken, New Jersey, pp 190–221
Thomas DJ, Griffin PM (1996) Coordinated supply chain management. Eur J Oper Res 94:1–115
Tragantalerngsak S, Holt J, Ronnqvist M (2000) An exact method for the two-echelon, single-source, capacitated facility location problem. Eur J Oper Res 13:473–489

A new binary model for university examination timetabling: a case study

Alireza Rashidi Komijan[1*] and Mehrdad Nouri Koupaei[2]

Abstract

Examination timetabling problem (ETP) is one of the most important issues in universities. An improper timetable may result in students' dissatisfaction as it may not let them study enough between two sequential exams. In addition, the many exams to be scheduled, the large number of students who have taken different courses, the limited number of rooms, and some constraints such as no conflict in a single student's exams make it very difficult to schedule experimentally. A mathematical programming model is required to formulate such a sophisticated problem. In this paper, a new binary model is developed for ETP. The novelty of the paper can be discussed in two directions. The first one is that a course can be offered more than once in a semester. If a course is requested by a few students, then it is enough to be offered once. If the number of students requesting a course is more than the maximum number of students who are allowed to attend a single class, then the course is multi-offered. The second novelty is that sharing a room for two simultaneous exams is allowed. Also, the model considers some hard and soft constraints, and the objective function is set in such a way that soft constraints are satisfied as much as possible. Finally, the model is applied in a sample department and is solved by GAMS.

Keywords: Binary programming, Mathematical modeling, Examination timetabling problem

Background

Timetabling is a common problem of educational institutions, e.g., schools and universities. It can be divided into three categories. The first one is school timetabling which includes scheduling of school classes. The second one is university course timetabling which determines which course should be assigned to which lecturer on which day and time slot. The third one is examination timetabling which schedules the examinations of given courses. Examination timetabling is a hard, complex, and time-consuming task. It is very confusing when developed experimentally. Examination timetabling problem (ETP) is to determine the exact date and time as well as the room in which each exam should be held. Such a problem can be formulated using mathematical models.

In this paper, a new model is presented for university examination timetabling problem (ETP). The paper has two main novelties. The first one is that it considers *multi-offered* or *multi-code courses*. At the beginning of a semester, the dean of a department may notice that a large number of students wish to take a specific course. The number of these students may be more than the capacity of a class. As a result, that course is offered in two or more classes. This is called the *multi-code course*. Obviously, these classes may be scheduled on different days of the week, but their exam should be held simultaneously. The second novelty of the paper is that *room split* is allowed. Naturally, each exam should be held in one room. If the number of students who have taken a course is larger than the capacity of a room, they should be divided into two rooms. Assume that the examination of courses A and B are held simultaneously. The number of students who took these courses is 40 and 45, respectively, and the capacity of each exam room is 30 students. It is possible to assign a room for course A (with a capacity of 30 students) and a room for course B (with a capacity of 30 students), and both courses share a room between the remaining students (10 students for course A and 15 students for course B). This is called *room split*.

The proposed model includes some hard and soft constraints. The objective function is set in such a way that

* Correspondence: rashidi@azad.ac.ir
[1]Department of Industrial Engineering, Firoozkooh Branch, Islamic Azad University, Firoozkooh, Iran
Full list of author information is available at the end of the article

soft constraints are satisfied as much as possible. The model is applied and tested in the Industrial Engineering Department of Islamic Azad University in Firoozkooh, Iran. In this department, 12 courses are offered in four terms for postgraduate students.

Timetabling problem in educational institutions is categorized in three groups: university examination, university course, and school timetabling. University examination timetabling defines the exact day, time slot, and room that each exam is held. In other words, which exam is scheduled on which day, the time slot, and in which building, floor, and room it is held are determined. Although there are some novel research that focused on the modeling and formulation of ETP, a large number of researchers are interested in the applications of heuristics to solve the problem. Here, a wide range of previous research both on modeling and solution approach are briefly presented.

A comprehensive survey on ETP was performed by Qu et al. (2009a). They considered both theoretical and practical research done in a 10-year period.

Dammak et al. (2006) focused on room assignment of exams and also on a heuristic to solve it. If the solution that resulted from the heuristic indicates that only one exam is held in a room, the problem is finished. Otherwise, the algorithm tries to reach a solution in which two exams are assigned to a single room at most. In a similar research, Ayob and Malik (2011) focused on room assignment for each exam in such a way that the total movement of students between rooms during two consecutive exams was minimized. Also, the model has another objective to minimize the number of rooms assigned to a single exam.

MirHassani (2006) developed a model based on a predefined exam timetable in order to maximize paper spread. Ayob et al. (2007a) used a new objective function for ETP. The objective function they presented was the penalty cost which was a modification of the objectives introduced by Carter et al. (1996) and Burke and Newall (1999).

Ayob et al. (2007b) solved a model with the aim of improving the quality of the timetable, which tried to minimize the number of students with two consecutive exams in a day. Burke et al. (2008) defined a model that included seven objectives. They grouped the objectives in such a way that each group satisfied the specific party (students, markers, invigilators, and estates).

Cheong et al. (2009) developed a model that minimized the length of timetable and prevented students to take exams in consecutive exams as much as possible. They formulated a multi-objective model and solved it using an evolutionary algorithm.

Sagir and Ozturk (2010) formulated the assignment of invigilators to exams as a multi-objective model and calculated the weights of objectives using the analytic network process. Kahar and Kendall (2010) applied a model on a real case regarding some new constraints such as distance between rooms. In the end, they used a heuristic to solve the problem. McCollum et al. (2011) considered an integer programming model with a cost penalty objective function. It tried to satisfy soft constraints as much as possible, and if it failed, a penalty was accounted.

Among the research done based on solution approach, Alvarez-Valdes et al. (1997) proposed a new algorithm that included several heuristics based on tabu search in order to solve an ETP in a Spanish university. The objective of the model was to reach the best schedule for all students. Dimopoulou and Miliotis (2001) focused on developing a course and exam timetable. At first, the course timetable was developed, and its solution was used to generate an initial exam schedule. The initial exam was further improved using a heuristic.

Mansour et al. (2011) applied the scatter search algorithm to generate a timetable for examinations and used it on a real case to show the efficiency of the heuristic. White et al. (2004) proposed a technique based on tabu search to solve an ETP. Naji Azimi (2005) applied the combination of simulated annealing, genetic algorithm, tabu search, and ant colony in order to solve an ETP.

Petrovic et al. (2007) used a case-based reasoning technique to choose the proper order of heuristics that should be applied to find an initial solution. Pillay and Banzhaf (2009) represented a new approach for heuristic combinations in which heuristics were applied simultaneously. Pillay and Banzhaf (2010) applied a genetic algorithm approach in order to get a set of solutions for an ETP in such a way that all hard constraints were met. It then improved the solutions in order to satisfy soft constraints as much as possible.

Modarres et al. (2010) used a new version of the fuzzy analytical hierarchy process to prioritize courses in a management school. They chose the best courses offered by ranking them.

Burke et al. (2010) used variable neighborhood search along with genetic algorithm for an ETP. Turabieh and Abdullah (2011) combined the principle of the electromagnetic-like mechanism with a great deluge algorithm to solve an ETP. Some algorithms based on graph coloring were applied by Malkawi et al. (2008) and Qu et al. (2009b) to solve ETP.

The structure of the paper is as follows. Problem assumptions are discussed in the 'Problem description' section. In the 'Methods' section, the mathematical model is described and is tested in a real case in the 'Results and discussion' section. Concluding remarks are in the 'Conclusions' section.

Problem description

In this paper, a binary model is represented to schedule examinations of postgraduate students of the Industrial Engineering Department of Islamic Azad University (Firoozkooh, Iran). Students were taking nine courses during the first three terms and worked on a thesis in the fourth term. In each term, they can take three courses. All exams should be held on the first and second floors of the department. Each floor has three available rooms for exams. Here is a list of educational rules and assumptions:

1. The postgraduate level includes four terms. Students take courses in the first three terms, with the last term dedicated to their thesis.
2. Students are categorized in three groups: beginners (students who are currently on their first term), juniors (students who have studied for one term and are currently on their second term), and seniors (students who have studied for two terms and are currently on their third one).
3. In total, 12 courses are offered by the department.
4. All 12 courses are categorized in three groups; each group includes four courses. The first, second, and third groups are offered to beginners, juniors, and seniors, respectively.
5. Each student should only take the courses offered for his/her group. For example, a junior student cannot take a course offered to beginners or seniors.
6. In each term, a student has to take three courses among the four ones offered to him. Thus, the student has options.
7. All courses are divided in to three groups: simple, medium, and hard.
8. All exams should be held on the first and second floors of the department.
9. On each floor, there are three rooms that can be used for exams.
10. The number of students registered for a course is given.
11. If the number of students who wish to take a specific course is more than the capacity of a class, the course can be offered more than once in the term. This is called the *multi-code course*.
12. The classes of a multi-code course may be scheduled on different days of the week, but their exam should be held simultaneously.
13. In order to use the capacity of a room efficiently, room split is allowed. It means that a room may be share between two courses whose examinations are held simultaneously. Assume that two exams are held simultaneously, and the number of students of each course is much more than the capacity of a room. In this case, two rooms may be assigned to these exams (one room to each exam) and a room for the remaining students of both exams (in total, three rooms are assigned).
14. All exams should be scheduled in a 6-day period. The planning period begins on a Saturday and ends on a Thursday.
15. Each day includes two time slots.

Methods

In this section, a binary model is developed based on the given rules and assumptions. There are some hard and soft constraints in the model. Obviously, hard constraints should be fully met. The objective function is set in such a way that soft constraints are satisfied as much as possible. In other words, the objective function minimizes the degree of soft constraint dissatisfaction. The model solution determines that each exam is held on which day and time slot and also in which building, floor, and room. The nomenclatures are as follows.

Indices and sets

Here are the indices and sets used:

I is the set of exams (courses).
i is the index of exam ($i \in I$).
J is the set of student groups (beginners, juniors, and seniors).
j is the index of student group ($j \in J$).
T is the set of time slots.
t is the index of time slot ($t \in T$).
R is the set of rooms.
r is the index of room ($r \in R$).
L is the set of days in the scheduled period length.
l is the index of day ($l \in L$).
B is the set of buildings. According to assumption 8, all exams should be held in one building. As a result, B includes only one building for the case under study. However, it is defined as we develop the general model.
b is the index of the building ($b \in B$).
F is the set of floors.
f is the index of the floor ($f \in F$).
c is the index of the code (refer to multi-code courses).
J_i is the set of student groups included in course i. According to assumption 4, each course is only for one group of students. As a result, J_i includes only one member for the case under study. However, it is defined as we develop the general model.
I_j is the set of courses offered to student group j.
$I_{m,j}$ is the set of medium courses offered to student group j.
$I_{h,j}$ is the set of hard courses offered to student group j.
T_l is the set of time slots available for day l.
C_i is the set of codes for course i (number of classes offered for course i). If course i is a multi-code course,

it is offered more than once due to the large number of students who wish to take it. For example, if a multi-code course is offered twice in a single term, C_i has two members (1 and 2). If a course is offered only once, C_i has only one member (1).
F_b is the set of floors in building b.
$R_{b,f}$ is the set of rooms located in floor f of building b.

Parameters
Here are the parameters used:

$S_{i,c}$ is the number of students registered at code c of course i.
Cap_{bfr} is the capacity of room r which is located on floor f of building b.
NC_i is the number of times that course i is offered in a semester. If i is not a multi-code course, NC_i is one.

Variables
Here are the variables used:

X_{ijlt} is a binary variable that has a value of 1 if exam i for student group j is scheduled on day l and time slot t; otherwise, it has a value of 0.
$Y_{icbfrlt}$ is a binary variable that has a value of 1 if the examination of code c of course i is scheduled on day l and time slot t and is held in room r of floor f of building b; otherwise, it has a value of 0.
H_{iclt} is a binary variable that has a value of 1 if code c of course i is scheduled on day l and time slot t; otherwise, it has a value of 0.
Z_{icbf} is a binary variable that has a value of 1 if the examination of code c of course i is held on floor f of building b; otherwise, it has a value of 0.
W_{bfrlt} is a binary variable that has a value of 1 if room r of floor f of building b is split on day l and time slot t; otherwise, it has a value of 0.

Constraints
All constraints are divided into hard and soft ones. They are as follows.

Hard constraints
There are some constraints that should not be violated. These are called hard constraints, and they are as follows:

1. All exams should be scheduled. For this, the following equation is used:

$$\sum\nolimits_{j\in J_i}\sum\nolimits_{l}\sum\nolimits_{t\in T_l} X_{ijlt} = 1\forall i. \tag{1}$$

2. No student should have two exams in a day. As beginners, juniors, and seniors, they can only choose among the courses offered to them. The constraint is formulated as follows:

$$\sum\nolimits_{i\in I_j}\sum\nolimits_{t\in T_l} X_{ijlt} \le 1\forall l, j. \tag{2}$$

3. All codes (classes) of a multi-code course should be scheduled simultaneously. If two or more classes are offered for a single course, their examination should be held on a single day and time slot. For this, the following equation is used:

$$\sum_{c\in C_i} H_{iclt} = NC_i X_{ijlt}\forall i, j \in J_i, l, t \in T_l. \tag{3}$$

4. At most, two rooms can be assigned to students of a class. For this, the following equation is used:

$$\sum\nolimits_{l}\sum\nolimits_{t\in T_l}\sum\nolimits_{r\in R_{bf}} y_{icbfrlt} \le 2Z_{icbf}\forall i, c \in C_i, b, f \in F_b. \tag{4}$$

5. All students of a class should be scheduled on the same floor. In other words, if the number of students of a class is more than the capacity of a room, they naturally should be split in two rooms. This constraint ensures that those two rooms should be in a single floor. For this, the following equation is used:

$$\sum\nolimits_{b}\sum\nolimits_{f\in F_b} Z_{icbf} = 1\forall i, c \in C_i. \tag{5}$$

6. All codes (classes) of a multi-code course should be scheduled on the same building. For this, the following equation is used:

$$\sum_{f\in F_b} Z_{icbf} = \sum_{f\in F_b} Z_{i\acute{c}bf}\forall i, c, \acute{c} \in C_i, c\neq\acute{c}, b. \tag{6}$$

7. Each room can be assigned to two exams at the most. For this, the following equation is used:

$$\sum_{i}\sum_{c\in C_i} y_{icbfrlt} \le 2 \quad \forall b, f \in F_b, r \in R_{bf}, l, t \in T_l. \tag{7}$$

8. Room capacity should not be violated. As shown in relation 8, half of the capacity of a room is decreased if it is split:

$$S_{ic}Z_{icbf} + (X_{ijlt} - 1)M \le \sum_{r\in R_{bf}} Cap_{bfr}\, y_{icbfrlt} - \frac{1}{2}\sum_{r\in R_{bf}} Cap_{bfr}\, W_{bfrlt} \quad \forall i, j \in J_i, c \in C_i, b, f, \in F_b, l, t \in T_l. \tag{8}$$

9. The following constraints relate variables $Y_{icbfrlt}$ and W_{bfrlt}. In other words, these constraints show the relation between $Y_{icbfrlt}$ and W_{bfrlt}. According to constraint 7, each room can be assigned to at most two exams. From constraint 9, it is clear that if a room is assigned to two exams, the left-hand side = 1. In that case, $W_{bfrlt} = 1$. On the other hand, relation 10 ensures that if $W_{bfrlt} = 1$, the room is split:

$$\sum_i \sum_{c \in C_i} y_{icbfrlt} - 1 \le W_{bfrlt} \ \forall b, f \in F_b, r \subset R_{bf}, l, t \in T_l, \tag{9}$$

$$2W_{bfrlt} \le \sum_i \sum_{c \in C_i} y_{icbfrlt} \ \forall b, f \in F_b, r \in R_{bf}, l, t \in T_l. \tag{10}$$

10. The following constraint relates $Y_{icbfrlt}$ and H_{iclt}. It shows that whenever exam i is scheduled on day l and time slot t, variable $H_{iclt} = 1$:

$$\sum_b \sum_{f \in F_b} \sum_{r \in R_{bf}} y_{icbfrlt} \le MH_{iclt} \forall i, c \in C_i, l, t \in T_l. \tag{11}$$

Soft constraints

Soft constraints are those that should preferably be satisfied. However, if a soft constraint is not met, the model will not be infeasible. In this section, soft constraints are introduced. The objective function is then set in such a way that deviations from soft constraint satisfaction are minimized. There are three soft constraints:

1. Each student should have at least one free day to study for a medium course. In other words, if a student takes the exam of a medium course on day l, he/she should not take any exam on day $l - 1$.

$$\sum_{s \in I_j, s \ne i} \sum_{t \in T_l} X_{sj\,l-1\,t} \le \left(1 - \sum_{t \in T_l} X_{ijlt}\right) M \quad \forall j, i \in I_{m,j}, l > 1. \tag{12}$$

2. There should be at least two free days to study for a hard course. This is actually shown in two constraints. Relation 13 shows that if a student takes a hard exam on the second day of the planning horizon (Sunday), he/she should not take any exam on the first day (Saturday). Relation 14 is written for $l > 2$.

$$\sum_{s \in I_j, s \ne i} \sum_{t \in T_l} X_{sj\,l-1\,t} \le \left(1 - \sum_{t \in T_l} X_{ijlt}\right) M \ \forall j, i \in I_{h,j}, l = 2, \tag{13}$$

$$\sum_{s \in I_j, s \ne i} \sum_{t \in T_l} X_{sjl-2\,t} + \sum_{s \in I_j, s \ne i} \sum_{t \in T_l} X_{sjl-1\,t} \le \left(1 - \sum_{t \in T_l} X_{ijlt}\right) M \ \forall j, i \in I_{h,j}, l > 2. \tag{14}$$

3. Hard courses should not be scheduled for the first two days of the planning period. According to relations 13 and 14, a hard course may tend to be scheduled for the first two days of the planning period. To avoid this and in order that students would have more time to study for a hard course, a soft constraint is applied to prevent hard courses to be scheduled on the first two days of the period.

$$\sum_j \sum_{i \in I_{h,j}} \sum_{l=1}^{2} \sum_{t \in T_l} X_{ijlt} \le 0. \tag{15}$$

Objective function

In order to make sure that soft constraints are satisfied as much as possible, a penalty value is related to a deviation from the ideal situation. As a result, relations 12, 13, 14, and 15 are rewritten as follows:

$$\sum_{s \in I_j, s \ne i} \sum_{t \in T_l} X_{sjl-1\,t} + g_{ijl} - gg_{ijl} = \left(1 - \sum_{t \in T_l} X_{ijlt}\right) M \forall j, i \in I_{m,j}, l > 1, \tag{16}$$

$$\sum_{s \in I_j, s \ne i} \sum_{t \in T_l} X_{sj\,l-1\,t} + k_{ijl} - kk_{ijl} = \left(1 - \sum_{t \in T_l} X_{ijlt}\right) M \forall j, i \in I_{h,j}, l = 2, \tag{17}$$

Table 1 Descriptions of courses

Course	Type of course	Students who can take the course	Number of registered students
A1	Simple	Beginners	30
A2	Medium	Beginners	25
A3-1	Hard	Beginners	30
A3-2	Hard	Beginners	30
A4-1	Hard	Beginners	25
A4-2	Hard	Beginners	25
B1-1	Simple	Juniors	30
B1-2	Simple	Juniors	25
B2	Medium	Juniors	30
B3	Medium	Juniors	25
B4-1	Hard	Juniors	25
B4-2	Hard	Juniors	25
C1	Hard	Seniors	25
C2	Medium	Seniors	25
C3	Hard	Seniors	25
C4	Simple	Seniors	25

Table 2 Room data

Floor	Rooms
First floor	111, 112, 113
Second floor	121, 122, 123

$$\sum_{s\in I_j, s\neq i}\sum_{t\in T_l} X_{sj\,l-2\,t} + \sum_{s\in I_j, s\neq i}\sum_{t\in T_l} X_{sj\,l-2\,t} + v_{ijl} - vv_{ijl} = \left(1-\sum_{t\in T_l} X_{ijlt}\right) M\ \forall j, i\in I_{h,j}, l>2, \quad (18)$$

$$\sum\nolimits_j \sum\nolimits_{i\in I_{h_j}} \sum\nolimits_{l=1}^{2} \sum\nolimits_{t\in T_l} X_{ijlt} + d - dd = 0. \quad (19)$$

With the aim of satisfying relations 12, 13 and 14, gg_{ijl}, kk_{ijl}, and vv_{ijl} are minimized. Also, both d and dd are minimized to ensure the satisfaction of relation 15. As the importance of all soft constraints is assumed the same, the objective function is written as follows:

$$\mathrm{Min}\sum_i \sum_{j\in J_i} \sum_l \left(gg_{ijl} + kk_{ijl} + vv_{ijl} + d + dd\right). \quad (20)$$

Results and discussion

The presented model was applied and tested in a real case, i.e., in the Department of Industrial Engineering of Islamic Azad University in Firoozkooh, Iran. We are going to develop the best examination timetable for postgraduate students of the department. The courses offered by the department are shown in Table 1. For example, it shows that course A1 is a fairly simple course that is offered to beginners. Also, it is not a multi-code course, while A3 is a hard course and is offered twice in the term (A3-1 and A3-2) due to the fact that a large number of students wish to take it.

According to assumption 8, all exams should be held on the first and second floors of the department. Thus, there is only one building in this case. According to assumption 9, there are three available rooms on each floor. Information about floors and rooms is shown in Table 2. The capacity of each room is 25 students.

The model is solved using GAMS 22.5, and the timetable is shown in Figure 1. The solution shows that the first soft constraint is satisfied: medium exams are not scheduled exactly after other exams. As shown in Figure 1, these exams (A2, B2, B3, and C2) are scheduled on the first day of the planning period. The second soft constraint is not fully satisfied. Although A3, B4, and C1 are hard exams, students do not have two free days to study for them. However, they have two free days to study for other hard exams (A4 and C1). The third soft

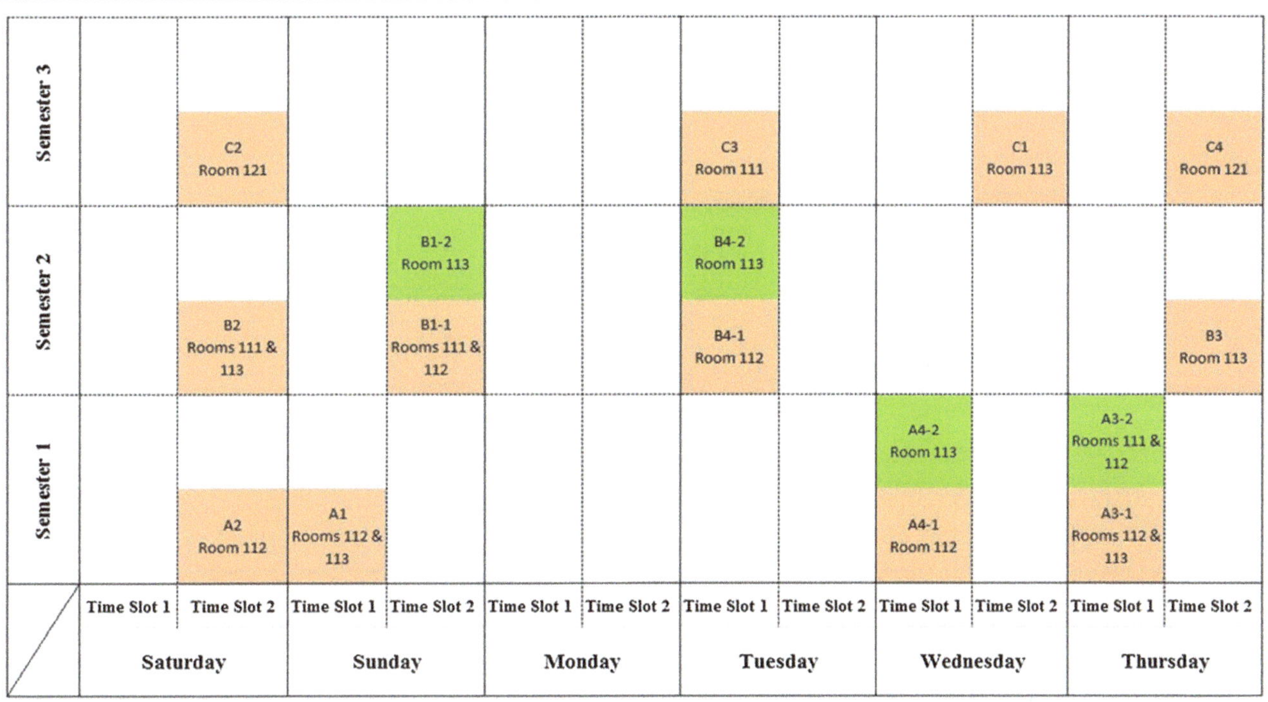

Figure 1 Examination timetable.

constraint is also fully satisfied. From Figure 1, it is clear that no hard exam is scheduled on the first two days of the planning period. The results also show that all hard constraints are fully met. For example, no student group takes two exams in a single day, no room is assigned to more than two exams, and all codes (classes) of a multi-code course are scheduled simultaneously.

Conclusions

In this paper, a new binary model is presented to generate a university examination timetable. The contributions of the model considered multi-code courses as well as room split. The model was applied in the Industrial Engineering Department of Islamic Azad University. The model includes some hard and soft constraints. The objective function is set in such a way that deviation from soft constraint satisfaction is minimized. The result shows that all soft constraints are fully satisfied, except for the one indicating that students have two free days for hard exams. There are totally five hard exams in the case. The result shows that students have two free days to study for just two hard exams. This is the best feasible solution as it is impossible to satisfy all soft constraints of the model.

Competing interests

The authors declare that they have no competing interests.

Authors' contributions

ARK presented the mathematical model, applied it in a real case, and solved the model by GAMS. MNK studied and categorized the literature. All authors read and approved the final manuscript.

Acknowledgments

This study was performed as a research plan and was financially supported by Islamic Azad University, Firoozkooh Branch.

Author details

[1]Department of Industrial Engineering, Firoozkooh Branch, Islamic Azad University, Firoozkooh, Iran. [2]Department of Management, Firoozkooh Branch, Islamic Azad University, Firoozkooh, Iran.

References

Alvarez-Valdes R, Crespo E, Tamarit JM (1997) A tabu search algorithm to schedule university examinations. Qüestiió 21(1–2):201–215

Ayob M, Malik A (2011) A new model for an examination-room assignment problem. International Journal of Computer Science and Network Security 11(10):187–190

Ayob M, Abdullah S, Malik A (2007a) A practical examination timetabling problem at the Universiti Kebangsaan Malaysia. International Journal of Computer Science and Network Security 7(9):198–204

Ayob M, Malik A, Abdullah S, Hamdan AR, Kendall G, Qu R (2007b) Solving a practical examination timetabling problem: a case study. Lecture Notes on Computer Science 4707:611–624

Burke EK, Newall JP (1999) A multistage evolutionary algorithm for the timetable problem. IEEE Transactions on Evolutionary Computation 3(1):63–74

Burke EK, Eckersley AJ, McCollum B, Petrovic S, Qua R (2010) Hybrid variable neighbourhood approaches to university exam timetabling. European Journal of Operational Research 206(1):46–53

Burke EK, McCollum B, McMullan P, Parkes AJ (2008) Multi-objective aspects of the examination timetabling competition track. In: Proceedings of PATAT 2008. Université de Montréal, Montreal. 18–22 August 2008

Carter MW, Laporte G, Lee SY (1996) Examination timetabling: algorithmic strategies and applications. J Oper Res Soc 47(3):373–383

Cheong CY, Tan KC, Veeravalli B (2009) A multi-objective evolutionary algorithm for examination timetabling. Journal of Scheduling 12(2):121–146

Dammak A, Elloumi A, Kamoun H (2006) Classroom assignment for exam timetabling. Advances in Engineering Software 37(10):659–666

Dimopoulou M, Miliotis P (2001) Implementation of a university course and examination timetabling system. European Journal of Operational Research 130(1):202–213

Kahar MNM, Kendall G (2010) The examination timetabling problem at University Malaysia Pahang: comparison of a constructive heuristic with an existing software solution. European Journal of Operational Research 207(2):557–565

Malkawi M, Al-Haj Hassan M, Al-Haj Hassan O (2008) A new exam scheduling algorithm using graph coloring. The International Arab Journal of Information Technology 5(1):80–87

Mansour N, Isahakian V, Ghalayini I (2011) Scatter search technique for exam timetabling. Applied Intelligence 34(2):299–310

McCollum B, McMullan P, Parkes AJ, Burke EK, Qu R (2011) A new model for automated examination timetabling. Annals of Operations Research 194 (1):291–315

MirHassani SA (2006) Improving paper spread in examination timetables using integer programming. Applied Mathematics and Computation 179(2):702–706

Modarres M, Sadi-Nezhad S, Arabi F (2010) Fuzzy analytical hierarchy process using preference ratio: a case study for selecting management short course in a business school. International Journal of Industrial Engineering Computations 1:173–184

Naji Azimi Z (2005) Hybrid heuristics for examination timetabling problem. Applied Mathematics and Computation 163(2):705–733

Petrovic S, Yang Y, Dror M (2007) Case-based selection of initialisation heuristics for metaheuristic examination timetabling. Expert Systems with Application 33(3):772–785

Pillay N, Banzhaf W (2009) A study of heuristic combinations for hyper-heuristic systems for the uncapacitated examination timetabling problem. European Journal of Operational Research 197(2):482–491

Pillay N, Banzhaf W (2010) An informed genetic algorithm for the examination timetabling problem. Appl Soft Comput 10(2):457–467

Qu R, Burke EK, McCollum B (2009a) Adaptive automated construction of hybrid heuristics for exam timetabling and graph colouring problems. European Journal of Operational Research 198(2):392–404

Qu R, Burke EK, McCollum B, Merlot LTG, Lee SY (2009b) A survey of search methodologies and automated system development for examination timetabling. Journal of Scheduling 12(1):55–89

Sagir M, Ozturk ZK (2010) Exam scheduling: mathematical modeling and parameter estimation with the analytic network process approach. Mathematical and Computer Modelling 52(5–6):930–941

Turabieh H, Abdullah S (2011) An integrated hybrid approach to the examination timetabling problem. Omega 39(6):598–607

White GM, Xie BS, Zonjic S (2004) Using tabu search with longer-term memory and relaxation to create examination timetables. European Journal of Operational Research 153(16):80–91

Bayesian change point estimation in Poisson-based control charts

Hassan Assareh[1*], Rassoul Noorossana[2] and Kerrie L Mengersen[3]

Abstract

Precise identification of the time when a process has changed enables process engineers to search for a potential special cause more effectively. In this paper, we develop change point estimation methods for a Poisson process in a Bayesian framework. We apply Bayesian hierarchical models to formulate the change point where there exists a step change, a linear trend and a known multiple number of changes in the Poisson rate. The Markov chain Monte Carlo is used to obtain posterior distributions of the change point parameters and corresponding probabilistic intervals and inferences. The performance of the Bayesian estimator is investigated through simulations and the result shows that precise estimates can be obtained when they are used in conjunction with the well-known c-, Poisson exponentially weighted moving average (EWMA) and Poisson cumulative sum (CUSUM) control charts for different change type scenarios. We also apply the Deviance Information Criterion as a model selection criterion in the Bayesian context, to find the best change point model for a given dataset where there is no prior knowledge about the change type in the process. In comparison with built-in estimators of EWMA and CUSUM charts and ML based estimators, the Bayesian estimator performs reasonably well and remains a strong alternative. These superiorities are enhanced when probability quantification, flexibility and generalizability of the Bayesian change point detection model are also considered.

Keywords: Bayesian hierarchical model; Change point; Control charts; Markov chain Monte Carlo; Poisson process

Background

Statistical process control charts are used to detect changes in a process by distinguishing between assignable causes and common causes of the process variation. When a control chart signals, process engineers initiate a search to identify and eliminate the source of variation. Knowing the time at which the process began to vary, the so-called change point, would help to conduct the search more efficiently in a tighter time-frame.

A Poisson process is often used to model the number of occurrences in an interval of time. In this regard, Poisson-based control charts have been developed and frequently applied in an industry context to monitor the number of defects and nonconformities in a product (Gardiner and Montgomery 1987; White et al. 1997) and in a health context to monitor patient mortality and spread of an infection in a hospital (Benneyan 1998; Limayea et al. 2008). The most commonly used control chart procedures adopted for Poisson-distributed data include c-charts (Shewhart 1926, 1927), cumulative sum of quality characteristic measurement (CUSUM; Page 1954, 1961; Brook and Evans 1972), and exponentially weighted moving average (EWMA; Roberts 1959; Trevanich and Bourke 1993; Borror and Rigdon 1998); see Woodall (1997) and Montgomery (2008) for more details. Furthermore, appropriate control charts and methods were developed in monitoring more complex Poisson data including correlated (Chiu and Kuo 2007; Niaki and Abbasi 2008; Niaki and Nafar 2008; Amiri et al. 2011) and auto-correlated observations (Weiß, 2007; Vermaat et al. 2008).

It has been shown that Poisson CUSUM and Poisson EWMA charts are more sensitive for detecting small shifts in the process parameters whereas a c-chart still remains efficient for the detection of large shifts (Montgomery 2008). However, upon signaling, none of them provide specific information regarding the time at

*Correspondence: h.assareh@unsw.edu.au
[1] Simpson Centre for Health Services Research, Australian Institute of Health Innovation, Faculty of Medicine, University of New South Wales, Sydney, NSW 2052, Australia
Full list of author information is available at the end of the article

which the process changed and the magnitude and the type of the change.

In recent years, statistical and machine-learning methods have been employed in the development of change point estimators for a broad range of processes and change types (Amiri and Allahyari 2012; Atashgar 2013). For Poisson processes, maximum likelihood (ML) methods were applied to estimate the true time of a step change (Samuel and Pignatiello 1998; Perry 2004) and a linear trend (Perry et al. 2006) in the Poisson rate. It was shown that more precise estimates were obtained when ML estimators were used in conjunction with Poisson control charts, compared to charts' signals and CUSUM (Page 1954) or EWMA built-in estimators (Nishina 1992). A confidence interval on the estimated change point was also constructed (Perry 2004; Perry et al. 2006). Furthermore, Perry et al. (2007a) challenged the underlying assumption of knowing the form of change types and derived a ML estimator for non-decreasing multiple step change points (unknown number of consecutive changes) using isotonic regression models. The estimator was reported a reasonable alternative for some magnitudes of the step and linear trend disturbances. In the presence of multiple change points, it was found to be the superior. ML estimators have also been extended for step change scenarios in correlated Poisson observations (Niaki and Khedmati 2012; 2013a; Sharafi et al. 2013). Similar methods were extended to other attributes including binary data (Perry et al. 2007b; Noorossana et al. 2009; Amiri et al. 2011; Hou et al. 2013; Niaki and Khedmati 2013b).

A Bayesian formulation has recently been proposed as an alternative in change point estimation within a clinical context (Assareh et al. 2011a). It can easily capture complexity of patient mix and provide highly informative and precise estimates for the true time of a step change (Assareh et al. 2011c) or linear trend (Assareh et al. 2011b) in the odds ratio of clinical outcomes or mean survival time following a clinical procedure (Assareh and Mengersen 2012). Application of the Bayesian framework to change point estimation provides a way of making a set of inferences based on posterior distributions for the time and the magnitude of a change as well as assessing the validity of underlying assumptions in the change point model itself (Gelman et al. 2004).

In this paper, we model the change point in a Poisson process using a Bayesian framework and compare the performance of the Bayesian estimator with ML estimators. We model and estimate change points assuming that the underlying change type is known. In this scenario, the changes are in the form of a step change, a linear trend and a multiple change with known number of changes. For each model, we analyze and discuss the performance of the Bayesian change point model through posterior estimates and probability-based intervals. The three models are demonstrated and evaluated in sections 'Bayesian change point model', 'Evaluation', and 'Performance analysis' and then compared with respect to goodness of fit in section 'Comparative performance and model selection'. We then compare the Bayesian estimator with ML estimators and others in section 'Comparison of Bayesian estimator with other methods' and summarize the study and obtained results in section 'Conclusion'.

Bayesian change point model

Statistical inferences for a quantity of interest in a Bayesian framework are described as the modification of the uncertainty about their value in the light of evidence, and Bayes' theorem precisely specifies how this modification should be made as below:

$$\text{Posterior} \propto \text{Likelihood} \times \text{Prior}, \tag{1}$$

where 'Prior' is the state of knowledge about the quantity of interest in terms of a probability distribution before the data are observed; 'Likelihood' is a model underlying the data, and 'Posterior' is the state of knowledge about the quantity after data are observed which also is in the form of a probability distribution. This structure is expendable to multiple levels in a hierarchical fashion, so-called Bayesian hierarchical models (BHM), which allows to enrich the model by capturing all kinds of uncertainties for data observed as well as priors. In complicated BHMs, it is not easy to obtain the posterior distribution analytically. This analytic bottleneck has been eliminated by the emergence of Markov chain Monte Carlo (MCMC) methods. In MCMC algorithms, a Markov chain, also known as a random walk, is constructed whose stationary distribution is the posterior distribution of the parameters. Samples generated from a long run of the Markov chain using a proposal transition density are drawn from posterior distributions of interest. Some common MCMC methods for drawing samples include Metropolis-Hastings and the Gibbs sampler (see Gelman et al. (2004) for more details).

Consider a Poisson process X_t, $t = 1, \ldots, T$, that is initially in-control, with independent observations coming from a Poisson distribution with a known rate λ_0. At an unknown point in time, τ, the Poisson rate parameter changes from its in-control state of λ_0 to λ_1, $\lambda_1 = \lambda_0 + \delta, \delta \neq 0$. The Poisson process step change model can thus be parameterized as follows:

$$p(x_t \mid \lambda_t) = \begin{cases} \exp(-\lambda_0)\lambda_0^{x_t}/x_t! & \textit{if} \quad t = 1, 2, \ldots, \tau \\ \exp(-\lambda_1)\lambda_1^{x_t}/x_t! & \textit{if} \quad t = \tau + 1, \ldots, T, \end{cases} \tag{2}$$

where δ is the magnitude of the step change and τ and T are the change time and the current time, respectively.

The departure from the in-control state may occur due to a non-constant change type scenario which can be explained by a linear trend model $\lambda_t = \lambda_0 + \beta(t - \tau)$ for $t > \tau$. In this model, β is the magnitude of the linear trend disturbance (slope) in the process parameter and its positive value implies an increasing trend in which $\lambda_t > \lambda_0$, while a negative β leads to a linear reduction of the Poisson rate and $\lambda_t < \lambda_0$ for $t = \tau + 1, \ldots, T$. The Poisson process linear trend change model can be modelled as follows:

$$p(x_t \mid \lambda_t) = \begin{cases} \exp(-\lambda_0)\lambda_0^{x_t}/x_t! & \text{if} \quad t = 1, 2, \ldots, \tau \\ \exp(-(\lambda_0 + \beta(t - \tau))) & \\ \quad (\lambda_0 + \beta(t - \tau))^{x_t}/x_t! & \text{if} \quad t = \tau + 1, \ldots, T. \end{cases} \tag{3}$$

In order to address the possibility of having change types other than step and linear trend forms (Perry et al. 2007a), we introduce a multiple change point scenario where the number of change points is known. This prior knowledge might have been obtained based on awareness and past experience of process engineers in factors such as changes in operators, materials, procedures, tools, and policies which may lead to increasing or decreasing step changes in the Poisson rate. Here, we consider the case of two sequential step changes. Other cases with more than two change points can be modelled in the same way. In this scenario, at an unknown point in time, τ_1, the Poisson rate parameter changes from its in-control state of λ_0 to λ_1, $\lambda_1 = \lambda_0 + \delta_1, \delta_1 \neq 0$. For a period of time, the process continues with the new parameter, λ_1, and then at an unknown point in time, τ_2, it changes to λ_2, $\lambda_2 = \lambda_0 + \delta_2, \delta_2 \neq \delta_1 \neq 0$. The Poisson process multiple change point model with two step changes can thus be parameterized as follows:

$$p(x_t \mid \lambda_t) = \begin{cases} \exp(-\lambda_0)\lambda_0^{x_t}/x_t! & \text{if} \quad t = 1, 2, \ldots, \tau_1 \\ \exp(-\lambda_1)\lambda_1^{x_t}/x_t! & \text{if} \quad t = \tau_1 + 1, \ldots, \tau_2 \\ \exp(-\lambda_2)\lambda_2^{x_t}/x_t! & \text{if} \quad t = \tau_2 + 1, \ldots, T. \end{cases} \tag{4}$$

Regarding above models to Equation 1, $p(. \mid .)$, is the likelihood that underlies the observations; and posterior distributions of the time (τ, τ_1, τ_2) and the magnitude of change (δ, β, δ_1, δ_2) will be constructed and investigated as they are the unknown parameters of interest in the change point analysis. Assume that the process X_t is monitored by a control chart that signals at time T. We assign a zero-mean normal distribution with a standard deviation of $6 \times \sqrt{\lambda_0}$ as a prior distribution for all change sizes (δ, β, δ_1, δ_2). This is a reasonably informative prior for the magnitude of the change in an in-control Poisson rate as the control chart is sensitive enough to detect very large shifts and estimate associated change points. Other distributions such as uniform or Gamma might also be of interest; see Gelman et al. (2004) for more details on selection of prior distributions. We place a uniform distribution on the range of $(1, T - 1)$ as a prior for the time of the change (τ, τ_1, τ_2). To avoid obtaining a negative value for process mean after a change, $\lambda_{1(2)}$, within MCMC, particularly when a drop has occurred, we added a constraint such that $\lambda_{1(2)}$ must be positive. Although other methods such as modelling the process on the log scale may be of interest, we do not pursue these here as we may lose simplicity and explicit or correct reflection of the Poisson process. See the Appendix for the change model codes in WinBUGS (Spielgelhalter et al. 2003).

Evaluation

We used Monte Carlo simulation to study the performance of the constructed BHMs in change estimation following a signal from c-, Poisson CUSUM, and Poisson EWMA control charts when a change (step, linear, multiple) is simulated to occur at $\tau = 100$. We generated 100 observations of a Poisson process with an in-control rate of $\lambda_0 = 20$. To investigate the behavior of the Bayesian estimators over the population for different change sizes, we replicated this simulation method 100 times. Simulated datasets that were obvious outliers were excluded. This setting allows us to have distribution of estimates with standard errors in orders of 10. The number of replication studies is a compromise between excessive computational time, considering MCMC iterations and sufficiency of the achievable distributions even for tails.

In the step of change scenario, we induced step changes of sizes $\delta = \{+2, +6\}$ as an example and $\delta = \{\pm 2, \pm 6, \pm 15\}$ for a replication study until the control charts signalled. In the linear trend model, changes of slopes $\beta = \{\pm 0.5, \pm 1.0, \pm 2.0\}$ were induced until the control charts signalled. For the multiple change point case, two consecutive changes are simulated to occur at $(\tau_1, \tau_2) = (100, 110)$. We induced two changes of sizes $(\delta_1, \delta_2) = \{(\pm 4, \pm 8), (\pm 4, \pm 12)\}$ as part of a replication study at the determined times of change (τ_1, τ_2) until the control charts signalled. In this scenario, the replication study was limited to c-chart, since other control charts mostly signalled prior to the induction of the second change point.

Because we know that the process is in-control, if an out-of-control observation was generated in the simulation of the early 100 in-control observations, it was taken as a false alarm and the simulation was restarted. However, in practice, a false alarm may lead to stopping the process and analyzing root causes. When no cause is found, the process would follow without adjustment. Furthermore, for the multiple change scenario, if in any simulation, the charts signalled earlier than simulating the second change, that simulation was terminated and not followed. The simulation was also repeated for rate parameters of 5 and 10 over equivalent change scenarios; since the results were similar to these obtained for $\lambda_0 = 20$, they are not reported here.

To construct control charts, we applied the Shewhart (1926, 1927), Brook and Evans (1972), and Trevanich and Bourke (1993) procedures for c-, Poisson CUSUM, and Poisson EWMA control charts, respectively. A Poisson CUSUM accumulates the difference between an observed value and a reference value k through $S_i^+ = \max\{0, x_i - k^+ + S_{i-1}^+\}$ and $S_i^- = \max\{0, k^- - x_i + S_{i-1}^-\}$, where $k^+ = (\lambda_1^+ - \lambda_0)/(\ln(\lambda_1^+) - \ln(\lambda_0))$ and $k^- = (\lambda_0 - \lambda_1^-)/(\ln(\lambda_0) - \ln(\lambda_1^-))$. If $S_i^\pm$ exceeds a specified decision interval $h^\pm$ then the control chart signals that an increase (a decrease) in the Poisson rate occurred. We calibrated the charts to detect a 25% shift in Poisson rates and have an in-control average run length ($\hat{ARL}_0$) of 370 approximately, close to a standard c-chart (see Woodall and Adams (1993)). The resultant Poisson CUSUM charts had $(k^+, h^+) = (22.4, 22)$ and $(k^-, h^-) = (17.4, 14)$. For simplicity, the values were rounded to one decimal place.

In a Poisson EWMA cumulative values of observations are obtained through $Z_i = r \times x_i + (r - 1) \times Z_{i-1}$, where $Z_0 = \lambda_0$, and plotted in a chart with UCL $= \lambda_0 + A^+\sqrt{\text{Var}Z_i}$ and LCL $= \lambda_0 - A^-\sqrt{\text{Var}Z_i}$. We let $r = 0.1$ and $A^\pm = 2.67$ to build a chart with an ARL_0 of 370, close to a standard c-chart.

All changes and control charts were simulated in the R package. To obtain posterior distributions of the time and the magnitude of the changes, we used the R2WinBUGS interface (Sturtz et al. 2005) to generate 100,000 samples through MCMC iterations in WinBUGS (Spielgelhalter et al. 2003) for all change point scenarios with the first 20,000 samples ignored as burn in. We then analyzed the results using the CODA package in R (Plummer et al. 2010). See the Appendix for the change point model codes in WinBUGS.

Performance analysis

Step change model

The posterior distributions for the time and the magnitude of a step change of size +6 are presented in Figure 1. For all control charts, posterior distributions of the change point concentrate on the 100th sample which is the real change point. Since the posteriors are asymmetric and skewed, particularly for the time of the change, the posterior mode is used as an estimator for change point model parameters (τ, δ).

Table 1 shows the posterior estimates for increases of sizes +2 and +6 in the process mean. The c-chart detects a fall of around half a standard deviation ($\delta = +2$) in the Poisson rate after 101 samples where the mode of the posterior distribution reports the 101st sample as the change point. For a medium-shift size, $\delta = +6$, around one-and-a-half standard deviation, the posterior mode concentrates on the 100th sample whereas the c-chart signals with 38 samples delay. The Poisson EWMA chart detects the shifts +2 and +6, after 42 and 13 samples, where the posterior distributions report the 103rd and 100th samples as the change points, respectively. This result implies that although the obtained posterior modes overestimate the change point for small shifts, they still perform relatively better than the Poisson EWMA chart. The resultant posteriors from a Poisson CUSUM are almost identical to those from Poisson EWMA. Bayesian estimates of the magnitude of the change tend to estimate small shifts almost precisely. However, the medium shift sizes are underestimated, although this slight bias must be considered in the context of their corresponding standard deviations.

Applying the Bayesian framework enables us to construct probability-based intervals around estimated parameters. A credible interval (CI) is a posterior probability-based interval which involves those values of highest probability in the posterior density of the parameter of interest. Table 2 presents 50% and 80% credible intervals for the estimated time and the magnitude of step changes in all three control charts. As expected, the CIs are affected by the dispersion and higher order behaviors of the posterior distributions. Under the same probability of 0.8 for the c-chart, the CI for the time of the step change of size $\delta = +2$ covers 53 samples around the 100th sample whereas it decreases to 6 samples for $\delta = +6$ due to the smaller standard deviation (see Table 1).

The comparison of the 50% and 80% CIs for the estimated time of a step change of size $\delta = +6$ in the Poisson EWMA chart reveals that the posterior distribution of the time is highly left skewed and the increase in the probability contracts the left boundary of the interval, from 96.9 to 88 in comparison with the shift in the right boundary. This investigation can be extended to other shift sizes and control chart scenarios for the time estimates. As shown in Table 1 and discussed above, the magnitude of the changes are not estimated as precisely as the time. However, Table 2 shows that in most cases for $\delta = +2$, the real size of change are contained in the respective posterior 50% and 80% CIs.

Having a distribution for the time of the change enables us to make other probabilistic inferences. As an example, Table 3 shows the probability of the occurrence of the change point in the last 10, 25, and 50 observed samples prior to signalling in the control charts. For a step change of size $\delta = +2$, since the c-chart signals very late (see Table 1), it is unlikely that the change point occurred in the last 10, 25, and even 50 samples. In contrast, in the Poisson EWMA and CUSUM charts, where they both signal earlier than the c-chart, the probabilities of occurrence in the last 10 samples are 0.55 and 0.59, then increase to 0.76 and 0.82, respectively as the next 15 samples are included. In

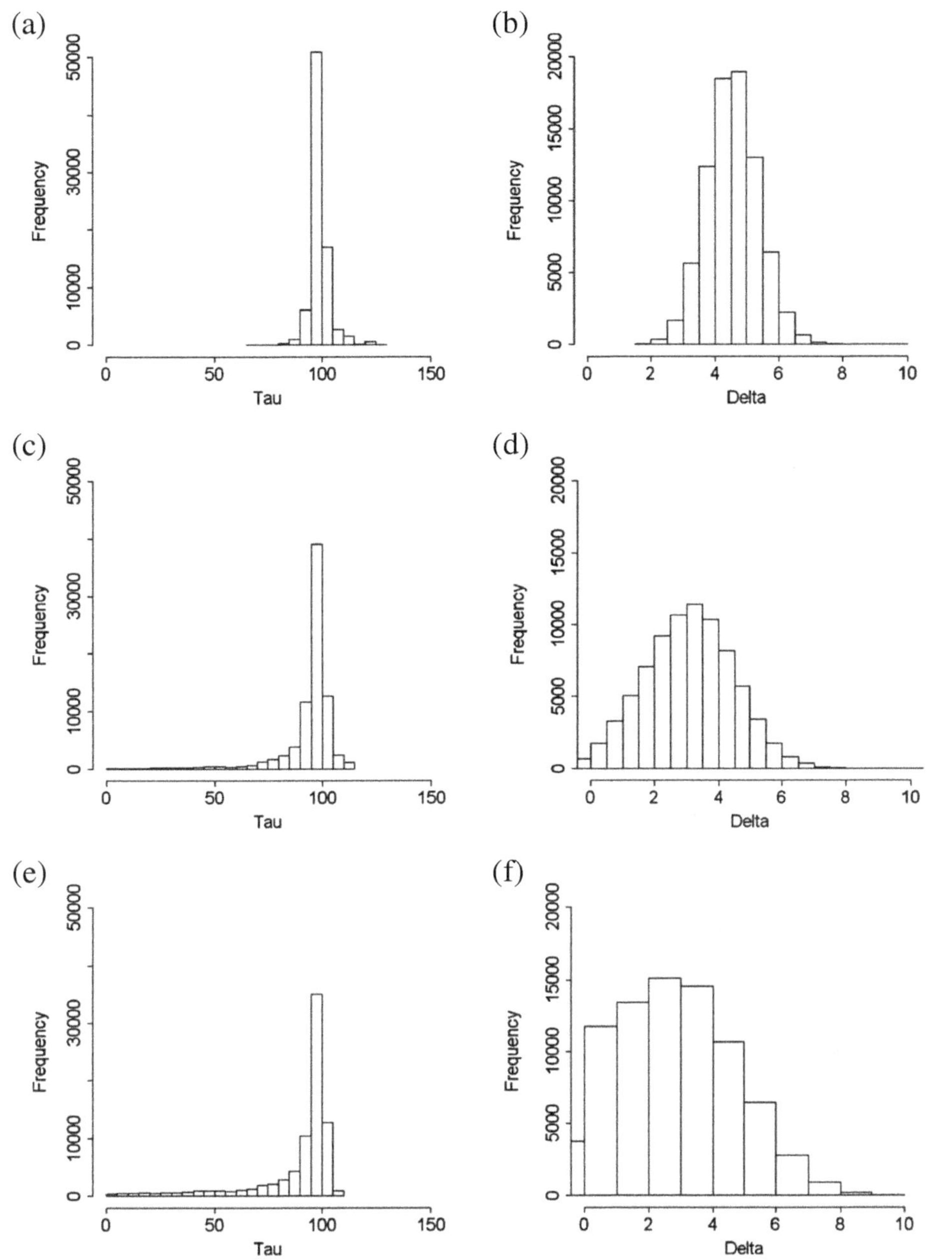

Figure 1 Posterior distributions of the time τ and the magnitude δ of a step change ($\lambda_0 = 20, \delta = +6, \tau = 100$) following signals from (a, b) c-chart, (c, d) Poisson EWMA and (e, f) Poisson CUSUM.

the case of $\delta = +6$, most of the probability density (0.98) is located between the last 25 and 50 samples for the c-chart, whereas with 0.80, it is between the last 10 and 25 samples for the Poisson EWMA chart and with probability 0.91, it is in the last 10 samples for the Poisson CUSUM chart. These kinds of probability computations and inferences can be extended to other change scenarios.

Table 4 shows the average of the estimated parameters obtained from the 100 replicated datasets. As seen, although the c-chart detects small to medium shifts with a large delay, it performs better where there exists a jump. Having a longer delay in the detection of a decrease in the Poisson rate in comparison with an increase of the same size in the c-chart is due to the equality of mean and the variance of the Poisson distribution. Therefore, a fall in the mean leads to less dispersed observations. The Poisson EWMA and CUSUM charts behave in the same manner.

For a step change of size around half a standard deviation ($\delta = \pm 2$) in the Poisson rate, the average of the modes, $E(\hat{\tau})$, reports the 103rd sample as the change point

Table 1 Posterior estimates (mode, SD) of step change point model parameters τ and δ following signals (RL)

δ	c-chart			Poisson EWMA			Poisson CUSUM		
	RL	$\hat{\tau}$	$\hat{\delta}$	RL	$\hat{\tau}$	$\hat{\delta}$	RL	$\hat{\tau}$	$\hat{\delta}$
+2	201	101.1 (16.3)	2.15 (0.46)	142	103 (24.1)	2.03 (0.92)	108	103.1 (16.6)	2.50 (2.1)
+6	138	100 (4.2)	4.5 (0.8)	113	100 (13.7)	3.1 (1.4)	106	100 (20.1)	5.8 (2.7)

From c-, Poisson EWMA ($r = 0.1$ and $A^{\pm} = 2.67$), and Poisson CUSUM charts $(k^+,h^+) = (22.4,22),(k^-,h^-) = (17.4,14)$, where $\lambda^0 = 20$ and $\tau = 100$. Standard deviations are shown in parentheses.

in all three control charts, whereas the charts detect the changes with delays greater than 17 samples, obtained in the Poisson CUSUM. This superiority persists where a medium shift of size $\delta = \pm 6$ has occurred in the process mean. As expected, for large-shift sizes ($\delta = \pm 15$), around three standard deviations, all control charts perform well, yet the mean of modes outperform them by a delay of less than one observation.

Table 4 reveals that in all three control charts, the variation of Bayesian estimates for time tends to reduce when the magnitude of shift in the process mean increases. However, by the nature of the Poisson distribution, for small to medium drops, $\delta = (-2, -6)$, the observed variation is less than those obtained in the estimation of jumps. The mean of the standard deviation of the posterior estimates of time, $E(\sigma_{\hat{\tau}})$, also decreases by moving for small-shift sizes to medium and large sizes in the Poisson EWMA and CUSUM charts. In contrast, the greatest variation is obtained for a large shift of size $\delta = \pm 15$ in the c-chart. This is due to the early detection of such shifts by the c-chart that leads to a very short run of samples after the change which then compresses the data and hence informs the MCMC algorithm.

The average of the Bayesian estimates of the magnitude of the change, $E(\hat{\delta})$, shows that the modes of posteriors for change sizes do not perform as well as the posterior distributions of the time across different shift sizes; however, promising results are obtained where a small shift, $\delta = \pm 2$, has occurred in the process mean. This estimator tends to underestimate the sizes, particularly where there exists a jump. This bias increases when the shift size increases since a very short run of samples coming from the out-of-control state of the process with a high variance was used. As seen in Table 4, the best estimates are obtained in Poisson EWMA cases. Having said that, Bayesian estimates of the magnitude of the change must be studied in conjunction with their corresponding standard deviations. In this manner, analysis of credible intervals would be effective.

Linear trend model

Posterior distributions and probabilistic intervals and inferences are obtainable for the time and the slope of the linear trend, similar to the step change scenario illustrated in section 'Step change model'. We, here, limited our results to the replication study over the simulated datasets outlined in section 'Evaluation'.

Table 5 shows the average of the estimated parameters over trends with slope sizes of $\beta = \{\pm 0.50, \pm 1.0, \pm 2.0\}$. For a linear trend with small slopes of size $\beta = \pm 0.5$ in the Poisson rate, the average modal value, $E(\hat{\tau})$, reports the 105th sample and less as the change point in all three control charts, whereas the charts detect the changes with delays greater than 10 samples, obtained in the Poisson CUSUM. This superiority also persists where a trend with larger slopes of size $\beta = \pm 1.0, \pm 2.0$ has occurred in the process mean. In these scenarios, the bias of the Bayesian estimator does not exceed two and one samples, where the minimum delays are seven and four samples, respectively.

Table 5 shows that in all three control charts, the variation of the Bayesian estimates for time tends to reduce when the magnitude of slope increases. The mean of

Table 2 Credible intervals for step change point model parameters τ and δ

δ		c-chart		Poisson EWMA		Poisson CUSUM	
		50%	80%	50%	80%	50%	80%
+2	$\hat{\tau}$	(101,105)	(65,118)	(96.6,114)	(71.2,125.8)	(98.2,105)	(65.2,108)
	$\hat{\delta}$	(2.1,3.2)	(1.9,3.2)	(1.41,2.65)	(0.76,3.08)	(0.12,2.50)	(-0.23,4.8)
+6	$\hat{\tau}$	(97.9,100)	(96,102)	(96.9,101)	(88,103)	(96,101)	(83,106)
	$\hat{\delta}$	(3.9,5.2)	(3.4,5.5)	(2.2,4.1)	(1.2,4.8)	(1.31,4)	(0.05,4.9)

Following signals from c-, Poisson EWMA ($r = 0.1$, and $A^{\pm} = 2.67$) and Poisson CUSUM charts ($(k^+,h^+) = (22.4,22),(k^-,h^-) = (17.4,14)$), where $\lambda_0 = 20$ and $\tau = 100$.

Table 3 Probability of occurrence of change point in last 10, 25, and 50 observed samples prior to signalling

δ	c-chart			Poisson EWMA			Poisson CUSUM		
	10	25	50	10	25	50	10	25	50
+2	0.00	0.00	0.01	0.55	0.76	0.86	0.59	0.82	0.91
+6	0.00	0.01	0.99	0.06	0.86	0.95	0.91	0.97	0.99

For c-, Poisson EWMA ($r = 0.1$ and $A^{\pm} = 2.67$), and Poisson CUSUM charts ($(k^+,h^+) = (22.4,22),(k^-,h^-) = (17.4,14)$), where $\lambda_0 = 20$ and $\tau = 100$.

posterior standard deviation for time, $E(\sigma_{\hat{\tau}})$, also decreases by moving for small slope sizes to medium and large sizes in both directions. However, the observed variation for estimation of a decreasing trend is less than those obtained for an increasing trend with the same slope size.

The average of the posterior estimates for the magnitude of the change, $E(\hat{\beta})$, shows that the modes of the posteriors for change sizes perform as well as the posterior estimates of the time, particularly, for the c-chart and Poisson EWMA chart. In the CUSUM chart, the posteriors tend to underestimate the slope sizes.

Multiple change model

We applied the multiple change point model following signals of the c-chart as the Poisson EWMA and CUSUM mostly signalled before simulating the second change in the process. Although results here are limited to the replication study, distribution of parameters and probabilistic inferences can easily be constructed.

As seen in Table 6 and discussed in section 'Step change model', the c-chart signals earlier when a larger shift, either an increase or decrease, has occurred in the second change; however, it performs better where there exists a jump, regardless of the direction of the first change. The chart alarmed after 38 samples when two consecutive drops of sizes around one and two standard deviations, $\delta_{1,2} = (-4, -8)$, occurred. Although this delay falls to 16 samples when the second change has happened in the opposite direction, the modes of posteriors for the time of the first change, $E(\hat{\tau}_1)$, outperform the chart. This superiority persists when the size of the second change increases to around three standard deviations, $\delta_2 = (\pm 12)$. The same results are also obtained, where the first change is an increase in the magnitude of one standard deviation, $\delta_1 = (+4)$.

Table 6 reveals that the Bayesian estimator tends to underestimate the time of the first change of two monotonic changes where the second change is of size $\delta_2 = (\pm 12)$. The associated variation, within replications, increases when the second step change increases in the same direction of the first change. The minimum variations of the posterior distributions for the time of the first change, $E(\hat{\sigma}_{\hat{\tau}_1})$, are obtained where there exist non-monotonic changes (see $\delta_{1,2} = (-4, +8)$ and $\delta_{1,2} = (+4, -8)$). This variation also increases when the second step change increases in the same direction of the first change.

The time of the second step change is estimated precisely by the posterior modes. Table 6 shows that

Table 4 Average of posterior estimates (mode, SD) of step change point model parameters τ and δ following signals (RL)

δ	c-chart				Poisson EWMA				Poisson CUSUM			
	$E(\text{RL})$	$E(\hat{\tau})$	$E(\sigma_{\hat{\tau}})$	$E(\hat{\delta})$	$E(\text{RL})$	$E(\hat{\tau})$	$E(\sigma_{\hat{\tau}})$	$E(\hat{\delta})$	$E(\text{RL})$	$E(\hat{\tau})$	$E(\sigma_{\hat{\tau}})$	$E(\hat{\delta})$
−15	101.17	100.45	22.48	−6.43	102.36	100.40	3.28	−11.49	101.13	100.46	23.21	−6.21
	(0.42)	(0.36)	(8.42)	(4.75)	(0.67)	(0.38)	(5.33)	(2.79)	(0.33)	(0.36)	(7.39)	(4.76)
−6	174.65	101.12	3.26	−5.90	106.43	100.72	14.92	-4.37	103.94	100.74	24.90	−2.10
	(66.38)	(1.72)	(4.48)	(1.06)	(2.84)	(0.76)	(7.79)	(2.47)	(2.36)	(0.76)	(5.31)	(2.13)
−2	663.24	103.05	21.33	−2.03	124.72	103.50	24.45	−2.11	127.54	103.23	27.13	−1.64
	(517.23)	(2.78)	(8.06)	(0.38)	(18.74)	(2.91)	(6.92)	(0.98)	(26.82)	(2.91)	(6.30)	(0.80)
+2	184.22	102.66	20.50	2.00	119.72	103.00	23.26	2.05	117.77	102.70	24.79	1.85
	(88.91)	(3.83)	(9.17)	(0.83)	(16.08)	(3.18)	(7.49)	(0.82)	(18.75)	(3.20)	(7.15)	(0.76)
+6	113.44	101.10	13.54	3.73	106.33	101.20	18.00	3.00	105.30	101.22	23.61	1.89
	(13.17)	(1.67)	(9.71)	(2.45)	(2.87)	(1.31)	(7.35)	(2.26)	(2.55)	(1.32)	(5.37)	(1.80)
+15	101.51	100.48	22.00	3.81	102.56	100.51	10.33	7.52	101.77	100.50	19.43	4.84
	(0.96)	(0.30)	(7.69)	(4.17)	(0.89)	(0.29)	(8.15)	(3.78)	(0.60)	(0.30)	(6.68)	(4.21)

From c-, Poisson EWMA ($r = 0.1$ and $A^{\pm} = 2.67$), and Poisson CUSUM charts ($(k^+,h^+) = (22.4,22),(k^-,h^-) = (17.4,14)$), where $\lambda_0 = 20$ and $\tau = 100$. Standard deviations are shown in parentheses.

Table 5 Average of posterior estimates (mode, SD) of linear trend change point model parameters τ and β following signals (RL)

β	c-chart				Poisson EWMA				Poisson CUSUM			
	$E(\text{RL})$	$E(\hat{\tau})$	$E(\sigma_{\hat{\tau}})$	$E(\hat{\beta})$	$E(\text{RL})$	$E(\hat{\tau})$	$E(\sigma_{\hat{\tau}})$	$E(\hat{\beta})$	$E(\text{RL})$	$E(\hat{\tau})$	$E(\sigma_{\hat{\tau}})$	$E(\hat{\beta})$
−2.0	106.48	100.83	2.27	−2.07	105.35	100.75	3.73	−1.72	104.07	100.92	5.29	−1.51
	(1.47)	(1.16)	(2.14)	(1.22)	(1.14)	(0.93)	(3.25)	(0.95)	(1.15)	(0.96)	(3.08)	(1.59)
−1.0	111.24	102.05	2.80	−1.35	108.01	102.14	5.96	−1.02	106.46	102.74	7.62	−0.78
	(2.56)	(2.36)	(1.73)	(0.75)	(1.76)	(2.07)	(5.33)	(0.71)	(1.92)	(2.18)	(4.82)	(1.06)
−0.5	120.08	102.96	4.19	−0.67	111.65	104.60	8.93	−0.50	109.67	104.70	9.40	−0.47
	(4.96)	(2.50)	(1.54)	(0.55)	(2.51)	(2.91)	(5.28)	(0.64)	(3.06)	(2.91)	(4.77)	(0.74)
+0.5	113.93	103.75	6.66	0.43	110.98	104.45	8.75	0.43	109.82	104.78	8.83	0.49
	(5.22)	(2.99)	(3.15)	(0.55)	(2.56)	(2.94)	(5.12)	(0.62)	(2.89)	(2.78)	(4.65)	(0.54)
+1.0	109.20	102.55	5.65	0.79	107.92	102.75	7.21	0.68	107.19	102.78	7.88	0.74
	(3.14)	(2.05)	(3.87)	(0.77)	(2.13)	(2.11)	(5.93)	(0.81)	(2.09)	(2.36)	(4.87)	(0.70)
+2.0	105.46	101.20	4.93	1.48	105.52	101.18	5.21	1.75	104.82	101.19	6.19	1.66
	(1.88)	(1.02)	(3.21)	(0.94)	(1.35)	(1.04)	(4.06)	(1.05)	(1.30)	(1.04)	(3.59)	(0.88)

From c-, Poisson EWMA ($r = 0.1$ and $A^{\pm} = 2.67$), and Poisson CUSUM charts ($(k^+, h^+) = (22.4, 22), (k^-, h^-) = (17.4, 14)$), where $\lambda_0 = 20$ and $\tau = 100$. Standard deviations are shown in parentheses.

the average, $E(\hat{\tau}_1)$, mostly concentrate on the 110th sample. Surprisingly, the variation between replications and also the variation of posterior distributions obtained for the time of the second change, $E(\hat{\sigma}_{\hat{\tau}_1})$, are less than those obtained for the first step change.

The average of the posterior estimates of the magnitude of the changes, $E(\hat{\delta}_1)$ and $E(\hat{\delta}_2)$, shows that the modes of the posteriors for change sizes do not perform as well as the posterior distributions of the time across different scenarios. The modes tend to underestimate the sizes, particularly, for jumps in either the first or the second step

Table 6 Average of posterior estimates (mode, SD) of multiple step change point model parameters $\tau_{1,2}$ and $\delta_{1,2}$ following signals (RL)

δ_1, δ_2	$E(RL)$	$E(\hat{\tau}_1)$	$E(\hat{\sigma}_{\hat{\tau}_1})$	$E(\hat{\delta}_1)$	$E(\hat{\tau}_2)$	$E(\hat{\sigma}_{\hat{\tau}_2})$	$E(\hat{\delta}_2)$
−4, −12	113.49	98.18	28.50	−1.17	109.26	7.08	−8.32
	(2.46)	(15.29)	(4.19)	(1.24)	(1.69)	(5.64)	(3.07)
−4, −8	138.94	101.04	28.84	−1.18	109.06	5.48	−7.69
	(25.94)	(7.60)	(5.31)	(1.33)	(2.06)	(3.36)	(1.65)
−4,+8	116.98	100.74	23.30	−2.48	110.49	8.14	4.69
	(5.75)	(6.52)	(9.20)	(1.80)	(0.60)	(8.90)	(3.63)
−4,+12	112.78	100.37	23.86	−2.21	110.30	11.08	5.42
	(2.29)	(6.75)	(8.27)	(1.75)	(1.06)	(8.90)	(4.87)
+4,−12	113.10	101.77	25.03	1.64	110.48	7.23	−9.34
	(2.50)	(3.40)	(7.67)	(1.74)	(0.34)	(8.91)	(3.28)
+4,−8	134.69	101.67	24.92	1.41	110.71	3.36	−7.7
	(22.74)	(7.04)	(6.78)	(1.38)	(0.73)	(5.59)	(2.03)
+4,+8	117.69	101.28	30.40	0.59	108.81	11.61	4.41
	(5.69)	(11.10)	(2.53)	(0.93)	(1.96)	(7.35)	(2.50)
+4,+12	112.23	98.32	29.93	0.09	108.37	11.90	4.17
	(2.50)	(15.32)	(3.19)	(1.13)	(2.00)	(8.15)	(2.71)

From c-, Poisson EWMA ($r = 0.1$ and $A^{\pm} = 2.67$), and Poisson CUSUM charts ($(k^+, h^+) = (22.4, 22), (k^-, h^-) = (17.4, 14)$), where $\lambda_0 = 20$ and $\tau = 100$. Standard deviations are shown in parentheses.

change. However, there still exists some gain in studying the estimated sizes and directions, particularly when the obtained standard deviations are also considered.

Comparative performance and model selection

We used Monte Carlo simulation to study the performance of the developed change point models in different change point scenarios following a signal from a c-chart. We generated 100 observations of a Poisson process with an in-control rate of $\lambda_0 = 20$. We then induced a step, a linear trend, and a multiple change in the Poisson rate. For each scenario the three change point models were applied and the time of the change was estimated. Based on the MCMC simulation, the Deviance Information Criterion (DIC) and related parameters, mean and variance of the posterior distribution of the deviance and the penalty value, were recorded. The DIC is a goodness of fit criterion which takes into account the deviance of the model, $-2\log(p(y \mid \theta))$, and a penalty for the model complexity, p_D (Spielgelhalter et al. 2002). To allow for asymmetry in the posterior distribution, seen in Figure 1, p_V was used as an alternative to p_D, where p_V is half of the variance of the posterior distribution of the deviance (Gelman et al. 2004).

Table 7 indicates that the Bayesian estimate of a step change outperforms other Bayesian estimates, linear and multiple, where there is a step change in the process parameter. It estimates 101.9 and 108.3 as the times of change of sizes $\delta = -4$ and $\delta = +4$, respectively, whereas the linear model underestimates the time with a bias of around 55 and 24 samples and the multiple model tends to overestimate it relative to the step model. According to the reported DICs, the DIC_V supports that the step model with values of 1,167 and 845.5 is a preferable fit where there exists either an increasing or a decreasing step change.

In the case of an occurrence of a linear trend shift in the Poisson rate, the Bayesian estimate of a linear trend change outperforms other Bayesian estimates in estimating the change point. The reported DIC_V is convincing that the linear model with values of 603.7 and 630.9 is also the best fit. These results can be extended to the multiple change scenario. Table 7 shows that the Bayesian estimate of a multiple change (two changes) outperforms other Bayesian estimates, step and linear, where there are two consecutive changes in the Poisson rate. Similarly, the reported DIC_V supports that the multiple model with values of 636.4 and 791.2 is also the best fit in this case.

Comparison of Bayesian estimator with other methods

To study the performance of the proposed Bayesian estimators in comparison with those introduced in

Table 7 Performance and goodness of change point models on different change types following signal

Change type	Change size	RL	Model	$\hat{\tau}$	Deviance	SD	p_D	DIC_D	DIC_V
			Step	101.9	1163.4	2.7	1.9	1165.3	1167
Step	$\delta = -4$	200	Linear	45.2	1168.9	2.1	2.1	1171	1171.1
			Multiple	102.5	1163.7	3.3	−0.8	1162.9	1169.1
			Step	108.3	842.2	2.6	0.8	843	845.5
Step	$\delta = +4$	148	Linear	86.3	843.7	2.1	2.1	845.8	845.9
			Multiple	108.5	841.7	3.2	−1.7	840	846.8
			Step	102.3	607	3.9	−2.1	604.9	614.6
Linear	$\beta = -1$	107	Linear	101.7	601.9	1.9	1.5	603.4	603.7
			Multiple	102.4	604.9	3.2	−0.5	604.4	610.0
			Step	102.1	631.5	3.8	0.9	632.4	638.7
Linear	$\beta = +1$	108	Linear	100.4	629.1	1.9	0.3	629.5	630.9
			Multiple	101.5	628.4	4.0	−1.2	627.2	636.4
			Step	109.5	788.8	2.7	−9.9	778.9	792.4
Multiple	$\delta_1 = -4$ $\delta_2 = -8$	138	Linear	88.1	788.2	3.3	1.9	790.1	793.6
			Multiple	100.1	784	3.8	0.8	784.9	791.2
			Step	100.5	723.1	2.8	0.0	723.1	727.0
Multiple	$\delta_1 = +4$ $\delta_2 = +8$	119	Linear	108.3	722.8	3.1	−0.7	658.9	727.6
			Multiple	100.3	722.6	2.9	1.7	722.3	726.8

From a c-chart where $\lambda_0 = 20$, $\tau_1 = 100$ and $\tau_2 = 110$.

Table 8 Average of estimated time of a step change in a Poisson process obtained by Bayesian estimators, CUSUM and EWMA built-in estimators, and ML estimator following signals (RL)

δ	c-chart			Poisson EWMA				Poisson CUSUM			
	$E(RL)$	$E(\hat{\tau}_{MLE})$	$E(\hat{\tau}_b)$	$E(RL)$	$E(\hat{\tau}_{EWMA})$	$E(\hat{\tau}_{\mathrm{MLE}})$	$E(\hat{\tau}_b)$	$E(RL)$	$E(\hat{\tau}_{CUSUM})$	$E(\hat{\tau}_{MLE})$	$E(\hat{\tau}_b)$
−15	101.17	99.97	100.45	102.36	95.42	99.98	100.40	101.13	99.56	99.96	100.46
	(0.42)	(0.22)	(0.36)	(0.67)	(10.12)	(0.20)	(0.38)	(0.33)	(0.89)	(0.24)	(0.36)
−6	174.65	100.19	101.12	106.43	96.73	99.65	100.72	103.94	100.08	97.78	100.74
	(66.38)	(1.72)	(1.72)	(2.84)	(6.41)	(2.19)	(0.76)	(2.36)	(1.69)	(13.16)	(0.76)
−2	663.24	93.16	103.05	124.72	97.86	102.70	103.50	127.54	122.70	103.56	103.23
	(517.23)	(19.76)	(2.78)	(18.74)	(17.80)	(17.91)	(2.91)	(26.82)	(26.51)	(15.62)	(2.91)
+2	184.22	94.20	102.66	119.72	100.87	96.75	103.00	117.77	109.12	96.89	102.70
	(88.91)	(22.15)	(3.83)	(16.80)	(13.51)	(17.12)	(3.18)	(18.75)	(19.63)	(21.09)	(3.20)
+6	113.44	100.55	101.10	106.33	95.94	99.31	101.20	105.30	99.75	99.29	101.22
	(13.17)	(2.65)	(1.65)	(2.87)	(10.04)	(7.81)	(1.31)	(2.55)	(2.36)	(7.79)	(1.32)
+15	101.51	99.95	100.48	102.56	94.95	99.51	100.51	101.77	98.92	99.51	100.50
	(0.96)	(0.45)	(0.30)	(0.89)	(9.09)	(4.02)	(0.29)	(0.60)	(2.32)	(4.02)	(0.30)

From c-, Poisson EWMA ($r = 0.1$ and $A^{\pm} = 2.67$), and Poisson CUSUM charts ($(k^+,h^+) = (22.4,22),(k^-,h^-) = (17.4,14)$), where $\lambda_0 = 20$ and $\tau = 100$. Standard deviations are shown in parentheses.

section 'Background', we run the alternatives, built-in estimators of Poisson EWMA and CUSUM charts and ML estimators, within replications as discussed in section 'Evaluation'.

Table 8 shows the mean of Bayesian estimates and detected change points provided by built-in estimators of EWMA (Nishina 1992) and CUSUM (Page 1954) charts and the ML estimator (Perry 2004) for a step change in a Poisson process.

Although the Bayesian estimator, $\hat{\tau}_b$, tends to overestimate the time of a step change of small sizes, $\delta = \pm 2$, with a delay of three samples, it outperforms the ML estimator, $\hat{\tau}_{\mathrm{MLE}}$, which underestimates the time by six samples following a signal from the c-chart. For step sizes of one and half and three standard deviations, the ML estimator performs slightly better than the Bayesian estimator; however considering the obtained standard deviations decreases this superiority, particularly where there exists a jump in the process mean.

Table 8 reveals that the EWMA estimator, $\hat{\tau}_{\mathrm{EWMA}}$, underestimates the change point when the size of shift increases for both directions where the Bayesian estimator tends to be more precise. $\hat{\tau}_b$ still remains the best estimator for small changes and shows acceptable performance in comparison with $\hat{\tau}_{\mathrm{MLE}}$ over larger shifts, particularly when the standard deviations are taken into account.

The CUSUM estimator, $\hat{\tau}_{\mathrm{CUSUM}}$, outperforms the equivalent estimators in EWMA for larger shifts, $\delta = (\pm 6, \pm 15)$; however, it overestimates the time of small shifts significantly. Similar to c-chart and EWMA cases, in CUSUM, the Bayesian estimator outperforms alternatives for small shifts and offers acceptable performance over other shift sizes, considering the obtained standard deviations over replications.

Table 9 shows the mean of the Bayesian estimates and detected change points provided by built-in estimators of EWMA (Nishina 1992) and CUSUM (Page 1954) charts and the ML estimator (Perry et al. 2006) for a linear trend change in a Poisson process. Application of the proposed ML estimator is restricted to trends with a positive slope as Newton's method is not tractable for decreasing trends in Poisson mean; see Perry et al. (2006) for more details.

The Bayesian estimator, $\hat{\tau}_b$, almost outperforms the built-in estimator of EWMA, $\hat{\tau}_{\mathrm{EWMA}}$, where there exists a decreasing trend. This superiority increases when the slope size raises, $\beta = -2$. The CUSUM estimator, $\hat{\tau}_{\mathrm{CUSUM}}$, estimates the change point more precisely than the EWMA; however the Bayesian estimator, $\hat{\tau}_b$, still remains the best alternative for detection of linear trends with negative slopes, when the variation of the estimates is taken into account.

Table 9 reveals that the Bayesian estimator, $\hat{\tau}_b$, is slightly outperformed by the ML estimator, $\hat{\tau}_{\mathrm{MLE}}$, across the charts when there exists an increasing linear trend in the process mean. Having said that, the Bayesian estimator can still be a reasonable alternative in light of the obtained standard deviations which are less than those observed from the ML estimator over replications.

The ML estimator proposed by Perry et al. (2007a) is suitable for monotonic consecutive changes. In contrast, the Bayesian estimator for a known number of change points proposed in section 'Multiple change model' can

Table 9 Average of estimated time of linear trend in Poisson process obtained by Bayesian estimator, CUSUM and EWMA built-in estimators, and ML estimator following signals (RL)

β	c-chart			Poisson EWMA				Poisson CUSUM			
	$E(RL)$	$E(\hat{\tau}_{MLE})$	$E(\hat{\tau}_b)$	$E(RL)$	$E(\hat{\tau}_{EWMA})$	$E(\hat{\tau}_{\text{MLE}})$	$E(\hat{\tau}_b)$	$E(RL)$	$E(\hat{\tau}_{CUSUM})$	$E(\hat{\tau}_{MLE})$	$E(\hat{\tau}_b)$
−2.0	106.48	-	100.83	105.35	97.59	-	100.75	104.07	100.36	-	100.92
	(1.47)	-	(1.16)	(1.14)	(6.11)	-	(0.93)	(1.15)	(1.82)	-	(0.96)
−1.0	111.24	-	102.05	108.01	97.34	-	102.14	106.46	102.49	-	102.74
	(2.56)	-	(2.36)	(1.76)	(9.92)	-	(2.07)	(1.92)	(2.74)	-	(2.18)
−0.5	120.08	-	102.96	111.65	97.61	-	104.60	109.67	104.94	-	104.70
	(4.96)	-	(2.50)	(2.51)	(12.03)	-	(2.91)	(3.06)	(3.74)	-	(2.91)
+0.5	113.93	103.55	103.75	110.98	99.37	102.02	104.45	109.82	104.00	102.12	104.78
	(5.22)	(3.48)	(2.99)	(2.56)	(9.12)	(9.23)	(2.94)	(2.89)	(3.30)	(11.68)	(2.78)
+1.0	109.20	102.70	102.55	107.92	97.13	101.08	102.75	107.19	101.07	101.57	102.78
	(3.14)	(3.19)	(2.05)	(2.13)	(9.70)	(12.42)	(2.11)	(2.09)	(3.01)	(3.59)	(2.36)
+2.0	105.46	100.23	101.20	105.52	96.35	100.57	101.18	104.82	99.61	100.59	101.19
	(1.88)	(2.81)	(1.02)	(1.35)	(8.80)	(4.07)	(1.04)	(1.30)	(3.47)	(3.81)	(1.04)

From c-, Poisson EWMA ($r = 0.1$ and $A^{\pm} = 2.67$), and Poisson CUSUM charts ($(k^+,h^+) = (22.4,22),(k^-,h^-) = (17.4,14)$), where $\lambda_0 = 20$ and $\tau = 100$. Standard deviations are shown in parentheses.

also be applied where there exists non-monotonic consecutive changes in the process mean. Therefore, the comparison study was not followed for the multiple change point case as there is no appropriate ML alternative against which to evaluate the Bayesian estimator. Similarly, the built-in EWMA and CUSUM estimators cannot be studied as they tend to signal before the second change point. In the case of signalling after the second change, they also failed as they tend to concentrate on the time of the latter step change as the change point in non-monotonic scenarios.

Apart from the accuracy and precision criteria used for the comparison study, the posterior distributions for the time and the magnitude of a change enable us to construct probabilistic intervals around estimates and probabilistic inferences about the location of change point as discussed in section 'Step change model' for the step change scenario. This is a significant advantage of the proposed Bayesian approach. Although similar results may be obtained when resampling in conjunction with ML methods, the inferential basics of this approach is more limited; see Bernardo and Smith (1994) for more details. Furthermore, flexibility of Bayesian hierarchical models, ease of extension to more complicated change scenarios such as combination of steps and linear and nonlinear trends, relief of analytic calculation of likelihood function, particularly for non-tractable likelihood functions and ease of coding with available packages should be considered as additional benefits of the proposed Bayesian change point model for monitoring purposes. This approach can be easily applied for other types of data and processes such as Bernoulli, normal and exponential family data and the mentioned advantages remain achievable.

The two-step approach to change point identification described in this paper has the advantage of building on control charts that may be already in place in practice. An alternative may be to retain the two-step approach but to use a Bayesian framework in both stages. There is now a substantial literature on Bayesian formulation of control charts and extensions such as monitoring processes with varying parameters (Feltz and Shiau 2001), over-dispersed data (Bayarri and Garcia-Donato 2005), start-up and short runs (Tsiamyrtzis and Hawkins 2005, 2008). A further alternative is to consider a fully Bayesian, one-step approach, in which both the monitoring of the in-control process and the retrospective or prospective identification of changes is undertaken in the one analysis. This is the subject of further research.

Conclusion

Identification of the time when a process has changed enables process engineers to pursue investigation of special causes more effectively. Indeed, knowing the change point restricts the search efforts to a tighter window of observations and related variables. In this paper, we modeled the change point estimation for a Poisson process in a Bayesian framework. We considered three scenarios of changes, a step change, a linear trend, and a multiple change when the number of changes is known. We constructed Bayesian hierarchical models and derived posterior distributions for change point estimates using MCMC. We compared the performance of the Bayesian estimators with c-, Poisson EWMA, and CUSUM control charts. The results showed that the Bayesian estimates outperform standard control charts in change estimation, particularly where there exists a small to medium size of step change(s) and a linear trend change with small to relatively large magnitude of slope. In comparison with built-in estimators of EWMA and CUSUM and ML-based estimators, the Bayesian estimator performs reasonably well and remains a strong alternative, particularly when other criteria such as probability quantification through credible intervals and probabilistic inferences, flexibility, generalization, and simplicity are taken into accounts.

The investigation on the performance of the Bayesian estimates over different change scenarios reveals that each Bayesian change point model outperforms other models where its underlying change type has occurred in the Poisson process. The results also support the idea of using DIC as a primary step in change point estimation which can direct process engineers to identify the appropriate change point model before making inferences about the derived underlying changes in the process.

Appendix

The following are the different change model codes:

1. Step change model

```
model {
for(i in 1 : RL_c){
xc[i] ~ dpois(lambda2[i])
lambda1[i]=lambda0+delta × step(i-change)
lambda2[i]=max(lambda1[i],0.000001)}
tau=1/(6 × sqrt(lambda0))
RL=RL_c − 1
delta ~ dnorm(0, tau)
change ~ dunif(1,RL)}
```

2. Linear trend change model

```
model {
for(i in 1 : RL_c){
xc[i] ~ dpois(lambda2[i])
lambda1[i]=lambda0+beta × (i-change) × step(i-change)
lambda2[i]=max(lambda1[i],0.000001)}
tau=1/(6 × sqrt(lambda0))
RL=RL_c − 1
beta ~ dnorm(0, tau)
change ~ dunif(1,RL)}
```

3. Multiple change model

```
model {
for(i in 1 : RL_c){
xc[i] ~ dpois(lambda2[i])
lambda1[i]=lambda0+delta1 × step(i-change1) ×
        step(change2-i)+delta2 × step(i-change2)
lambda2[i]=max(lambda1[i],0.000001)}
tau=1/(6 × sqrt(lambda0))
RL=RL_c − 1
delta1 ~ dnorm(0, tau)
delta2 ~ dnorm(0, tau)
change1 ~ dunif(1,change2)
change2 ~ dunif(change1,RL)}
```

Competing interests

The authors declare that they have no competing interests.

Authors' contributions

HA contributed to the conception, design, and implementation of statistical analysis and writing and modification of the manuscript. RN and KM contributed to the conception, design, and modification of the manuscript. All authors read and approved the final manuscript.

Authors' information

HA is a research fellow at the University of New South Wales. His primary research interests include Bayesian methods and engineering statistics and its applications in statistical quality control. RN is Professor of Industrial Engineering at Iran University of Science and Technology. His primary research interests include statistical process control, process optimization, Six Sigma, and statistical analysis. KM is Professor of Statistics at Queensland University of Technology. Her research interests are Bayesian statistics, mixture models, hierarchical modelling and meta-analysis.

Acknowledgements

The authors gratefully acknowledge the financial support from Queensland University of Technology and St Andrews Medical Institute through an ARC Linkage Project.

Author details

[1]Simpson Centre for Health Services Research, Australian Institute of Health Innovation, Faculty of Medicine, University of New South Wales, Sydney, NSW 2052, Australia. [2]School of Industrial Engineering, Iran University of Science and Technology, Tehran, Iran. [3]Discipline of Mathematical Sciences, Science and Engineering Faculty, Queensland University of Technology, Brisbane, QLD 4001, Australia.

References

Amiri A, Allahyari S (2012) Change point estimation methods for control chart postsignal diagnostics: a literature review. Qual Reliability Eng Int 28(7): 673–685. http://dx.doi.org/10.1002/qre.1266

Amiri A, Koosha M, Azhdari A (2011) Profile monitoring for Poisson responses In: IEEE International Conference on Industrial Engineering and Engineering Management (IEEM), Singapore, Singapore, 6-9 Dec 2011.

Assareh H, Mengersen K (2012) Change point estimation in monitoring survival time. PLoS ONE 7(3):e33630. http://dx.doi.org/10.1371/journal.pone.0033630

Assareh H, Smith I, Mengersen K (2011a) Bayesian change point detection in monitoring cardiac surgery outcomes. Qual Manag Health Care 20(3): 207–222

Assareh, H, Smith I, Mengersen K (2011b) Bayesian estimation of the time of a linear trend in risk-adjusted control charts. IAENG Int J Comput Sci 38(4): 409–417

Assareh H, Smith I, Mengersen K (2011c) Change point detection in risk adjusted control charts. Stat Methods Med Res. http://dx.doi.org/10.1177/0962280211426356

Atashgar K (2013) Identification of the change point: an overview. Int J Adv Manuf Technol 64(9-12):1663–1683. http://dx.doi.org/10.1007/s00170-012-4131-2

Bayarri MJ, Garcia-Donato G (2005) A Bayesian sequential look at u-control charts. Technometrics 47(2):141–151

Benneyan JC (1998) Statistical quality control methods in infection control and hospital epidemiology, Part I: introduction and basic theory. Infect Control Hosp Epidemiol 19(3):194–241

Bernardo JM, Smith AFM (1994) Bayesian theory. Wiley, New York

Borror C C and Champ, Rigdon S (1998) Poisson EWMA control charts. J Qual Technol 30(4):352–361

Brook D, Evans D (1972) An approach to the probability distribution of CUSUM run length. Biometrika 59(3):539–549

Chiu JE, Kuo TI (2007) Attribute control chart for multivariate Poisson distribution. Commun Stat Theory Methods 37:146–158
Feltz CJ, Shiau JJH (2001) Statistical process monitoring using an empirical Bayes multivariate process control chart. Qual Reliability Eng Int 17(2): 119–124
Gardiner J S Montgomery, D C (1987) Using statistical control charts for software quality control. Qual Reliability Eng Int 3:15–20
Gelman A, Carlin J, Stern H, Rubin D (2004) Bayesian Data Analysis. Chapman & Hall/CRC, London
Hou CD, Shao YE, Huang S (2013) A combined MLE and generalized P chart approach to estimate the change point of a multinomial process. Appl Math Inf Sci 7(4):1487–1493
Limayea SS, Mastrangeloa CM, Zerrb DM (2008) A case study in monitoring hospital-associated infections with count control charts. Qual Eng 20(4): 404–413
Montgomery DC (2008) Introduction to Statistical Quality Control. Wiley, New York
Niaki S, Khedmati M (2012) Detecting and estimating the time of a step-change in multivariate Poisson processes. Scientia Iranica 19(3):862–871
Niaki S, Nafar M (2008) An artificial neural network approach to monitor and diagnose multi-attribute quality control processes. J Ind Eng Int 4(7):10–24
Niaki SA, Abbasi B (2008) A transformation technique in designing multi-attribute C control charts. Scientia Iranica 15:125–130
Niaki STA, Khedmati M (2013a) Estimating the change point of the parameter vector of multivariate Poisson processes monitored by a multi-attribute T2 control chart. Int J Adv Manuf Technol 64(9-12):1625–1642
Niaki, S T A, Khedmati M (2013b) Identifying the change time of multivariate binomial processes for step changes and drifts. J Ind Eng Int 9:1–11
Nishina K (1992) A comparison of control charts from the viewpoint of change-point estimation. Qual Reliability Eng Int 8(6):537–541
Noorossana R, Saghaei A, Paynabar K, Abdi S (2009) Identifying the period of a step change in high-yield processes. Qual Reliability Eng Int 25(7):875–883
Page ES (1954) Continuous inspection schemes. Biometrika 41(1/2):100–115
Page, E S (1961) Cumulative sum charts. Technometrics 3:1–9
Perry MB (2004) Robust change detection and change point estimation for Poisson count processes. PhD thesis, Florida State University, USA
Perry MB, Pignatiello JJ, Simpson J (2006) Estimating the change point of a Poisson rate parameter with a linear trend disturbance. Qual Reliability Eng Int 22(4):371–384
Perry MB, Pignatiello JJ, Simpson J (2007a) Change point estimation for monotonically changing Poisson rates in SPC. Int J Prod Res 45:1791–1813
Perry MB, Pignatiello JJ, Simpson JR (2007b) Estimating the change point of the process fraction non-conforming with a monotonic change disturbance in SPC. Qual Reliability Eng Int 23(3):327–339
Plummer M, Best N, Cowles K, Vines K (2010) The Coda package: Output Analysis and Diagnostics for MCMC, R Package Version 0.13-2. Available: CRAN. R-project. org/package= coda (April 2010). Citeseer
Roberts S (1959) Control chart tests based on geometric moving averages. Technometrics 1:239–250
Samuel T, Pignatiello J (1998) Identifying the time of a change in a poisson rate parameter. Qual Eng 10(4):673–681
Sharafi A, Aminnayeri M, Amiri A (2013) An MLE approach for estimating the time of step changes in Poisson regression profiles. Scientia Iranica 20(3): 855–860
Shewhart W (1926) Quality control charts. Bell Syst Tech J 5:593–602
Shewhart, W (1927) Quality control. Bell Syst Tech J 6:722–735
Spielgelhalter D, Best NCB, Van Der Linde A (2002) Bayesian measures of model complexity and fit. J R Stat Soc Series B (Methodological) 64(4):583–639
Spielgelhalter D, Thomas A, Best N (2003) WinBUGS version 1.4. Bayesian inference using Gibbs sampling. MRC Biostatistics Unit. Institute for Public Health, Cambridge, United Kingdom
Sturtz S, Ligges U, Gelman A (2005) R2WinBUGS: a package for running WinBUGS from. R. J Stat Softw 12(3):1–16
Trevanich A, Bourke P (1993) EWMA control charts using attributes data. Statistician 42(3):215
Tsiamyrtzis P, Hawkins DM (2005) A Bayesian scheme to detect changes in the mean of a short-run process. Technometrics 47(4):446–456
Tsiamyrtzis P, Hawkins DM (2008) A Bayesian EWMA method to detect jumps at the start-up phase of a process. Qual Reliability Eng Int 24(6):721–735
Vermaat MB, Does RJMM, Bisgaard S (2008) EWMA control chart limits for first- and second-order autoregressive processes. Qual Reliability Eng Int 24(5): 573–584. http://dx.doi.org/10.1002/qre.922
Weiß CH (2007) Controlling correlated processes of Poisson counts. Qual Reliability Eng Int 23(6):741–754. http://dx.doi.org/10.1002/qre.875
White CH, Keats JB, Stanley J (1997) Poisson CUSUM versus c-chart for defect data. Qual Eng 9(4):673–679
Woodall DH (1997) Control charts based on attribute data: bibliography and review. Journal Qual Technol 29(2):172–183
Woodall WH, Adams BM (1993) The statistical design of CUSUM charts. Qual Eng 5(4):559–570

27

Resource leveling scheduling by an ant colony-based model

Mohsen Garmsiri* and Mohammad Reza Abassi

Abstract

In project scheduling, many problems can arise when resource fluctuations are beyond acceptable limits. To overcome this, mathematical techniques have been developed for leveling resources. However, these produce a hard and inflexible approach in scheduling projects. The authors propose a simple resource leveling approach that can be used in scheduling projects with multi-mode execution activities. In the mentioned approach, an ant algorithm determines the execution mode of each activity so that resource leveling index and project time become optimum. In the model, some visibility functions (defined in accordance with problem characteristics) are utilized, and the best, which return the best result, is selected for the model.

Keywords: Ant colony optimization, Resource leveling, Visibility function

Background

Resource availability constraints and resource leveling are typical real-life scheduling problems. It is, thus, imperative that the development of project schedules includes not only activity precedence relations and duration assignments but also resource assignments for each activity in the network. This issue has been studied extensively. For example, Easa (1989) developed an improved tabu search model to solve resource leveling problems. Kastor and Sirakoulis (2009) evaluated the effectiveness of resource leveling tools of three popular packages by comparing the results when leveling two real construction projects as case studies. Leu et al. (2000) proposed an algorithm that can effectively provide the optimal or near-optimal combination of multiple construction resources as well as starting and finishing dates of activities subjected to the objective of resource leveling. Jiang and Shi (2005) presented a new algorithm called the enumerative branch-and-cut procedure for minimizing the total project duration of a construction project under multiple resource constraints based on an enumeration tree. Kim and Graza (2005) evaluated the resource-constrained critical path method, which some writers had proposed. Najafia et al. (2009) proposed a genetic algorithm method to solve a resource investment problem with discounted cash flows (a project-scheduling problem in which (a) the availability levels of the resources are considered decision variables and (b) the goal is to find a schedule such that the net present value of the project cash flows optimizes).

Moumene and Ferland (2009) represented list representation for a generalization of the resource-constrained project scheduling problem. Chtourou and Haouari (2008) presented a two-stage algorithm for a robust resource-constrained project scheduling. Christodoulou et al. (2009) proposed a method for resource allocation and scheduling of resource-constrained projects by use of a measure of the projects' entropy. Ranjbar and Kianfar (2007) used a genetic algorithm-based model to schedule each activity in one of its modes in order to minimize the project makespan.

There are also some researches about resource scheduling under multi-project scheduling. For example, Kim et al. (2005) proposed a hybrid genetic algorithm with fuzzy logic controller to solve the resource-constrained multiple project scheduling problem.

There are also researches about project scheduling under situations where availability levels of the resources are variable. For example, Buddhakulsomsiri and Kim (2006) studied about the multi-mode resource-constrained project scheduling problem when activities can be split during scheduling under situations where resources may be temporarily not available.

* Correspondence: mohsen.garmsiri@yahoo.com
MBA Program, Payame Noor University, Tehran, Iran

There are also researches about project scheduling for various stages. For example, Shi and Deng (2000) developed a construction scheduling and planning method, named object-oriented resource-based planning method, for meeting the different requirements at various planning stages.

For many real-life projects, each activity can be performed using alternative modes that define the subset of the resources needed, and an activity may share different resources simultaneously. The problem is to select a single mode for each operation and, accordingly, to schedule the resources while optimizing a target. This issue has been studied extensively. For example, (Peteghem and Vanhoucke 2009) presented a genetic algorithm for the multi-mode resource-constrained project scheduling problem in which multiple execution modes are available for each of the activities of the project. Wuliang and Chengen (2009) used genetic algorithm to solve a multi-mode resource-constrained discrete time/cost trade-off problem. Chen and Weng (2009) proposed a two-phase genetic algorithm (GA) model in which both the effects of time–cost trade-off and resource scheduling are taken into account. Lova et al. (2009) proposed a hybrid genetic algorithm for finding the start times and execution modes for the activities of a project that minimize project makespan while verifying a set of precedence and resource constraints. (Damaka et al. 2009) proposed a differential evolution algorithm for solving multi-mode resource-constrained project scheduling problems. He et al. (2009) developed a model to assign activities' modes and progress payments so as to maximize the net present value of the contractor under the constraint of project deadline. Ranjbar and Kianfar (2007) used a genetic algorithm-based model to schedule each activity in one of its modes in order to minimize the project makespan.

This paper proposes a new heuristic approach for scheduling each activity of a project in one of its defined modes subject to precedence constraints in order to minimize the project makespan. Since employing time–cost trade-off analysis usually causes great fluctuations in the daily resource requirements by mixing compressed-duration activities of high resource demand with others of low resource demand, therefore, the need for resource management techniques becomes inevitable to ensure efficient utilization of resources. So in addition to project time, resource leveling index is considered as a part of scheduling objectives. Ant colony optimization algorithms are applied because they have been applied to many combinatorial optimization problems ranging from quadratic assignment to protein folding or routing vehicles, and a lot of derived methods have been adapted to dynamic problems in real variables, stochastic problems, multi-targets and parallel implementations. It has also been used to produce near-optimal solutions to the traveling salesman problem. They have an advantage over simulated annealing and genetic algorithm approaches of similar problems when the graph may change dynamically; the ant colony algorithm can be run continuously and adapt to changes in real time.

The rest of this paper starts with a definition of the project scheduling problem being solved. The focus then moves to an introduction of ant colony optimization algorithm and the details of the proposed algorithm. The application of the proposed algorithm is demonstrated through an example project, which is followed by analyses on the results. Conclusions and future developments appear in the last section.

Results and discussion

Resource leveling scheduling by an ant colony-based model

Objective function

The objective function includes two objectives:

Objective 1 Is to determine the minimal duration for a construction project, that is, it is better if the project completion time is shorter. Minimize 'T' where T is the set of the maximum finish time of all activities in the project.

Objective 2 The second objective is defined to minimize deviation of resource usage amount. There are some resource leveling indexes (Elazouni and Metwally 2007; Leu et al. 2000). In this research, objective 2 implemented the minimum moment algorithm to make the daily resource consumption (r_{tj}) as smooth as possible. Also, objective 2 uses an amplification factor, which is the difference between the daily resource consumption (r_{tj}) and the resource consumption of the last day $r_{(t-1)j}$. This difference is used to select the optimal daily resource consumption and enable it to reach the optimum use of resources. In the paper, also, considering the multi-type resource allocation problem, the weight of every kind of resource is generally added wherein the weighted value W_j represents its importance. For example, whether a resource type is important or not could depend on its unit price. The higher the unit price of this resource type, the more important is this resource type.

Minimize

$$\sigma = \sum_{j=1}^{m} \left\{ \left[\sum_{t=1}^{n} r_{ij}{}^2 + \sum \left(r_{tj} - r_{(t-1)j} \right)^2 \right] \times W_j \right\} \tag{1}$$

where r_{tj} is the consumption of type j at time t in the project, W_j is the weight of the resource j, n is the number of optimal project duration on the condition of the resource constraints, and m is the number of total type resource project days (Pan et al. 2008).

Total objective

A compound of project cost index and resource leveling index makes the total objective:

$$\text{Total objective} = \mathrm{IL}_T \times \frac{T}{T_{basis}} + \mathrm{IL}_\sigma \times \frac{\sigma}{\sigma_{basis}} \tag{2}$$

where IL_T is the importance level of the project time index, IL_σ is the importance level of the resource leveling index, T is the project time, σ is the resource leveling index, T_{basis} is the basis amount of the project time index, and σ_{basis} is the basis amount of the resource leveling index.

Ant colony optimization system

Ant colony optimization (ACO) is a metaheuristic technique for hard discrete optimization problems that was first proposed in the early 1990s. In ACO, blind insects like ants can manage to establish the shortest route paths from their colony to feeding sources and back. When searching for food, ants initially explore the area surrounding their nest in a random manner. As soon as an ant finds a food source, it evaluates the quantity and quality of the food and carries some of the found food to the nest. During the return trip, the ant deposits a chemical pheromone trail on the ground. The quantity of pheromone deposited, which may depend on the quantity and quality of the food, will guide other ants to the food source (Blum and Dorigo 2004; Gambardella et al. 1997). While an isolated ant moves essentially at random, an ant encountering a previously laid trail can detect it and decide with high probability to follow it, thus reinforcing the trail with its own pheromone. The collective behavior that emerges is a form of autocatalytic behavior where the more ants follow a trail, the more attractive that trail becomes for being followed. The process is characterized by a positive feedback loop, where the probability with which an ant chooses a path increases with the number of ants that previously chose the same path. Also, pheromone concentration vanishes with time, and then, the less-used path will be much lower in pheromone concentration. The indirect communication between the ants via the pheromone trails allows them to find the shortest paths between their nest and food sources (Bonabeau et al. 1999).

ACO will have some major differences with a real (natural) one:

(1.) Artificial ants will have some memory.
(2.) They will not be completely blind.
(3.) They will live in an environment where time is discrete.

The idea is that if, at a given point, an ant has to choose among different paths, it does this based on two main components:

(1.) Pheromone trails that were heavily chosen by preceding ants (that is, those with a high trail level) are chosen with higher probability. Furthermore, high trail levels (pheromone values) are synonymous with short paths.
(2.) Heuristic, which is visibility, is defined as the inverse of the distance, and so, closer nodes have a higher chance of being next.

Two parameters (α and β) are introduced to control the relative weight of pheromone trail intensity and visibility. At run-time, ACO algorithms try to update the pheromone values in such a way that the probability to generate high-quality solutions increases over time. The pheromone values are updated using previously generated solutions. The reinforcement of solution components depending on the solution quality is an important ingredient of ACO algorithms. It implicitly assumes that good solutions consist of good solution components. To learn which components contribute to good solutions can help to assemble them into better solutions (Dorigo et al. 1996).

The basic ACO algorithm involves the following: ants choose the trail with the largest concentration of pheromone τ with a certain probability p. The shorter path will have more pheromone concentration. For a whole colony following this behavior, the shortest path will be much higher in pheromone concentration than the other paths. Then, the probability of choosing any other paths will be very small. Also, the less-used path will be much lower in pheromone concentration due to pheromone volatility with time. Let an ant k reside in a node i at time t. Assume that one of the possible trails for its next step is (i, j) connecting node i to node j. Then, the probability of choosing node j as the next node is given by

$$P_{(k,i,j)} = \begin{cases} \dfrac{\tau(i,j)^\alpha \eta(i,j)^\beta}{\sum\limits_{r\in\Gamma} \tau(i,j)^\alpha \eta(i,j)^\beta} & r\in\Gamma \\ 0 & r\notin\Gamma \end{cases} \tag{3}$$

where the pheromone trail update is given by $\tau\ (i,j)\ (t+1) = (1-\nu)^*\ \tau\ (i,j)\ (t) + \nu^*\ \delta\ (k,i,j)$.

ν ⊠ (0, 1) is the evaporation coefficient. $\delta(k, i, j)$ is the pheromone released by the ant k on the trail (i, j).

$$\delta(k,i,j) = \begin{cases} Q/L & \textit{if ant selected trail} = j \\ 0 & \textit{if ant selected trail} \neq j \end{cases} \tag{4}$$

where Q is a positive constant, and L is the total path time. Γ is a tabu list, which contains all the trails that

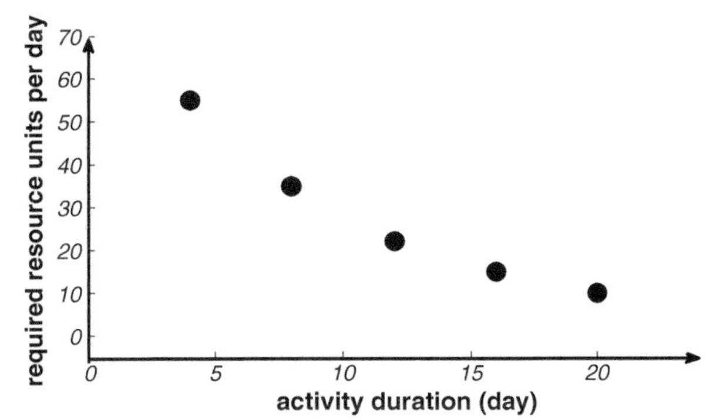

Figure 1 Execution mode selection and route establishment by ants.

the ant has already passed and must not be chosen again; this acts as the memory of the ant. η(*i*, *j*) is the visibility function, which is usually selected as the inverse of the distance for *i* to *j*, i.e.,

$$\eta(i,j) = \gamma / d(i,j).$$

Where γ is the visibility parameter, α and β express the relative weight between the importance of τ and η (Abdallah 2009).

Proposed ACO technique for solving resource leveling scheduling

The ant selects an execution mode (from available execution modes) for each activity and establishes a route from the initial activity until the last. A schedule is established based on the route and as-soon-as-possible scheduling approach. A route that has optimal objective function with respect to all other routes is the best route. Figure 1 demonstrates the execution mode selection and route establishment by ants.

In the algorithm proposed here, each feasible solution is demonstrated by the mode string. The mode string *ms* can be regarded as a vector. There are as many parts in the vector as there are activities. The situation that an activity a_j performed in mode *m* can be represented by $Ms(j) = m$, where $1 \le j \le N$ and $1 \le m \le M_j$, respectively. Figure 2 demonstrates a feasible solution determined by ant 1. In this figure, the third number of the mode string is 1, indicating that the execution mode of activity 3 is 1.

In the proposed algorithm, the visibility function η(i, j), which is usually selected as the inverse of the distance from *i* to *j*, is proposed here to be proportional to the (1) inverse of the cost, (2) inverse of the cost multiplied by the duration, (3) inverse of the duration, (4) duration, and (5) inverse of the number of required resource per day. So, some models for different visibility functions are established.

For the first run, each ant is given a random guide to the execution modes for each activity. In other words, the ants choose one of the available execution modes for each activity. They put their pheromone trace on their path. A tabu list, which contains all the execution modes that the ant has already selected, is made. The tabu list acts as the memory of the ant. The pheromone level is also updated. For other runs, ants choose one of the available execution modes for each activity guided by the pheromone trail and visibility level, putting its pheromone trace on this path. The pheromone level is also updated. The mentioned procedure is repeated, until certain stopping criterion is met (e.g., a fixed number of iterations).

Computational study

A simple project is used to demonstrate the proposed model. The data of each activity and the relationship between activities are shown in Table 1. Resource cost rates are 20 and 15 units per day.

We use procedure with ACO-related input data as follows: number of ants = 50 and iteration number = 1,000. The proposed algorithms were implemented in MATLAB language. The running time using Intel (R) Pentium D, 3 GHz, 160 MB RAM is approximately 60 s.

In the previous researches, the effects of some parameters, including *v*, α, and β, have been analyzed. In this research, in addition to afore-stated parameters, the impact of different definitions of visibility function is analyzed. The normalized different definitions of visibility

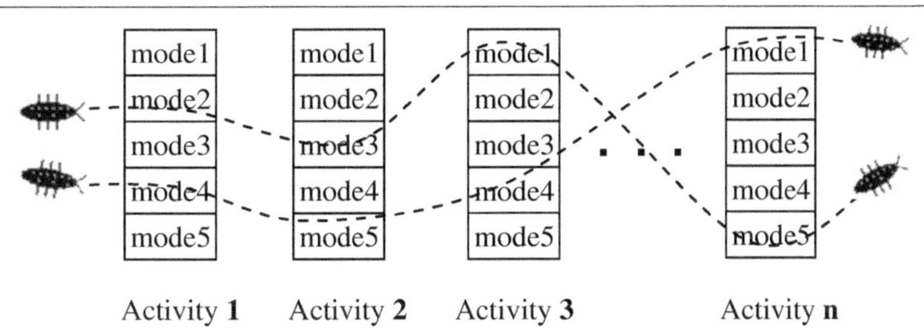

Figure 2 A feasible solution.

Table 1 The activity data

Activity	Precedence	Option	Duration	Resource 1 requirement	Resource 2 requirement
1	-	1	25	5	2
		2	15	7	4
		3	10	10	8
2	1	1	35	10	3
		2	22	15	7
		3	18	25	15
3	1	1	22	8	5
		2	15	15	9
		3	12	25	20
4	1	1	25	10	7
		2	18	18	15
		3	14	30	25
5	2	1	13	10	15
		2	10	15	28
		3	8	25	40
6	3, 4	1	20	18	20
		2	18	30	30
		3	14	40	45
7	2, 4	1	18	10	14
		2	12	20	18
		3	8	25	35
8	6	1	15	5	8
		2	12	10	15
		3	8	15	25
9	7	1	30	10	15
		2	18	1	20
		3	15	25	30
10	8	1	25	10	5
		2	20	15	10
		3	15	25	20
11	8	1	24	5	10
		2	20	10	15
		3	18	15	25
12	8	1	20	10	10
		2	16	15	20
		3	14	20	25

function for different execution modes of activities are presented in Tables 2, 3, 4, 5 and 6.

The values of w_T and w_σ are fixed to 0.5. The values of T_{basis} and σ_{basis} are equal to the project time and resource leveling index of the project schedule in which the first execution mode is selected for each activity.

With each parameter and visibility function combination, 100 runs of experiments on the project example are performed so that the stochastic characteristic in ACO can be reflected. Results from the runs are shown in Table 7. The results of model running based on different values of v, α and β and different definitions of visibility function are presented in Table 7.

In each cell of Table 7, the first number is the total objective function. The second number is the iteration number in which the ants select a route as the best, in other words, model result does not change after the mentioned iteration number. The third number is the project time of the solution, and the fourth number is the resource leveling index of the solution. The best solutions that have minimum total objective function are in bold.

Comparing different combinations that have different parameters values and definitions for visibility function, it is found that a big α and small β lead to better solutions because using extremely big α and small β encourages the ants to follow the preceding ants' pheromones. Using small α and big β encourages the ants to choose the next node as having the largest trail time. This is clear from Equation 3. In this case, the ants are not heavily attracted to the pheromones of preceding ants. This may lead to inaccurate identification of the best solution.

More than that, using high evaporation coefficient leads the ants to make all paths randomly, and the ant does not influence the succeeding ones, and evaporation coefficient leads to fast convergence to the preceding path and may lead to immature convergence due to the lack of exploration of other paths. However, results of the project show that changing the evaporation coefficients causes little difference in the ACO performance. It can be concluded that the initial adoption of the ants are good solutions.

As it can be seen in Table 7, the first definition for visibility function gets good result with respect to all other definitions. Project time and resource leveling index optimization are the objectives. In addition, the first definition is an inverse of cost, and cost is a function of activity time and number of required resource per day. Thus, the first definition produces the best results.

Another consideration that is dependent on the model results is the rapidity of getting result. As it can be seen in Table 7, the combinations which get good result get the result more rapidly. As a conclusion of the mentioned fact, the amounts of v, α, β and the definition for visibility function affect the quality and rapidity of getting result.

Conclusions

An ACO-based method including its corresponding framework is proposed for solving the project scheduling

Table 2 Normalized first definitions of visibility function for different execution modes of activities

(1)	Mode 1	Mode 2	Mode 3
A1	0.32	0.35	0.33
A2	0.38	0.37	0.25
A3	0.43	0.34	0.23
A4	0.44	0.32	0.24
A5	0.42	0.32	0.26
A6	0.52	0.25	0.23
A7	0.36	0.33	0.32
A8	0.44	0.29	0.27
A9	0.42	0.33	0.25
A10	0.43	0.33	0.25
A11	0.45	0.32	0.22
A12	0.42	0.31	0.27

problem in terms of minimizing project duration and resource leveling index. An ant representation of activity execution mode, based on which a feasible schedule (including duration and resource allocation for all activities and, in result, project duration and resource leveling index) can be definitely produced, is adopted for the ACO-based approach. The ACO artificial agent seems to provide a powerful alternative to traditional methods in terms of construction scheduling since it is able to provide an efficient and easy-to-implement alternative to analyze and achieve the project scheduling problem and, in result, efficiently construct the best solutions in acyclic (unidirectional) network topologies.

The proposed model is applied to a project comprising 15 activities. For the sample project, the proposed model has shown that despite the seemingly iterative approach of the ACO algorithm, the method exhibits quick convergence to the final solution and, thus, low computational time. Furthermore, the model also assists project managers in making decision about resources.

Ongoing work on the applicability of the ACO artificial agent addresses the inclusion of resource-based scheduling techniques to account for resource leveling in multi-project scheduling. More than that, the proposed ACO method will be developed into a user-friendly system so that the practitioners can utilize it in solving the project scheduling problems.

Table 3 Normalized second definitions of visibility function for different execution modes of activities

(2)	Mode 1	Mode 2	Mode 3
A1	0.19	0.34	0.48
A2	0.26	0.40	0.34
A3	0.32	0.37	0.31
A4	0.33	0.34	0.33
A5	0.33	0.33	0.34
A6	0.46	0.25	0.29
A7	0.23	0.31	0.46
A8	0.34	0.27	0.39
A9	0.37	0.33	0.3
A10	0.34	0.33	0.33
A11	0.4	0.34	0.26
A12	0.35	0.32	0.33

Table 4 Normalized third definitions of visibility function for different execution modes of activities

(3)	Mode 1	Mode 2	Mode 3
A1	0.19	0.33	0.48
A2	0.22	0.35	0.43
A3	0.23	0.34	0.43
A4	0.24	0.33	0.43
A5	0.25	0.33	0.43
A6	0.28	0.31	0.4
A7	0.21	0.32	0.47
A8	0.24	0.3	0.45
A9	0.29	0.32	0.39
A10	0.26	0.32	0.43
A11	0.28	0.34	0.38
A12	0.27	0.34	0.39

Methods

Problem statement

In project scheduling, if resources fluctuations are beyond acceptable limits, actions must be taken to move activities or resources around in order to level out the resource-loading graph. For example, it is bad for employee morale and public relations when a company has to hire and lay people off indiscriminately. Proper

Table 5 Normalized fourth definitions of visibility function for different execution modes of activities

(4)	Mode 1	Mode 2	Mode 3
A1	0.5	0.3	0.2
A2	0.47	0.29	0.24
A3	0.45	0.31	0.25
A4	0.44	0.32	0.25
A5	0.42	0.32	0.26
A6	0.39	0.35	0.27
A7	0.47	0.32	0.21
A8	0.43	0.34	0.23
A9	0.38	0.34	0.28
A10	0.42	0.33	0.25
A11	0.39	0.32	0.29
A12	0.4	0.32	0.28

Table 6 Normalized fifth definitions of visibility function for different execution modes of activities

(5)	Mode 1	Mode 2	Mode 3
A1	0.49	0.32	0.20
A2	0.52	0.31	0.17
A3	0.55	0.30	0.16
A4	0.55	0.28	0.17
A5	0.51	0.30	0.20
A6	0.57	0.25	0.18
A7	0.50	0.30	0.20
A8	0.54	0.28	18
A9	0.46	0.33	0.21
A10	0.51	0.31	0.18
A11	0.51	0.3	0.19
A12	0.49	0.29	0.22

resource planning will facilitate a reasonably stable level of the work force. Other advantages of resource leveling include simplified resource tracking and control, lower cost of resource management and improved opportunity for learning. For leveling resources, there are some tools such as Primavera and MS Project. Each tool uses a special approach to level resources.

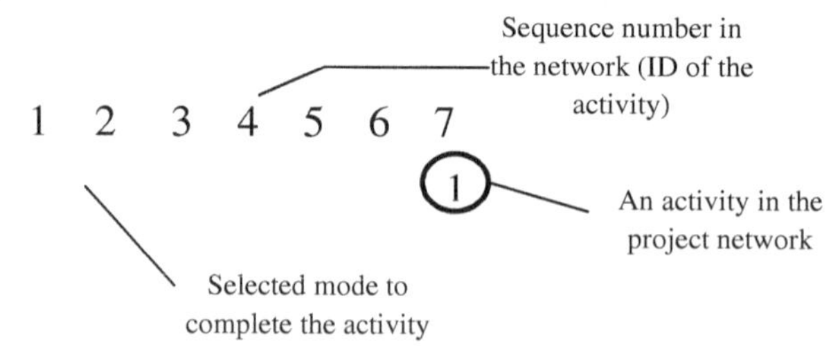

Figure 3 A typical activity time-resource curve.

These scheduling tools largely rely on heuristics in leveling resource in forward-pass scheduling; thus, in general, they do not yield the optimum solution in terms of the shortest project duration. In addition, the proposed resource leveling methods, though, do not necessarily optimize the allocation of resources so as to minimize the project duration. For example, a resource leveling may leave the project schedule unchanged even if a different resource allocation may result in a shorter or longer project schedule. Therefore, it is essential to use a more appropriate method, which considers project time, for leveling resources. In other words, both of the objectives, project time minimization and resource leveling, should be considered. To this purpose, a resource leveling index

Table 7 Result of the model

	$v = 0.5, \alpha = 1, \beta = 0.1$	$v = 0.5, \alpha = 0.1, \beta = 1$	$N = 0.5, \alpha = 0.5, \beta = 0.5$	$v = 0.9, \alpha = 1, \beta = 0.1$	$v = 0.9, \alpha = 0.1, \beta = 1$	$v = 0.9, \alpha = 0.5, \beta = 0.5$	$v = 0.1, \alpha = 1, \beta = 0.1$	$v = 0.1, \alpha = 0.1, \beta = 1$	$v = 0.1, \alpha = 0.5, \beta = 0.5$
1	1.3	1.56	1.53	1.28	1.51	1.47	1.37	1.53	1.42
	97	200	200	45	200	200	200	200	200
	66	69	50	59	63	66	70	56	64
	2,157,355	2,923,620	3,488,240	2,333,190	2,963,125	2,724,050	2,257,260	3,260,540	2,625,010
2	1.46	1.57	1.49	2	1.6	1.4	1.41	1.46	1.49
	194	200	200	1	200	200	200	200	200
	66	78	68	99	63	56	59	62	66
	2,694,100	2,636,525	2,735,690	3,391,750	3,270,925	2,824,710	2,757,290	2,814,450	2,779,960
3	1.39	1.63	1.52	2	1.57	1.65	1.36	1.49	1.64
	154	54	200	1	200	200	200	200	200
	59	**66**	**75**	**99**	**78**	**61**	**57**	**54**	**64**
	2,691,540	3,264,390	2,601,240	3,391,750	2,636,525	3,505,895	2,653,155	3,085,215	3,353,520
4	1.43	1.44	1.55	1.38	1.35	1.52	1.34	1.38	1.45
	37	200	200	96	200	200	200	200	200
	75	75	60	70	67	60	61	72	64
	2,266,535	2,303,975	3,215,840	2,274,415	2,289,725	3,109,700	2,452,995	2,208,825	2,739,430
5	1.37	1.46	1.46	1.3	1.38	1.48	1.34	1.46	1.46
	94	199	200	109	200	200	200	200	200
	70	70	61	70	61	67	75	61	60
	2,257,260	2,540,170	2,868,195	1,994,880	2,598,640	2,726,395	1,987,000	2,868,195	2,888,185

The best solutions that have minimum total objective function are in bold.

should be used in project scheduling, and in addition to project time, the mentioned index should be optimized.

In addition, if several resources need to be leveled simultaneously, the objective can be defined as the sum of the multiplication of the index for each resource and its corresponding weighting score. The weighting score is a man-made value, and its value, ranging from 0 to 1, depends on the importance of the resource.

For many real-life projects, it is possible to execute each activity in one of several alternative modes, which usually represent relations between the consumed resources and the duration of the activity, leading to some problems. The mentioned problems involve the selection of a set of execution modes in order to achieve a certain objective. The study involves the selection of a set of execution modes in order to level resources. In other words, each activity can be performed using alternative modes that define the subset of the resources needed. A typical activity discrete time-resource curve is shown in Figure 3.

In project scheduling, at first, the execution mode of each activity is selected so that objectives, project time minimization and resource leveling index optimization, are considered. Then, noncritical activities are shifted within their available slacks so that the resource usage curve became smoother. So, the execution mode of each activity determination is under the research problem.

In reference to the project duration, the first criterion set is that the project completion date must be, as much as possible, equal to or earlier than the specified required project completion date. The required project completion date is up to the project manager to be specified and may depend on several considerations. This requirement is such that any project completion later than this has low value. Thus, the required project completion date serves as a limit or deadline on how late the completion of the project can be. In reference to the fluctuations in the daily resource requirements, the second criterion set is that the fluctuations in the daily resource requirements should be minimized. Thus, time and fluctuations in the daily resource requirement optimization should arrive at the set of shifting amount of activity that results in a valid project completion date and have the minimum fluctuations in the daily resource requirements.

The project scheduling and resource leveling problem under study is based on the following assumptions: (1) some activities compose a project; (2) each activity can be performed using alternative modes that define the subset of the resources needed and its duration; (3) all predecessors must be finished before an activity can start (i.e., precedence constraints); (4) resources are renewable from period to period; and (5) activities are non-preemptive, that is, cannot be interrupted when in progress.

Competing interests

Both authors declare that they have no competing interests.

Authors' contributions

MG drafted the manuscript, participated in the design of the study, and performed the statistical analysis. MRA conceived of the study and participated in its design and coordination. All authors read and approved the final manuscript.

Acknowledgements

The authors are grateful to the editor miss Fatemeh Jazebi for her insightful and helpful comments and suggestions.

References

Abdallah H, Emara HM, Dorrah HT, Bahgat A (2009) Using ant colony optimization algorithm for solving project management problems. Expert Syst Appl 36:10004–10015

Blum C, Dorigo M (2004) The hyper-cube framework for ant colony optimization. Transactions on Systems, Man and Cybernetics – Part B. Cybernetics 34 (2):1161–1172

Bonabeau E, Dorigo M, Theraulaz G (1999) Swarm intelligence from natural to artificial systems. Oxford University Press, Oxford

Buddhakulsomsiri J, Kim DS (2006) Properties of multi-mode resource-constrained project scheduling problems with resource vacations and activity splitting. Eur J Oper Res 175:279–295

Chen P-H, Weng H (2009) A two-phase GA model for resource-constrained project scheduling. Autom Constr 18:485–498

Chtourou H, Haouari M (2008) A two-stage-priority-rule-based algorithm for robust resource-constrained project scheduling. Comput Ind Eng 55:183–194

Christodoulou S, Ellinas G, Pooyan A (2009) Entropy-based scheduling of resource-constrained construction projects. Autom Constr 18:919–928

Damaka N, Jarbouia B, Siarryb P, Loukil T (2009) Differential evolution for solving multi-mode resource-constrained project scheduling problems. Comput Oper Res 36:2653–2659

Dorigo M, Maniezzo V, Colorni A (1996) Ant system: Optimization by a colony of cooperating agents. IEEE Transactions on Systems, Man, and Cybernetics – Part B :29–41

Easa SM (1989) Resource leveling in construction by optimization. J Constr Eng Manag 115:302–316

Elazouni AM, Metwally FG (2007) Expanding finance-based scheduling to devise overall-optimized project schedules. Journal of Construction Engineering and Management 133:86–90

Gambardella LM, Taillard E, Dorigo M (1997) Ant Colonies for QAP. Technical Report Number: IDSIA 97–4, IDSIA, Lugano, Switzerland

He Z, Wang N, Jia T, Xu Y (2009) Simulated annealing and tabu search for multi-mode project payment scheduling. Eur J Oper Res 198:688–696

Jiang G, Shi J (2005) Exact algorithm for solving project scheduling problems under multiple resource constraints. Journal of Construction Engineering and Management 131:986–992

Kastor A, Sirakoulis K (2009) The effectiveness of resource levelling tools for resource constraint project scheduling problem. Int J Proj Manag 27:493–500

Kim K, de la Garza JM (2005) Evaluation of the resource-constrained critical path method algorithms. Journal of Construction Engineering and Management 131:522–532

Kim KW, Yun YS, Yoon JM, Gen M, Yamazaki G (2005) Hybrid genetic algorithm with adaptive abilities for resource-constrained multiple project scheduling. Comput Ind 56:143–160

Leu S-S, Yang C-H, Huang J-C (2000) Resource leveling in construction by genetic algorithm-based optimization and its decision support system application. Automation in Construction 10:27–41

Lova A, Tormos P, Cervantes M, Barber F (2009) An efficient hybrid genetic algorithm for scheduling projects with resource constraints and multiple execution modes. Int J Production Economics 117:302–316

Moumene K, Ferland JA (2009) Activity list representation for a generalization of the resource-constrained project scheduling problem. Eur J Oper Res 199:46–54

Najafia AA, Niaki STA, Moslem S (2009) A parameter-tuned genetic algorithm for the resource investment problem with discounted cash flows and generalized precedence relations. Comput Oper Res 36:2994–3001
Pan N-H, Hsaio P-W, Chen K-Y (2008) A study of project scheduling optimization using tabu search algorithm. Eng Appl Artif Intel 21:1101–1112
Ranjbar MR, Kianfar F (2007) Solving the discrete time/resource trade-off problem in project scheduling with genetic algorithms. Appl Math Comput 191:451–456
Shi JJ, Deng Z (2000) Object-oriented resource-based planning method (ORPM) for construction. Int J Proj Manag 18:179–188
Van Peteghem V, Vanhoucke M (2009) A genetic algorithm for the preemptive and non-preemptive multi-mode resource-constrained project scheduling problem. Eur J Oper Res 201:409–418
Wuliang P, Chengen W (2009) A multi-mode resource-constrained discrete time–cost trade-off problem and its genetic algorithm based solution. Int J Proj Manag 27:600–609

Permissions

All chapters in this book were first published in JIEI, by Springer; hereby published with permission under the Creative Commons Attribution License or equivalent. Every chapter published in this book has been scrutinized by our experts. Their significance has been extensively debated. The topics covered herein carry significant findings which will fuel the growth of the discipline. They may even be implemented as practical applications or may be referred to as a beginning point for another development.

The contributors of this book come from diverse backgrounds, making this book a truly international effort. This book will bring forth new frontiers with its revolutionizing research information and detailed analysis of the nascent developments around the world.

We would like to thank all the contributing authors for lending their expertise to make the book truly unique. They have played a crucial role in the development of this book. Without their invaluable contributions this book wouldn't have been possible. They have made vital efforts to compile up to date information on the varied aspects of this subject to make this book a valuable addition to the collection of many professionals and students.

This book was conceptualized with the vision of imparting up-to-date information and advanced data in this field. To ensure the same, a matchless editorial board was set up. Every individual on the board went through rigorous rounds of assessment to prove their worth. After which they invested a large part of their time researching and compiling the most relevant data for our readers.

The editorial board has been involved in producing this book since its inception. They have spent rigorous hours researching and exploring the diverse topics which have resulted in the successful publishing of this book. They have passed on their knowledge of decades through this book. To expedite this challenging task, the publisher supported the team at every step. A small team of assistant editors was also appointed to further simplify the editing procedure and attain best results for the readers.

Apart from the editorial board, the designing team has also invested a significant amount of their time in understanding the subject and creating the most relevant covers. They scrutinized every image to scout for the most suitable representation of the subject and create an appropriate cover for the book.

The publishing team has been an ardent support to the editorial, designing and production team. Their endless efforts to recruit the best for this project, has resulted in the accomplishment of this book. They are a veteran in the field of academics and their pool of knowledge is as vast as their experience in printing. Their expertise and guidance has proved useful at every step. Their uncompromising quality standards have made this book an exceptional effort. Their encouragement from time to time has been an inspiration for everyone.

The publisher and the editorial board hope that this book will prove to be a valuable piece of knowledge for researchers, students, practitioners and scholars across the globe.

List of Contributors

Reza Tavakkoli-Moghaddam
Department of Industrial Engineering, College of Engineering, University of Tehran, Tehran, Iran

Mehdi Alinaghian
Department of Industrial and Systems Engineering, Isfahan University of Technology, Isfahan, Iran

Alireza Salamat-Bakhsh
Department of Industrial Engineering, Tehran South Branch, Islamic Azad University, Tehran Iran

Narges Norouzi
Department of Industrial Engineering, College of Engineering, University of Tehran, Tehran, Iran

Masoud Yaghini
School of Railway Engineering, Iran University of Science and Technology, Tehran 16846-13114, Iran

Masoud Seyedabadi
School of Railway Engineering, Iran University of Science and Technology, Tehran 16846-13114, Iran

Mohamad M Khoshraftar
School of Railway Engineering, Iran University of Science and Technology, Tehran 16846-13114, Iran

Siamak Noori
Department of Industrial Engineering, Iran University of Science and Technology, Narmak, Tehran 16846-13114, Iran

Seyed Farid Ghannadpour
Department of Industrial Engineering, Iran University of Science and Technology, Narmak, Tehran 16846-13114, Iran

Payel Ghosh
Department of Mathematics, Adamas Institute of Technology, Barasat, North 24 Parganas 700126, West Bengal, India

Tapan Kumar Roy
Department of Mathematics, Bengal Engineering and Science University, Shibpur, Howrah 711103, West Bengal, India

Ramin Sadeghian
Islamic Azad University, South Tehran Branch, Industrial Engineering Faculty,Tehran, Iran

Babak H Tabrizi
Department of Industrial Engineering, Faculty of Engineering, University of Tehran, P.O. Box: 11155-4563, Tehran, Iran

Jafar Razmi
Department of Industrial Engineering, Faculty of Engineering, University of Tehran, P.O. Box: 11155-4563, Tehran, Iran

Seyed Taghi Akhavan Niaki
Department of Industrial Engineering, Sharif University of Technology, P.O. Box 11155-9414, Azaadi Ave., Tehran 1458889694, Iran

Majid Khedmati
Department of Industrial Engineering, Sharif University of Technology, P.O. Box 11155-9414, Azaadi Ave., Tehran 1458889694, Iran

GVSS Sharma
Department of Mechanical Engineering, GMR Institute of Technology, GMR Nagar, Rajam-532127, Andhra Pradesh State, India

P Srinivasa Rao
Department of Industrial Engineering, GITAM Institute of Technology, GITAM University, Visakhapatnam - 530045 Andhra Pradesh State, India

Meghdad HMA Jahromi
Department of Industrial Engineering, Khomein Branch, Islamic Azad University, Khomein, Iran

Reza Tavakkoli-Moghaddam
Department of Industrial Engineering, College of Engineering, University of Tehran, Tehran, Iran

Ahmad Makui
Department of Industrial Engineering, Iran University of Science and Technology, Tehran, Iran

Abbas Shamsi
Department of Industrial Engineering, Khomein Branch, Islamic Azad University, Khomein, Iran

Ghasem Tohidi
Assistant professor Department of Mathematics, Islamic Azad University, Central Tehran Branch, Tehran, Iran

Shabnam Razavyan
Assistant professor Department of Mathematics, Islamic Azad University, South Tehran Branch, Tehran, Iran

A Narvand
School of Industrial Engineering, South Tehran Branch, Islamic Azad University, Tehran, Iran

P Soleimani
School of Industrial Engineering, South Tehran Branch, Islamic Azad University, Tehran, Iran

Sadigh Raissi
School of Industrial Engineering, South Tehran Branch, Islamic Azad University, Tehran, Iran

Seyed Taghi Akhavan Niaki
Department of Industrial Engineering, Sharif University of Technology, P.O. Box 11155-9414 Azaadi Ave, Tehran 1458889694, Iran

Saeid Hoseinzade
Department of Industrial Engineering, Sharif University of Technology, P.O. Box 11155-9414 Azaadi Ave, Tehran 1458889694, Iran

Arvind K Lal
School of Mathematics and Computer Applications, Thapar University, Patiala, PB 147004, India

Manwinder Kaur
School of Mathematics and Computer Applications, Thapar University, Patiala, PB 147004, India

Sneh Lata
University Institute of Engineering, Chandigarh University, Gharuan-Mohali, PB 140413, India

Mahdi Bashiri
Department of Industrial Engineering, Faculty of Engineering Shahed University, Tehran, Iran

Amir Moslemi
Department of Industrial Engineering, Faculty of Engineering Shahed University, Tehran, Iran

Department of Industrial Engineering, Amirkabir University of Technology (Tehran Polytechnic), Hafez, Tehran, Iran

Hossein Karimi
Department of Industrial Engineering, Shahed University, Tehran 3319118651, Iran

Arokiasamy Mariajayaprakash
Department of Mechanical Engineering, Rajiv Gandhi College of Engineering and Technology, Kirumabakkam, Puducherry 607402, India

Thiyagarajan Senthilvelan
Department of Mechanical Engineering, Pondicherry Engineering College, Pillaichavady, Puducherry 605014, India

Krishnapillai Ponnambal Vivekananthan
TENNECO Automotive India private Limited, Mettupalayam, Puducherry 605009, India

Bahman Naderi
Department of Industrial Engineering, Faculty of Engineering, University of Kharazmi, Karaj, Iran

Esmaeil Najafi
Department of Industrial Engineering, Science & Research Branch, Islamic Azad University, Tehran, Iran

Mehdi Yazdani
Department of Industrial Engineering, Qazvin Branch, Islamic Azad University, Qazvin, Iran

Zeinab Hosseini
Department of Industrial Engineering, Alzahra University, Tehran, Iran

Reza Ghasemy Yaghin
Department of Industrial Engineering, Amirkabir University of Technology, Tehran, Iran

Maryam Esmaeili
Department of Industrial Engineering, Alzahra University, Tehran, Iran

Kailash Lachhwani
Department of Mathematics, Government Engineering College, Bikaner 334004, India

Mahaveer Prasad Poonia
Department of Mechanical Engineering, Government Engineering College, Bikaner 334004, India

Charan Jeet Singh
Department of Mathematics, Guru Nanak Dev University, Amritsar 143005, India

Madhu Jain
Department of Mathematics, Indian Institute of Technology, Roorkee 247667, India

Binay Kumar
Department of Mathematics, M.L.U. DAV College, Phagwara 144402, India

Seyed Saeid Hashemin
Department of Industrial Engineering, Ardabil Branch, Islamic Azad University, Ardabil, 56157-31567, Iran

Seyed Mohammad Taghi Fatemi Ghomi
Department of Industrial Engineering, Amirkabir University of Technology, Tehran, 15988-46611, Iran

Madhu Jain
Department of Mathematics, IIT, Roorkee 247667, India

Richa Sharma
Department of Mathematics, JK Lakshmipat University, Jaipur 302026, India

Gokul Chandra Sharma
Department of Mathematics, I.B.S., Khandari, Agra 282002, India

Arman Izadi
Department of industrial engineering, Amirkabir university of technology, Hafez street, NO 424, 15875-4413, Tehran, Iran

Ali mohammad Kimiagari
Department of industrial engineering, Amirkabir university of technology, Hafez street, NO 424, 15875-4413, Tehran, Iran

Alireza Rashidi Komijan
Department of Industrial Engineering, Firoozkooh Branch, Islamic Azad University, Firoozkooh, Iran

Mehrdad Nouri Koupaei
Department of Management, Firoozkooh Branch, Islamic Azad University, Firoozkooh, Iran

Hassan Assareh
Simpson Centre for Health Services Research, Australian Institute of Health Innovation, Faculty of Medicine, University of New South Wales, Sydney, NSW 2052, Australia

Rassoul Noorossana
School of Industrial Engineering, Iran University of Science and Technology, Tehran, Iran

Kerrie L Mengersen
Discipline of Mathematical Sciences, Science and Engineering Faculty, Queensland University of Technology, Brisbane, QLD 4001, Australia

Mohsen Garmsiri
MBA Program, Payame Noor University, Tehran, Iran

Mohammad Reza Abassi
MBA Program, Payame Noor University, Tehran, Iran